SPACECRAFT POWER SYSTEMS

SPACECRAFT POWER SYSTEMS

Mukund R. Patel

CRC PRESS

Boca Raton London New York Washington, D.C.

Cover photo source: Dornier Satellitensysteme GmbH, Daimler-Benz Aerospace, Germany. With permission.

Library of Congress Cataloging-in-Publication Data

Patel, Mukund R., 1942-
Spacecraft power systems / Mukund R. Patel
p.cm.
Includes bibliographical references and index.
ISBN 0-8493-2786-5 (alk. paper)
1. Space vehicles – Auxiliary power supply. I. Title

TL1100.P38 2004
629.47′44–dc22 2004054509

International Standard Book Number 0-8493-2786-5
Library of Congress Card Number 2004054509
Printed in the United States of America 1 2 3 4 5 6 7 8 9 0
Printed on acid-free paper

. . . dedicated
to my late sister, Bhanumati Amin,
who left loving memories for me to cherish;
and
to my parents, Shakariben and Ranchhodbhai Patel,
who still inspire me in their nineties.

Preface

The first man-made satellite was *Sputnik I,* launched and operated by Russia in the low Earth orbit in 1957. In the decade that followed, the American space exploration programs launched many Earth orbiting satellites. The first commercial geosynchronous satellite, *Intelsat-1,* was placed in orbit in 1965, and, in 1969, NASA's *Apollo-11* became the first manned spacecraft to land on the moon. Since then, many countries have successfully undertaken small and large space programs. In 2003, a civilian became the first space tourist and China became the third nation after Russia and the United States to put a human in space. In 2004, U.S. President Bush announced a new space initiative for manned landing on the moon by 2015, and then onward to Mars. In the same year, both China and India announced their own plans to send unmanned spacecraft to the moon by 2010. On the commercial side, the recent worldwide communications technology boom has established the satellite as an integral part of the national infrastructure. Today, many countries have the capability of placing and operating satellites in orbit.

In the past, the U.S. government has provided a large share of the space industry revenue, and telecommunications satellites dominated the non-government market. The 1990s saw milestones in the development of personal communications systems, expansion of the remote sensing industry and the approval of sale of spy-quality data. The 2000 world market for commercial communications satellites was 28 satellites at a procurement cost of 12 billion dollars, about 50% of which went to U.S. companies. Added to this was the insurance cost for launch and 1-year on-orbit operation ranging from 7 to 15% of the launch vehicle and satellite costs. According to Aerospace America, the total number of launches in 2003 was 90, and is projected to be 150 in 2012. These endeavors symbolize the increasing expansion of a commercial market. Many changes occurring in the industry continue the evolution of this dynamic sector of the technology. New commercial ventures are rapidly expanding. It is on these ventures that the future expansion of the space industry rests.

The GPS III of the U.S. Air Force with the first launch targeted in the 2012–2015 time frame may be among the largest defense satellite programs during the next 20 years. It will have some 30 satellites at an estimated cost of $5 billion. New large-scale commercial satellites, such as Europe's *Galileo* navigation program, will involve many countries, possibly including China. Science missions also continue to provide new knowledge of our universe. In the spring of 2001 alone, our horizon was broadened by the discovery of 11 new planets by Keck telescopes in Hawaii, the Haute-Provence Observatory in France, and the European Southern Observatory in Chile, bringing the total to about 100 planets in other solar systems. And, recently, scientists have discovered a solar system just like ours. In our own solar system, 39 moons of Jupiter, 30 moons of Saturn and a massive reservoir of water ice just below the Mars surface have been discovered.

Now that commercial and science satellites have established their strong foothold in the modern world, it is important to optimize their technical performance and ultimately the rate of return for the investors. The satellite has certain resources — instruments in science satellites and bandwidth in communication satellites — all driven by electrical power. This book focuses on the design, performance, and operation of electrical power systems for all types of spacecraft, Earth-orbiting, interplanetary, and deep space for science, commercial, and defense applications.

The design of spacecraft power systems has been advancing at a rapid rate, along with improvements in the system components. The book provides the most modern in-depth data and procedures for designing the electrical power systems to meet the mission requirements at a minimum cost and weight. The launch cost in 2002 U.S. dollars averaged about $10,000 per kilogram for placing a satellite in low Earth orbit, and about $50,000 per kilogram for placing one in a geosynchronous orbit. The power system weight is therefore a significant design driver. The design must be totally optimized to the limit such that even a small change in any one component would result in a penalty of some sort at the satellite level. Moreover, the engineer must incorporate new technologies as they become available. However, the satellite is like a spider's web — touch one thing and everything moves. It is for this reason that the power system engineer must be cognizant about more than the solar array and the battery to optimally design even a conventional satellite. This becomes even more important in view of the fact that one out of four commercial GEO satellites launched in the 5-year period 1998 to 2002 has experienced electrical power problems. Together, they make 60% of the industry-wide insurance claims for in-orbit operational defects.

The author, a recognized expert in the industry, has written this book to provide reliable information on a variety of subjects to engineers and managers involved in spacecraft power systems design, development, testing, and operation. The details covered in the book would help industry engineers develop and maintain a leading edge in this rapidly growing and evolving industry.

Yardley, Pennsylvania

Acknowledgment

The emerging technologies in the spacecraft power systems cannot possibly be documented in a book without help from many sources. I have been extremely fortunate to receive full support from many organizations and individuals in the field. The book is a direct result of my teaching a continuing education course to engineers working at many prime contractors and NASA centers. I am grateful to Marshall Kaplan, president of Launchspace Inc., for inviting me to develop and teach the course since 1995 under the initial sponsorship of the U.S. Space Foundation.

Several organizations provided current data and reports on the space power technologies. They are NASA, the European Space Agency, the U.S. Department of Energy, and spacecraft manufacturers worldwide. Many expert individuals at those organizations gladly provided all the help I asked for. My special gratitude goes to Jim Haines, head of the Power Systems Section at the European Space Agency, and Bob Beaman, senior staff engineer at NASA Goddard Space Flight Center, for reviewing the book proposal and providing valuable comments.

Professional engineers attending my courses often knew more in their specialties than I did. Their stimulating discussions have continuously sharpened my knowledge on space power. Several of them not only encouraged me to write the book on this timely subject, but also provided valuable suggestions and comments during the development of the book.

I am also grateful to Professor Jose Femenia, Engineering Department Head; Dr. Warren Mazek, Dean; and Vice Admiral Joseph Stewart, Superintendent; of the U.S. Merchant Marine Academy, Kings Point, NY, for supporting my research and publications, including a sabbatical leave for writing this book.

I heartfully acknowledge the valuable support from all.

Mukund R. Patel

About the Book

This book has evolved from a three-day continuing education course taught by the author on spacecraft electrical power systems over the last ten years. The material covered is based on the author's deep involvement in many design and development programs at the General Electric Space Division and the Lockheed Martin Space Systems.

Spacecraft power systems have undergone significant new developments in the last decade and will continue to do so even at a faster rate in the current decade. The design and analyses methods used by the engineers have evolved as needed in every corporation. Except for limited publications in bits and pieces at various conferences, and a few books having short sections with sketchy coverage on this vast subject, there is no single book available as of today that covers the entire scope of spacecraft power systems.

All aspects of the electrical power system design, analyses, and operation are covered in this book in details not found in any other book. The wide coverage of the fundamentals of energy conversion, energy storage, power conditioning, energy management, and the operational aspects will help engineers maintain a leading edge in designing and operating a variety of spacecraft power systems.

The book is written to benefit the spacecraft power system design and development engineers, mechanical and aeronautical engineers, chief engineers and program managers. It can serve as a ready reference book. It assumes a college level background in physics and mathematics, and can also be used as a textbook for senior or graduate level university courses. Although the designs are always mission specific, the design numbers or curves are given at many places as examples or typical values for that class of satellites only.

The book covers all aspects of spacecraft power systems in 27 chapters divided in four major parts:

Part A: Satellite overview, space environment and its effects, power system options, PV-battery system, power system requirements, and design trades.

Part B: Solar array, battery, power electronics and magnetics, distribution harness, protection, and ancillary components.

Part C: Energy management, dynamic performance and stability, EMI/EMC, ESD, reliability, derating, and integration and testing.

Part D: Special systems such as interplanetary and deep space, nuclear RTG, thermodynamic system with alternator, high-power and high-voltage systems, electric propulsion, flywheel and superconducting energy storage, fuel cell, and microwave beam power satellite.

Both systems of units — International and British — are used in the book to present data, as they came from various sources. An extensive conversion table is therefore given in the front, and so is a list of acronyms commonly used in the spacecraft industry and also in the book.

About the Author

Mukund R. Patel, Ph.D., PE, is a research engineer, educator and consultant in electrical power systems. He has served as Senior Staff Engineer at Lockheed Martin Space Systems, Principal Engineer at General Electric Space Division, Fellow Engineer at Westinghouse Research Center, Space Products Manager at Induction General Inc. and 3M Distinguished Visiting Professor at the University of Minnesota, Duluth. Presently, he is a Professor at the U.S. Merchant Marine Academy, Kings Point, NY.

Dr. Patel earned his M.S. and Ph.D. degrees from the Rensselaer Polytechnic Institute, Troy, NY, in 1967 and 1972, respectively. He is a Fellow of the Institution of Mechanical Engineers (UK), Associate Fellow of the American Institute of Aeronautics and Astronautics (AIAA), Senior Member of the Institute of Electrical and Electronic Engineers (IEEE), Professional Electrical Engineer in Pennsylvania, Chartered Mechanical Engineer in the UK, and Member of Eta Kappa Nu, Tau Beta Pi, Sigma Xi and Omega Rho. He has published over 50 papers at national and international conferences, holds several patents, and has earned NASA recognition for exceptional contribution to the power system design for UARS. He has taught IEEE workshops and in-house courses at GE. Since 1995, he has been teaching the Spacecraft Power System Design Course to professional aerospace engineers across the United States.

He lives in Yardley, PA, with his wife, Sarla. They have three grown up children, Ketan, Bina, and Vijal; and two grandchildren, Rayna and Dhruv. He can be reached at patelm@usmma.edu.

Acronyms and Abbreviations

ACS	attitude control system
ADE	array drive electronics
AE	autumnal equinox
AIAA	American Institute of Aeronautics and Astronautics
AMTEC	alkaline metal thermoelectric converter
ANSI	American National Standards Institute
AO	atomic oxygen
APL	Applied Physics Laboratory (Johns Hopkins University)
APS	auxiliary power supply
a-Si	amorphous silicon (PV cell)
AU	astronomical unit
AWG	American wire gage
BCC	battery charge controller
BDCU	bi-directional battery charge–discharge unit
BCN	battery charge network
BCU	battery charge unit
BCR	battery charge regulator
BCVM	battery cell voltage monitor
BDN	battery discharge network
BDR	battery discharge regulator
BDU	bus data unit
BIU	battery interface unit
BJT	bipolar junction transistor
BLU	battery letdown unit
BOL	beginning of life
BPC	battery power converter
BPM	Battery pressure monitor
BSR	back surface reflector
BU	backup unit
BVR	bus voltage regulator
CASI	Canadian Aeronautics and Space Institute
CD	command directory

CDR	critical design review
C/D	charge/discharge
CDU	command decoder unit
CIGS	copper indium gallium diselenide
CMD	command
C&DH	command and data handling
CNES	Center National d'Etudes Spatiales (France)
COTS	commercial off-the-shelf (equipment)
CPV	common pressure vessel (with two battery cells)
c-Si	crystalline silicon (PV cell)
DAR	double anti-reflective (coating)
DARPA	Defense Advanced Research Projects Agency
DCA	Defense Communication Agency
DDCU	d.c. to d.c. converter unit
DET	direct energy transfer
DEU	deployment electronics unit
DIPS	dynamic isotopes power system
DMS	defense meteorological satellite
DOD	depth of discharge (of battery)
DoD	Department of Defense (U.S.)
DoE	Department of Energy (U.S.)
DPV	dependent pressure vessel (battery)
DPU	deployment power unit
DRL	Deuche (German) Aerospace Center
DSCS	defense system communications satellite (< 1 kW load, 28 ± 0.28 V, GEO bus)
EED	electro-explosive device
EHD	electro-hydrodynamic
EHT	electro-thermal hydrazine thruster
ELF	extra-low frequency
EMC	electromagnetic compatibility
EMF	electromotive force
EMI	electromagnetic interference
EMP	electromagnetic pulse
EMW	energy-momentum wheel
EOL	end of life
EOS-AM	Earth observing system (3-kW load, 120 V ± 5 V, LEO bus)
EPS	electrical power system
EQX	equinox
ESA	European Space Agency
ESD	electrostatic discharge
ESR	equivalent series resistance

FBA	fuse board assembly
FESS	flywheel energy storage system
FIT	failure rate in time (10^6 hours)
FMECA	failure mode and effect analysis
FPGA	field programmable gate array
GN&C	guidance, navigation and control
GEO	geosynchronous Earth orbit
GPS	Global Positioning System (of U.S. Air Force, NAVSTAR) (800-W load, 28-V Regulated, 10900 nm, 63° inclined, MEO bus)
GRC	Glenn Research Center (of NASA at Lewis Field)
GSE	ground support equipment
GSFC	Goddard Space Flight Center (of NASA)
HAJ	hydrazine arcjet
HEO	highly elliptical orbit (Molniya)
HTSC	high-temperature superconductor
HV	high voltage
IECEC	Intersociety Energy Conversion Engineering Conference
IEEE	Institute of Electrical and Electronic Engineers
IEMS	integrated energy momentum wheel system
IGBT	insulated gate bipolar transistor
INM	Inmarsat satellites (3500-W load, 28-V sun-regulated, GEO bus)
IMC	inter-module connector
I/O	input/output
IPS	ion propulsion system
IPV	individual pressure vessel (for one battery cell)
IR	infrared
Isp	specific impulse (per kg of fuel)
ISRO	Indian Space Research Organization
ISS	International Space Station (105 kW, 160/120 V regulated, LEO bus)
ISO	International Standards Organization (Geneva, Switzerland)
I–t	current versus time
I & T	integration and testing
ITO	indium tin oxide
I–V	current versus voltage
JPL	Jet Propulsion Laboratory
KOH	potassium hydroxide (electrolyte)

LCP	light-concentrating panel
LEO	low Earth orbit
Li-Ion	lithium ion
Li-Poly	lithium polymer (solid electrolyte)
LN	liquid nitrogen (−193°C boiling point)
LISN	line impedance stabilization network
LV	low voltage
LPC	load power conditioner
MEO	mid Earth orbit
MBSU	main bus switching unit
MEP	main enabling plug
M-G	motor–generator
MHD	magneto-hydrodynamic
MIL-SPEC	military specifications (U.S.)
MIL-STD	military standard (U.S.)
MMW	multimegawatts
MOSFET	metal-oxide semiconducting field effect transistor
MPG	multiple point ground
MTBF	mean time between failures
MWA	momentum wheel assembly
NASA	National Aeronautics and Space Administration
NASDA	National Space Development Agency (of Japan)
NiCd	nickel cadmium battery
NiH_2	nickel hydrogen battery
NiMH	nickel metal hydride
N-S	north-south
NSSK	north-south station keeping
NOAA	National Oceanic and Atmospheric Administration
NSTS	National Space Transportation System
OBC	on-board computer
OC	ordnance controller, also open circuit
PCDU	power control and distribution unit
PCE	payload control electronics
PDP	power distribution point
PCU	power control unit
PDU	power distribution unit
PEM	proton exchange membrane (fuel cell)
PEMS	power and energy management software
PM	permanent magnet
PMAD	power management and distribution
PMCS	power management and control software

PME	propulsion module electronics
P-P	peak-to-peak (ripple)
PPT	peak power tracking
PRI	primary unit
PRP	power return point
PRU	power regulator unit
PPR	prime power return
PRA	pyrotechnic relay assembly
PV	photovoltaic
P–V	power versus voltage
PWM	pulse width modulation
PYRO	pyrotechnic (ordnance)
RBI	remote bus isolator
R-DET	regulated-direct energy transfer
RFI	radio frequency interference
RFC	regenerative fuel cell
rms	root mean squared
rpm	revolution per minute
RPC	remote power controller
rss	root sum squared
RTG	radioisotope thermoelectric generator
RTN	return
RUL	remaining useful life (of battery)
RUE	remaining useful energy (of battery)
RWA	reaction wheel assembly
SA	solar array
S&A	safe and arm
SAD	solar array drive
SAPS	solar array position sensor
SAS	solar array switch
SBA	shunt boom assembly
SBR	space-based radar
SC	short circuit
SD	shunt dissipator
SDI	Strategic Defense Initiative
SDV	shunt drive voltage
SESB	single-event secondary breakdown
SEU	single event upset
SEL	single event latch-up
SLA	stretched lens array
SMC	safe mode controller
SOC	state of charge (of battery)
SPA	solar panel assembly

SPE	solid polymer electrolyte (fuel cell)
SPG	single point ground
SPS	solar power satellite
SPT	stationary plasma thruster
SPV	single pressure vessel (for all battery cells)
SP-100	Space Power-100 (DoD Program)
SR-DET	sunlight regulated direct energy transfer
SS	summer solstice
SSA	sun sensor assembly
SSU	sequential shunt unit
SSM	switched shunt module
SSPA	solid state power amplifier
SSPC	solid state power controller
SSS	solid state switch
TCS	thermal control system
TDMA	time division multiple access
TDRS	tracking and data relay satellite (1700-W load, 28-V regulated, GEO bus, 0° inclination)
TE	thermoelectric
TEC	thermoelectric converter
TF	transfer function
TLM	telemetry
TPR	thrust to power ratio
TPV	thermo-photovoltaic
TRDF	total ripple distortion factor
TSS	tethered satellite system
TT&C	tracking, telemetry and command
TWTA	traveling wave tube amplifier
USRA	Universities Space Research Association
USSF	U.S. Space Foundation (Colorado Springs, CO)
VE	vernal equinox
V/f	voltage to frequency ratio
$V–t$	voltage versus time
$V–T$	voltage versus temperature
XIPS	xenon ion propulsion system
WCBV	worst case bus voltage in normal operation
WS	winter solstice

Suffixes

mp	maximum power point
oc	open circuit
op	operating point
sc	short circuit

Trade names

A2100	Lockheed Martin Space Systems, Sunnyvale, CA
BSS-702 and BSS-601	Boeing Satellite Systems, El Segundo, CA
Bussmann	Cooper Bussmann Inc., St. Louis, MO
Littelfuse	Littelfuse Corporation, Des Plaines, IL
Microsheet	Corning Inc., Corning, NY
Nomex, Mylar, Kapton	E.I. Dupont Company, Wilmington, DE
Kevlar, KelF, Teflon	
SuperNiCd	Eagle Picher Technologies, LLC, Joplin, MO
P600L Fuse	Mepcopal Corporation, San Diego, CA
PADS A/D	Mentor Graphics, Wilsonville, OR
PSPICE	Intusoft Corporation, Gardena, CA

Systems of Units and Conversion Factors

	SI unit =	English unit ×	Factor below
Length	meter	foot	0.304 8
	mm	inch	25.4
	micron	mil	25.4
	km	mile	1.609 3
	km	nautical mile	1.852
	m^2	foot squared	0.092 9
Area	μm^2	circular mil	506.7
	liter (dm^3)	cubic foot	28.316 8
Volume	liter	cubic inch	0.016 39
	cm^3	cubic inch	16.387 1
	m^3/s	cubic foot/hr	0.028 31
	liter	gallon (US)	3.785 3
	liter/s	gallon/minute	0.063 09
Mass	kg	pound mass	0.453 59
	kg	Slug	14.593 9
Density	kg/m^3	pound/mass/ft^3	16.020
	kg/cm^3	pound mass/in^3	0.027 68
Force	N	pound force	4.448 2
Pressure	kPa	pound/in^2 (psi)	6.894 8
	kPa	Bar	100.0
	kPa	std atm (760 torr)	101.325
	kPa	1mm Hg at 20°C	0.132 84
Torque	N·m	pound-force foot	1.355 8
Power	W	foot pound/sec	1.355 8
	W	Horsepower	745.7
Energy	J	foot pound-force	1.355 8
	kJ	Btu International	1.055 1
	MJ	horsepower hour	2.684 5
	MJ	Therm	105.506
Temperature	°C	°F	(°F − 32)·5/9
	K	R	(°F + 459.67)·5/9
Heat	W	Btu(Int.)/hour	0.293 1
	kW	Ton Refrigeration	3.517

	W/m^2	Btu/(ft^2 hr)	3.154 6
	W/(m^2·°C)	Btu/(ft^2 hr) °F)	5.678 3
	MJ/(m^3·°C)	Btu/(ft^3 °F)	0.067 1
	W/(m·°C)	Btu inch/(ft^2 hr °F)	0.144 2
	W/(m·°C)	Btu ft/(ft^2 hr °F)	1.730 4
	J/kg	Btu/pound	2.326
	MJ/m^3	Btu/ft^3	0.037 3
	J/(kg·°C)	Btu/(pound °F)	4.186 8
Velocity	m/s	foot/sec	0.304 8
	m/s	mile/hour	0.447 04
Magnetics	Weber	line	10^{-8}
	wb/m^2 (tesla)	kiloline/inch^2	0.0155

Other Units and Conversions

1 e = -0.1592×10^{-18} coulomb
1 eV = 0.1592×10^{-18} J
1 MeV = 0.1592×10^{-12} J
1 calorie = 4.1868 J
1 kg cal(Int.) = 4.1868 kJ
1 erg = 0.1 μJ
1 dyne = 10 μN
1 tesla = 10,000 gauss (lines/cm^2)
1 micron = 1 μm
1 nautical mile = 1.15081 mile
1 rad = 100 erg/g
1 rad(Si) = 1 krad = 10 J/kg
1 rad(Si) generates 1.7×10^{13} electron hole pairs in 1 g of silicon
Absolute zero temperature = −273.16°C = −459.67 °F
Boltzmann constant = 1.38×10^{-23} J/K
Stefan–Boltzmann constant = 5.67×10^{-8} W/m^2·K^4

Space Flight Constants

Acceleration due to Earth gravity	$9.806\ 7\ \mathrm{m/s^2}$ ($32.173\ 5\ \mathrm{ft/s^2}$)
Earth equatorial radius (mean)	6378.140 km
Geosynchronous orbit altitude	35,786 km
Obliquity of ecliptic at Epoch 2000	23.439 291°
Precession of the equinox	1.396 971 3°/century
Earth-sun distance (AU)	149.597 870 million km
Mean lunar distance	0.384 401 million km
Sidereal year [a,b,c,d]	$3.158\ 814\ 955 \times 10^7$ seconds
Solar year	365 days, 6 hours, 9 minutes, 9.54 s
Solar flare maxima year	1990 + 11.0 n (n = 1,2,...)
Solar flux radiation	$1358 \pm 5\ \mathrm{W/m^2}$ at 1 AU
Earth annual mean albedo	0.30 (fraction of the solar flux reflected by the Earth)
Earth's long wavelength radiation[e,f,g]	$237 \pm 7\ \mathrm{W/m^2}$
Earth's effective black body temperature	206 to 262 K
Permittivity of free space, μ_o	$4\pi \times 10^{-7}$ henry/m
Permittivity of free space, ε_o	8.85×10^{-12} farad/m

Velocity of light in free space $V_o = \dfrac{1}{\sqrt{\mu_o \varepsilon_o}} = 299.863 \times 10^{-6}$ m/s

a. Solar year is measured with respect to the sun. It is the time the Earth takes to complete one revolution around the sun, which is 356 d, 6 h, 9 m, 9.54 s.
b. Sidereal year is the time interval between two successive passages of the vernal equinox over the meridian.
c. Solar day is the time the Earth takes to complete a rotation, relative to the sun, measured from noon to noon at the prime meridian. It is divided into 24 solar hours.
d. Sidereal day is the time the Earth takes to complete one revolution around the imaginary axis through the North and South Poles. It is measured with respect to any fixed start in the sky, and is equal to 23 h, 56 m, 4.09 s, about 4 m shorter than the solar day. It is divided into 24 sidereal hours.

e. Long wavelength radiation is a minimum when albedo is a maximum
f. Due to orbital orientation, peak solar flux, albedo or long wavelength radiation may not be incident on a given surface
g. Diurnal (daily) variation can be 36 W/m^2.

List of Chapters

Part A: Power System Fundamentals

Part B: PV–Battery System

Part C: Power System Performance

Part D: Special Power Systems

Contents

Part C: Power System Performance

Part A
Power System Fundamentals

Chapter 1

Satellite Overview

1.1 Introduction

A satellite consists of various systems designed to meet the mission specific requirements. All but the simplest satellites require a common set of systems shown by the solid lines in Figure 1.1. Complex satellites require additional systems shown by the dotted lines. The systems are classified into two groups, the payload and the bus. The payload consists of the communications equipment in commercial satellites or science instruments in research satellites. The bus consists of all remaining equipment grouped into several functional systems that support the payload. The power system is one of the bus systems that consist of the solar array, battery, power electronics, distribution harness, and controls. Other essential bus systems are the communications and data handling system to receive commands and return information, telemetry sensors to gage the satellite state, and a

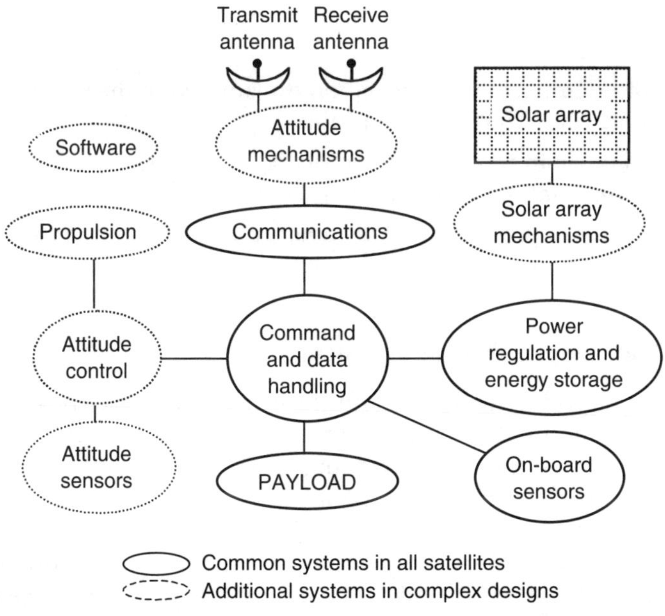

FIGURE 1.1 Satellite systems.

central computer to coordinate and control activities of all the systems. Satellites with complex missions also require systems to determine the spacecraft attitude and orbit orientation, and propulsion to control both.

Satellite design is generally optimized to the fullest extent such that any change would result in a higher cost. The total mission cost, however, is a complex function of many variables, and so is the power system cost. The cost, C, in dollars per watt of power generated can be expressed as a function of four major variables as follows:

$$C = f(X_1, X_2, X_3, X_4) \tag{1.1}$$

where X_1 = cost per kg of the power system mass launched,

X_2 = cost per liter of the power system volume launched,

X_3 = cost per watt of the power system generation capacity, and

X_4 = cost of altitude control related with power system components.

1.2 Satellite Systems

The typical communications satellite bus consists of the following systems.

1.2.1 Communications and Data Handling

The communications and data handling system performs three independent functions:

It receives and demodulates information transmitted to the satellite from the ground station via command links.

It transmits data, both recorded (remote) and real-time, from the satellite to the ground station via data links.

It transmits bus equipment and other telemetry data from the satellite to the ground station via telemetry links.

1.2.2 Attitude and Orbit Control System

The attitude and orbit control system determines the exact position of the satellite with respect to the local vertical, thus providing precise pointing of the communications antennas, imaging sensors, and any other mission sensors. The attitude control function accepts error signals from which the basic or the precision attitude determination function provides 3-axis attitude control using three reaction wheels. The basic attitude determination function obtains the pitch and roll data from the Earth sensor and yaw data from the gyroscopes with updates from the sun sensor to provide a basic 3-axis pointing within 0.1° accuracy. The precise attitude determination function uses three gyroscopes with updates from a star sensor to provide a 3-axis pointing within 0.01° accuracy.

1.2.3 Tracking, Telemetry, and Command System

The tracking, telemetry, and command (TT&C) system accepts analog, discrete, and digital data from various systems of the spacecraft, and processes them into a continuous data stream for direct transmission to the ground or for on-board storage for later transmission. These data are analyzed and evaluated on the ground to determine the spacecraft state of health and the operational configuration. The command and control function is all digital. It provides ascent guidance from the booster separation through transfer orbit, and controls the satellite attitude and operating modes on orbit. The control is exercised in accordance with commands and data received from the ground station, supplemented by signals and data supplied by other systems of the satellite. This system also provides the error correction coding, a key function of the system.

1.2.4 Electrical Power System

The electrical power system generates, stores, conditions, controls, and distributes power within the specified voltage band to all bus and payload equipment. The protection of the power system components in case of all credible faults is also included. The basic components of the power system are the solar array, solar array drive, battery, battery charge and discharge regulators, bus voltage regulator, load switching, fuses, and the distribution harness. The harness consists of conducting wires and connectors that connect various components together.

In the Earth orbiting satellite, the solar array is rotated once per orbit by the solar array drive to track the sun at or near normal angle. The rotation is rate-servo controlled. The body information and position errors are computed by the satellite computer to derive rate control signals. The nominal rate of rotation is 0.06° per s. Using slip rings and carbon brushes is one way of providing the rotary joint between the rotating array and the satellite body. The control signals for required rotation rate come from the TT&C system, which also selects the rotation direction.

1.2.5 Thermal Control System

The thermal control system maintains the temperature of all equipment within the specified limits during normal and abnormal operations. It provides both passive and active cooling as needed. Typical components of this system are fixed radiators, louvers, multi-layer blankets, coatings, tapes, heaters, thermostats, temperature sensors, and control electronics. Thermistors are widely used as temperature sensors. The components are sized for the average electrical power dissipation, the external heat input from the sun, the Earth's reflected sunlight (albedo), and the long wavelength (infrared) heat radiated from the Earth.

1.2.6 Structure and Mechanisms System

The structure and mechanisms system primarily provides a frame for mounting and linking various mechanical components together. Mechanisms for deploying the booms, the solar array and other components after orbit injection are often included in this system. The deployment power circuits and devices are well shielded for electromagnetic interference (EMI) to prevent unspecified deployment. The deployment motion is derived by spring-loaded rotary mechanisms. A rotary vane damper filled with viscous silicone fluid governs the deployment rate. The solar array deployment generally involves cable cutting and/or rod cutting. Structures are often made of magnesium and aluminum. Composites are also common. Some steel and occasionally beryllium are used where needed.

1.2.7 Propulsion System

The propulsion system provides propulsion torque for 3-axis control during ascent and for maintaining the satellite momentum below a specified maximum level during the mission. It often uses a mixed system of high-pressure regulated helium or nitrogen and liquid hydrazine. The gaseous helium or nitrogen and hydrazine are carried in high-pressure cylindrical titanium alloy tanks. The propulsion systems also provides the translational delta-V for orbit changes and orbit trims

Figure 1.2 depicts the anatomy of the Global Positioning Satellite (GPS), a mid-Earth orbit communications satellite fleet of the U.S. Air Force. The payload antenna is facing the Earth. The solar array panels are oriented north–south on two booms running through the solar array drive. Mounted inside the north and south panels of the spacecraft body are the batteries, power regulators, and the control electronic boxes (not shown).

1.3 Earth Orbit Classification

The Earth is a sphere with a slight flatness at the top. Its diameter is 12,713.54 km at the poles and 12,756.32 km at the equator, the difference being 42.78 km. Air surrounds the Earth and extends up to 160 km above the surface, beyond which the atmosphere gradually fades into space. The satellites orbiting the Earth are classified into several groups with their typical parameters listed in Table 1.1. The orbits are often described by the following abbreviations:

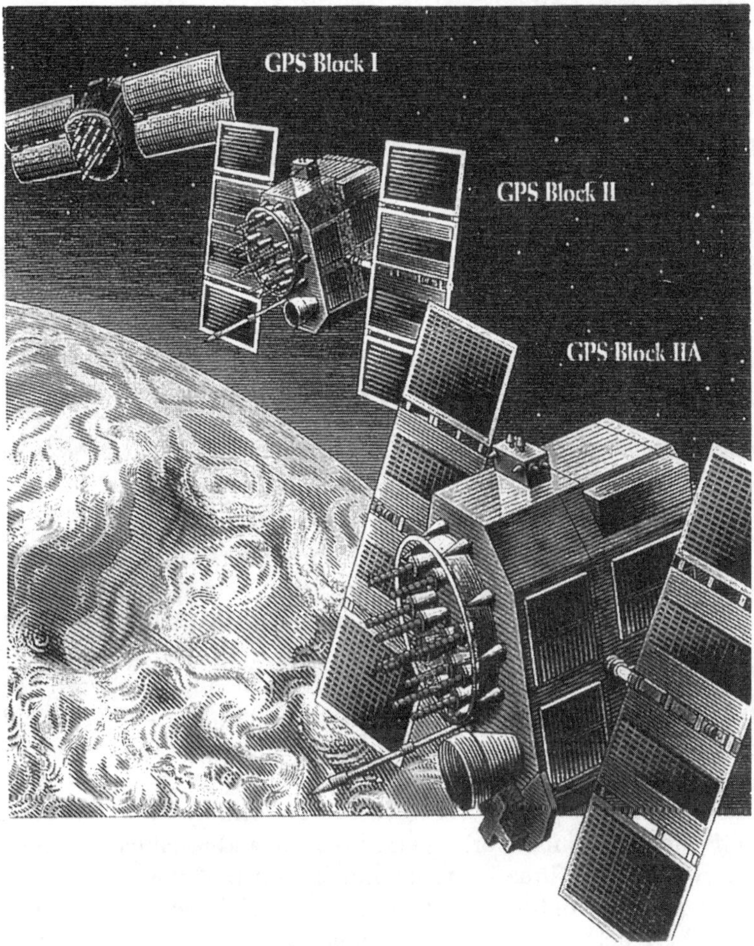

FIGURE 1.2 Configuration and evolution of GPS, mid-Earth orbit navigation satellites of the U.S. Air Force.

GEO: geosynchronous Earth orbit, circular at 35,786-km altitude
MEO: mid Earth orbit, circular at 2000 to 20,000-km altitude
LEO: low Earth orbit, generally circular at 200 to 2000-km altitude
HEO: highly elliptical orbit, such as Molniya

A communications satellite provides interconnectivity over a large area for point-to-point, point-to-multi-point, and broadcast communications. It can serve fixed as well as mobile terminals anywhere — on land, on the sea, in air or in space. A typical satellite transponder receives an uplink signal from a ground station, frequency converts, amplifies and retransmits it to the ground.

Table 1.1 Classification of the Earth's orbits

Orbit type	Apogee (km)[a]	Perigee (km)[b]	Eccentricity[c]	Inclination[d] (degrees)	Period[e]
Geostationary	35,786	35,786	0	0	1 sidereal day
Geosynchronous	35,786	35,786	Near 0	0-90°	1 sidereal day
Elliptical Molniya	39,400	1000	High	62.9°	½ sidereal day
Low Earth	Various	Various	0 to high	0 to 90°	>90 min

[a]Closest distance from the Earth surface.
[b]Farthest distance from the Earth surface.
[c]Ratio of difference to sum of apogee and perigee radii.
[d]Angle between orbit plane and equatorial plane.
[e]One sidereal day is 23 h, 56 min, 4.09 s.

In the case of LEO and MEO, the orbit parameters are chosen to avoid the radiation belts that surround the Earth at altitudes of 1.3 to 1.7 and 3.1 to 4.1 Earth radii. A typical LEO satellite has an altitude of 500 to 1500 km, an orbit period of 1.5 to 2 h, and is visible to a given Earth station for only a few minutes in every orbit period. A typical MEO satellite is between 5000 and 12000 km altitude with orbit period of several hours. In a highly elliptical inclined orbit, it can see the polar regions for a large fraction of its orbit period.

A GEO satellite moving west to east at an altitude of 35,786 km (22,237 miles) results in a nominal orbit period of 24 h, and remains stationary with respect to the Earth. Three such satellites spaced 120° apart in the equatorial plane can provide continuous coverage of the globe except near the poles. The launch vehicle booster and its upper stages deliver the satellite in the transfer orbit, which is an elliptical orbit with the Earth at one of its foci and the apogee at the geosynchronous orbit. An apogee kick motor is then fired to circularize the orbit at the geosynchronous height. The primary features of various orbits are described below.

1.3.1 Geostationary Orbit

The geostationary orbit is a very special geosynchronous orbit, and, in fact, is unique. It is exactly circular with a radius of 42,164 km in the Earth's equatorial plane with zero degree inclination and zero eccentricity. A satellite placed in this orbit is synchronized with the Earth's rotation rate and direction (eastward). It does not move with respect to the Earth, and sees the same object on the Earth steadily. The orbit period is the same as that of the Earth's rotation, i.e., 23 h 56 m 4.09 s. As a result, the satellite's beam-to-Earth and the ground station's beam-to-satellite are steady in position. This simplifies the design and operating requirements of both the satellite and the ground station. However, it takes more fuel to reach and maintain the geostationary orbit than any other orbit around the Earth at that altitude. Numerous satellites already placed there make it difficult to get a desirable location in this orbit that would avoid radio frequency interference from neighboring satellites. The Tracking and Data Relay Satellite (TDRS) of the U.S. Department of Defense is an example of a geostationary satellite. A satellite placed in this orbit tends to drift away from its assigned station. Hence, a periodic station-keeping operation is required.

The time in space is kept in sidereal time, which measures the rotation of the Earth in relation to a fixed star. Solar time is used on Earth to measure the Earth's rotation in relation to the sun. The same star is not in the same place at the same solar time, but is at the same place at the same sidereal time from day to day. A sidereal day consisting of 24 sidereal hours is the time the Earth takes to rotate once on its axis past an imaginary line from the Earth's center to any star. Thus, the sidereal time is measured from a point in the sky called the vernal equinox, although no bright start marks this point.

The geostationary orbit period is exactly 1 sidereal day. It is slightly shorter than the mean solar day of 24 h because of the sun's apparent motion resulting from the Earth's rotation around the sun, which is 360° in 365.24 days, i.e., 0.9856° per day. By the time the Earth has rotated once in relation to a distant star, it has moved westward along its orbit, as depicted in Figure 1.3. The sun is then 0.9856° east of its position at the start of the Earth's rotation. The Earth needs additional time to rotate eastward to come back in line with the sun. The Earth must thus rotate a total of 360.9856° in 1 mean solar day so that the meridian will align itself with the sun from one noon to the next in exactly 24 h (86,400 s). The time for the Earth to rotate 0.9856° past one rotation is $86{,}400/(0.9856/360.9856) = 235.91$ s. The sidereal period of rotation is therefore $86{,}400 - 235.91 = 86{,}164.09$ s, or 23 h, 56 m, 4.09 s, which is shorter by 3 m and 55.91 s than the mean solar day.

1.3.2 Geosynchronous Orbit

Most commercial communications satellites operate in numerous geosynchronous orbits. The geostationary orbit described above is one unique orbit in the entire class of the geosynchronous orbit. The distinction between the two is minor and fine, but is important. The geosynchronous orbit is similar to the geostationary orbit, except that its inclination can be any value between 0 and 90°. Inclinations other than 0° requires ground station tracking antennas. Sometimes, that may not be a disadvantage because the ground stations require tracking antennas for other reasons. Mobil platforms, such as planes and ships, also require tracking antennas. The geosynchronous orbit is chosen for fuel-efficient launch and orbit

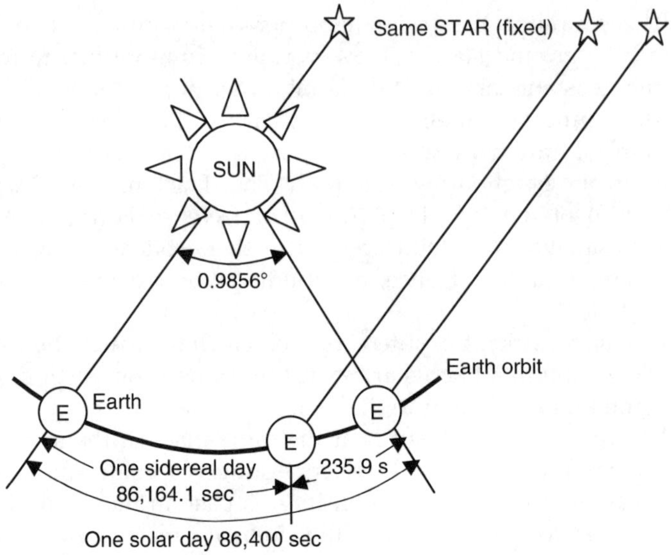

FIGURE 1.3 Sidereal day and mean solar day for Earth.

maintenance. If a satellite is placed in i° inclination orbit, the point directly below the satellite oscillates between i° north and i° south every day, and appears to drift to the north and south in a figure eight as shown in Figure 1.4. The angular height of this figure is just the magnitude of $\pm\pi i^\circ/180$ radians. This motion away from the equator induces a longitudinal difference between the ideal and actual satellite points. The difference

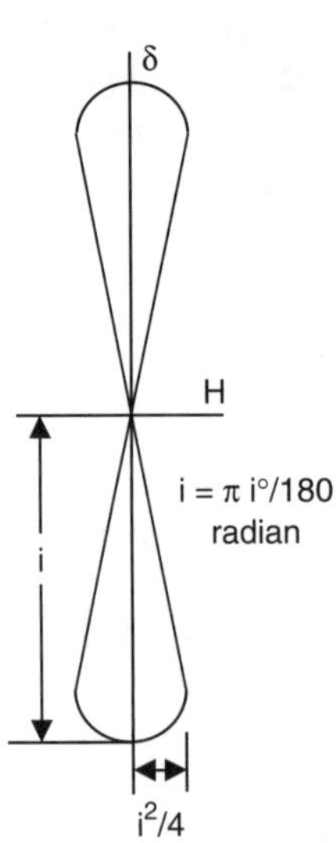

FIGURE 1.4 Satellite motion in geosynchronous orbit with i° inclination.

appears as the satellite moves towards the equator. The maximum longitudinal deviation is $(\pi i^\circ/180)^2/4$. The variation in distance is $\pm(\pi i^\circ/180)R_o$, where R_o is the orbit radius. The ground stations must be aligned with the north–south (N–S) motion of the satellite in an inclined orbit.

When the sun and the moon are not in the equatorial plane, the N–S components of their combined gravity force change the orbit inclination of the geostationary and geosynchronous satellites at a rate about 0.85° per year. Station-keeping maneuvers are required to compensate for such orbit drift. This consumes some fuel, which must be carried on board to last for the mission duration. The use of arc jets minimizes the fuel requirement by increasing the propulsion efficiency. The arc jets and station-keeping fuel requirement can be eliminated if 0.85° drift per year is acceptable, or can be accommodated in the mission design, as in TDRS. In exchange, the satellite would require several degrees of yaw maneuvering on daily basis to remain pointed at the ground station.

1.3.3 Highly Elliptical Orbit

Among highly elliptical orbits, Molniya is one specific orbit named after a Russian communications satellite with a 1000-km perigee and a 39,400-km apogee. Having the period of ½ sidereal day in this orbit, the satellite comes to the same longitude on every other apogee. The advantage of the Molniya orbit is good coverage of the entire northern hemisphere. The disadvantage is no coverage over the southern hemisphere. Moreover, it requires more satellites and needs two tracking antennas at each ground station. GPS, although not in a Molniya orbit, has an orbit period of ½ sidereal day because of its selected MEO location in a circular orbit. Some US military satellites have used the elliptical Molniya orbit with 63.4° inclination to cover Russia for 10 h out of a 12-h period.

1.3.4 Low Earth Orbit

This is approximately a circular orbit at low altitude. The International Space Station (ISS) and NASA's space shuttle orbiter operate in low Earth orbit. Most communications satellites operate in GEO, but some newer constellations are planned and/or placed in LEO between 500 and 2000-km altitudes and 30 to 90° (polar) inclinations. Being closer to the Earth, smaller and simpler satellites can be used in this orbit. Also, two-way communications introduces a time delay of only 0.02 s versus 0.5 s in geosynchronous orbits. On the negative side, LEO communications satellites require tracking of omni-directional antennas, and many birds are needed for wide coverage.

1.3.5 Sunsynchronous Orbit

A satellite in this orbit maintains a constant angle between the sun's direction and the orbit plane, and always sees the sun at the same angle. It is used for special applications.

1.4 Orbit Mechanics

Kepler's three laws of planetary motion, based on Newton's laws, apply to a satellite orbiting a planet. They are as follows:

First law: The satellite orbit is an ellipse with the planet at one focal point.

Second law: The line joining the planet and the satellite sweeps equal areas in equal times. If the time intervals $\triangle t_1$ and $\triangle t_2$ in Figure 1.5 are equal, then the swept areas A_1 and A_2 are also equal.

Third law: The square of the orbit period is proportional to the cube of the semi-major axis,

i.e.,
$$T_o^2 = \frac{4\pi^2 a^3}{\mu} \tag{1.2}$$

where a is the semimajor axis of the orbit, and μ is the gravity constant of the planet. For the Earth, μ is $3.986 \times 10^{14}\ \text{m}^3/\text{s}^2$ or $3.986 \times 10^5\ \text{km}^3/\text{s}^2$.

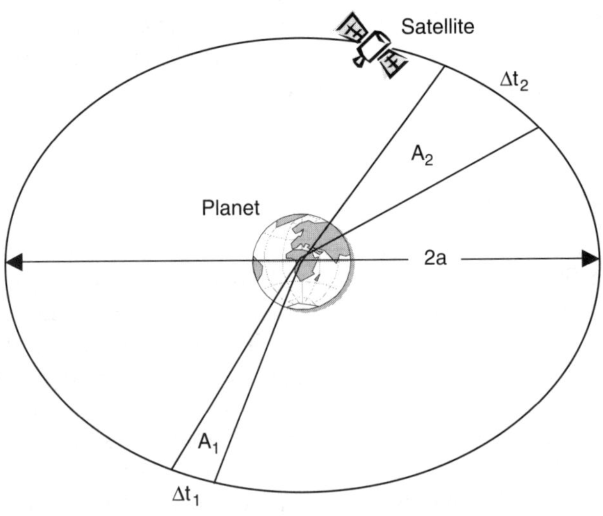

FIGURE 1.5 Kepler's second law of planetary motion.

For circular orbits, the third law gives the orbit period T_o in terms of the orbit radius R_o,

$$T_o = \frac{2\pi R_o^{1.5}}{\sqrt{398600}} s, \quad \text{or} \quad T_o = 2.7644 \times 10^{-6} R_o^{1.5} h \tag{1.3}$$

For the orbit period to be 1 sidereal day of 86,164.09 s, the orbit radius must be 42,164 km. Deducting the mean radius of the Earth surface 6378 km, we get the geosynchronous satellite altitude of 35,786 km above the Earth surface. This altitude is about six times the Earth's radius.

The satellite velocity in circular orbit is given by

$$v = \frac{2\pi R_o}{T_o} \tag{1.4}$$

which is 3.075 km/s in GEO orbit. In comparison, the Earth travels in its orbit around the sun at speed of 30 km/s, about ten times faster.

1.5 Satellite Stabilization Methods

The satellite stabilization in orbit is achieved by either an active or passive method described below.

1.5.1 Gravity Gradient Method

Gravity gradient is a passive method, sometimes used in small LEO satellites. The difference in the attractive force of gravity on the parts closest and farthest from the Earth creates a moment that maintains the satellite aligned with the local vertical. This method requires long booms in order to have adequate moment, and it does not work in geosynchronous orbits because of near-zero gravity there.

1.5.2 Magnetic Damping

Magnetic damping is another passive method. It uses long booms with magnets that interact with the Earth's magnetic dipole field to produce the stabilization moment.

1.5.3 Spin Stabilization

Spin stabilization is an active method that has been commonly used in most satellites until recently. It is still used for small satellites. The spinning bicycle wheel and the spinning top are stable above certain minimum spin rate. The satellite stability can be similarly maintained by storing angular momentum in a spinning body on board the satellite. For spin stabilization, the moment of inertia about the desired spin axis must be greater than that about any orthogonal axis. Small satellites are spun in entirety. Large satellites using complex antennas are split in two sections, the despun antenna section and a spinning cylindrical body. The solar cells are mounted on the spinning body as shown in Figure 1.6. A typical spin rate is 30 to 60 revolutions per minute. Heavier satellites require higher spin rates for stability. Spin-stabilized satellites are also called dual-spin or gyrostat satellites.

1.5.4 Three-axis Stabilization

Three-axis or body stabilization is another active method commonly used in modern satellites. In this method, the satellite uses several spinning momentum wheels located inside the body as shown in Figure 1.7. The orientation is automatically maintained by servo-control that adds or subtracts momentum from the spinning momentum wheels. Thrusters are used periodically to maintain the orientation as necessary. The satellite usually has a box-shaped body with flat solar panels (wings) extending from the north and south faces. Table 1.2 compares key features of the spin-stabilized and 3-axis stabilized satellites. The 3-axis stabilization generally results in lower dry mass for satellites with solar array power exceeding a few hundred watts. For this reason, it is widely used in modern large high-power satellites.

All spacecraft, once disturbed from their stable position, may oscillate for a long time in characteristic modes which may be close to being unstable. It is important that these modes are identified and suitable damping is

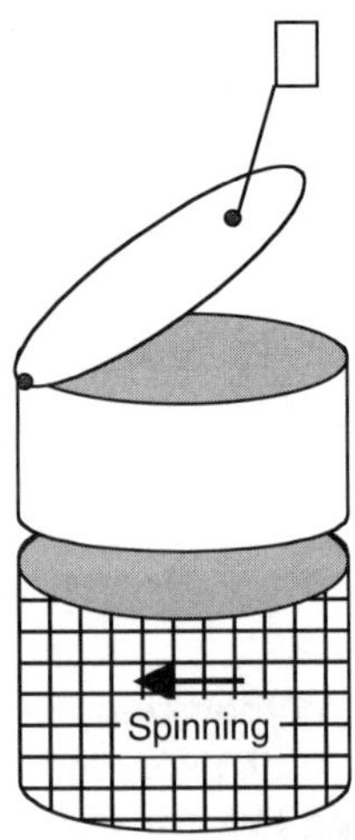

FIGURE 1.6 Spin-stabilized gyrostat satellite.

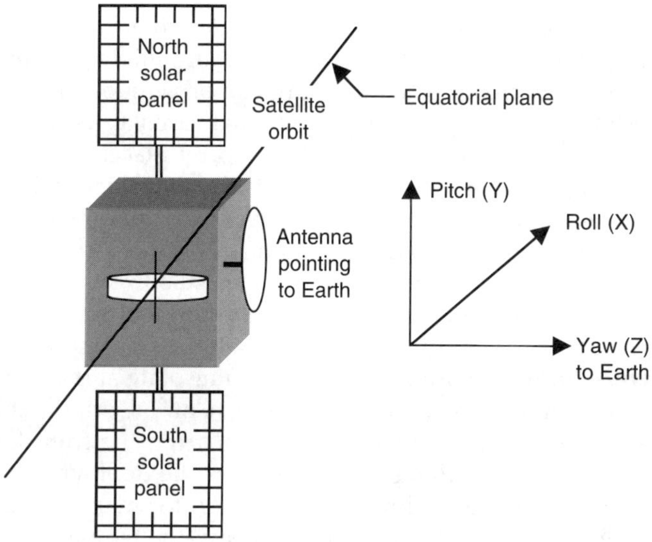

FIGURE 1.7 Three-axis body-stabilized satellite with definition of attitude axes.

introduced by the attitude and/or orbit control system to restore the vehicle to the stable position. Fuel movement in the tanks may also add in to the oscillations, but it is normally controlled by baffles. There are five points within the reference frame in space at which a stationary body will be in equilibrium. All these points are in the plane in which the dominant masses rotate. They are referred to as the Lagrangian or Libration points, and are of potential use for the spacecraft in the Earth–Moon type systems.

1.6 Launch and Transfer Orbits

The communications satellite is placed in a geosynchronous orbit in two main steps. The launch vehicle places the satellite first into a low Earth

Table 1.2 Key features of spin-stabilized and 3-axis stabilized satellites

Spin-stabilized	3-axis stabilized
Inherently stiff due to rotational inertia	Bias or zero momentum maintains the stability
Simple mechanical structure	Complex attitude control
Only 1/3rd of the solar array generate power at any time	Full solar array generates power all the time
Power limited by body size that fits the launch vehicle	Can have high power by adding solar panels
Less flexibility in design	Great flexibility in design
Suitable for small satellites	Suitable for large satellites

circular orbit, called the parking orbit. Then, the so-called Hohmann transfer takes the satellite to the final orbit using minimum fuel. The first velocity increment changes the low circular orbit into a highly elliptical transfer orbit with perigee that of the final circular orbit. The second velocity increment at the apogee of transfer orbit places the satellite in the final circular orbit. When the perigee and the apogee kick motors are fired, some sort of stabilization is needed because the thrust would tumble the satellite and cause incorrect orbit injection.

The fully deployed satellite, which is 3-axis stabilized in the operational orbit, can use spin stabilization in the transfer orbit when the solar panels are stowed into a box shaped body. The satellite is despun by applying reaction wheel torque to bring to a non-spinning state at the end of transfer orbit. The de-spinning operation takes about 10 min. Until the solar array is fully deployed, the sunlit panel radiates heat from the front face only, as opposed to both the front and back faces after the deployment. Moreover, the exposed panel is oriented normal to the sun for maximizing the power generation except during maneuvering. To keep the temperature of the sun side panel from rising above the tolerance limit, the satellite is spun at a low rate, such as $1/10^{th}$ to 1 revolution per minute. Spinning at such a slow barbecue rate is merely for thermal reasons even when the spinning is not required for stability. The spin rate is gyro controlled. One can deploy the array in the transfer orbit, but it adds a mechanism and structural complexity, resulting in added mass, low reliability, and difficult transfer orbit maneuvers.

1.7 Operational Orbit

As the satellite revolves around the Earth in operational orbit at inclination θ measured from the equatorial plane, it changes its orientation with respect to the sun with seasons as depicted in Figure 1.8. The north of the Earth rotation axis is inclined 23.45° towards the sun on the summer solstice day, and 23.45° away on the winter solstice day. On the autumnal and vernal equinox days, the axis inclination is zero, resulting in equal day and night.

1.8 Eclipse due to Earth

Since the ecliptic and equatorial planes are inclined to each other by 23.45°, the angle of incidence of the sunlight on the solar arrays varies from 66.55 to 90°. The corresponding incident solar flux varies from 91.75% on a solstice day to 100% on an equinox day. However, the satellite on equinox days encounters longest eclipse once per day when the Earth blocks the sunlight from illuminating the satellite.

When the satellite is in shadow of the Earth, the solar array power generation ceases and its temperature drops sharply. Predicting the eclipse

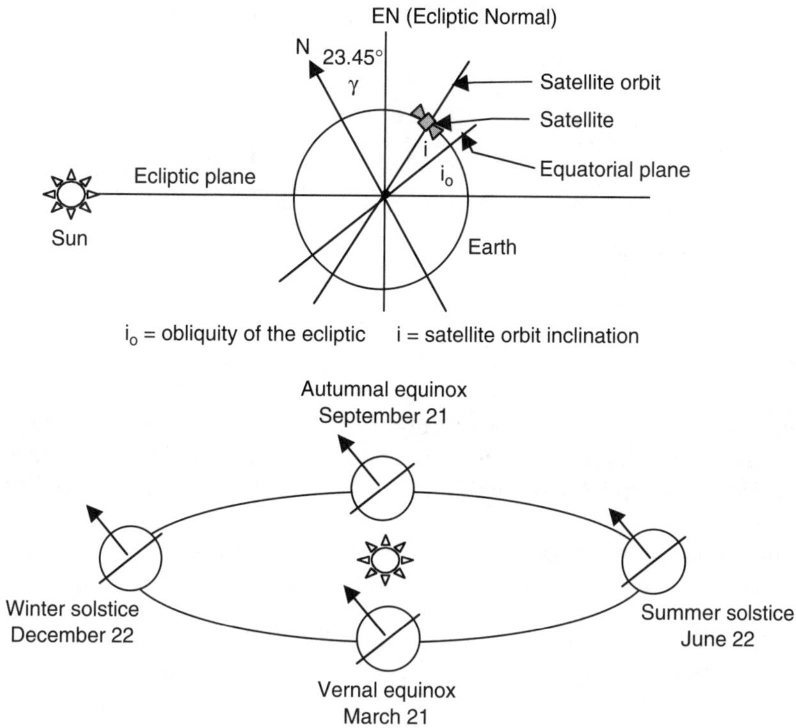

FIGURE 1.8 Satellite in Earth's orbit with seasonal variations.

duration is, therefore, important for the spacecraft power system design. For the geosynchronous satellite, the longest eclipse occurs on the vernal and autumnal equinoxes when the sun is in the equatorial plane as shown in Figure 1.9. The duration for which the entire sun is blocked is called the umbra (total eclipse marked by dotted arc). The total arc when the sun is fully or partially blocked is called the penumbra (arc *a–b*). It is proportional to the mean solar day accounting for the Earth's orbital motion during the eclipse. The umbra duration varies with the seasons, the longest being 69.4 min occurring around March 21 and September 21. From the geometrical considerations of the geosynchronous orbit in Figure 1.9, the penumbra duration is 73.7 min (1.228 h) and the umbra is 4.3 min shorter than the penumbra. Since the solar array output voltage and current during this 4.3 min would not meet the requirement for the power system operation, penumbra is taken as the eclipse duration for the power system design.

As the sun moves above or below the equator after an equinox, the eclipse duration becomes shorter and shorter, and finally becomes zero when the inclination of the sun becomes high enough (Figure 1.10). The number of days the geosynchronous satellite sees an eclipse, and the eclipse

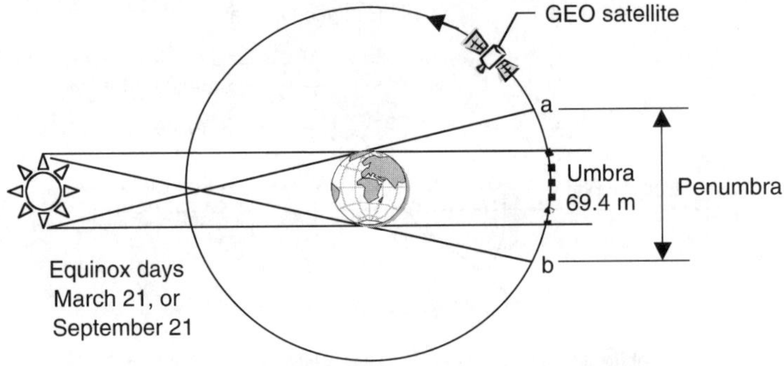

FIGURE 1.9 Eclipse in geosynchronous orbit — umbra and penumbra.

duration on that day, are shown in Figure 1.11. The eclipse onset time on a particular day is of interest to the satellite design engineer, because it determines the required services and the battery requirement onboard the satellite.

In near-equatorial, circular, low Earth orbits, eclipses of approximately equal duration occur once every orbit period. The eclipse duration is

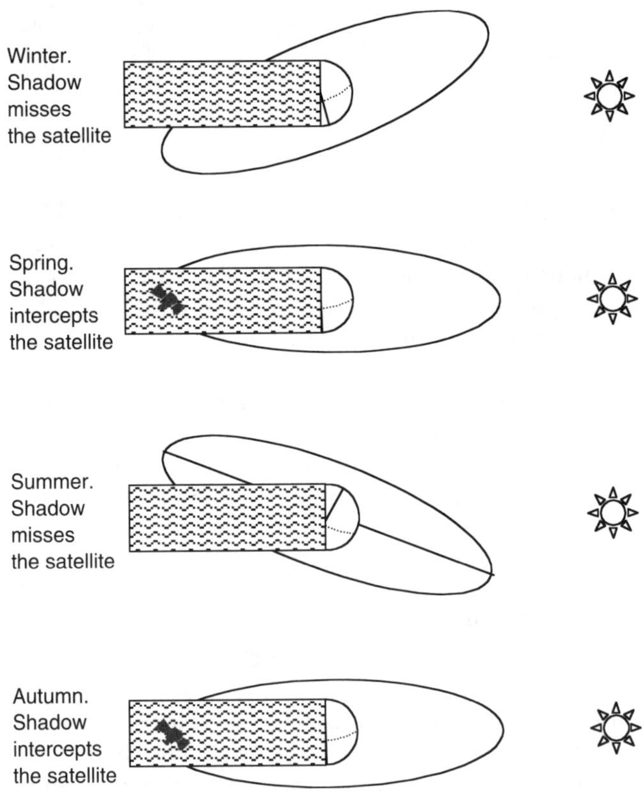

FIGURE 1.10 GEO eclipse, once per orbit in spring and autumn seasons only.

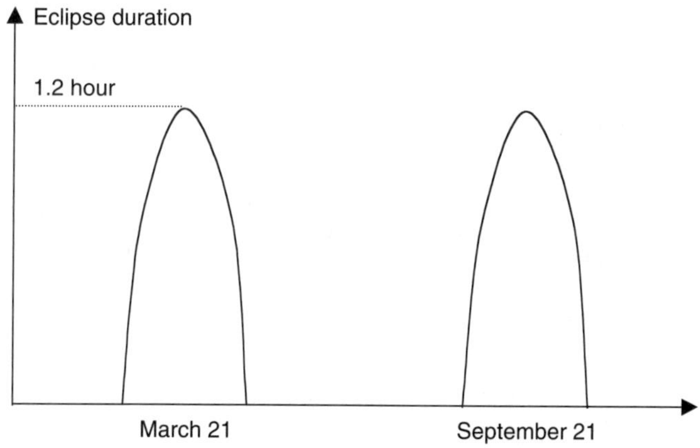

FIGURE 1.11 GEO eclipse duration longest on vernal and autumnal equinox days.

dependent on the orbit altitude, inclination, and the sunlight incidence angle on the orbit plane (Figure 1.12). It can vary by a factor of two in LEO.

For a circular orbit, the eclipse duration (in h) is given by

$$T_e = \frac{1}{2} + \frac{1}{\pi}\sin^{-1}\left\{\frac{\left(1 - \left(\frac{R_{\text{Earth}}}{R_{\text{orbit}}}\right)^2\right)^{1/2}}{\cos\beta}\right\} \tag{1.5}$$

where β = angle of the sunlight incidence on the orbit plane, i.e., the angle between the Sun–Earth line and the local normal of the orbit plane.

The β angle varies seasonally between $\pm(i + \gamma)$, where i = orbit inclination with respect to the equator and γ = angle between the sun line and the ecliptic plane (23.45°). As β increases, the eclipse duration decrease, which improves the load capability of the electrical power system. At certain value of high β, no eclipse occurs. There are polar and near-polar low Earth orbits

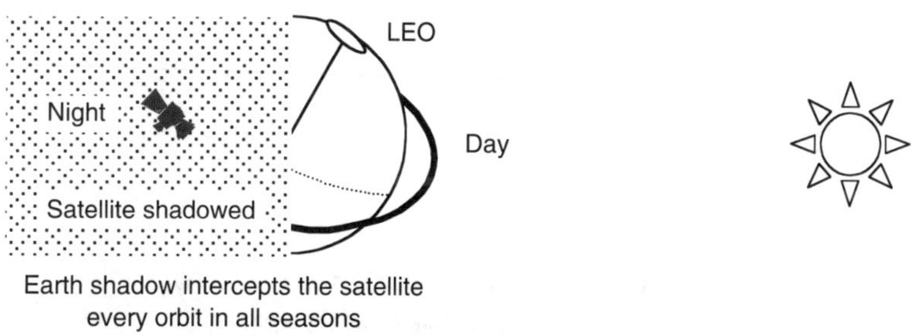

FIGURE 1.12 Eclipse in near-equatorial LEO, once per orbit in all seasons.

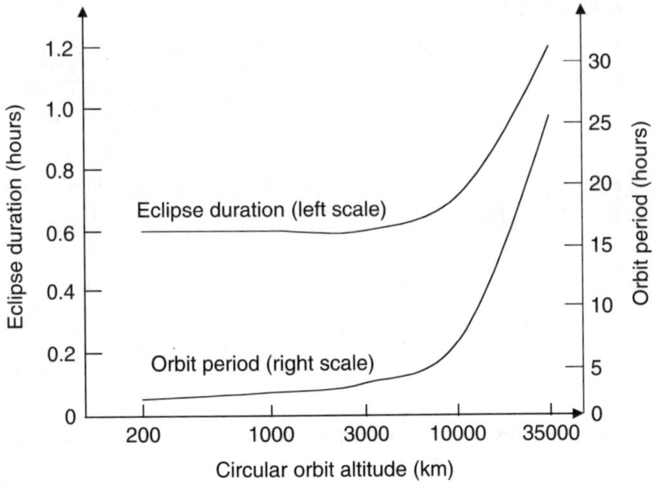

FIGURE 1.13 Eclipse duration and orbit period vs circular orbit altitude.

that never have an eclipse of the sun. On the other hand, the longest eclipse occurs at $\beta = 0$.

1.8.1 Example

For a satellite in a 6343-mile radius and 20° inclination orbit, the above equation gives $T_e = 0.63\,\text{h}$ or 38 min long eclipse (note that the argument of the $\sin^{-1}$ must be in radians).

For circular orbits, the eclipse duration and the number of eclipses per year are plotted in Figures 1.13 and 1.14, respectively. The ratio of the maximum eclipse to minimum sunlight duration is an indicator of a

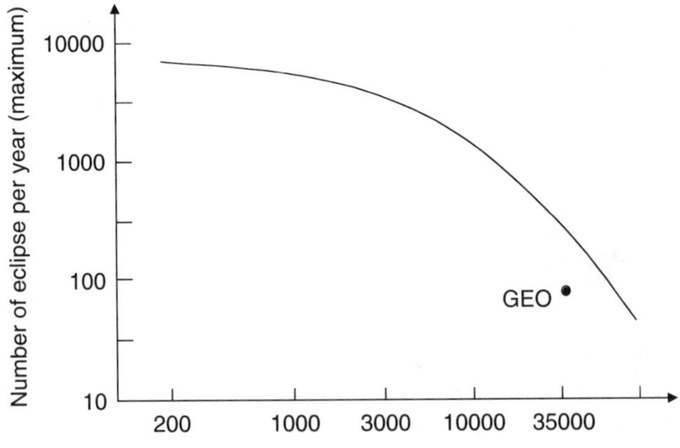

FIGURE 1.14 Maximum number of eclipse per year vs orbit altitude.

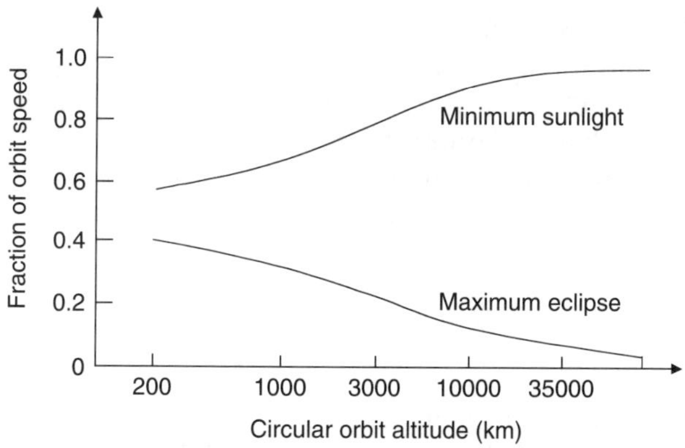

FIGURE 1.15 Minimum sun time and maximum eclipse duration in circular Earth orbits.

challenge posed to the power system design engineer. The greater the ratio, the heavier the battery requirement to power the load during an eclipse. It also requires a larger solar array to capture the energy required during shorter periods of sunlight, and to divert a higher fraction of it to charge the battery while simultaneously supplying full load to the payload. Such a demand is greater on a low Earth orbit satellite, as seen in Figures 1.15 and 1.16.

1.9 Eclipse due to Moon

In addition to eclipses due to the Earth's shadow on the satellite, the moon can also obstruct the sun as seen by the satellite. Such eclipses are irregular in occurrence, varying from 0 to 4 per year, with an average of two. Usually they are spaced far apart, but in the worst case two eclipses can occur in 24 h. The duration can vary from several minutes to more than 2 h, with an

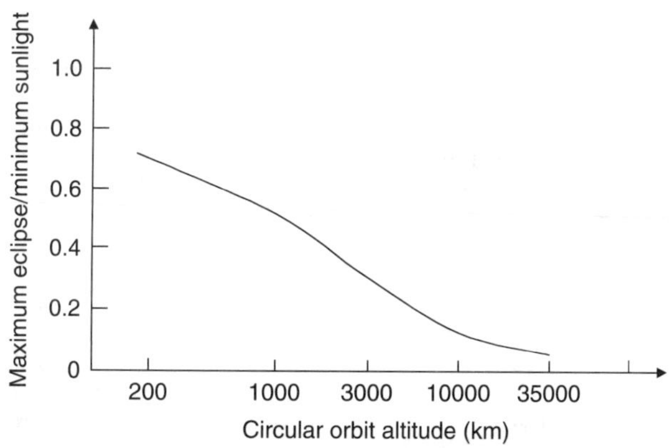

FIGURE 1.16 Ratio of maximum eclipse to minimum sun time in circular Earth orbits.

average of 40 min. The satellite may experience additional battery depth of discharge and fall in temperature in the case where an eclipse due to the moon occurs adjacent to an eclipse due to the Earth. In most missions, however, eclipses due to the moon impose no additional design requirement, but that has to be ascertained. Otherwise, shedding a noncritical load temporarily during the worst moon eclipse often circumvents the situation.

1.10 Solar Flux

The energy received from the sun in space varies with the distance squared. The Earth's orbit around the sun is approximately circular with a slight eccentricity of 0.01672. The distance, therefore, varies within ± 0.01672 times the average distance between the sun and the Earth, which is 149.6 million km, defined as one astronomical unit (AU) of distance. Thus, the solar flux varies over $(1 \pm 0.01672)^2$ or 1 ± 0.034 of the yearly average. Within these small variations, the Earth is closest to the sun around January 2 (perihelion), and farthest around July 2 (aphelion). For many years, the average solar radiation in the Earth orbit has been taken as $1358 \pm 5\,W/m^2$ on the surface normal to the sun. Measurements reported by Frohlich[1] show a higher average value of $1377 \pm 5\,W/m^2$, which is now generally accepted. However, the conservative number of $1358 - 5 = 1353\,W/m^2$ continues in wide use.

Table 1.3 shows the ratio of the seasonal flux over the yearly average flux on the equinox and solstice days. It also shows the seasonal variations in the orbit inclination. Satellites using solar arrays with single-axis tracking gimbals generate less power in accordance with the cosine of the sun angle. The last column gives the combined effect of the cosine and the flux variation due to eccentricity on the solar array output power. The output on the summer solstice day is 11% lower than that on the vernal equinox day.

Table 1.3 Variation of solar flux and angle of incidence with the seasons

Day of the year	Seasonal flux/ yearly average	Angle of incidence with single-axis tracking gimbals	Flux on array/yearly average on sun-tracking surface
Vernal equinox, March 21	1.001	0°	1.001
Summer solstice, June 22	0.967	23.45°	0.887
Autumn equinox, September 21	0.995	0°	0.995
Winter solstice, December 22	1.034	23.45°	0.949

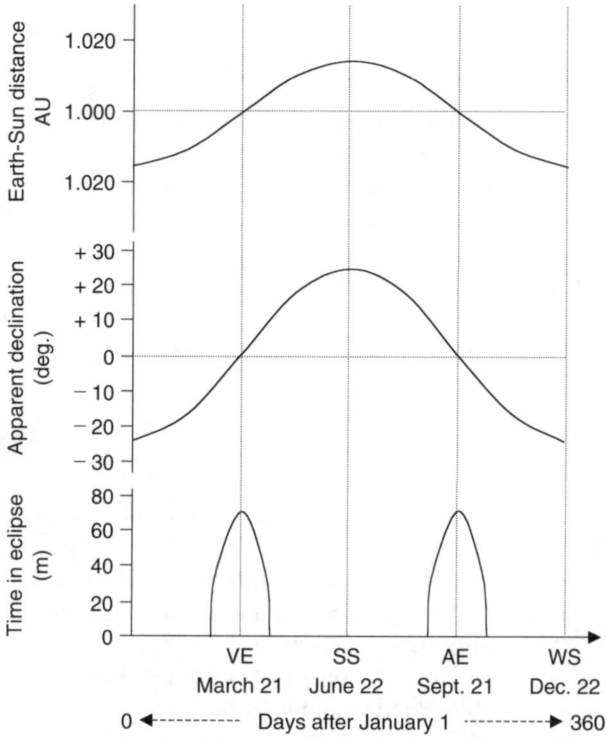

FIGURE 1.17 Seasonal variations in solar parameters over 1 year in geosynchronous orbit.

The seasonal variations in solar parameters in geosynchronous orbit are collectively depicted in Figure 1.17.

1.11 Beta Angle

The beta angle, β, is defined as the angle between the Earth–Sun line and the orbit plane when the spacecraft is closets to the sun (orbit noon). It varies seasonally between 0° and $(i + i_o)^\circ$, where i = orbit inclination, and i_o = angle between equator and ecliptic plane, which is 23.45°.

The value of $\beta = 90^\circ$ results in the greatest solar flux on the satellite body, and $\beta = 0^\circ$ gives zero flux. In most satellites with the array always pointed to the sun by the solar array drive motor, the β angle has an insignificant effect on the generation of electric power. However, the thermal control system is impacted by the β angle. Low β may require additional heaters, while high β may require additional cooling. The β angle impacts the solar array temperature, which in turn has a small secondary effect on the power generation. The most significant effect of the β angle on the power system design comes from the eclipse duration as given by Equation 1.5. As β

increases, the eclipse duration decreases, which consequently requires a smaller battery and less charging power during sunlight.

The β angle is not to be confused with the sun angle, θ, often used to define the sunlight incidence angle on the solar array, which could be canted to catch the sun normally in case β is not 90°. The sun angle is defined as the angle between the solar array plane and the sun-pointing vector. The power generation is proportional to cos θ, so θ of 90° results in the maximum power generation, and 0° gives zero power.

The power system engineer starts with the orbit parameters specified by the customer, which primarily set the orbit period, eclipse duration and the β angle. These parameters in turn have the greatest impact on the power system design.

1.12 Spacecraft Mass

Hundreds of small spin-stabilized and large 3-axis stabilized satellites have been built and deployed in various orbits. The spacecraft dry mass and the corresponding electrical power requirements for GEO satellites have been steadily rising as seen in Figure 1.18. The new technological developments in solar cells and evolutionary design improvements have jointly contributed to the higher solar array output power per kilogram for modern communications satellites. For large 3-axis body stabilized GEO satellites with total eclipse load of 5 to 10 kW, the electrical power system (EPS) mass is about 30% of the total spacecraft dry mass, the payload mass is also about 30%, and the remaining 40% is in the structure and all other spacecraft systems.

The average mass of 16 commercial satellites contracted in 2002 was about 3600 kg each, which is considered the mass of medium size satellites. On the large side, Thuraya for the United Arab Emirates will have lift off mass of 5200 kg.

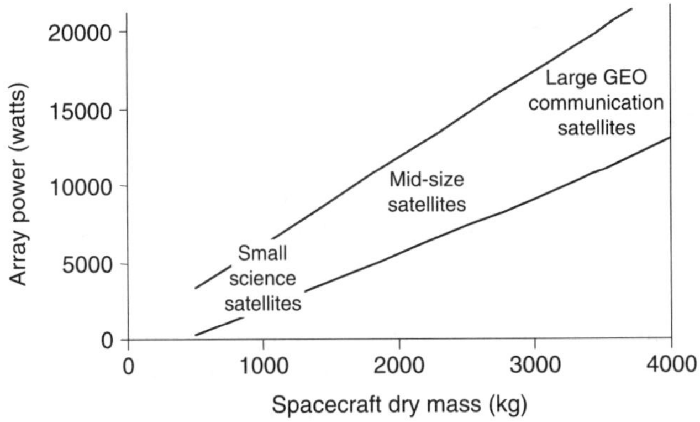

FIGURE 1.18 Power vs dry mass of GEO communications satellites.

Reference

1. Frohlich, R.C., *Contemporary Measures of the Solar Constant: The Solar Output and its Variations*, Colorado Associated University Press, Boulder, CO, 1977, pp. 93–109.

Chapter 2

Near-Earth Space Environment

2.1 Introduction

The near-Earth space environment, i.e., thermosphere and ionosphere, presents a substantial risk to space systems. The spacecraft power system must withstand the space environment and meet full performance specifications over the entire mission life, starting from the launch phase to disposal at the end of life. General design criteria required for this purpose are studied and documented by space research organizations such as NASA.[1] As the solar array is directly exposed to the space environment, it is particularly vulnerable to damage. In fact, the rate of damage on a solar array determines the spacecraft life in many cases. The environmental factors also affect the overall design of all other components of the power system.

2.2 Launch and Transfer Orbit Environment

High levels of acceleration, shock, and vibration are present during lift-off and during pyro-driven deployments at various stages. The resulting stress levels, varying with the launch vehicle, have effects on the power system design, particularly on the solar panels. For example, the solar panels are required to withstand the launch acceleration of about $3g$ on the space shuttle *Orbiter*, and about $10g$ on the launch vehicle *Saturn*, where g is the acceleration due to gravity on the Earth's surface. The pyro shock levels during deployments can be extremely high for a few milliseconds. The shock spectrum is generally characterized by high-frequency components. The peak shock level in *Ariane* during payload separation, for example, could be about $2000g$ at frequencies above 1.5 kHz.

In transfer orbit, the solar panel is still folded, but must withstand the accelerating force at perigee and the braking force at apogee. Thermally, the outer solar panels must withstand the Earth's heat radiation and albedo, in addition to the solar radiation, without exceeding the specified temperature limit.

2.3 On-orbit Environment

The power system design is influenced primarily by the following on-orbit environmental factors.

2.3.1 Lack of Gravity and Atmosphere

The lack of gravity and atmosphere in space has significant effects on the power system design. Zero gravity and vacuum jointly deprive all spacecraft equipment from the natural convection cooling which is normal on Earth. The internal thermal design in space, therefore, depends primarily on conduction cooling and, to some extent, on radiation cooling. The heat rejection to the outer space, however, must depend only on radiation cooling.

The vacuum causes sublimation and outgassing, more in some materials than others. The subsequent condensation of the sublimed vapor on cold surfaces can cause short circuits in electrical parts. Since zinc has a high sublimation rate, and some polymers have high outgassing rates, their use in space is restricted. The solar array current collecting slip rings and brushes can experience cold welding under high contact pressure in a vacuum. The use of lubricants having low sublimation rates, or keeping them in sealed pressurized enclosures, is therefore important.

2.3.2 Magnetic Field

The interaction of the Earth's magnetic field, B, and the magnetic dipole moment, M, created by an electrical current loop produces torque, $\boldsymbol{T}$, on the spacecraft. The cross-vector product of M and B gives this torque (in Nm), i.e.,

$$T = M \times B \tag{2.1}$$

where M is expressed in $A{\cdot}m^2$ and B in tesla. The magnetic moment is produced by current loops in the spacecraft power circuits. The moment (in $A \cdot m^2$) is defined as the vector product of the current and the loop area, i.e., $M = I \times A$. Its direction is perpendicular to the loop area as shown in Figure 2.1.

In the geostationary orbit, the normal component of B is constant around $0.104\,\mu T$, and the radial component varies between $\pm 0.042\,\mu T$ as the satellite revolves around the Earth. In orbits at other altitudes, the magnetic field varies inversely with the orbit radius cubed. The normal component of B produces a torque in the equatorial plane, averaging to zero over the orbit period. Both components of the torque affect the satellite's attitude. The power system contribution to the magnetic torque comes from the solar panel, battery, and wires connecting various system components.

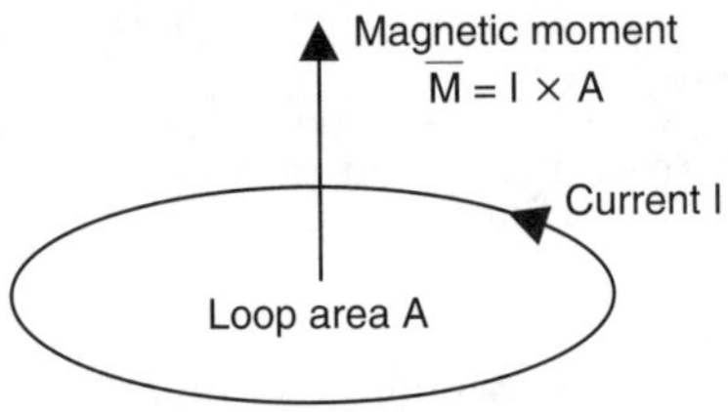

FIGURE 2.1 Definition of magnetic moment.

The magnetic moment of the spacecraft is minimized by design and compensation in the following ways:

- Lay out the current carrying conductors such that they form the smallest possible loops.
- Make two adjacent loops in opposite directions, so that they compensate each other.
- Twist wires wherever possible to neutralize the moments of adjacent twists.

After using one or more of the above design means, the net residual magnetic moment is determined by tests after the satellite is assembled. A moment that would limit the resulting torque below 100 μNm in the operational orbit is generally considered acceptable.

2.3.3 Meteoroids and Debris

Impact of solid objects can cause damage to the solar array. Small particles not large enough to cause immediate damage can cause gradual degradation in output power over time. The particle mass and the hit rate (flux) depends on the orbit. Several NASA studies conducted for predicting the meteoroid flux in the Earth's orbit have given consistent results. They show that the average number varies inversely with the meteoroid mass as given by the following simple expression:

$$\text{meteoroid flux} = \frac{1}{m} \tag{2.2}$$

where the flux is measured in number of particles of mass greater than m grams per square kilometer per year. The most common meteoroids are small, having masses between 0.1 and 100 mg, generally called micrometeoroids. The probability of having n impacts with particles of mass between m_1 and m_2 on a given surface area during y years is estimated using Poisson's probability distribution function. In the absence of accurate information, their mass density is assumed to be 0.5 g/cm^3 and a mean

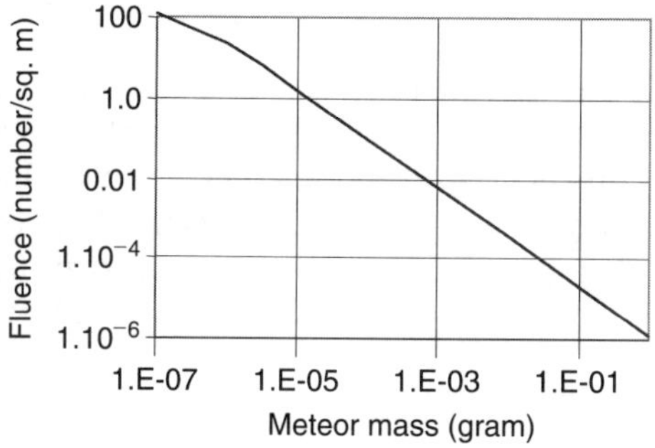

FIGURE 2.2 TDRS meteoroid environment fluence over 15 years.

impact speed of 20 km/s. The actual fluence varies with the orbit. For example, in the geostationary orbit of TDRS, the estimated meteoroid fluence for a 15-year mission is shown in Figure 2.2. The yearly fluence rate in other orbits is shown in Figure 2.3.

In addition to the natural micrometeoroids, the man-made debris can vary from 1 μm to 10 mm in diameter. They have an average density equal to that of aluminum, the most widely used material for satellite and launch vehicle components. Their relative velocity with respect to the spacecraft varies from zero to twice the orbital velocity, with an average around 10 km/s. The impact energy of large meteoroids can instantly damage the solar array protective glass and cells. Micrometeoroids, on the other hand, erode the glass surface gradually with subsequent degradation in the power output over time.

2.3.4 Atomic Oxygen

Atomic oxygen is present in low Earth orbit. It severely erodes some materials, such as silver, which is extensively used in the solar array

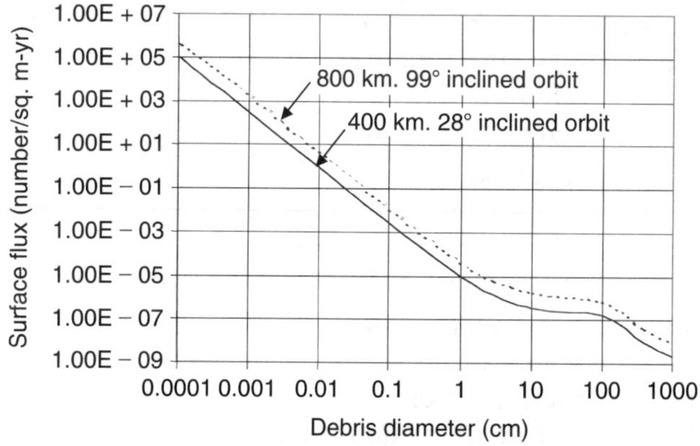

FIGURE 2.3 Debris flux rate estimate in various orbits.

construction. Similar erosion is also seen in some electrical insulation such as Kapton™ and silicon rubber. The erosion comes not only from the chemical activity, but also in large part by the atoms travelling at several km/s velocity relative to the spacecraft. In addition to the erosion (surface recession), atomic oxygen can also form stable oxides on the metal surface. For these reasons, exposed bare silver in solar arrays and silicon or Kapton insulation on electrical wires are undesirable. This topic is dealt with in Chapter 5.

2.3.5 Charged Particles

The sun continuously radiates energy in several forms — visible light, invisible infrared, ultraviolet, X-rays, gamma rays, radio waves, electrons, protons, and plasma (electrically charged hot gas). The abundance of charged particles that come with the sun's radiation makes space a hostile environment. The energy in these particles impinging on a surface causes damage, which accumulates over a period of time. The following terms are widely used in discussing the radiation of charged particles in the space environment.

One million electron volts (1 MeV) is the unit of the equivalent energy level of various charged particles. It is defined as the energy released by the total charge of one million electrons falling through one volt in the electric potential field. With the electron charge of 0.1592×10^{-18} C, 1 MeV equals 0.1592×10^{-12} J.

Flux density expresses the number of charged particles present per unit volume.

Flux is the term that expresses the rate of flux hitting a surface. It is given as the number of particles impinging upon a unit surface area per unit time, i.e., in number/m^2s or in MeV/m^2s to account for their total energy level. The flux varies as the spacecraft moves around the orbit. It is greater on the sun-side of the Earth than on the farther side. Often the orbit average flux is quoted.

Integrated flux or fluence expresses the accumulated number of particles impinging upon a unit surface area over some time duration in orbit. Its unit is number/m^2 year or often MeV/m^2 over the mission life to account for the total energy level the surface must withstand.

Dose expresses the amount of energy absorbed per unit mass of the subject matter. It is measured in rad or rad (Si). Since silicon makes up the bulk of micro-electronic devices, it is taken as the reference material for comparing the radiation energy. One rad is equal to 100 ergs of absorbed energy in 1 g of target material. One rad (Si) is the unit of energy that will generate 1.7×10^{13} electron–hole pairs in 1 g of silicon.

The common sources of charged particles in the space environment are as follows:

Solar radiation, consisting of mostly electrons and protons coming from the sun.

Solar wind brings bursts of charged particles in addition to the normal radiation from the sun. It is primarily made of protons and some electrons radiating from the sun. Their flux and energy level around the Earth's orbit are low during normal solar wind. The mean proton flux is about 200 million protons/cm^2 with a mean energy level of several keV. However, the energy levels can reach 100 MeV during solar flares, and even 1 GeV during intense flares.

Cosmic radiation comes mostly from the outer space and some probably from the sun. It consists of about 85% proton, 12% alpha particles and 3% electrons. The proton energy level is high in GeV, but the flux rate is low: in the order of a few particles/cm^{2}s.

2.4 Van Allen Belts

The Earth's magnetic field covers the magnetosphere, a doughnut shaped region of space influenced by the Earth's magnetic field. It acts on the electrons and protons coming from the space. The Van Allen radiation belts are parts of the magnetosphere that contain a large number of particles. The magnetosphere normally shields the Earth from these particles. However, when a disturbance on the sun radiates an abundance of particles, some reach the Earth's atmosphere near the magnetic poles and cause a glow known as an aurora, popularly known as the northern lights (or the southern lights, in the southern hemisphere).

The Earth's magnetic field varying with the radius cubed has a radial gradient, and it converges near the magnetic poles a shown in Figure 2.4. The solar wind brings charged particles to the Earth's magnetic field. These particles are deflected by the Lorentz force, F, given by the following cross-vector product:

$$F = q\boldsymbol{V} \times \boldsymbol{B} \tag{2.3}$$

where q is the particle charge, V is the particle velocity, and B is the Earth's magnetic field intensity. The particle moving radially towards the Earth at velocity V generates a circumferential force that will impart velocity V_ϕ. The resultant velocity is then V_R, as shown in the plan view. The radial gradient

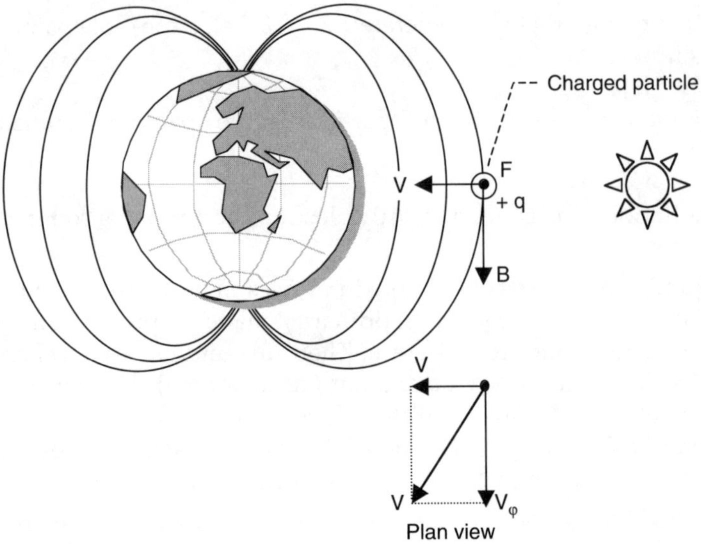

FIGURE 2.4 Lorentz force on a charged particle moving in Earth's magnetic field.

in the field imparts a spiral motion to the particle. The spiraling particle comes to rest once all the energy is absorbed, and then springs back and forth as explained below.

A particle moving in a converging magnetic field generates a force that pushes the particle into a weak magnetic field (Figure 2.5), causing it to drift at velocity V_z. The V_z decreases as the particle moves towards the equatorial plane of weak magnetic field; and V_ϕ increases to keep the kinetic energy constant. The particle eventually comes to a stop in the z-direction and is reflected back into the weaker field. Since the field is again converging on the other side, the particle is reflected back again. Thus, the converging field confines the charge particle between two magnetic mirrors. This way, most

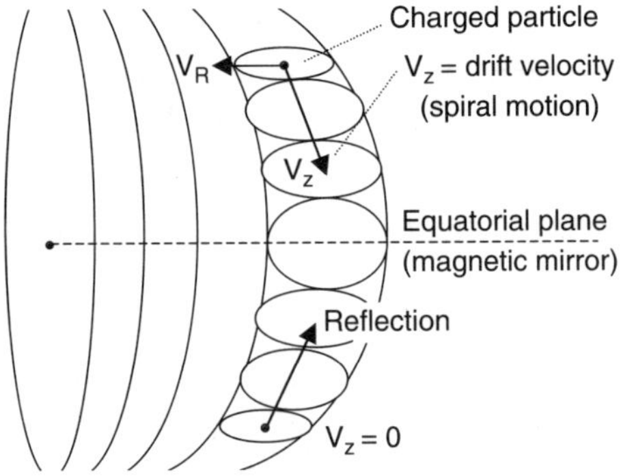

FIGURE 2.5 Charged particle moving in a convergent magnetic field.

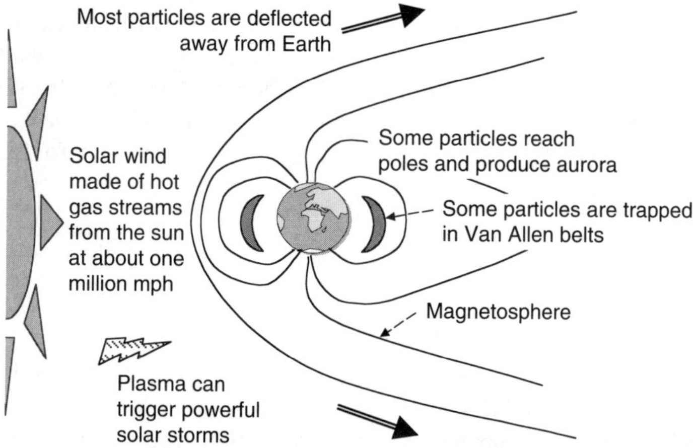

FIGURE 2.6 Van Allen radiation belts.

particles remain trapped in a certain belt all the time. However, particles having sufficiently high axial kinetic energy do not get reflected, and escape the confinement belt.

Most particles become trapped in two doughnut-shaped radiation belts known as the Van Allen belts, which are a part of the magnetosphere. The belts are stronger on the sun-side than on the side away from the sun, as shown in Figure 2.6. The trapping is concentrated in two belts shown in Figure 2.7. The electrons get trapped in a belt spanning 2 to 5 Earth radii with a heavy concentration between 3 and 4 Earth radii. The high-energy protons penetrate closer to the Earth and get trapped in another belt spanning around 1 to 2 Earth radii with a heavy concentration around 1.5 Earth radius. Since the charged particle flux varies over the satellite orbit, the integrated radiation flux (fluence) is used in the power system design.

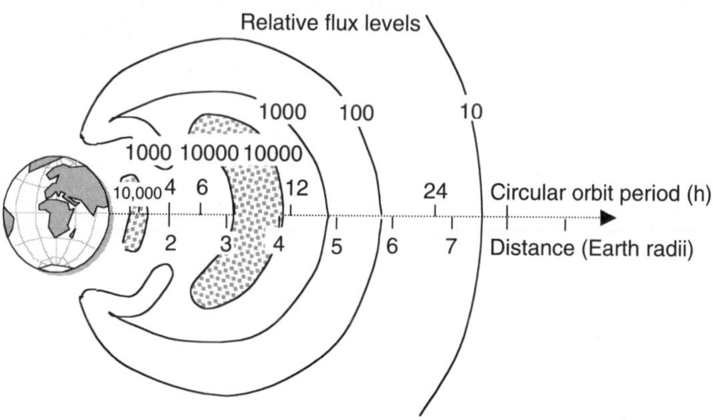

FIGURE 2.7 Extent of Van Allen radiation belts around Earth per NASA's AP-8 Environment Model.

The solar array, in particular, requires protection against radiation damage. The geosynchronous satellites orbiting at 6.6 times the Earth radius are way outside the inner proton belt and at the outer edge of the outer electron belt. Therefore, they are exposed to only a small part of the outer electron belt. However, they are more vulnerable to high-energy proton generated by solar flares.

2.5 Solar Wind and Solar Flare

A continuous stream of protons, electrons, and ionized gas leaving the sun and sweeping the solar system is called the solar wind. It is very thin, containing about 5 charged particles per cubic centimeter of space. The solar wind streams constantly, but not evenly. These particles travel past the Earth at speeds of about 480 km/s. In comparison, the Earth travels in its orbit at 30 km/s. When these particles flow past the Earth, some get trapped in the Earth's magnetosphere.

A solar flare is a gush of ionized gas with charged particles spewed out by the corona, the sun's glowing outer shell. It is associated with dark sunspots bounded by intense magnetic fields that come and go in 11-year cycles. The flare can cause auroras and disturb radio signals on Earth. The ionized gas is extremely hot (>1 million °C), such that atoms of hydrogen and helium are homogenized into dilute plasma, composed mainly of negatively charged electrons and positively charged protons. Although this constitutes a million tons of matter moving at a million miles per hour, its density is so low that it is still considered a vacuum. However, its effects can be devastating as depicted in Figure 2.8.

Solar flares occur in 11-year cycles. The last major period was November and December 2000. Several flares occur during each solar cycle in which 7 years are active and 4 inactive. The intensity levels of the flares vary over a wide range. Therefore, their energy levels and the corresponding probabilities that must be considered in designing the power system are mutually agreed upon by the satellite customer and the manufacture. The relatively strong 1989–1990 flares were extensively analyzed by spacecraft engineers because of the quantity and quality of data obtained and their impact on many spacecraft in orbit at the time. The newly developed model suggests that the proton fluences below 10 MeV were significantly underestimated in the past. This causes more degradation of the photovoltaic (PV) cells in the I_{sc}, although not much in V_{oc}. From this study, it is estimated that a solar array design based on the old King model[2] of 1974 would have about 20% probability of missing the power output requirements by a few percent compared to the new Feynman model[3] of 1989–1990 flares.

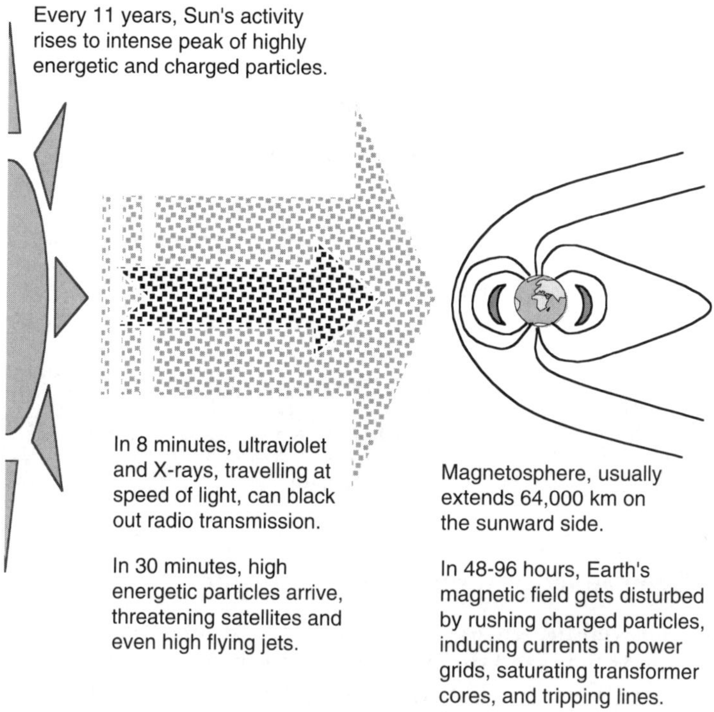

FIGURE 2.8 Solar flare and effects on Earth.

Table 2.1 gives the total annual fluence in a typical geosynchronous orbit. The integrated flux of solar flare protons per 11-year cycle based on NASA/JPL model with 80% confidence level is given in Table 2.2.

Spacecraft going beyond the solar system, such as *Pioneer* and *Voyager*, experience the so-called termination shock wave at the edge of the solar system. The termination shock wave is a zone around the solar system where the solar wind crashes into the tenuous gas and dust that fills interstellar space.

Table 2.1 Total annual fluence in a typical geosynchronous orbit (number/cm^2)[a]

Particle type	Normal solar flares	Intense solar flares
Trapped electrons	$\sim 10^{14}$	$\sim 10^{14}$
Trapped protons	Negligible	Negligible
High energy solar flare protons	$\sim 10^{7}$	$\sim 10^{10}$

[a]The energy level of the electrons varies from a few keV to several MeV, and that of protons in several hundreds MeV range.

Table 2.2 Integrated flux of solar flare protons per 11-year solar cycle[a]

Energy (>MeV)	Solar protons (number/cm^2)
1	5.3×10^{11}
5	2.0×10^{11}
10	1.0×10^{11}
30	2.3×10^{10}
50	9.7×10^{9}
70	6.6×10^{9}
100	2.5×10^{9}

[a]Based on the NASA/JPL model with 80% confidence level.

2.6 Geomagnetic Storm

There is a well-recognized link between solar activity, disturbances in the geomagnetic field, and disruptions to man-made systems such as satellites, communications networks, defense systems, and even the sturdy power grids on Earth. Analysis suggests that these vulnerabilities to magnetic disturbances will continue and perhaps even increase as these systems continue to evolve and grow.

What could happen to the spacecraft power system under the geomagnetic storm can perhaps be appreciated by examining what has happened to the power grids on Earth in the past. Under such storms, the magnetic field of the Earth is disturbed, causing the induced currents in overhead transmission lines. The high-voltage lines, due to their long length and large loop area with the Earth's surface, can induce a high current. This low frequency d.c.-like geomagnetically induced current then enters the high voltage transformers, driving them in deep magnetic saturation, causing the voltage collapse and severe overheating. The great storm of March 13, 1989 affected large areas of North America with a power blackout for 6 million Hydro Quebec customers in Canada for a long time. Voltage regulation was affected, undesired relay operations occurred on system equipment, and new areas of vulnerability were exposed due to unanticipated common mode currents saturating the transformer cores, drawing high currents and tripping off the transmission lines. Less severe storms in September 1989, March 1991, and October 1991 reinforced the fact that geomagnetic disturbances can hamper utility operation. The October 2003 storm knocked out satellites (e.g., Japan's Earth observation satellite *ADEOS-11*), hampered radio communication and caused an electrical blackout in Sweden. To protect airline passengers against the charged particles, the U.S. Federal

Aviation Administration warned pilots to stay below 25,000 feet when flying near the poles in order to avoid excess radiation. The storm arrived within a few hours of NOAA's prediction time.

The space research models of plasma population and circulation, monitoring the interplanetary real-time solar wind and the geomagnetically induced currents, and tracking the auroral zone current can provide timely information for the next solar maxima, which begins in 2010–2011.

Major effects of solar flares and geomagnetic storms on the spacecraft power system are:

- Rapid degradation of the solar array power generation. This is a life-shortening, not a life-threatening, effect.
- Common-mode EMI noise entering the electrical power system components through exposed cables. This can upset the power system possibly causing a permanent damage. In the 1990s, the power systems of a few communications satellites (e.g., *Anik-E1* of Canada and *Intelsat-K*) were damaged, and a few were suspected to have failed under intense solar flares. An in-depth investigation indicated that a spark caused by geomagnetic storm shorted out connections between the satellite solar panels and several dozens of relays.

2.7 Nuclear Threat

Some defense satellites are required to withstand a man-made threat of nuclear detonation of a specified intensity. The energetic particles released during a nuclear event are fission electrons, neutrons, gamma rays and X-rays. Their energy levels depend solely on the destructive power of the nuclear device deployed. The threat level specified for the design is determined from the probability and the consequence considerations. The level of nuclear threat, however, is always a classified number under the U.S. Department of Defense (DoD) secret classification system.

2.8 Total Radiation Fluence

The energy levels of various sources of the natural radiation in the Earth's orbit are as follows:

- Trapped electrons 0.1 to 7 MeV
- Trapped protons 0.1 to 100 MeV
- Solar flare protons 1 to 200 MeV
- Solar flare alphas 1 to 300 MeV
- Galactic cosmic rays >1 GeV

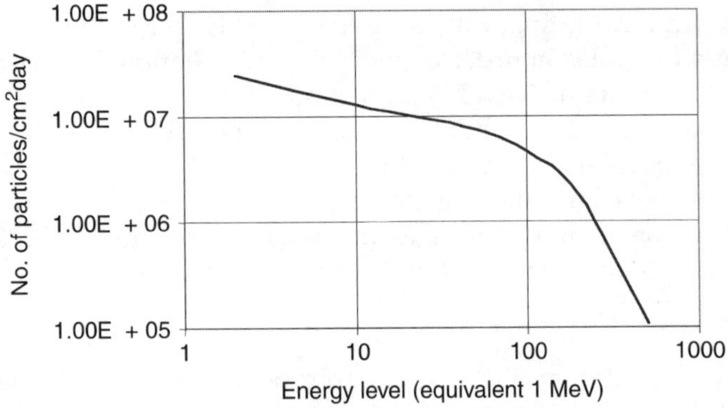

FIGURE 2.9 Trapped proton environment in 852-km, 99°-inclined circular orbit. (Source: NASA/GSFC.)

The total fluence is calculated for a given mission using the environmental model specified by the customer. As an example, the orbit integrated flux of trapped proton and electron environment in a 852-km, 99°-inclined circular orbit is shown in Figures 2.9 and 2.10, where the y-axis is the fluence with energy levels on the x-axis.[4]

The total fluence in equivalent MeV/cm^2 day is the area under such a number (n) versus energy level (E) curves for the electrons, protons, solar flares, and any other source of radiation, natural or man-made. That is,

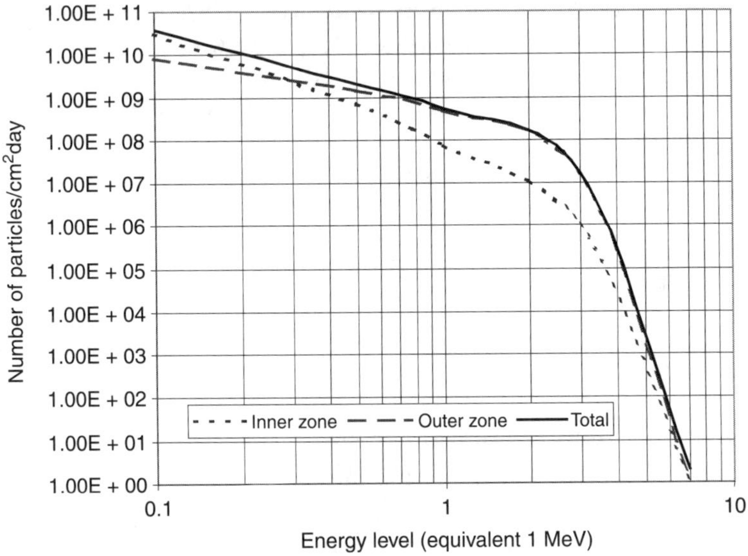

FIGURE 2.10 Trapped electron environment in 852-km, 99°-inclined circular orbit. (Source: NASA/GSFC.)

$$\text{fluence} = \int_0^\infty n\,\mathrm{d}E \text{ electrons} + \int_0^\infty \text{protons} + \int_0^\infty \text{flares} + \int_0^\infty \text{nuclear} + \int_0^\infty \text{any other} \tag{2.4}$$

The integrated fluence determined by Equation 2.4 is then used to determine the damage on the solar array and other components during the mission life.

Several recently completed cross-disciplinary scientific and engineering research projects are significantly improving space environment characterization. Operational users of space systems are the direct beneficiaries of these research advances.

References

1. NASA, "*Space Vehicle Design Criteria* (*Environment*)," Technical Report No. SP-8005, 1980.
2. King, J.H., "Solar proton fluence for 1977–1983 space missions," *Journal of Spacecraft and Rockets*, 11, 401–407, 1974.
3. Feynman, J., "New interplanetary proton fluence model," *Journal of Spacecraft*, 27(4), 403–408, 1990.
4. Stassinopoulos, J.M., Barth, J.M., and Smith, R.L., "*METSAT Charged Particle Environment Study, Revised Edition, Method 2*," NASA Report No. GSFC X-600-87-11, 1987.

Chapter 3

Power System Options

3.1 Introduction

Power requirements in very early satellites were several watts. In today's communications satellites, it is several kilowatts and is growing. Some strategic defense spacecraft power requirements are estimated to be in hundreds of kilowatts and some concepts require hundreds of megawatts of burst power. Solar radiation is the only external source of energy available in space. Any power system not using solar energy must carry its own source of energy on board, such as the primary battery, fuel cell, nuclear or chemical fuel.

The basic components of the spacecraft power system are shown in Figure 3.1. They are the primary energy source, energy conversion, power regulator, rechargeable energy storage, power distribution and protection, and power utilization by the user's equipment (loads). Candidates for the primary energy source include solar radiation, radioisotopes, nuclear reactors, and electrochemical and/or chemical fuel. The energy conversion may be photovoltaic, thermoelectric, dynamic alternator, fuel cell, or thermionic. The energy storage has been primarily electrochemical, although flywheel technology is under development at NASA Glenn Research Center (GRC).

From the available options that are compatible with a given mission and its environment, the satellite level optimization study is conducted to select the best combination of energy source, energy conversion, and energy storage technologies. Final selection must meet multiple criteria, but the

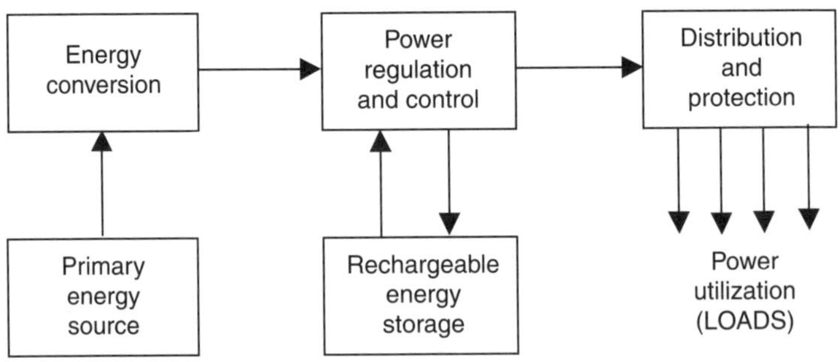

FIGURE 3.1 Basic components of a spacecraft power system.

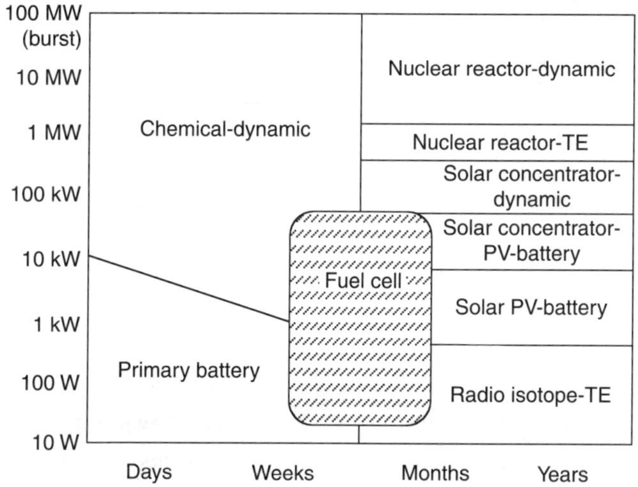

FIGURE 3.2 Optimum energy sources for various power levels and mission durations.

primary criteria are always low mass and low life-cycle cost. Such selection is largely influenced by the product of the power level and the mission duration as shown in Figure 3.2. The dividing lines among various options are only approximate and have large overlaps. The following sections briefly describe these options with their optimum application ranges. The detailed description and performance of often-used options are covered in separate chapters.

3.2 Primary Battery

A primary battery can economically power a small elementary spacecraft requiring only several watts over several days. Early short mission spacecraft flew with primary batteries such as AgZn and NaS. Even today, low power short life satellites carrying instruments with low duty ratio may be designed using a primary battery such as LiCFx as the only power source, thus eliminating the solar panel and battery charge electronics.

The battery cell consists of two electrode plates submersed in an electrolyte as shown in Figure 3.3(a). The electrochemistry of the cell generates an electrical potential difference between the electrodes, which can drive electrical current through an external load circuit. Thus, the battery converts the stored chemical energy between the electrode plates into direct current electricity. The cell can deliver only a certain amount of charge, measured in ampere-hours (Ah), before all of its energy is depleted. The cell voltage decays with the Ah discharged as shown in Figure 3.3(b). The primary battery has nonreversible electrochemistry. It cannot be

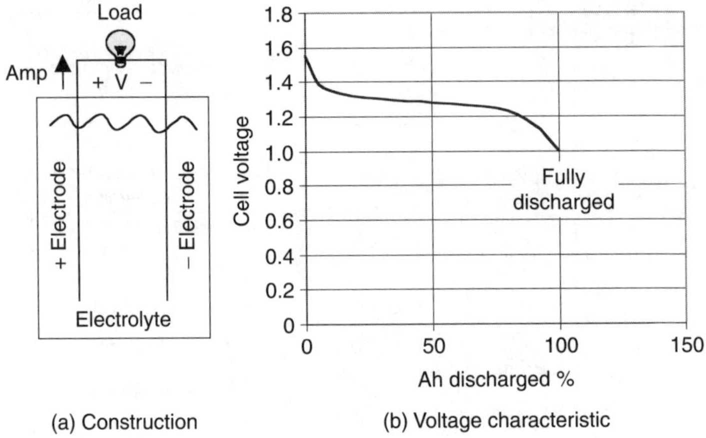

FIGURE 3.3 Battery cell construction and voltage characteristics.

recharged once its Ah capacity has been drained. It is then discarded, often jettisoned from the spacecraft to shed mass.

3.3 Fuel Cell

Powering loads greater than several watts for more than a few days or a few weeks is not practical using a battery, but is easily done with a fuel cell. The fuel cell, developed as an intermediate-term power source for space applications, was first used in a moon buggy and continues being used to power NASA's space shuttles. It also finds other niche applications at present. A fuel cell converts chemical energy in the fuel, such as hydrogen and oxygen, into electricity (Figure 3.4(a)). Since the fuel continuously

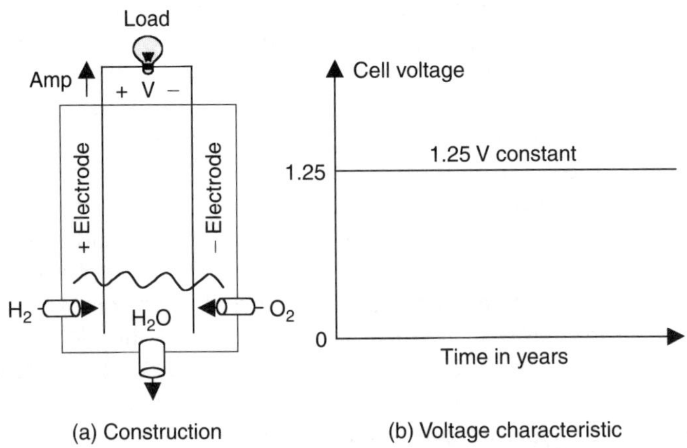

FIGURE 3.4 Fuel cell construction and voltage characteristics.

refurbishes the energy, the cell does not run out of energy. Hence, the fuel cell is not rated in terms of the Ah capacity, but in terms of the power generation rate. The cell voltage remains constant (Figure 3.4(b)) as long as the fuel is supplied at the required rate. Therefore, the fuel cell can be an optimum choice for supplying hundreds or thousands of watts over a few weeks using the on-board fuel.

A fuel cell is a static electrochemical device that generates direct current electricity by a chemical reaction without altering the electrodes or the electrolyte materials. In concept, hydrogen and oxygen are combined to produce electricity and water, which is the reverse of the electrolysis of water. The crew in manned missions can use this water. The fuel does not burn as in an internal combustion (IC) engine. Thus, the fuel cell operates differently from both the electrochemical battery and the IC engine. The conversion efficiency of the fuel cell is not limited to that of the Carnot cycle, because the fuel cell bypasses the thermal-to-mechanical conversion and its operation is isothermal. That is why its efficiency can be, in principle, much greater than that of the IC engine. The efficiencies of some commercial fuel cells approach 70 to 80%, about twice the combustion engine efficiency. The space qualified fuel cell efficiency, however, is around 10% at present, but has a potential for a significant increase.

3.4 Solar PV–Battery

One of the most valuable breakthroughs in the space industry was probably the photovoltaic (PV) cell used to convert sunlight into electricity for Earth-orbiting satellites. Today, it is the most widely used energy conversion technology in the industry that has fueled the information revolution using high-power communications satellites. Power requirements in tens of watts to several kilowatts over a life ranging from a few months to 15 to 20 years can be met with an array of photovoltaic cells. Satellites requiring continuous load power even during an eclipse must use a rechargeable battery along with the PV array. The battery is charged during sunlight and discharged to power the load during an eclipse. A power regulator and control circuits are used as required for the mission.

The general layout of the PV–battery power system is shown in Figure 3.5. All components other than the solar array are generally located inside the satellite body. The orientations of the core body and the solar array are maintained relative to the sun and the Earth. The core body is normally maintained in a near constant orientation relative to the Earth, while the α drive and the β gimbals orient the solar array to the sun. The α drive rotates 360° once per orbit as the satellite revolves around the Earth. The β gimbals rotate $\pm\ \beta^\circ$ to compensate for the variation in the solar β angle and also to prevent array shadowing if applicable. Not all satellites have β gimbals, but almost all using the solar energy for power generation have an α drive. The

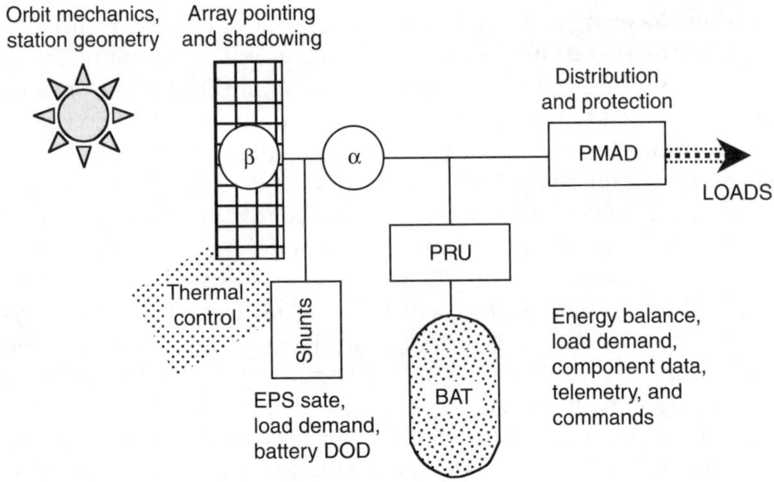

FIGURE 3.5 Solar photovoltaic–battery power system configuration.

most common form of α drive is a slip ring assembly with a solar array drive in 3-axis stabilized satellites, and a rotary power transfer assembly in gyrostats. Angular errors induced by the structural distortions are often compensated by the α and/or β drive settings.

The seasonal variations of the β angle and the eclipse duration over 1 year for the International Space Station in 400-km (220-n.m.) altitude and 51.6° inclination orbit are shown in Figure 3.6.[1] For a given system design, the

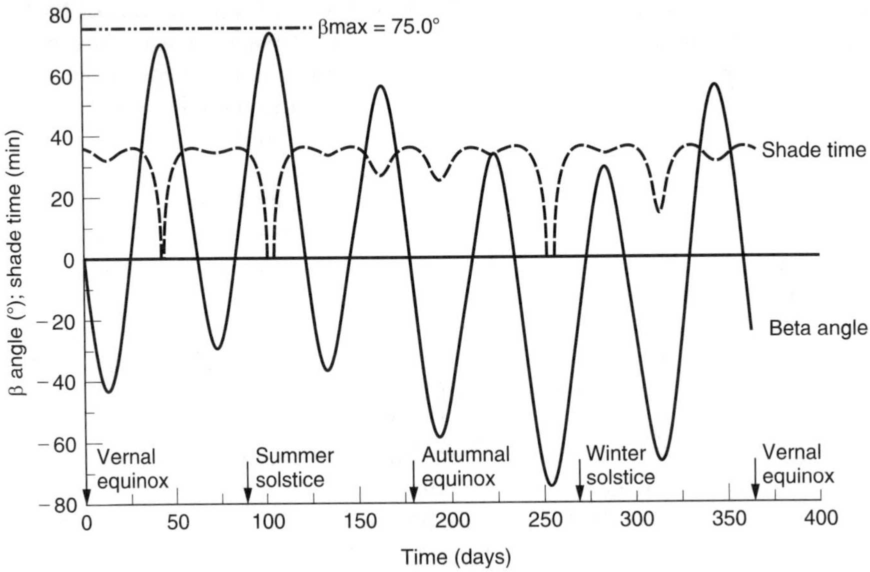

FIGURE 3.6 Beta angle and eclipse duration variation with season for the International Space Station. (Source: NASA Glenn SPACE Team/J. Hojnicki.)

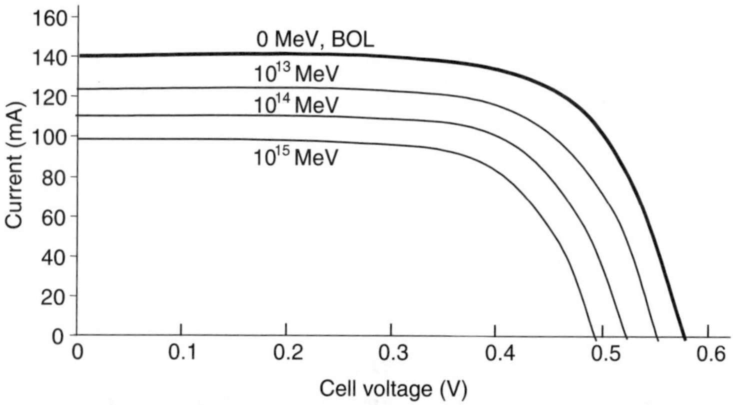

FIGURE 3.7 Degradation in *I–V* characteristics of a typical PV cell under radiation.

power available to the load varies over the year due to seasonal variation in the β angle. At high $\pm\beta$ when the eclipse duration is zero, the load capability of the electrical power system would be the greatest, as no battery charge power is required. For the ISS, there would be no eclipse at all for $\beta > 71^\circ$, making the orbit sun-synchronous.

The PV cell has been a building block of space power systems since the beginning. The cell is a diode-type junction of two crystalline semiconductors, which generates electricity under sunlight. Its performance at the beginning of life (BOL) is characterized by the output voltage and current at its terminals as shown by the heavy line in Figure 3.7. The two extreme points on this curve, namely the open circuit voltage, V_{oc}, and the short circuit current, I_{sc}, are often used as the performance indicators. The maximum power a cell can generate is the product of V_{oc}, I_{sc}, and a factor that is approximately constant for a given junction. The *I–V* characteristic of the PV cell degrades as shown by thin lines with the increasing fluence of charged particles on the solar array in the space environment. Such degradation results in decreasing power generation with time. With the combination of seasonal variations of β angle and yearly degradation of charged particles, the power generation of the solar array over the mission life varies as shown in Figure 3.8.

3.5 Solar Concentrator–Dynamic Power System

Solar energy can be used in systems other than photovoltaic cells. For example, the sun's energy is collected in the form of heat using a concentrator. The heat, in turn, is used to generate steam and drive a rotating turbo-generator or a reciprocating alternator: either way uses a thermodynamic energy converter.

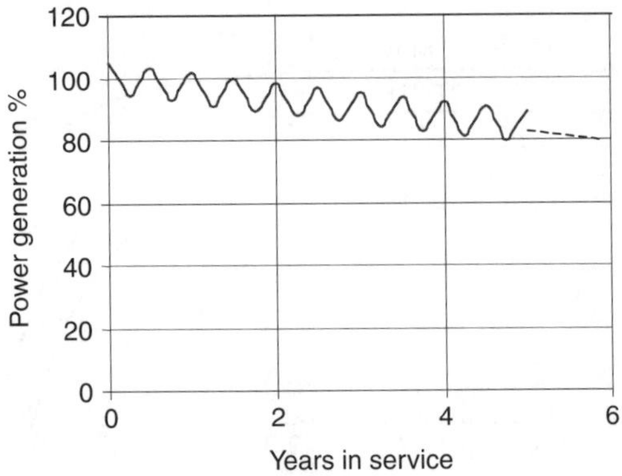

FIGURE 3.8 Degradation of solar array output power versus service years.

The dynamic power system was a primary candidate for early space station design, having an estimated power requirement of 300 kW. The system configuration is shown in Figure 3.9. A parabolic concentrator focuses the sun's heat on to a receiver, which boils a fluid. The fluid can be a suitable liquid or even a liquid metal, such as potassium chloride. High-pressure steam produced in the receiver drives a steam turbine based on the Rankine cycle. The fluid can also be a gas, such as a mixture of helium and

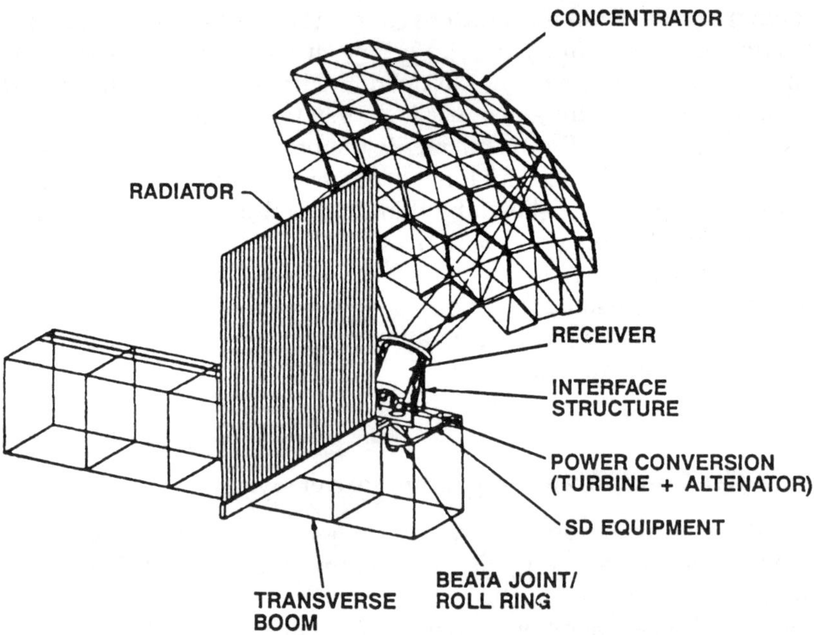

FIGURE 3.9 Solar concentrator–dynamic system. (Source: NASA Glenn Research Center.)

xenon, having a molecular weight around 40. The heated gas drives a turbine working on the Brayton cycle. The gas-based system, however, minimizes erosion and the problem of sloshing when transporting a liquid. In either case, the high-pressure high-temperature fluid drives the turbine, which in turn drives an electrical generator. The energy conversion efficiency is about twice that of the photovoltaic system. This minimizes the deployed collector area and the aerodynamic drag in the low Earth orbit. An indirect advantage is that the energy storage is interwoven in the system at no extra cost. It primarily resides in the form of latent heat with the phase change at high temperature around 1000 K.

The usable energy extracted during a thermal cycle depends on the working temperatures. The maximum thermodynamic conversion efficiency that can be theoretically achieved with the hot side temperature, T_{hot}, and the cold side temperature, T_{cold}, is given by the Carnot cycle efficiency, which is

$$\eta_{Carnot} = \frac{T_{hot} - T_{cold}}{T_{hot}} \tag{3.1}$$

where the temperatures are in degrees absolute. The higher the hot side working temperature and lower the cold side exhaust temperature, the higher the efficiency of converting the captured solar energy into electricity. The hot side temperature, T_{hot}, however, is limited by properties of the working medium. The cold side temperature, T_{cold}, is largely determined by the cooling method and the environment available to dissipate the exhaust heat.

The dynamic power system incorporates the thermal energy storage for hours with no degradation in performance, or for longer duration with some degradation. This feature makes the technology capable of producing high-value electricity for meeting peak demands. Moreover, compared to the solar–photovoltaic system, the solar–thermal system is economical, as it eliminates the costly PV cells and battery. The solar concentrator–dynamic system with a turbo-alternator also offers significant advantage in efficiency and weight, and hence the overall cost over solar PV technology. The efficiency advantage comes from the higher efficiency of the engine (about 30%) as compared to silicon solar cells (about 15%), and higher efficiency of thermal energy storage of the receiver (about 90%) as compared to the battery efficiency (about 75%).

The concept is sufficiently developed for use in the future, particularly in high-power LEO missions. It may also find applications in high-power defense spacecraft where large solar arrays can make the mission nonmaneuverable and vulnerable to enemy detection and attacks. The higher efficiency requiring less solar collection area results in reduced drag and less concern regarding station dynamics, approach corridors, and experimental viewing angles. The reduced drag is particularly important because it allows lower flight altitudes within given constraints of drag-makeup fuel and orbit decay time. At high power approaching the 100-kW range for space-based radar (SBR), the solar array collector area becomes prohibitive. The solar dynamic system can be extremely cost effective over a wide range of power between a few kilowatts and hundreds of kilowatts. It was considered for the dynamic isotopes power system (DIPS) in the 5 to 10 kW power range, and the space station in the 200 to 500 kW power range.

3.6 Nuclear–Thermoelectric

Interplanetary and deep space missions far away from the sun cannot be effectively designed using photovoltaic power generation because of weak solar flux. The spacecraft must therefore carry on board a primary energy source, such as a radioactive isotope or a nuclear reactor. In the first alternative, the radioisotope heats a thermoelectric (TE) material such as lead telluride, which generates electrical potential (Figure 3.10). The principle is similar to that in a thermocouple, only with higher conversion efficiency. The radioisotope thermoelectric generator (RTG) is being routinely used for interplanetary missions requiring power levels of several hundred watts. The reactor, on the other hand, has been considered for high power in the 30 to 300 kW range. Both power sources have the advantage of supplying power all the time, thus eliminating the need for a battery in base load systems having no peak power requirement. The obvious disadvantage is the heavy radiation shielding required around the electronic components. Also the nuclear fuels that are safe and easy to handle with little shielding, such as curium-244 and plutonium, are expensive. Inexpensive easily available fuels, such as strontium-90, are unsafe.

High-energy particles emitted from the radioactive isotope material are the primary energy source, which heats the absorbing material. The thermal power radiation decreases proportionally with the remaining mass. The mass of the isotope material decay exponentially at a rate characterized by the half-life, $T_{1/2}$, which is long. Therefore, the power generation essentially remains constant for decades. This makes the nuclear energy source ideal

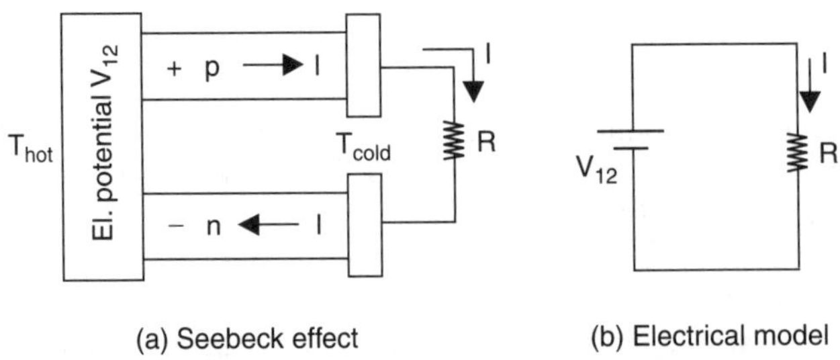

FIGURE 3.10 Thermoelectric converter.

for long-life interplanetary missions; many of them require several hundred watts power. However, it has also been used for some defense missions requiring a high degree of radiation hardness under nuclear threat, such as in the Space Power-100 (SP-100) program having power requirements ranging from 30 to 300 kW. Fairly advanced development has been carried out under the SP-100 funding, which is described in detail in Chapter 22.

3.7 Nuclear or Chemical–Dynamic

Hundreds of kilowatts or multi-megawatts of electrical power require a chemical or nuclear reactor as the primary energy source and rotating a.c. generator. A short-life burst power mission may consider carrying on-board chemical fuel. On the other hand, a long-life mission may require a nuclear reactor. The primary source of power in such a reactor is nuclear fission, just like that in a ground-based nuclear power plant, only much smaller in scale. A fissile material such as uranium-235 works as the heat source to vaporize a fluid, typically liquid metal such as mercury or sodium potassium. The vapor then drives a turbine-generator using the Brayton or the Rankine cycle. The Stirling cycle is suitable in the 50 W to 50 kW power range for long life space missions, while the Brayton cycle is suitable in the 50 kW to 10 MW power range for advanced electric propulsion. A variety of concepts using a nuclear–dynamic system was considered for the strategic defense initiative (SDI) programs in the late 1980's and the early 1990s. The SDI conceptual design power levels were in the multi-megawatt range between 30 and 300 MW. The basic requirement for such missions is to generate high power in the pulse or burst mode for a short duration of time, such as engagement at full power for about 15 min for a few times over the mission life. For the remaining time during the entire mission, the system must coast consuming minimum quiescent power. The design driver for such a power system is large-scale energy storage with a slow charge rate, and energy discharge at a high rate. For energy storage, both the flywheel and the superconducting magnet have been considered along with other options. The multi-megawatt nuclear dynamic systems are further discussed in Chapter 22. A nuclear–dynamic system for a few hundreds watts of power for interplanetary and space defense missions has been developed by NASA GRC, and is being space qualified. It uses a radioisotope as the heat source, a Stirling thermodynamic engine, a reciprocating electrical alternator, and a.c.–d.c. power converters if needed for d.c. loads. It is intended to replace the presently used RTGs. It works at temperatures of 1000 °C on the hot side and 500 °C on the cold side of the fluid. Compared to RTGs, it has a higher system efficiency (20%), higher specific power, and is scalable to higher power levels. The space-worthy technology has been demonstrated on a NASA test bed for applications up to 10 kW by building and testing a 2 kW unit. The technology is similar to that used for balloon flights where a parabolic solar concentrator at the top of the balloon collects the energy, boils a liquid, and drives a Stirling engine at the base of the balloon.

3.8 Other Systems

There are other power system options under various stages of development. They may find applications in some special niche missions. In evaluating potential applications of any new technology, the prime considerations are always the mass, cost, reliability, and the technology risk. For example, a system using any fluid has a potential for leakage and sloshing. On the other hand, a system having no moving parts have three inherent advantages: high reliability, no vibration, and no torque on the platform.

3.8.1 Thermo-Photovoltaic

In the thermo-photovoltaic (TPV) scheme, the radioisotope or solar heat is directed at the PV cells as opposed to the present use of RTGs for generating electricity. It has a significantly higher conversion efficiency. Therefore, it may find applications as the mission-enabling technology where its high cost may be acceptable. The system can have a cylindrical or a flat configuration as shown in Figure 3.11. A heated surface radiates infrared heat onto an array of photovoltaic cells sensitive in the infrared range. A part of the energy is converted into d.c. electricity; and some is reflected back and dissipated as heat. The energy conversion process is different to that in the conventional PV cell. The efficiency varies with the radiator temperature. It ranges from 10% at 800 °C to 12% at 1100 °C based on the absorbed heat, but is only 3 to 4% based on the total incident energy.

Most current TPVs use low bandgap PV cells such as 0.55 eV InGaAs or 0.73 eV GaSb to optimize the cell response to energy sources in the 1 to 2 μm range. The low bandgap of these cells leads to low open circuit voltage (about 0.25 to 0.45 V) and poor fill factor caused by high intrinsic carrier

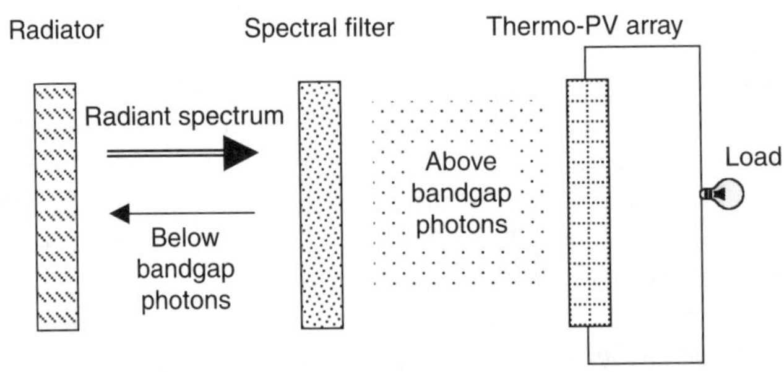

FIGURE 3.11 Thermo-photovoltaic converter.

concentration. Thus, the cell must be operated at low temperatures, generally below 60 °C to have adequate output voltage. As a result, a large radiator area is required. The temperature has considerable influence on the PV cell performance. The bandgap decreases with increasing temperature and the photocurrent increases with temperature by absorbing extra longer wavelength photons from the emitter. However, the open circuit voltage of the cell decreases linearly with increasing cell temperature due to the exponential increase in the saturation current.

The TPV concept has recently seen renewed interest with new developments in semiconductor technology. Recent advances have produced low bandgap (0.50 to 0.55 eV) material that is lattice-matched to GaSb substrates. These developments enable TPV systems to achieve reasonable efficiency and power density with radiators operating at about 1000 °C. At relatively low radiator temperatures, there are many viable options for the heat source and a number of applications become attractive. A conversion efficiency of 11.7% has been reported[2] at a radiator temperature of 1076 °C and a module temperature of 30 °C. This is the highest achieved efficiency for an integrated TPV system. It is the ratio of the peak (load-matched) electrical power output and the heat absorption rate. The TPV cell degrades like the GaAs cell, hence it is suitable for application in Earth orbits passing through high radiation belts.

3.8.2 Solar–Thermoelectric

The power system for a near-sun probe operating at high temperature cannot effectively use PV cells because of severe temperature degradation in performance. In such a mission, the collected solar heat can be directed at a pile of thermo-electric converters (TECs). The energy conversion details in such systems are common with the RTGs covered in Chapter 20. The only difference between the RTG and the solar–TE is the source of heat. The RTG heat source is the nuclear reaction in a suitable radioisotope, while heat in the solar–TE scheme comes from the sun.

3.8.3 Thermionic

In this conversion process, the thermal energy is converted into electricity by using the electrons released from a hot body, known as thermionic emission, or the Edison effect. The electrons released from the cathode are collected at the anode, and a closed path through a load is established to complete the circuit back to the cathode. It has no moving parts. The concept is an old one, but is attracting new interest due to advances made in high-temperature materials. The thermionic converter is basically a heat engine with electrons as the working fluid, and is subject to the Carnot efficiency limitation. For this reason, it operates at much higher temperatures with a hot side around 1800 to 2000 K and heat rejection around 800 to

1000 K. This imposes limits on candidate materials and life. Also, the thermionic converter produce a very low voltage requiring extensive power conversion and large power conditioning loss. Further developments are under way with the DoD funding at present. A typical example of development is the thermionic fuel element that integrates the converter and nuclear fuel for space power in the kilowatts to megawatts range for long missions. Converters filled with ionized gas, such as cesium vapor, in the inter-electrode space, yield higher specific power due to space charge neutralization.

A 100-kW thermionic power system with a nuclear reactor has been built and tested in Russia, which is believed to be scalable to 300 kW and perhaps to a megawatts power level. A solar concentrator–thermionic power system may be practical up to 100 kW. This technology is not suitable for power levels lower than several kilowatts. General Atomic has designed a 50 kW unit in which solar heat is concentrated directly on the cathode, thus eliminating a working fluid. The conversion efficiency is around 10% using a high-side temperature of 1800 °K, and 5% using the high-side temperature just below 1000 °K. The specific power can approach 700 to 1000 W/kg.

3.8.4 Alkaline Metal Thermal to Electric Converter

The alkaline metal thermal to electric converter (AMTEC) is a direct thermal to electric energy conversion system. It operates at temperatures around 1000 K on the high side and 600 K on the low side, and yield efficiency of 18 to 20%. It is a static system suitable for power levels below 100 W. It uses a unique characteristic of the alkali metal conducting ceramic β-alumina.[3] It is a solid electrolyte that conducts sodium ions, but is an electron insulator. An electrochemical potential is generated when sodium is present at two different pressures separated by an electrolyte. The sodium is circulated through the converter using a porous stainless steel wick that uses capillary forces to transport the sodium from the low-pressure region to the high-pressure region. It has no moving parts and develops low voltage requiring a high level of power conditioning. A commercial version of this technology with titanium nitride electrodes is used for small down-hole instruments in the oil industry. Developments are under way for space-worthy AMTEC using a radioisotope as the heat source for interplanetary and deep space missions requiring power in the few hundred watts range.

3.9 Technology Options Compared

The practical limit and performance characteristics of major power system options are summarized in Table 3.1. Although the PV–battery power system is the most widely used system for satellites, a variety of alternative power system technologies have been flown, developed, and are being

Table 3.1 Practical limit and performance comparison of various power system options

Power system option	Practical power limit (kW)	Net system efficiency (%)	Specific power (W/kg)
Solar–PV	20	15–30	5–10
Isotope–TEC	1	7–15	7–15
Nuclear–TEC	100	7–15	–

developed for various missions. Table 3.2 summarizes the present state of the spacecraft technology options[4] along with current development work that has promising applications in the near future. It also includes some concepts that potentially may find beneficial applications in the future.

Development engineers continuously evaluate new technologies for possible incorporation into new designs as they become available from research laboratories. Many times, these changes bring incremental benefits. Quantum benefits can be achieved by incorporating several changes simultaneously into a new design. For example, DoD and NASA have funded development and testing of protoflight solar array designs which could yield specific power over 100 W/kg, a factor of 3 greater than the state of the art, and a factor of 5 greater than the state of the practice. The design under consideration integrates three promising technologies: a flexible copper indium diselenide thin-film PV cell, smart mechanisms using shape memory metal, and a multifunctional lightweight structure.[5] An important criterion in the application of a new technology is the development and flight qualification status. New components must be subjected to time-consuming and expensive testing to prove their ability to withstand launch and space environments.

3.10 System Voltage Options

Early spacecraft with loads of a few hundred watts used 28 V d.c., primarily based on the product specifications readily available for the aircraft power system at the time. Since then, the power levels have increased significantly. As power is the product of voltage and current, high power requires a high-voltage bus in order to keep the current at a reasonable level. Otherwise, the excessive power loss in switching devices and the I^2R loss in conductors reduce the system efficiency considerably. Today's spacecraft bus voltages, somewhat standardized by the product lines of various manufacturers and government agencies, are 28 V, 50 V, 70 V, 100 V, 120 V, and 160 V, as shown in Figure 3.12. The 160-V limit comes primarily from the bare conductor interaction with space plasma, particularly in the low Earth orbit. Above

Table 3.2 Technology options and status of various power system components

Technology option	Technology status
PV cell: converts sunlight into electricity	Used in most long-mission spacecraft for power generation. The mission design life is generally limited to about 15 years due to degradation in the Van Allen radiation belts.
TE cell: converts radioisotope heat into electricity	Well developed and flight-proven for a few hundred watts. It is expensive. Solar heated TE may be considered in regions of intense Van Allen radiation belts and/or missions with nuclear threats.
AMTEC: converts isotope heat into electricity using an alkali metal TE converter	Under development. Suitable up to a few hundred watts systems for interplanetary and deep-space missions, rovers, etc.
Thermionic: converts heat into electricity	Solar and nuclear heated thermionic converters have been built and tested in the laboratory. Suitable for hundreds of kilowatts. Have high specific power.
Nuclear reactor: converts heat into electricity	US prototype designs have been developed, but not proven in space. Russian space programs have flown many such units. Scalable to the megawatt power range.
Fuel cell: converts a fuel's chemical energy into electricity	Used routinely for the space shuttles. Was considered for the ISS. May find more applications in the future.
Battery energy storage	Most spacecraft use secondary batteries, and some short mission spacecraft use primary battery. Fully developed.
Flywheel energy storage	Under active development for installation on ISS in 2006–2007 time frame. Targeted to replace the battery.
Thermo–PV: converts heat into electricity using PV type cells	Under development. Low specific power, but may find niche application.

160 V, the solar array current leakage to plasma increases exponentially with potential sparking above 180 to 200 V.

The plasma current is collected at exposed conductors having a bias voltage. With an insulated conductor, parasitic power loss due to plasma current at pinholes or coating defects have been shown to be small in ground plasma chamber testing at NASA GRC. Tests on several cable-insulating materials have shown that the collected plasma current remains negligible below a few hundred volts positive. However, at voltages of 200 V and above, insulation pinholes showed snap-over effects. If conductors are at high positive potentials relative to the plasma, snap-over may greatly increase the electron current and the resulting power drain. Still, regardless of the spacecraft grounding scheme, it is unlikely that more spacecraft will be at potentials greater than +100 V above the plasma potential, so snap-over is unlikely to occur. An exception is the 160-V ISS

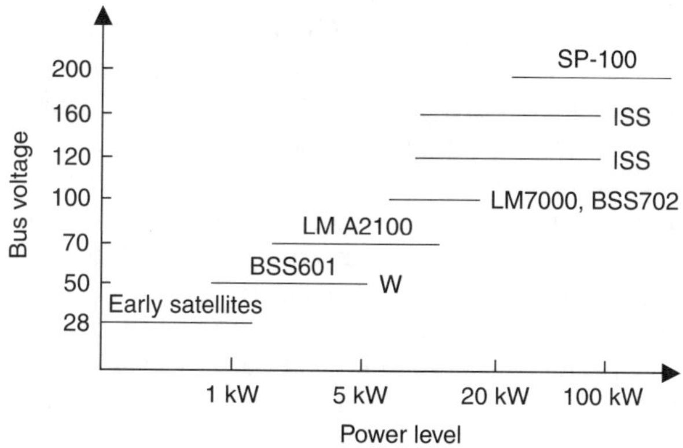

FIGURE 3.12 Optimum voltage for various power levels.

solar array, where the plasma contactor keeps the structure close to the plasma potential. However, the most positive end of the solar array being more than 100 V positive, the solar cell edges would collect more current than otherwise, essentially increasing the power drain and demanding a high current capacity for the plasma contactor. For other high-voltage LEO spacecraft, such as the 200-V SP-100, analysis has shown that the power loss from currents to other surfaces is small compared to the total delivered current, and thus the percentage efficiency loss is also small.

A rule of thumb is that for every square meter of exposed conductor in LEO at 100 V positive, a parasitic structure current of about 1 mA may be expected. Thus, for a payload of about 100 m^2 surface area on a 100 V bus, for instance, only 100 mA of structure current may drain from the power system capacity. This is negligible compared to 100 A that a 10-kW power system would deliver at 100 V.

Voltages higher than 160 V can be used in low Earth orbit with insulated cables covered in a shielded enclosure as shown in Figure 3.13, and by

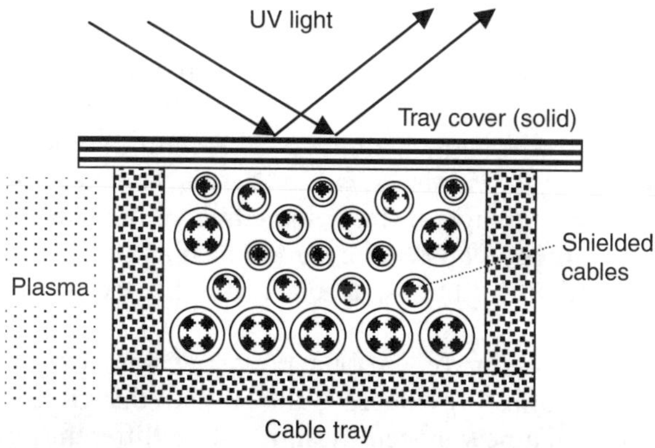

FIGURE 3.13 Covered cable tray protection against voltage breakdown.

encapsulating all connectors and circuit boards. NASA has selected a 120-V distribution system for the ISS with necessary step-down converters for existing 28-V hardware. Early in the ISS design, 270 V d.c. and 440 V 20 kHz a.c. were considered, but, finally, a 160-V solar array voltage and 120-V distribution voltage were selected. Alternating current was seriously considered for the space station design in the 1980s for multiple benefits, but was dropped as the power requirement was significantly scaled down and high development costs were projected.

High-voltage design, however, impacts upon the component selection. Moreover, the space environment considerations also limit the voltage to a certain level for a given mission. The factors that influence the voltage selections are:

- Power level being the primary driver
- Space environment and space plasma
- Paschen minimum breakdown voltage between bare conductors
- Human safety
- Availability of components, such as semiconductor devices, power distribution and protection devices, tantalum capacitors, etc.

Design issues at much higher voltages up to several hundred kilovolts are discussed in Chapter 22.

3.11 Scaling for Power Level

Selecting the bus voltage from available options is a matter of the technical factors discussed above and in Chapter 22, taking into consideration the corporation's in-house design heritage. In the case of a new development characterized by an absence of the heritage design, the following guideline may be used:

$$\text{optimum voltage} = 0.025 \times \text{power requirement} \tag{3.1}$$

For example, a 5000-W payload power system would have a mass-optimized design at $0.025 \times 5000 = 125\,\text{V}$. This voltage is then shifted to the nearest standard voltage. If the voltage falls in the bands given in Table 3.3, the design engineer would have a good availability of parts which have been widely used in other spacecraft design, both commercial and government.

For top level screening of various power system options for a given mission, the mass of a previously built similar design can be scaled to estimate the mass of a new system design with a different power level. The following is a rough empirical scaling law:

Table 3.3 Voltage bands

	High voltage (V)	Low voltage (V)	Medium voltage (V)
Distribution voltage	28	70	120
Solar array output	38	80	160

$$\text{mass of new design} = \text{mass of similar design} \times \left(\frac{\text{new power requirement}}{\text{power in similar design}}\right)^{0.7} \tag{3.2}$$

For example, the mass of a new system with 10 times the power requirement is 5 times the similarly built system using the same technology. This equation suggests that specific power doubles for every 10-fold increase in the power level.

For a.c. systems, the power system mass depends both on the power level and also on the frequency. The design and fabrication experience of commercial and aircraft industry suggest the following rule of thumb.

$$\text{mass of a.c. system} = \left(\frac{\text{kW}}{f}\right)^{\alpha} \tag{3.3}$$

where the exponent α is equal to 0.5 for small systems in hundreds of watts, and 0.75 for large systems in hundreds of kilowatts.

References

1. Hojnicki, J.S. et al. "*Space Station Freedom Electrical Performance Model,*" NASA Glenn Research Center, Report No. TM-106395, 1993.
2. Brown, E.J. et al. "Measurements of conversion efficiency for a flat plate thermo-photovoltaic system using a photonic cavity test system," in *Proceedings of the 35th Intersociety Energy Conversion Engineering Conference,* AIAA, 2000, Paper No. 3029.
3. Pantalin, J.E. et al. "Advanced AMTEC converter development," *Proceedings of the 35th Intersociety Energy Conversion Engineering Conference,* ASME, 2001, pp. 519–524.
4. Hyder, A.K. et al. "*Spacecraft Power Technologies,*" Imperial College Press/World Scientific Publishing Co, London, 2003.
5. Marshall, C.G. et al. "Example of a prototype lightweight solar array and the three promising technologies it incorporates," *Proceedings of the 35th Intersociety Energy Conversion Engineering Conference,* SAE, 1999, Paper No. 01-2550.

Chapter 4

Photovoltaic–Battery System

4.1 Introduction

Photovoltaic conversion of the sun's energy is the most common source of electrical power in space. An array of photovoltaic cells powers the load and charges a battery during sunlight. The battery powers the load during an eclipse. If the solar array, the battery and the load were operated at the same constant voltage, no voltage regulator would be needed. All the equipment could be wired to the same bus. However, the solar array output voltage is higher at the beginning of life, and when the array is cold for several minutes after each eclipse. Also, the battery has a lower voltage during discharge than during charge. Since the system is required to provide power to the load at a voltage regulated within specified limits, a voltage regulator is always needed to match voltages of various power components during the entire orbit period.

The photovoltaic power system, therefore, primarily consists of a solar array, a rechargeable battery, and a power regulator which regulates power flow between various components to control the bus voltage. Other components such as various sensors are also required to make the array and the battery work together. The total power system is thus coordinated internally as well as externally through interfaces with other systems of the spacecraft. The top-level performance characteristics of the basic PV–battery system are described in this chapter, leaving the component level details for later chapters.

4.1.1 Solar Array

The solar array is made of numerous PV cells stacked in series–parallel connections to obtain the desired voltage and current from the assembly. It converts the incident photon energy into d.c. voltage, which drives current through the external load circuit. The solar array works more like a constant current source over its normal operating range. Its terminal voltage versus current characteristic, referred to as the *I–V* curve, is shown in Figure 4.1. The end of life (EOL) curve must meet the performance requirement. The characteristic curve changes significantly with temperature and the radiation dose of charged particles as shown in the figure. The power, being the product of the voltage and current, varies as in Figure 4.2. The

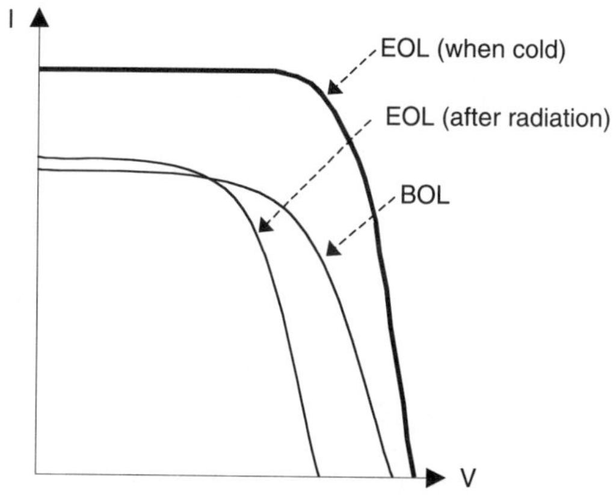

FIGURE 4.1 Current versus voltage characteristics of a photovoltaic cell.

power output of an array is maximum at the knee-point voltage. The system produces less power when operating at any other voltage. Also, the power output gradually degrades with accumulated radiation dose. At the end of life, the array generates less power than at the beginning of life.

4.1.2 Battery

The battery is made of rechargeable electrochemical cells connected in a series–parallel combination to obtain the desired voltage and current. Its terminal voltage depends primarily on the state of charge (SOC), and to some extent on the operating temperature. The battery charge is measured in terms of the ampere-hours stored between the positive and negative plates. The voltage is highest when the battery is fully charged, and the lowest when it is fully discharged. Since the battery works more like a constant voltage source over the normal operating range, its terminal characteristic is generally expressed in terms of the battery voltage versus the state of charge. Figure 4.3 depicts the voltage of one fully charged cell as it discharges and then gets recharged. The voltage scale in the figure

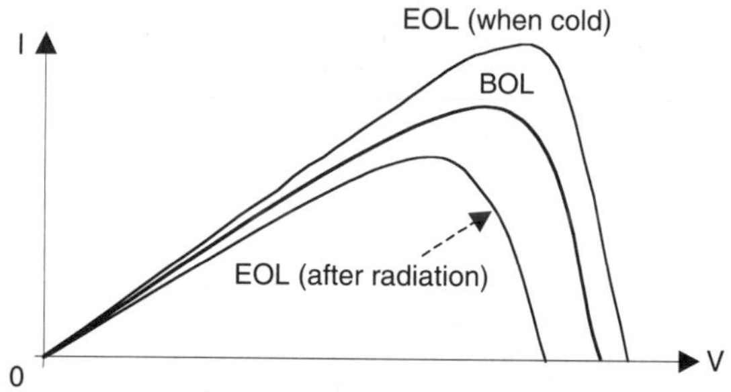

FIGURE 4.2 Power versus voltage characteristics of a photovoltaic cell.

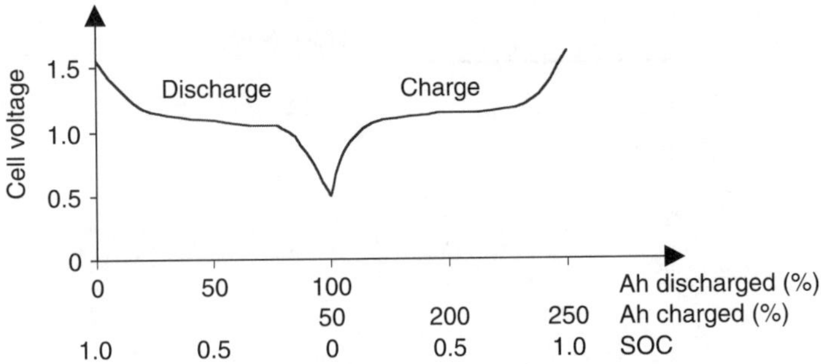

FIGURE 4.3 Battery cell voltage versus state of charge in one complete cycle.

represents both the NiCd and the NiH_2 cells. The battery voltage drops significantly with increasing discharge, and then rises during charge. The average voltage during charge is higher than that during discharge.

4.1.3 Power Regulation

Power regulation is primarily accomplished by battery charge and discharge converters, a shunt dissipator, and a mode controller that responds to the bus voltage error signal. The shunt dissipator is necessary to control the bus voltage during sunlight as described hereafter. The solar array (source) and the constant power load have their own *I–V* characteristics as shown in Figure 4.4. The system can operate at either of the two intersection points A or B. However, point A is inherently unstable because the load slope is less than the source slope. Point B, on the other hand, is inherently stable. Without a shunt control, the system would operate at point B, producing a lower power. With shunts designed to regulate the

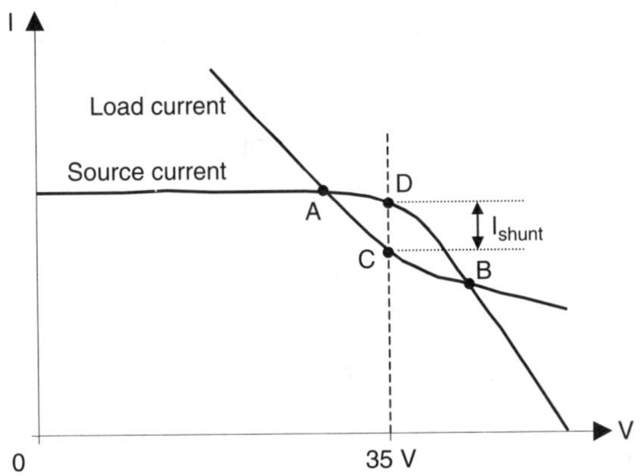

FIGURE 4.4 Stability of operating point and shunt control during sunlight.

sunlight voltage at say 35 V; the system will pull back from point B to point C by shunting the excess current to the ground. The shunt current in this operating mode would be I_{shunt}, the difference between the source current at D and the load current at C.

The mode controller sets the system's operating mode in response to the error signal, which is the difference between the actual bus voltage and the reference voltage representing the required bus voltage. Depending on the error signal value and its polarity (positive or negative), the mode controller sends a control signal to either the shunt regulator, or the battery charge regulator, or to the battery discharge regulator (Figure 4.5). One of these three sub-regulators, in turn, maintains the bus voltage within the specified limits. Details of the next level of the mode controller during the battery discharge regulation are shown in Figure 4.6.

External interfaces of the power system with other systems of the spacecraft are mission specific, and difficult to describe in general. However, Figure 4.7 depicts key external interfaces in a typical communications satellite.

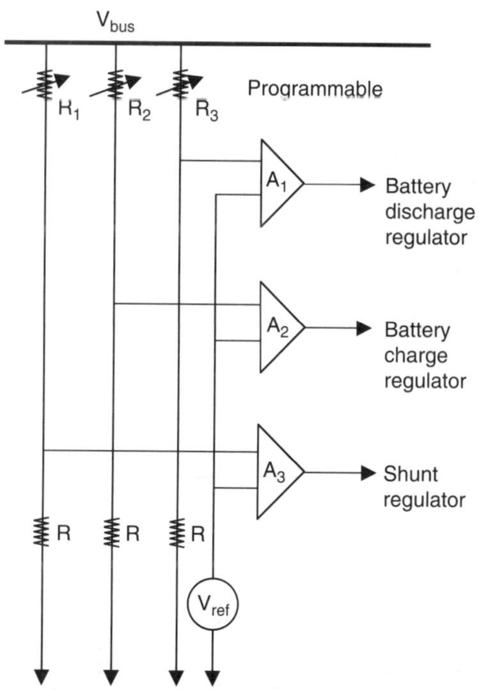

FIGURE 4.5 Mode controller error signal routing concept.

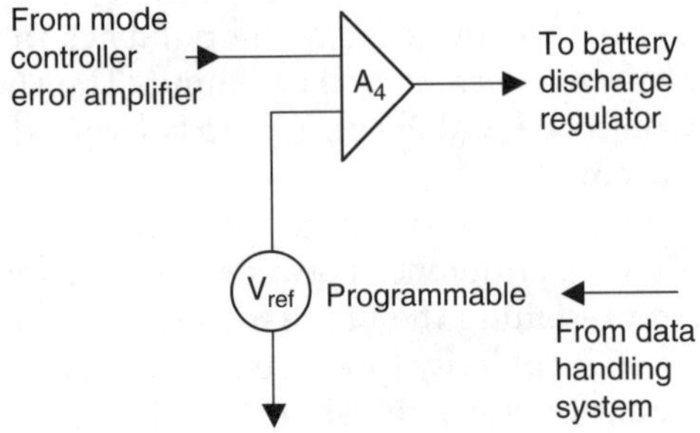

FIGURE 4.6 Mode controller error signal to battery discharge regulator.

4.2 Power System Architectures

The solar array, battery, and the shunt characteristics described above, along with the load voltage requirement, are extremely important in selecting the power system architecture that is most suitable for the mission. The mission-imposed or self-derived requirements on the power system also have bearings on the architecture selection. The PV–battery system is configured in one of the following architectures that would optimize the system performance for a given mission.

4.2.1 Direct Energy Transfer

Direct energy transfer (DET), in which the solar power is transferred to the loads with no series component in between. The necessary exceptions are: (1) the slip rings to provide a rotary joint between the Earth-facing

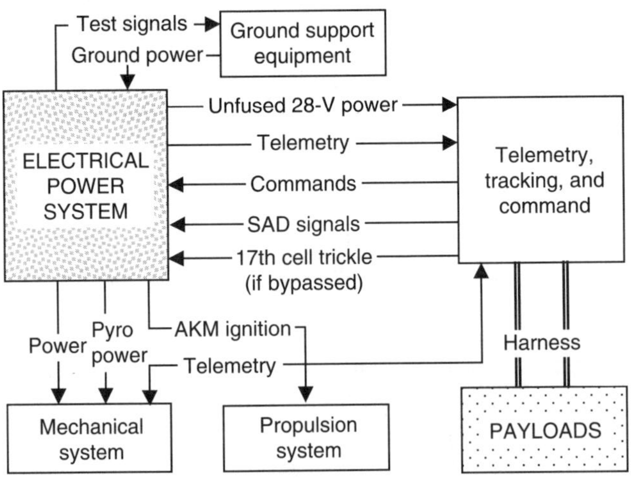

FIGURE 4.7 External electrical interfaces in a typical power system.

spacecraft body and the sun-facing solar array, and (2) the power distribution unit consisting of load switching relays and fuses to protect the power system from faults in the load circuits. The DET can be further subdivided into two classes: (1) the fully regulated bus, and (2) the sunlight regulated bus. The components and operation of these two busses are similar, except that the latter has no battery discharge converter in the power regulator unit (PRU).

4.2.2 Peak Power Tracker

Peak power tracker (PPT), in which the solar array output voltage is always set at the value which results in the maximum power transfer from the array to the load. A series power converter between the array and the load matches the load voltage requirement and the array output voltage. For this architecture to be cost effective, the power loss in the PPT converter must be less than the gain in operating the system at the peak power point all the time.

4.3 Fully Regulated Bus

In a fully regulated direct energy transfer bus, commonly known as the *regulated bus*, the bus voltage is controlled within a few percent during the entire orbit period. The typical bus voltage variation is ±2 to 5% of the nominal voltage. The architecture of this bus is shown in Figure 4.8. It has the following components.

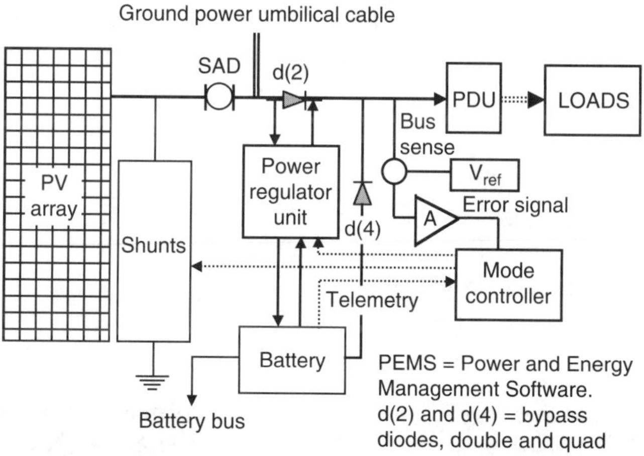

FIGURE 4.8 Fully regulated direct energy transfer architecture.

4.3.1 Solar Array

The solar array (SA) is an array of PV cells that convert sunlight into electricity. The large array is generally divided into many parallel circuits (strings), each with isolation diodes so that a failed string would not draw power from the healthy ones. The ungrounded end of each section and individual string is isolated from the main bus and battery by diodes. These diodes may be located inside the PRU for protection, or on the back of the solar array exposed to the environment. In one case, the slip rings are involved, but not in the other. The electrostatic discharge induced transient voltage suppressing snubber capacitors may be located inside or outside. The array negative terminal is usually grounded. If the positive were grounded, the power system would work just as well, except that we must use p–n–p semiconductors instead of traditional n–p–n devices in the PRU. The power converter design would change accordingly. One advantage of the positive ground is that the structure becomes negatively charged, and attracts atomic oxygen ions at lower speed, causing less sputtering of the anodized aluminum commonly used in the spacecraft structure. The bus diodes pair d(2), typically Schottky diodes, prevents loss of the bus voltage in case the battery charger gets an internal fault, which may clear after some time. Such diodes — appropriately called the DET diodes — have been used in many defense meteorological satellites (DMSs). They are not absolutely needed, and would reduce efficiency. In its absence, adequate fusing should be used to protect the bus from faults in a battery charger. The current required to clear a charger fault would come from the solar array whether or not d(2) is present.

4.3.2 Solar Array Drive

The solar array drive (SAD) consists of slip rings, a motor, and motor drive electronics. It continuously orients the solar array to face the sun for generating the maximum possible power during the entire sunlight period of the orbit.

4.3.3 Shunt Dissipator

During sunlight, particularly in the beginning of life, this component dissipates power that is unwanted after meeting the load power and the battery charge power requirements.

4.3.4 Battery

The battery stores energy in an electrochemical form to supply power to the loads during eclipse periods over the entire mission life. NiCd secondary batteries were extensively used in satellites until the mid 1980's. In the newer designs, however, they have been replaced by NiH_2 for better performance and lower mass.

4.3.5 Power Regulator Unit

The power regulator unit (PRU) provides an interface between the solar array bus and the battery. The battery voltage varies widely with the cell voltage varying from 1.0 V when fully discharged to 1.55 V when fully charged. The discharge converter in the PRU boosts the battery voltage to the bus voltage during an eclipse and the charge converter bucks the array voltage to the battery voltage during sunlight. For GEO missions, the charge converter rating is much smaller than the discharge converter rating. In such situations, designing the charge and discharge converters separately would be beneficial, as their individual designs can be better tuned to their own ratings and requirements. For LEO missions, on the other hand, a bi-directional converter may be beneficial, as the charge converter and the discharge converter ratings are comparable. The following approximations may help trading various options in conceptual designs. A 1-kW battery discharge converter may weigh 2 to 3 kg. A 1-kW PRU designed to operate between –25 and 60 °C with a reliability of 0.99975 may weigh 6 to 7 kg including the battery charge converter and dedicated support electronics.

4.3.6 Power Distribution Unit

The power distribution unit (PDU) ensures that all loads, except critical and essential loads, are powered through switches and fuses. The fuses are not to protect the loads as much as to protect the power system from faults in the user equipment.

4.3.7 Bus Voltage Controller

The bus voltage controller consists of the bus voltage sensor, the reference voltage and the error signal amplifier. The amplified error signal output of the bus controller enters the mode controller, which in turn sends command signals needed to regulate the bus voltage within required limits.

4.3.8 Mode Controller

The mode controller automatically changes the EPS mode in response to the error signal as follows. The mode selector switch is typically a magnetic latching relay.

4.3.8.1 *Shunt mode*

During sunlight, if the solar power exceeds the load and battery charge requirements, the mode controller turns on the shunts to dissipate excess power, else the bus voltage will rise above the allowable limit. During this mode, the battery is charged at full, partial or trickle charge rate, as required.

4.3.8.2 *Charge cut-back mode*

When the battery is approaching full charge, the charge rate is cut back to control the battery temperature. This mode includes trickle charge rate when the battery reaches a 100% state of charge. The charge cut-back mode is also activated in case the solar power exceeds the load requirement, but not enough to supply the required charge current to the battery. In such a case, the battery charge power is cut back to the level that will maintain the bus voltage within the allowable limits.

4.3.8.3 *Discharge mode*

In the absence of solar power during an eclipse, the battery is discharged to maintain the bus voltage. The battery voltage falls with the decreasing state of charge. The discharge converter, therefore, must increase the boost ratio accordingly. The PRU does this automatically by increasing the duty ratio of the discharge converter as the battery voltage decays.

4.3.8.4 *PRU bypass mode*

In case a fault occurs in any of the loads, the fuse must be blown as quickly as possible in order to minimize the bus voltage decay. The delay in PRU control loop response and the additional impedance introduced by the PRU would significantly extend the fuse blow time. To circumvent this, the battery is instantly connected to the bus by the bypass diode quad d(4), which quickly delivers the battery energy to the fuse in case of a fault, and blocks power flow to the battery at all other times.

4.3.9 Battery Bus

The battery bus is essentially a tap point directly off the battery. During the launch and ascent phases of the mission, the PV array is not deployed, and the battery meets all the energy needs. For example, all electro-explosive devices (EEDs) for deployment are powered directly from the battery bus for high reliability and low bus impedance (as there is nothing between the power source and the load).

4.3.10 Power and Energy Management Software

Although the power and energy management software (PEM) is part of the software system, it is dedicated to the EPS performance, health monitoring, control, and protection. In case of emergencies, or during planned mission operations, PEMS sheds loads in a preset sequence if and when the battery state of charge cannot support all loads. The battery telemetry consists of the battery voltage, current, temperature, individual cell voltages and the internal pressures of selected cells. Some of these telemetry readings also go to PEMS.

4.3.11 Loads

The term *load* includes all loads, i.e., the payloads (transmitters, receivers, science instruments, etc.) as well as the bus system loads. Most loads in satellites are constant power loads. Some loads may have a low duty ratio. For all loads, the orbit average power requirement is taken into account in sizing the PV array and the battery.

4.3.12 Ground Power Cord

To preserve the battery power during pre-launch testing and final checks before lift-off, the on-board system uses external ground power via an umbilical cord. To further preserve the battery, power transfer is scheduled as late as possible in the countdown.

4.4 Bus Voltage Control

The regulated bus in normal operation maintains the bus voltage between the specified upper limit and the lower limit. The mode control scheme for GPS IIR type 28-V regulated MEO bus with about 1000 W load, may be set up as follows:

Shunt mode regulates the bus voltage by using a shunt dissipator. In this mode, the batteries are fully charged and the solar array power generation exceeds the spacecraft's needs, so the excess current from the solar array is shunted. This mode is established when the shunt drive voltage signal is greater than 2.5 V d.c. and the battery charge currents are not responsive to the shunt drive voltage.

Battery charge control mode regulates the bus voltage using the battery chargers as linear shunting circuits. In this mode, the batteries are charged and the spacecraft loads are met by the solar array. If the solar array power capability is only slightly above the system load, the battery charge current must be limited. This mode is established when the shunt drive voltage is 1.0 to 2.5 V d.c. and the battery charge current is linearly responsive to the shunt drive voltage and is between 0 and 3.6 A.

Discharge mode regulates the bus by the battery discharge converter. In this mode, solar array current is not adequate for spacecraft bus loads, so the batteries are in discharge to provide the power needed for the bus. The solar array power capability is less than the bus load demand. This mode is established when the shunt drive voltage is less than 1.0 V d.c.

Reference and error amplifiers measure the difference between the 28-V volt bus and reference voltage. Depending on the magnitude and sign of the error signal, signals are sent to activate the shunt dissipators, the battery charge controllers or the battery discharge controllers. These modes are mutually exclusive and easily identified by monitoring the shunt drive voltage telemetry signal. The PRU performance is significantly different for each mode but varies only slightly for various input conditions within each mode.

In another example, in an EOS-AM type fully regulated 120-V LEO bus with a 3000-W load, the bus voltage is regulated within ±5 V to remain between 115 and 125 V as shown in Figure 4.9. This regulation band is further subdivided into five sub-bands. The voltage control is implemented using a mode controller, which places the power system in one of the following operating modes:

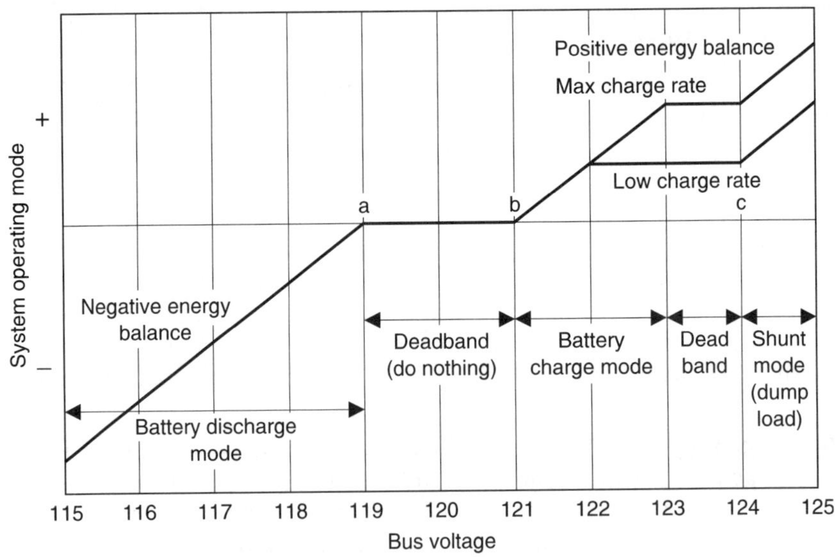

FIGURE 4.9 Bus voltage versus operating mode bands in a 120 ± 5-V regulated bus.

- Discharge mode when the bus voltage falls below the specified limit (point a)
- Dead band (do nothing band) when the voltage is between the dead-band limits (a and b)
- Charge mode when the bus voltage rises above the dead-band limit (point b)
- Shunt mode when the battery is fully charged and solar array output power exceeds the load power requirement (point c)

As a refinement, the control scheme may have two dead bands between the charge mode and the shunt mode, one for placing the battery in high charge rate, and the other for a low charge rate, as shown between the 122 and 124-V bands in Figure 4.9.

The bus voltage is regulated at a specified bus sense point, generally defined inside the power regulator unit. For dynamic stability of the feedback control loop, the mode controller amplifier provides a minimum phase margin of 45° and gain margin of 10 dB. The bus is generally designed to accommodate payload shedding in case the bus voltage drops to 90–95%, and is capable of powering the essential spacecraft functions to 75–80% of the nominal bus voltage.

4.5 Control Circuit

The control circuit may be analog, digital or a hybrid, as described below.

4.5.1 Analog Control

Analog control has been used in spacecraft for decades. A fully analog control loop for the main bus voltage regulator gives much better dynamic response to the power bus transients. The bandwidth of this control loop is kept high to optimize the bus regulation while reducing the noise.

4.5.2 Digital Control

Digital control is an alternative to the traditional analog control. A digital shunt regulator uses relatively small shunt dissipators to provide small signal control of the bus. As the analog shunt approaches its limit, it would clock up/down counter, which controls the number of shunt switches in use. While it may present challenges with respect to dynamic response of the system controller, it offers flexibility in tailoring the system to multiple missions. For example, by simply changing the gain constant in the software table, one can adjust the system's transient response. One can also incorporate a number of different battery charge regimes, and then adjust the charge rate in orbit with single command from the ground. This arrangement is adaptable to a variable number of solar array segments and to a number of array configurations. The result is more flexibility in using standard modules of the solar array and the battery, which translates into cost reduction.

4.5.3 Analog–Digital Hybrid Control

In a hybrid control scheme, a high bandwidth analog control loop for the main bus voltage regulator assures better dynamic response to the power bus transients while keeping the EMI low. The digital control loop is used for managing the lower functions where a low bandwidth is acceptable. A field programmable gate array (FPGA) may be used to command solar array shunts on and off the bus and for commanding the battery charge states. The FPGA samples the power system parameters, calculates the averages and monitors large rate changes in order to make adjustments needed in the solar array shunts and battery charge state. Such a hybrid control system results in good bus transient response with flexibility, reprogramming capability, bandwidth, and design simplicity.

4.6 Sun-Regulated Bus

If the design objective is to minimize complexity, an obvious approach is to distribute power from both sources — the solar array and the battery — directly to the load. Although such a direct energy transfer bus is sometimes known as *unregulated bus*, the bus voltage *is* regulated by shunt control during sunlight, and is unregulated only during an eclipse. For this reason, such a bus is also known as the partially regulated bus, sunlight regulated bus, or *sun-regulated bus*. Typical sun-regulated bus architecture is depicted in Figure 4.10. The difference in the sun-regulated bus and the fully regulated bus is only in the PRU. The sun-regulated bus has the usual battery charge regulator to regulate the battery charge rate during sunlight as commanded, but does not have the battery discharge converter. Instead, the battery discharges directly to the bus during eclipse through diode 'd', which is not called the 'bypass diode' any more, but is now called the 'battery discharge diode'. It only allows discharge from the battery, but blocks any uncontrolled charge current coming directly to the battery, leaving the charging function to the charge regulator. Thus, the battery is disconnected from the bus during sunlight when the shunt controller is regulating the bus voltage. This architecture would be most beneficial in multiple battery systems in GEO where the sunlight duration is long and the eclipse is short.

In this architecture, the voltage regulation during sunlight is achieved by shunt control circuits. Since the battery is directly connected to the bus without a discharge converter in between, the bus voltage is the same as the

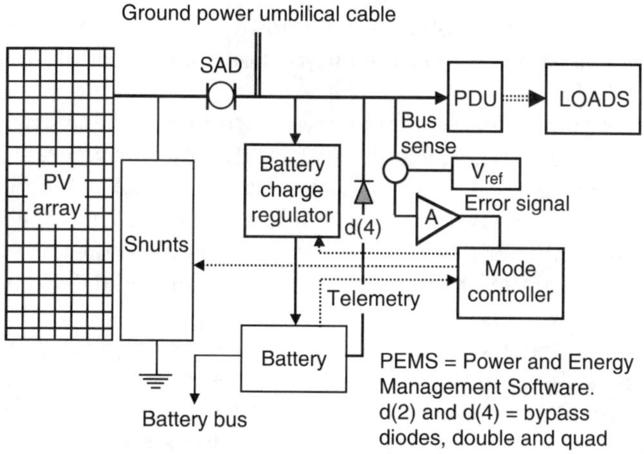

FIGURE 4.10 Sun-regulated direct energy transfer architecture.

battery voltage. Therefore, the bus voltage falls as the battery discharges during eclipse, and rises as it gets recharged during sunlight. The bus voltage variations are the same as the battery voltage. A nominal 28-V bus voltage typically varies from 22 to 35 V during the orbit period. In the absence of shunt regulation, the solar array output voltage would settle at the naturally stable operating point B in Figure 4.4, which would be too high particularly in the beginning of life and for several minutes after each eclipse when the array is cold. The maximum to minimum bus voltage ratio without the shunt control may approach 3 in some cases during mission life, versus about half as much with the shunt control.

The mode controller controls the battery charger and the shunts as needed. The power and energy management software maintains the energy balance. Non-critical loads are powered through switches and fuses.

4.7 Fully Regulated Versus Sun-Regulated Bus

Power system design must give equal importance to conditioning and distributing power from both the solar array during sunlight and from the battery during eclipse. This poses a difficult design problem since both power sources have fundamentally different characteristics. The solar array is inherently a constant current source whose output voltage is limited by the forward junction voltage of the PV cell. On the other hand, the battery is inherently a constant voltage source of low internal resistance giving a discharge voltage regulation of about 10%. It has high current and power capability and its size is determined by the energy requirement during eclipse. However, the fully regulated bus finds application in spacecraft having the load power requirements above 3 kW, typically in GEO. The sun-regulated bus, on the other hand, finds applications mostly in satellites having the load power requirement of less than a couple of kilowatts, typically in LEO.

The sun-regulated bus is simple and hence more reliable. However, the equipment connected to the bus must be designed to operate over wide variations in voltage up to ±25% around the nominal value. Some loads, such as traveling wave tube amplifiers (TWTAs) may accept such variations, but most others require a voltage regulator at the point where the power is distributed to the equipment. The sun-regulated bus saves the cost and the power loss in the battery discharge converter. However, the power loss in a battery discharge converter in the regulated bus and the battery bypass diode in the sun-regulated bus compensate each other.

The fully regulated bus offers certain advantages at the cost of complexity and potentially lower reliability. Low system impedance can cause voltage variations due to cross-coupling between items of equipment. This may increase the electromagnetic interference particularly in the time division multiple access (TDMA) mode. However, the advantages of a fully regulated bus are as follows:

- Low battery mass and cost, as it allows use of fewer higher capacity battery cells.
- Low solar array mass and cost due to not having a battery latch-up problem, which requires over-sizing the array as described in Chapter 13. The saving is typically 7 to 10%, and can be as high as 15 to 20% in some cases.
- Lighter power distribution harness mass, since the main bus works at a constant voltage. In contrast, the sun-regulated bus requires heavier wires to carry high current at the lowest voltage level at the end of eclipse.
- Lightweight load power converters, since they are supplied with constant voltage input.

In the sun-regulated bus, if the battery latch-up occurs at the eclipse exit, the resulting bus current can be in theory 60% higher than that in the fully regulated bus. However, due to the reduced thermal load in eclipses and the availability of charge power, latch-up in equinox is not a design driver. Summer solstice is more critical period, when the bus latches onto a fully charged battery, limiting the array over-sizing to 10 to 15%. The over-sizing can be further reduced by temporarily reducing the heater load.

The harness connecting two pieces of equipment is usually designed to limit the voltage drop under maximum load current. The specified allowable voltage drop primarily depends on the bus voltage, and can be from 100 to 500 mV in a 28-V bus, and 1 to 2 V in a 120-V bus. The sun-regulated bus requires thicker conductors to limit the voltage drop at high current at the end of eclipse. For a 5-kW load power capability, the power distribution harness, including the metallic cable shield and various connectors, could weigh approximately 20 to 30 kg in a sun-regulated bus and 15 to 20 kg in a fully regulated bus.

The load power conditioner (LPC) provides interface between the EPS bus and the radio frequency (RF) amplifier of the communication loads. The output ratings of LPC vary in the 50 to 500-W range. Generally, it has a pre-regulator and a switch mode converter if the bus is sun-regulated, or a linear regulator with fully regulated bus. The LPC efficiency ranges from 90 to 98% depending on the design. However, for comparable power ratings, its efficiency is about 2% higher for a fully regulated bus than that for a sun-regulated bus, and the LPC mass is lower by 2 to 4 g/W of output.

A fully regulated bus in general provides straightforward power system specifications, simple design interface, in-orbit operating flexibility, autonomous overload and eclipse entry control, and simple testing at the user end. At power levels above a few kilowatts, it also provides high efficiency, low mass and low overall cost at the power system level and/or the satellite level. On the other hand, it needs a battery discharge converter. This can be expensive, as each item of equipment has some base cost associated with the design, documentation, manufacture, quality control, and testing.

The fully regulated bus also offers great flexibility in battery cell selection. Fewer high capacity cells can be selected, since the battery voltage can be any value within the charge and discharge converter duty ratio limit, This generally reduces the battery cost and footprint by selecting one of the standard ampere-hour capacity cells to match the exact requirement of the bus. On the other hand, the required cell ampere-hour rating in a sun-regulated bus may fall between two standard capacity cells, forcing the design engineer to select the higher capacity cell, costing more in mass and money.

4.8 Peak Power Tracking Bus

The solar array generates more power at higher voltage at the beginning of life and when cold coming out of an eclipse. The maximum power can be extracted only if the bus voltage is varied with years in service and with temperature. However, the load must be supplied power at the same voltage, generally lower than the maximum power producing voltage of the solar array. A suitable switching regulator between the solar array and the load, as shown in Figure 4.11, remedies this disparity between the maximum power producing voltage and the constant load voltage. The series regulator input voltage is then maintained at the maximum power producing level by the peak power tracker, and the output voltage is stepped down to the constant load voltage by varying the duty ratio as required. The peak power tracking is activated only when the battery needs

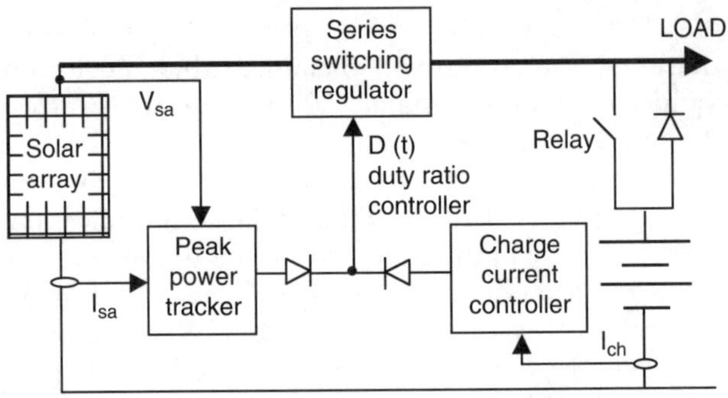

FIGURE 4.11 Peak power tracking architecture for a mission with wide variations in solar flux and temperature.

charging or the load demand exceeds the solar array output. Otherwise, the excess power is left on the array raising the array temperature. The battery relay is opened up when the battery is fully charged.

The peak power tracking (PPT) electronic controller senses the maximum power point in one of the following ways:

- The solar array output power — the product of voltage and current — is continually computed and fed to the peak power tracker. The array operating voltage is changed until the peak is detected.
- As seen in Chapter 8, the bus dynamic and static (a.c. and d.c.) impedances are equal in magnitude at the peak power point. A ripple is injected into the solar array bus, and the dynamic impedance dV/dI and the static impedance V/I are continuously measured. The bus voltage is adjusted such that both impedances are equal.
- The ratio of the V_{mp} to the V_{oc} for any solar array is approximately constant, say K (typically 0.70 to 0.75). The V_{oc} of a solar cell coupon maintained in the same environment as the main array is continuously monitored. The operating voltage of the main array is then adjusted to $K{\cdot}V_{oc}$ to extract the maximum power.
- The inner voltage control loop regulates the solar array output voltage to the reference value from the PPT controller. By changing this reference value at regular intervals, the PPT controller moves the operating point of the solar array. In each time interval, the PPT controller calculates the solar array power slope by multiplying the sensed solar array voltage and current. If this power slope is positive, the PPT controller increases the reference value until the sensed power slope is negative, and vice versa. Thus, the operating point of the solar array is located near the peak power point where the power slope is zero. The algorithm can be written as

$$V_{ref}(n+1) = V_{ref}(n) + K\frac{\Delta P}{\Delta V} \tag{4.1}$$

where K is a suitable constant. The peak power voltage thus derived is fed to the series switching regulator, which then converts the array voltage to the load voltage. Figure 4.12 shows a complete PPT assembly.

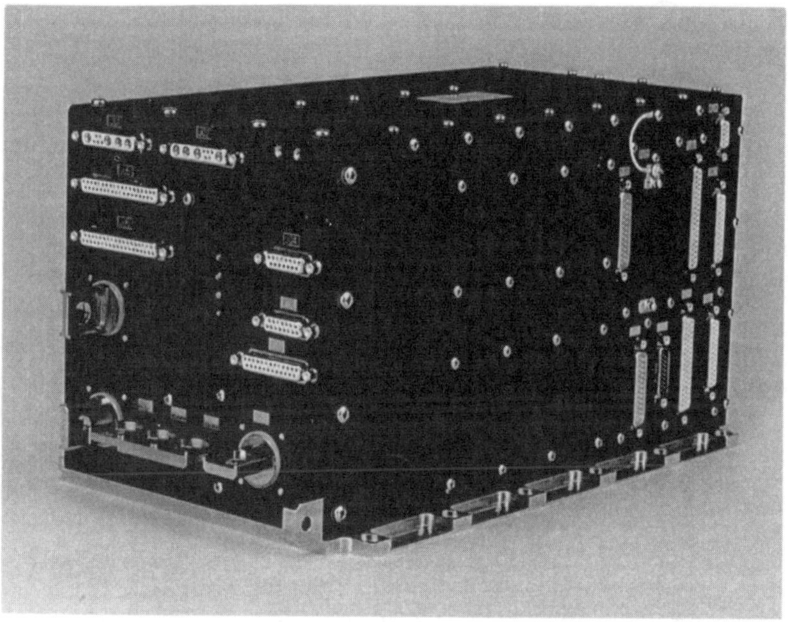

High Performance SA Power Regulation

- Peak Power Tracking at <1 % of Pmax
- Regulation Efficiency ≥ 96 %
- Power Accomodation > 1,500 W
- Bus Load Variation 300 to 2,200 W

Charge Control for Extended Battery Life

- Autonomous by V/T-Method
- 16 Commandable EOC Voltage Limits (Temperature & Current Compensated)

Comprehensive Redundancy

- 4 for 3 Power Regulators
- Power & Charge Control
- Pyro Electronics
- RS 485 Serial Data Bus

FIGURE 4.12 Power regulator unit with peak power tracker.
(Source: Dornier Satellitensysteme Gmbh, Daimler-Benz Aerospace. With permission.)

The peak power tracking architecture is particularly useful in the following applications, where the additional weight, power loss, and cost of adding such an assembly can be justified:

- Small satellites having no pointing gimbals, such that the solar array is not always oriented towards the sun
- Satellites having the solar radiation and array temperature varying over a wide range, indirectly varying the array voltage

In low Earth orbit, where the battery must be charged in a short period. The PPT allows maximum power to be captured for several minutes after each eclipse when the array is cold. Architecture without the PPT feature, such as a DET bus, would waste a significant amount of power as shown in Figure 4.13. If a DET system were designed to deliver the required power at one-half the illumination at EOL, the power waste would be CD watts at EOL full sun, BD watts at BOL full sun, and AD watts at EOL full sun on a cold array. The PPT design eliminates this waste by utilizing all the power that can be generated.

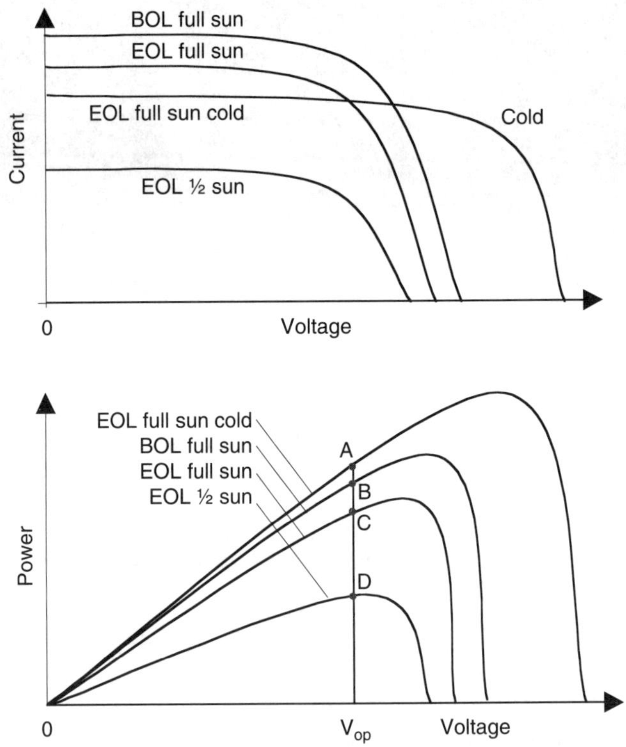

FIGURE 4.13 Power wasted in direct energy transfer architecture in certain conditions.

The main advantages of the peak power tracking are that it maximizes the solar array output power all the time, and it does not require the shunt regulator and the battery charge regulator. On the other hand, it results in poor system efficiency due to power loss in the peak power tracking converter. Moreover, since this loss is dissipated inside the spacecraft body, it negatively impacts the thermal system.

The PPT can have three configurations: series, parallel, and series–parallel, as shown in Figure 4.14.[1] The series–parallel configuration yields better system efficiency because the input and the output power conversion is processed by a single converter for all operating modes, as seen in Table 4.1.

A PPT algorithm has been developed[2] and tested using only the solar array voltage information, giving the tracking control without using current sensors. This results in low ripple current, hence lighter bus filter capacitors.

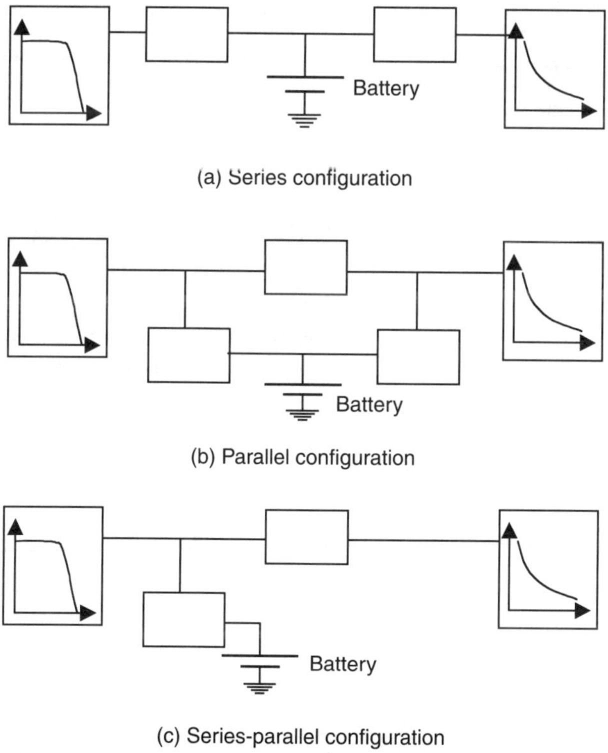

FIGURE 4.14 Battery charge and discharge options in peak power tracking architecture.

Table 4.1 Operating modes of series–parallel battery discharge regulators in the PPT system

Mode	Series regulator	Battery charger	Battery discharger
Eclipse	Off	Off	Regulated bus
PPT discharge (partial sun)	PPT	Off	Regulates bus
PPT charge (full sun)	Regulates bus	PPT	Off
Trickle charge	Regulated bus	Trickle charge	Off

4.9 Architecture Trades

A primary criterion of choosing between alternative architectures is the overall mass, efficiency and cost of the system. Table 4.2 summarizes the trades between various architectures with their best applications. The final selection of the architecture depends on the mission specific details. However, past experience indicates that peak power tracking architecture

Table 4.2 Pros and cons of various architectures and their best applications

System	Sun regulated	Fully regulated	Peak power tracking
Pros	High power transfer efficiency from solar array and battery to load. Fewer power system components.	Well regulated input voltage to all loads. Simpler, lighter and more efficiency load converters.	No need for shunt regulator and battery charge regulator.[a] Makes the maximum use of the incident solar energy.
Cons	More complex load converters. Battery latch-up concern. Larger solar array.	Needs more power converters. Series power loss between battery and load.	Lower efficiency than DET at EOL in many cases. More heat dissipation inside the spacecraft body.
Best application in missions with these features	Small load variations. Small variations in illumination for most of the sun period.	Loads requiring close regulation. Large solar array output voltage variations.	Large variation in solar array input energy (illumination) throughout the mission.

[a]True for a single battery bus with the battery connected directly to the bus. For a multiple battery system, or a fully regulated bus, each battery must have its own charger for effective battery charge management.

is generally advantageous for small satellites in low or irregular orbits having a power requirement of less than 500 W. Between 1000 and 3000 W, the sun-regulated direct energy transfer architecture would most probably be advantageous. For power levels exceeding 5000 W, the regulated direct energy transfer architecture is generally found advantageous. Although the regulated bus requires additional equipment, the added mass is compensated by the elimination of converters and regulators in the loads. Moreover, decoupling the battery design from the solar array voltage, and operating the system at a constant voltage optimizes the operating point of the solar array. The battery voltage is then chosen to optimize the number of battery cells in the most economical capacity.

The PPT architecture is more suitable for LEO satellites having relatively short periods of sunlight. It is also attractive for missions with large variations in solar flux, solar array temperature, and sun angle in satellites with no sun tracking gimbals. It makes the best use of the solar array all the time in such missions.

In addition to the DET and PPT architectures described above, some spacecraft use hybrid architecture, such as two power busses, one regulated and the other sun-regulated. Charging the battery from a dedicated solar array section rather than from the main bus is another possibility.

The architectures of some of the spacecraft busses flying at present are described in the next sections.

4.10 The International Space Station 160- to 120-V Bus

The International Space Station is truly the largest and most complex space structure ever built, with 16 international partners. The complete assembly shown in Figure 4.15 weighs one million pounds and the total interior space of its six laboratories equals two Boeing 747 aircraft. It is taller than a 30-story building and wider than the length of a football field. The attitude is controlled within 1° stabilized at a rate less than 0.1°/s. It is in 335 to 500-km low Earth, 51.6° inclined orbit with a 90-min orbit period and 35-min eclipse. The experiments planned for its on-board laboratories are targeted to enhance our understanding in:

- The Earth, by viewing 75% of its surface
- Long term exposure of micro-gravity on humans
- Producing new medicines and materials

Like any other large space programs, such as the manned landing on Moon, the engineering and technology spin-offs on our everyday lives on Earth are the side benefits.

The ISS power system generates 105 kW using a solar array with an area of nearly 1 acre. The loads are powered during an eclipse by thirty-eight

NASA artist's conception

FIGURE 4.15 International Space Station with solar array and other modules in view. (Source: NASA.)

81-Ah NiH_2 batteries[3] via bi-directional battery charge–discharge units (BDCUs), autonomously controlled by the bus voltage set point. It has two interconnected power systems; the 160/120-V US built system, and 120/28-V dual voltage Russian built system to power the US, European, and Japanese modules. The two systems are generally independent, but are interconnected via d.c. converters to allow bi-directional transfer of power. Figure 4.16 is the functional block diagram of the US system, including the converter at the interface with the Russian power source. The US system is described below.[3]

The solar array is made of four modules (wings) for a total of 76 kW power under nominal conditions, and more during favorable conditions. Each wing consists of two thin blankets held under tension on each side of a central collapsible mast. The entire assembly turns on a β gimbal, which provides one axis of rotation for sun pointing. A second orthogonal axis rotation is provided at the α gimbal, where the entire solar array connects to the rest of the truss structure of the station. The 76 kW from the US system added to the 29 kW from the Russian system makes the 105 kW capability of the complete station assembly. Each of the four solar modules is made in 82 strings of eighty 8 × 8 cm PV cells and coverglass for protection against space charged particles. The PV cells are crystalline silicon with 14.5% EOL efficiency over a 15-year life. The module printed circuit is Kapton/copper/Kapton laminate welded to each cell to provide a series interconnection for

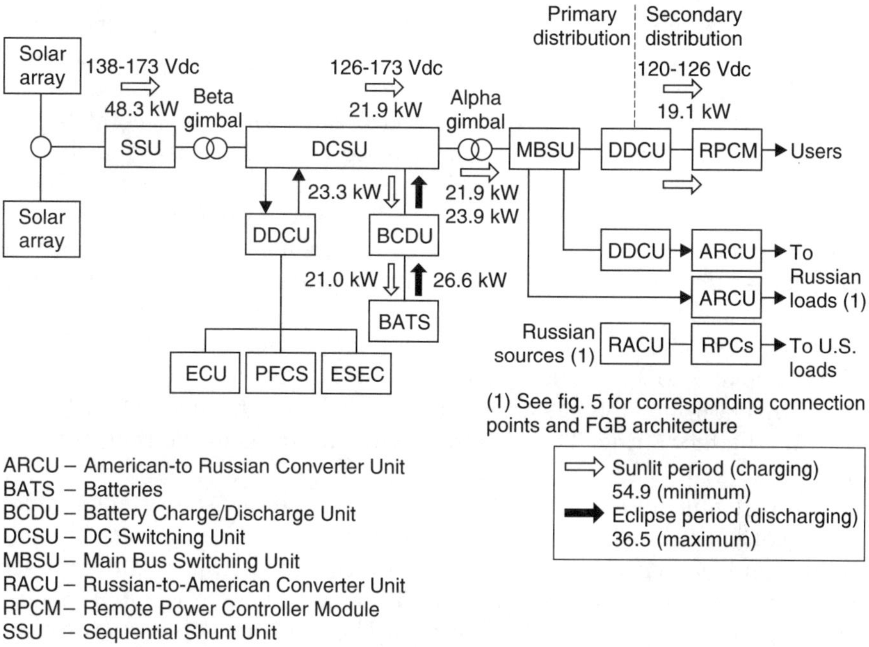

FIGURE 4.16 International Space Station single channel power flow diagram. (From E.B. Gietl et al. NASA Report No. 210209, 2000.)

the assembly. Parallel bypass diodes are used every eight cells for reliability in case of cell damage and to avoid reverse current heating during prolonged shadows. Sequential shunt units (SSUs) on the solar arrays operate at 20-kHz switching frequency.

The seasonal sun-pointing is done by β gimbals and the orbit sun-following by α drive and roll rings. For currents of the space station magnitudes, the roll rings provide superior power transfer performance over the slip rings with rubbing contacts, as described further in Chapter 22.

The solar array output voltage is 160 V, which is the highest voltage that can be practically used in a low Earth (high plasma) orbit in view of potential plasma arcing and/or leakage current concerns. The 160 V is stepped down to 120 V using d.c.–d.c. converter units (DDCUs), each rated at 4.25 kW, for utilization inside the user modules. The DDCUs provide 150% current limiting capability and 20 dB isolation between the generation point and the distribution points for personnel safety.

The solar array area and the operating voltage are greater than in any other spacecraft flown before. Therefore, the nature of the single point ground on the ISS in high plasma in LEO poses an arcing problem. To preclude such arcing, a device called the plasma contactor located on the truss creates a plume of ionized xenon gas, which acts as a conductive bridge between the station and the plasma. It protects the array and other conductive surfaces of the station from arcing, pitting, and erosion.

The battery is made of 48 battery packs, each with thirty-eight 81-Ah IPV NiH_2 cells, designed for 5 years (40,000 charge/discharge cycles) at 35% depth of discharge (DOD), although the actual operational DOD is around 15%. With two packs connected in series, the battery voltage varies between 95 and 115 V. Battery replacements are planned every 5 years to assure the power margin. With 15 years life of the station, the battery replacements every 5 years constitute significant recurring cost. The maximum charge rate is 50 A, which tapers down to 40, 27, 10 and 5 A, and finally to 1 A trickle charge rate. No bypass diodes are used in either direction, but the cells are closely matched. An open cell will lose the entire battery string. The operating temperature is in the range of 0 to 10 °C.

The switching and fault protection is achieved by solid-state remote power controllers (RPCs) in six ratings from 3.5 to 65 A in both current limiting and non-current limiting designs. The RPCs trip at different set points of over-current, over-voltage, and under-voltage to isolate faults as close as possible to the faulted equipment. The solid state power controllers (SSPCs) provide switching and protection. They reset automatically for critical loads and remain tripped for non-critical loads. The upstream coordinated fault protection in the 160-V segment is achieved by larger main bus switching units (MBSUs) containing a remote bus isolator (RBI). The RBIs are essentially large relays capable of interrupting up to 350 A of d.c. fault current.

The power system stability is also a serious concern because the loads on the station are constantly changing as new scientific experiments are brought on board. As a result, the output impedance of the DDCUs and the input characteristic of the loads have been specified to ensure stability with any combination of the expected loads. The stability criteria under such diverse load combinations are discussed in Chapter 14.

A hierarchy of redundant computers linked via the MIL-STD-1553 bus controls the ISS power system. The computers autonomously control many functions such as sun tracking, battery energy storage, and thermal control. The power flow balance between major segments of the station is coordinated by the on-board command and control systems, which also provide interface control between the segments.

Future plans for new technology developments on the ISS that are related to the electrical power systems include: (1) replacing the silicon solar cells with GaAs panels to increase conversion efficiency, (2) using high discharge cycle NiH_2 batteries, and (3) using flywheel energy storage to significantly increase the specific energy.

4.11 Large Communications Satellite Buses

In view of reducing cost and delivery time, all prime manufacturers have developed standards busses, which can meet various power requirements without making many changes. The standard bus, however, has a degree of flexibility to tweek a little if and when needed to meet a customer's specifications. A few such standard busses presently available for communication satellites are briefly described below.

4.11.1 100-V Bus

A standard fully regulated DET bus operating at 100 V d.c. is offered by Boeing Satellite Systems under the trade name BSS-702™. Its power system architecture is shown in Figure 4.17.[4] The BSS-702 bus is a dual voltage bus, primary 100 V d.c. for high power equipment and secondary 30 V d.c. for low power component. It uses a NiH_2 battery and silicon triple junction solar cells. A xenon electric ion-propulsion system is used for N–S station keeping, which is powered from the 100-V bus.

After each discharge of the battery, the basic function of any battery charge management is to return the battery to the highest achievable SOC with the minimum amount of overcharge. In the BSS-702 bus, this is achieved by performing the following health monitoring and maintenance functions:

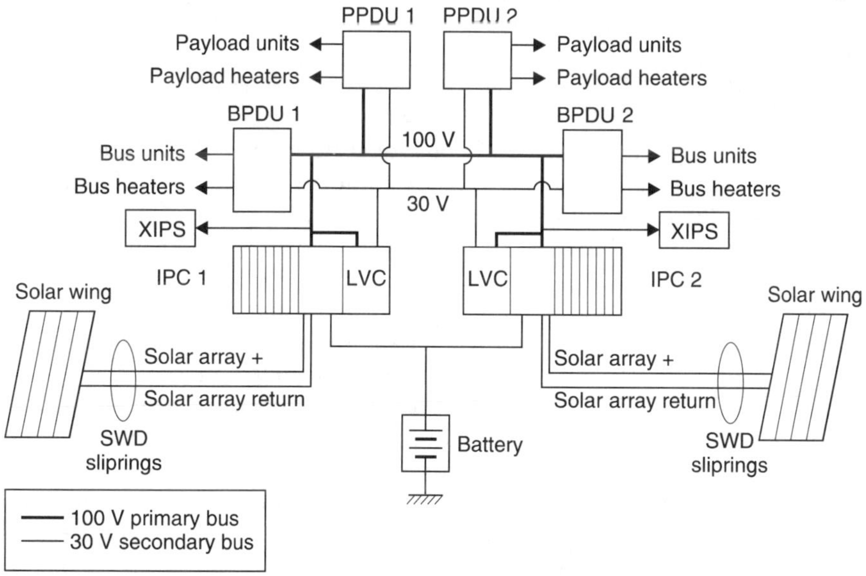

FIGURE 4.17 Fully regulated 100-V bus architecture.
(From R. Hill, Proc. 36th IECEC, ASME, p. AT-59, 2001. With permission.)

- Monitor and process the battery voltage, current, temperature, and pressure sensor readings.
- Control over-temperature to prevent inadvertent overcharge.
- Autonomously control the charge rate by switching between high, low, and trickle charge rates as needed in different seasons.
- Adjust battery bias to account for voltage degradation in NiCd and pressure growth in NiH_2 cells.
- Detect and report low SOC conditions.

Each battery is built in multiple packs to distribute the mass and heat as needed to balance the spacecraft layout. Each pack has redundant heaters, reconditioning circuit health sensors, and cell bypass diodes in case of cell failure. The sensor monitors the cell voltage, pressure, temperature, and bypass and reconditioning circuit status telemetry. The cells are 5.5-inch diameter NiH_2 and are stacked vertically with the electrodes on two separate sides.

The solar array panel tracks the sun in one axis only, so the seasonal variations in power generation can be more than 10% over the year. The original solar array design was a concentrator type with channeled solar reflector panels along both sides of the wing with GaInP2/GaAs/Ge PV cells mounted on a graphite face sheet on an aluminum honeycomb substrate. After suspecting some plume related problems, the array was replaced with a traditional flat panel array.

Loads are either unfused, fuse protected, or switched by relay or transistor, depending on the current magnitude and the nature of the load.

4.11.2 70-V Bus

A standard fully regulated DET bus operating at 70 V is offered by Lockheed Martin Space Systems under trade name A2100™. Of the 16 commercial satellites contracted in 2002 and the first half of 2003, five were for A2100. It is available up to 15-kW power level, scalable up to 20 kW using multi-junction GaAs PV cells. The power system architecture is shown in Figure 4.18.[5] The solar array is made of flat panels with Si, GaAs or multi-junction PV cells as needed for the required power level. The PV cell coverglass is coated with indium tin oxide (ITO) to eliminate arcing due to electrostatic discharge (ESD). The honeycomb substrate with a graphite epoxy composite face sheet saves mass over that with aluminum. The booms and yokes are made from aluminum or graphite. One solar array drive (SAD) per solar array wing with redundant motor winding provides a large torque margin in case jerking is needed. Power slip rings and signal slip rings are in separate sections.

The batteries are made of NiH_2 cells assembled in two batteries for the required energy storage. The battery power converters in the PRU can be

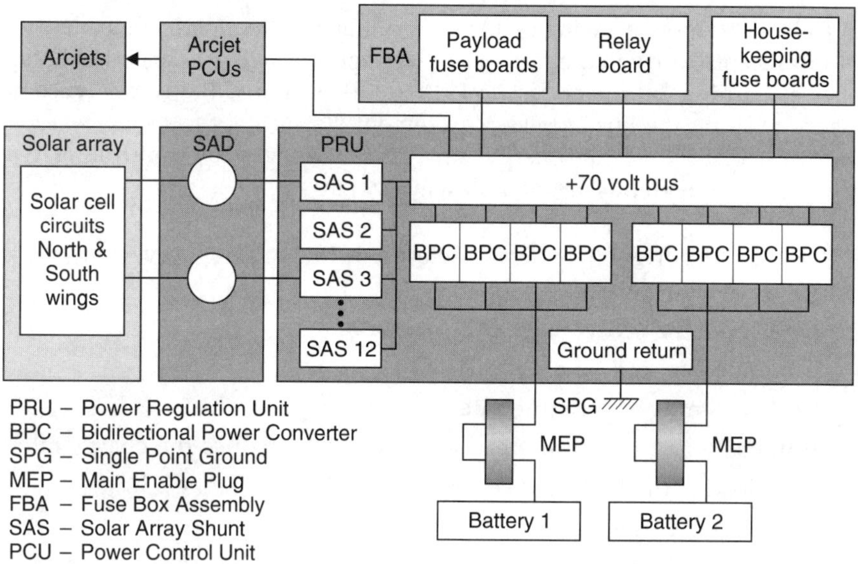

FIGURE 4.18 Fully regulated 70-V bus architecture.
(From A.A. Salim, Proc. 35th IECEC, AIAA, p. 2809, 2000. With permission.)

directional (buck and boost separately), or bi-directional if it is beneficial. The charge converters have multiple charge rates. The batteries have multiple heater control zones for maintaining desired temperature with redundant temperature and pressure sensors. For open cell protection, a thermally activated switch is preferred over the bypass diode, as the former results in lower loss of power. Reconditioning is considered unnecessary for the NiH_2 batteries. However, if required by the specifications, resistances are connected across the entire battery, or across a battery section, or across each cell. In the first option, which is the most mass effective, two parallel resistors are first commanded across the battery. As the battery voltage approaches 1.0 per cell, the discharge rate is cut back by commanding one of the resistors off to avoid cell voltage reversal. The charge rate is also cut back in case of any peaking power on the bus sensed by a voltage dip.

The battery nominally operates between –5 and 0 °C during the charge period. However, its temperature rises to about 20 °C at the end of the longest eclipse. The allowable cell-to-cell and battery-to-battery gradients are 5 to 10 °C. Locating the battery outside the spacecraft body can be advantageous in maintaining the 0 to 5 °C temperatures during the charge periods. The MEP is the module-enabling plug for the battery. The electro-explosive devices (EEDs) for deployment are powered directly from the battery at the 22nd cell tapping, and are protected by fuses in the EED enable plugs.

The main power bus assembly provides the load distribution, current sensors, voltage sensors, and bus filter. The fuse box houses protective fuses and also load power switches for controlling the bus load when needed. The transient over-voltage and under-voltage following a large load step, such as turning on an arc jet (>4 kW total), is individually analyzed to ensure meeting the bus ripples requirement.

In the A2100 bus, the energy balance is managed by controlling two batteries independently in the following steps:

- Using the current telemetry, determine the total battery charge and discharge current
- Determine the cell voltage and pressure
- Compute the state of charge
- Control the battery charge rate
- If the battery temperature exceeds a set limit, reduce the charge rate
- Shed load if the battery discharge falls below the set level
- Provide for parallel or sequential battery charge
- Prevent the disconnection of all battery power converters
- Provide ground override capability

The bus under-voltage protection under a set limit is provided by commanding and latching the payload power converter off. The shaded loads can be turned on only by a ground command.

A shunt is connected across each solar circuit. The shunt switch is either completely turned on or off. The on-switch shunts the power to the ground, while the off-switch feeds the power to the bus. The desired bus regulation with fine resolution is maintained by using one active shunt. Such configuration also minimizes the bus voltage ripple.

4.11.3 Under-50-V Buses

British Aerospace's standard 42.5-V fully regulated and 28 to 42.5-V sun-regulated busses have been used on many small satellites. US manufacturers, on the other hand, offer 28-V fully regulated and 22 to 35-V sun-regulated busses for small satellites requiring load power under a couple of kilowatts. One such bus offered by Daimler-Benz Aerospace is depicted in Figure 4.19.

4.12 Small Satellite Bus

The following power system features are typical in small satellites:

- A solar array covering a wide range of solar flux and temperature with wide swings in the *I–V* characteristics.

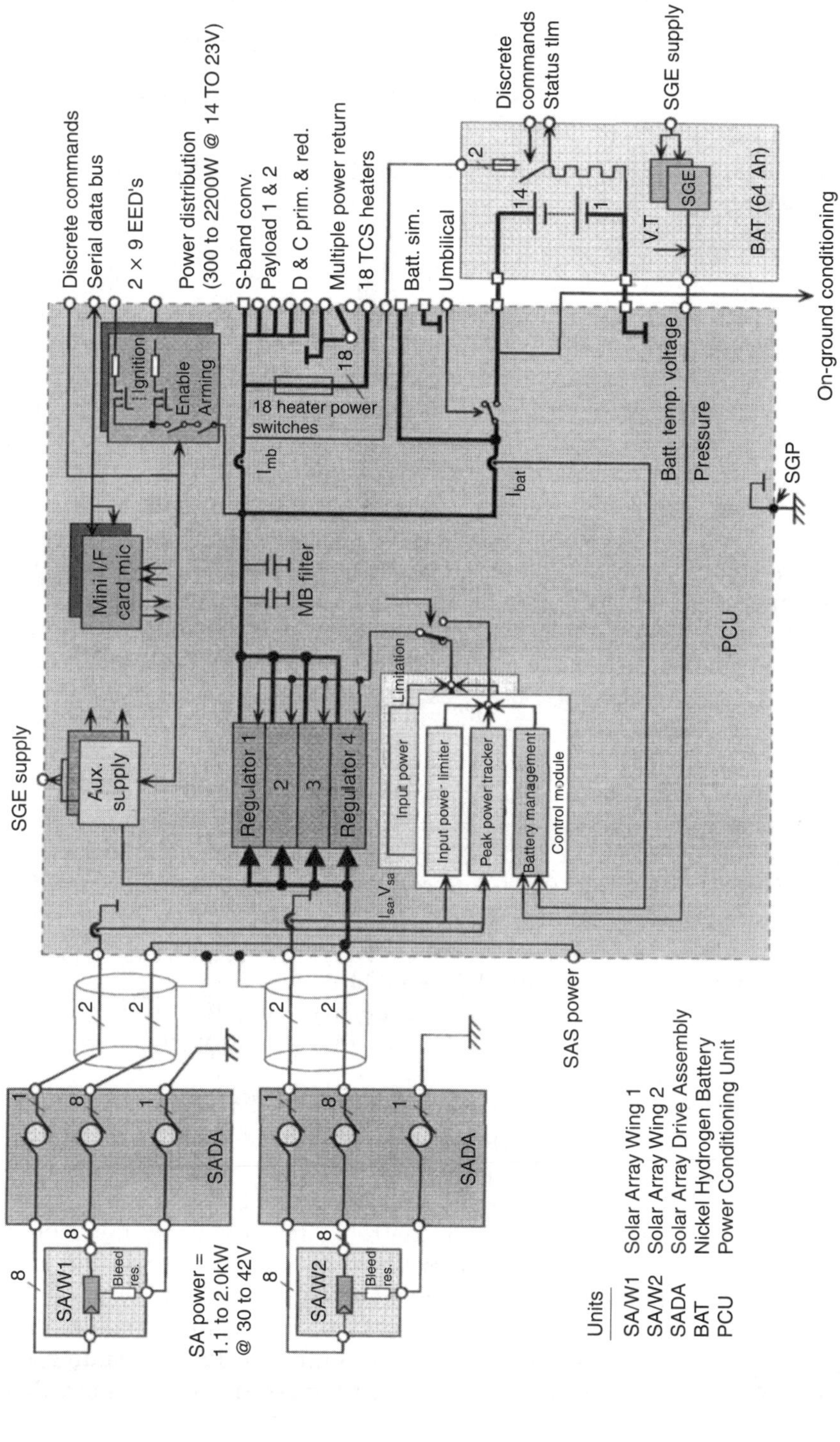

FIGURE 4.19 Sun-regulated low voltage bus architecture. (Source: Dornier Satellitensysteme Gmbh, Daimler-Benz Aerospace. With permission.)

- Peak power tracking architecture and no sun-tracking gimbals.
- Body mounted solar array or three to four flat panels on 1 to 2 mm aluminum substrate with insulation and optimized coverglass. Or, flexible solar array with cloth-type substrate weighting about 5% of the rigid substrate.
- One battery, if permissible; otherwise two; NiCd or NiH_2; both to have a reliable flight history.

Early small satellites were spin stabilized, with their drum-like body covered with PV cells all around. Such configuration used the cells ineffectively; because it uses less than one third of the PV cells for power generation at any given time. Today's spacecraft have 3-axis active attitude control with solar cells installed on flat panels pointed normally to the sun.

The PPT is more suitable for being one common architecture for a number of small satellites for a variety of missions. The design can be modular and adaptable without component redesign, thus being cost effective. The power control unit maximizes the energy delivered to the bus by driving the solar arrays to operate at their maximum power point over a wide range of temperature, sun inclination, and illumination intensity. The tracker system, in addition to its flexibility, does not dissipate power associated with the shunt regulation inside the spacecraft. It adjusts the operating point of the solar array as needed, thereby leaving the excess power on the array.

A PPT bus that has flown successfully had the following design features, which can be easily changed to accommodate other missions.

The solar array continuously tracked the sun using a solar array drive. The PV cell could be either 8 mil or 2 mil thick silicon or 5.5 mil thick GaAs. The NiH_2 battery used 22 cells. The maximum battery DOD for LEO was nominally 30% and 35% with one cell failed. For GEO, it was nominally 70% and 75% with one cell that failed. The charge/discharge ratio was maintained at 1.05 for LEO and 1.2 for GEO.

The peak power tracker received telemetry from the battery charge control loop, which varied the modulation of the control signal operating the tracker in each PRU.

The PRU worked as the interface between the solar array and the bus, designed to deliver the total power generated by the solar array to the spacecraft bus. It contained a peak power tracker comprised of a buck regulator with pulse width modulation (PWM) control and the required battery voltage–temperature (V–T) control. After receiving the battery telemetry and comparing with the selected V–T charge curve, the PRU varied the duty ratio of the buck converter. This changes the solar array operating point thus reducing the power to the bus. All power regulation functions were monitored and controlled within the PRU, including the battery charge control. Separate PRUs were used for each wing if more than one solar array wing under different operating conditions were needed for some satellites in the standard product line. This ensures that the EPS design remained modular to accommodate all possible mission scenarios with little or no design changes.

4.13 Micro-Satellite Bus

Very small satellites with load power requirements in watts can use simple and lean architecture shown in Figure 4.20. The solar array, the battery, and the loads are permanently connected in parallel. The battery feeds to the loads automatically during an eclipse, and recharges itself during sunlight by imposing its drained voltage on the bus. Once fully charged, the battery voltage is relatively constant, and the battery works as the buffer. Any excess current is absorbed by the shunt resistance connected in parallel with the battery by means of a switch.

Two simple architectures for satellites that require power only during sunlight are shown in Figure 4.21(a) and (b). The shunt voltage regulator makes the power source behave like a current source. On the other hand, the series regulator makes the source behave like a voltage source. The disadvantage of the series approach is that, without the shunt regulator, the solar array output voltage in a typical GEO satellite would exceed 150% of its EOL voltage. For a few minutes after exiting the eclipse, the cold array can produce as much as 300% of the nominal bus voltage.

Low voltages (7 to 15 V) have been considered for satellites with load in tens of watts. However, such low voltage is likely to end up in high cost because it requires components other than standard 28-V class components. For example, ST5 spacecraft designed by NASA GSFC[6] for a few watts load used a battery dominated low voltage bus depicted in Figure 4.22. To minimize the number of components for high reliability and low mass, it used a body mounted solar array with triple-junction cells, one lithium-ion battery, and simple power electronics. The electronics that traditionally

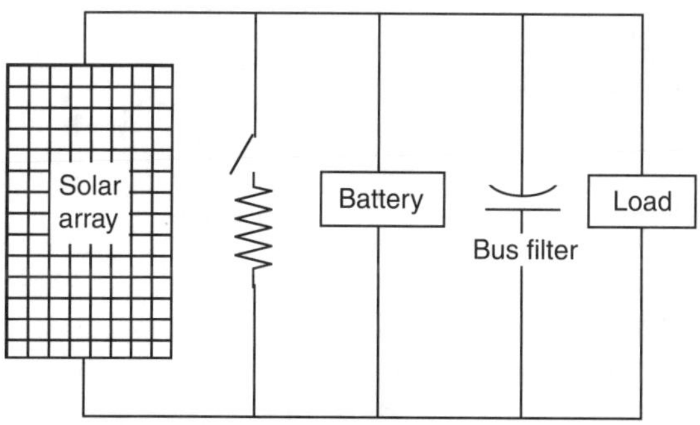

FIGURE 4.20 Simple power system for a micro-satellite.

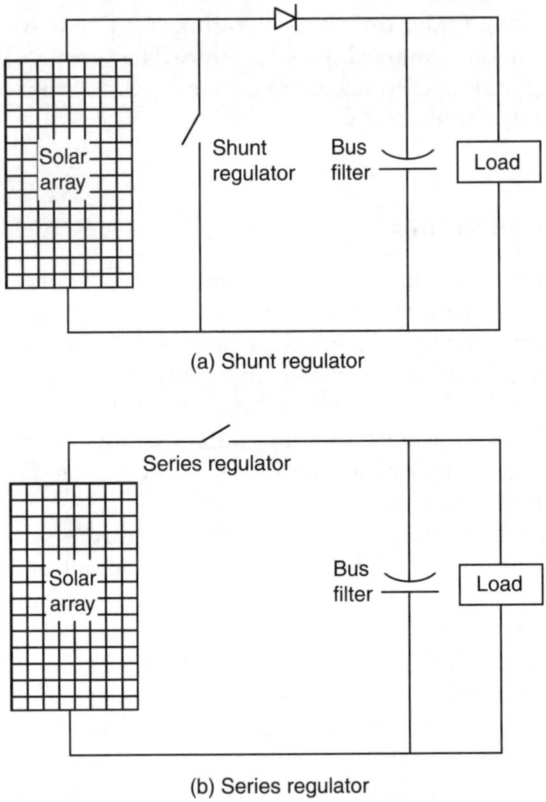

FIGURE 4.21 Micro-satellite power system with shunt regulator and series regulator.

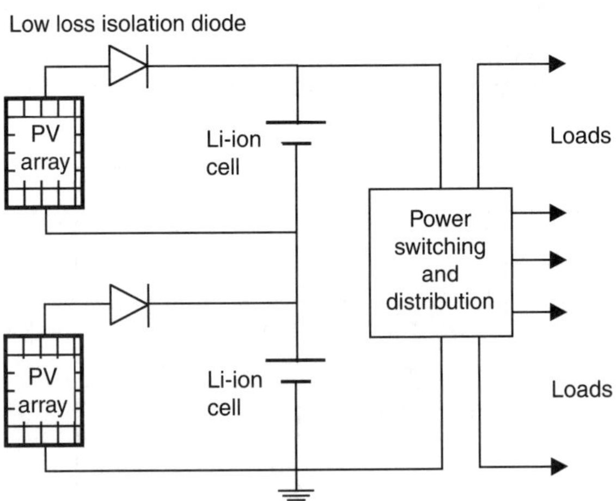

FIGURE 4.22 Micro-satellite power system with no regulator.

provide the solar array regulation and battery charge control are eliminated. Low loss diodes are used in series with a solar array for fault isolation. The bus voltage is the same as the battery voltage, which may vary from 2.7 to 4.0 V per cell. The design can be modular in that for a higher power microsat, two modules can be connected in series as shown in the figure.

References

1. Cho, Y.J. and Cho, B.H., "A novel battery charge–discharge of the regulated peak power tracking systems," in *Proceedings of the 34th Intersociety Energy Conversion Engineering Conference*, SAE, 1999, Paper No. 01-2445.
2. Veerachary, M., Senjyu, T., and Uezato, K., "Voltage-based maximum power point tracking control of PV system," *IEEE Transactions on Aerospace and Electronics Systems*, 38(1), 262–267, 2002.
3. Gietl, E.B., Gholdston, E.W., Manners, B.A., and Delventhal R.A., "Electrical power systems of the international space station — A platform for power technology development," in *Proceedings of the 35th Intersociety Energy Conversion Engineering Conference*, AIAA, 2000, Paper No. 35-AP-SS-1. Also NASA Report No. TM-210209, June 2000.
4. Hill, R., "Boeing Satellite Systems 702 Electrical Power System," in *Proceedings of the 36th Intersociety Energy Conversion Engineering Conference*, ASME, 2001, Paper No. AT-59.
5. Salim, A.A., "In-orbit performance of Lockheed Martin's Electrical Power System for A2100 Communication Satellite," *Proceedings of the 35th Intersociety Energy Conversion Engineering Conference*, AIAA, 2000, Paper No. 1-AP-SP-1
6. Castell, K. and Wingard, R., "Recent advances in power system design at GSFC," *Proceedings of the 34th Intersociety Energy Conversion Engineering Conference*, SAE, 1999, Paper No. 01-2534.

Chapter 5

Environmental Effects

5.1 Introduction

The radiation and other aspects of the space environment that have damaging effects on various power system components are discussed in this chapter. The charged particles impinging on the spacecraft surface slow down by collision and lose their energy. The deposited energy in the material causes damage by atomic excitation in metals and semiconductors, and ionization in plastic materials. Various particles lose energy at different rates as they penetrate the surface and cause damage at different rates. The higher the rate, the more damage they produce. The damage is most severe in the Van Allen belts because of the high level of radiation. Although the belt region, which is between the low Earth orbit and the crowded geosynchronous orbit, is potentially a valuable piece of orbital real estate, the region has remained relatively unexplored because of the practical and economic difficulties of shielding conventional satellites from charged particles. NASA has measured charged particles in many orbits to develop and validate a computer model. In 2002, nine astronauts aboard the ISS each wore a 0.04-inch square silicon chip dosimeter that measured the levels of radiation they were exposed to during space walks. On Earth, similar devices are used in hospitals worldwide to measure the radiation given to cancer patients.

5.2 Solar Array Degradation

The solar array is the most affected component in the power system under space radiation. The degradations are listed in Table 5.1 in order of their importance. The array performance degrades with the increasing level of fluence, but not in a linear relationship. The damage is more in the initial years than in the latter years. The impinging particles produce defects in the crystalline structure of the PV cells. The resulting defects degrade the voltage and current outputs of the cell. Low-energy particles create damage close to the surface, and therefore lower the open circuit voltage. On the other hand, high-energy particles penetrate deeper in the base and lower the lifetime of electron hole pairs, thus decreasing the short circuit current. The power generation capability continues to degrade as the radiation dose accumulates over the years. When it falls below the specified requirement,

Table 5.1 Environmental causes and effects on solar array performance

Environment	Effect
Ionized particles	Damage to the p–n junction
Micrometeoroids and debris	Damage to cell active area and interconnects Increased cell shunt resistance
Contamination from out-gassing and propulsion plume	Degradation in optical properties of the coverglass
Space plasma	Increased leakage current
Thermo-elastic cycling	Cracks in solder joints of the interconnects
Bypass diode failure	Shorts in cell blocks and reduced string current
Ultraviolet radiation	Degradation in optical properties of the coverglass

the solar array is considered to have reached the end of life. Figure 3.7 and Figure 3.8 showed the gradual degradation of the *I–V* curves under increasing fluence levels, which result in continuously decreasing power output with years in service. With traditional silicon cells, the average power degradation is about 1% per year in LEO and less in GEO. Because the radiation damage to the n-type crystalline lattice is less than that on the p-type, the n–p type PV cells are more radiation resistant than the p–n type used in early satellites. Today, almost all cells used in the industry are n–p type. A coverglass, such as fused silica or ceria doped Microsheet^{TM}, over the PV cell is used for protection against radiation. The degree of radiation protection provided by the coverglass comes from both the density and thickness of the material. For example, 12-mil fused silica coverglass provides the same protection as 10-mil ceria Microsheet. Normally, the coverglass is coated with anti-reflective material, such as silicon monoxide. The coating minimizes reflection of the sunlight, thus enhancing the solar energy absorption and the power output of the cell. Susceptibility to radiation can be further enhanced using radiation-hard solar cells. Higher resistivity of the cell substrate (base), while initially less efficient, may provide higher EOL power in a long mission spacecraft.

5.3 Electrostatic Discharge in the Solar Array

The arcing in a solar array due to electrostatic discharge (ESD) is known to have caused power loss in both LEO in GEO satellites. Small ESD discharges do not have enough energy to cause damage. However, the risk is increased in a high-voltage array, particularly in the low Earth orbits. Bus voltages greater than 120 V inherently increase such a risk because of current leakage and arc discharges. The mechanism of damage is as follows. An anamolously large charge collection from the space environment can occur on positively biased array components, and the negatively biased portions may be involved in arc discharges. These discharges may damage solar cells and produce electromagnetic interference. The vast majority of arcing would be of a microsecond duration. The arcs are generally localized near the coverglass edges and are believed to involve electron currents from the conductor. ESD arcing between adjacent PV cells produces intense low impedance plasma, leading to an array-driven current coupling between adjacent cells as shown in the electrical circuit model in Figure 5.1. The resulting ohmic heating damages the conducting path, possibly causing a permanent short in the affected string. The plasma charge eventually discharges in the space and reaches a balanced voltage. In a high plasma orbit, this voltage is high.

Flight experiments have shown strong dependence of the arc rate on bias voltage and plasma density and the existence of a threshold bias voltage. Dielectric materials such as the coverglass or adhesive must be near a highly negative conductor or semiconductor, and both components should be in contact with space plasma. The charging rate in GEO is believed to be determined primarily by the flux of electrons with energy in the order of 30 keV. The arcing rate is sensitive to temperature, and most GEO arcing occurs just after coming out of an eclipse. The primary mechanism of arcing in high voltage solar arrays is believed to involve differential charging. It triggers an arc followed by a more energetic discharge between cells that is sustained by the array power itself. If Kapton insulation on the array substrate is carbonized, a permanent short and a loss in power generation can occur. This risk can be mitigated by grouting adjacent cells to increase the trigger threshold of the sustain discharge.

Figure 5.2 from Crofton and Francis[1] shows data from a simulated test on a solar array under a 20-keV electron gun and sunlight. As expected, the substrate and coverglass could become charged to kilovolt levels. The data

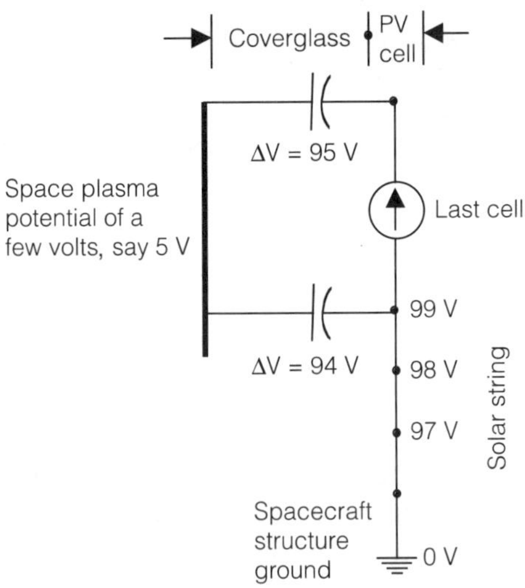

FIGURE 5.1 Electrical circuit model of solar array string in plasma.

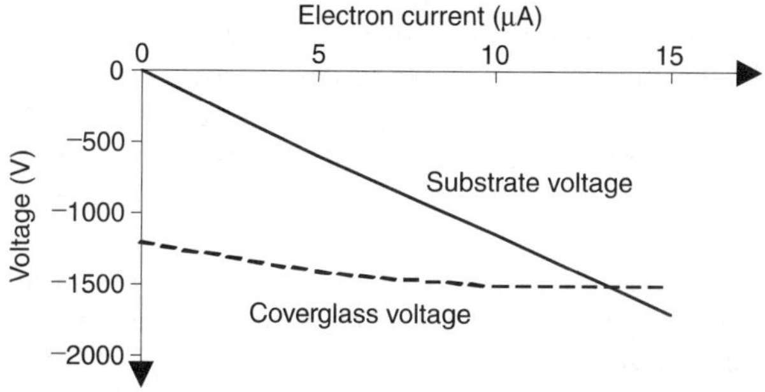

FIGURE 5.2 Aluminum substrate and coverglass charging under 20-keV electron gun in sunlight.

were obtained by starting at the highest electron beam current, and reducing it in steps at intervals of several minutes each. The conducting substrate voltage decayed quickly with time, but the coverglass potential was sustained due to the long time constant of the resistive coverglass once the charging source was removed. Such differential voltage can arise because the flux of electrons and protons in GEO can vary minute by minute.

The spacecraft charging environment in LEO is primarily due to ions. It consists of low energy plasma of less than 1 eV with a density ranging from 10^5 to 10^6 particles per cm^3. The floating potential of the spacecraft is affected by the bus voltage and the area of the solar array. The large area of the coverglass floats near the plasma potential of a few volts. The spacecraft structure, on the other hand, is driven negative to approximately the bus voltage. Under this condition, the cells near the structure ground have high voltage gradient across the coverglass and the cell, leading to arcing across a small clearance (<1 mm) between the coverglass and the cell interconnects. The resulting relatively dense and highly conductive plasma plume expands over the array and sweeps past the cells. The charge held in the coverglass-to-cell capacitance is released in the conductive plume, causing current coupling directly between the cells. The cell-to-cell coupling is driven by current and voltage produced by the array strings, and can last much longer than the initiating arc between the coverglass and the cell interconnects. If sufficient string current is available, such sustained arc can cause heating damage, leading to a permanent short circuit and power loss. NASA tests have shown a sustained cell-to-cell arcing threshold of about 60 V bus voltage and 1 A string current. The solar array arc discharge mitigation options are discussed in Chapter 8.

5.4 Damage to the Power Electronics

The charged particles can interact with sensitive regions of semiconductor devices and degrade the electrical performance and even trigger catastrophic failures. The photons, protons, and electrons cause gradual degradation in performance with increasing radiation dose. The space environment also contains high-energy protons, neutrons, and heavy ions that can cause single-event displacement damage by introducing high-energy transients in the sensitive region of the power MOSFET. The effect can be a single-event burnout, single-event gate rupture or single-event latch-up. The burnout causes permanent short circuit between the source and the drain, rendering the MOSFET useless. The gate rupture causes a permanent short between the gate and the drain. The latch-up can also be caused by electrical transients on input or output lines, high temperature or improper sequencing of power supply biases. Various mechanisms of such damage are as follows:

Ionization damage: The electron–hole pairs are produced by ionization in the device insulation or protective coatings. The charge thus trapped degrades the electrical properties of the device.

Dose-rate damage: The electron–hole pairs produced by ionization over a short duration produces a current transient. The resulting voltage transients can upset logic or memory devices and can induce latch-up.

Displacement damage: High-energy particles collide with silicon atoms and displace atoms from their crystal lattice, producing defects leading to degradation in electrical performance of the device.

The above three types of damage do not occur one after another in steps as the radiation dose increases in a device. All three types may occur simultaneously in a single device. The results of the damage can be classified into two categories:

- Failure under single-event radiation from cosmic rays
- Gradual degradation under the accumulated dose of radiation

The worst-case cosmic ray radiation can cause the following types of failure in semiconducting devices:

- Single-event upset (SEU). This causes a change in the state of the stored memory (bit-flip) in the semiconductor device. However, it does not result in a damaged device.
- Single-event latch-up (SEL). This causes a high current condition with loss of stored data, inability to function, and even burnout. The device must be powered off to restore the function.

Table 5.2 Types of damage in semiconducting devices due to charged particles

Device	Ionization damages	Displacement damage
Diodes	Decrease in forward voltage drop	Decrease in forward voltage drop
	Increase in leakage current	Increase in leakage current
	Decrease in Zener voltage	Increase in breakdown voltage
Bipolar transistors	Decrease in gain	Decrease in gain
	Increase in leakage current	Increase in leakage current
		Increase in saturation voltage

- Single-event secondary breakdown (SESB). This is caused by a very high current due to secondary breakdown of parasitic transistors in power MOSFETs.

The types of damage on various semiconducting devices under a dose of charged particles are summarized in Table 5.2. Under radiation effects, the diode forward voltage drop can decrease by 50 mV maximum, and the leakage current can increase by 10 μA maximum. Generally, diodes used in a metal enclosure do not pose radiation damage risk. However, they may be damaged when placed unprotected outside the spacecraft, such as on a solar array for cell bypass or string isolation purposes.

Many d.c.–d.c. converters use a power MOSFET because of its higher switching speed relative to the bipolar transistor. The MOSFET also has a lower potential of thermal runaway and has a simpler input drive requirement. However, the power MOSFET is more susceptible to degradation and catastrophic failure under radiation.

The radiation studies performed by Attia et al.[2] are summarized in Table 5.3 and Table 5.4. The study showed that the effect of total dose radiation on the Buck, Boost and Cuk converters were insignificant. However, all converters lose their performance function under the worst-case single-event failure. For these reasons, space qualified converters must use radiation hardened power MOSFETs, which cost much more than their

Table 5.3 Reduction in converter output voltage at various radiation doses

Radiation dose (krad)	Buck	Boost	Buck–Boost	Cuk
0	0	0	0	0
4	negligible	negligible	0	negligible
8	negligible	negligible	69%	negligible
12	negligible	negligible	100%	negligible

Table 5.4 Converter output voltage after a single-event failure

Converter	Output voltage
Buck	Same as the supply voltage
Boost	0
Buck–Boost	0
Cuk	0

commercial varieties. However, improved circuit designs that can reliably use commercial MOSFETs in space can yield significant cost benefits.

Given that the SEU bit-flips in computer memory, it induces data error that propagates through the run-time operational flight program, causing erroneous outputs from the flight computer system.

5.5 Effects on other Components

Ultraviolet radiation modifies the surface resistivity, opacity, and thermal characteristics of the coverglass and its adhesive used in the solar array construction. The result is reduced illumination and increased temperature, both of which degrade the solar array output power. Since charged particles also darken the coverglass, it is a usual design practice to allow a couple of percent power degradation due to coverglass darkening.

Space missions to Jupiter involve radiation exposure exceeding 1 Mrad, which corresponds to the intense radiation in the Van Allen belts. The RTG power source used for Jupiter missions cannot deliver the peak power needed during certain science-acquisition events, so batteries are used in 2 to 10 Ah ratings. In the NiH_2 battery, the hydrogen electrode polymeric compression seal contains Teflon™, which has poor radiation tolerance, and hence is not suitable for such a mission. Super NiCd™ batteries have high radiation tolerance, and have been tested up to 20 Mrad.

The glass dielectric capacitors are high-reliability components suited for any environment where radiation or high temperature (200 °C) could alter or destroy the circuit performance. They have been tested under neutron and gamma radiation environments and have excellent transient and long-term radiation performance characteristics. They are radiation hardened and qualified to MIL-C-23269 and MIL-C-11272 for withstanding nuclear radiation and high-voltage transients with little or no degradation. Their performance relative to other types of capacitors in fast neutron flux fields is shown in Figure 5.3. As the bar graph illustrates, glass capacitors can operate in fields ten times more intense than any other capacitor type tested. Furthermore, when exposed to nearly 10×10^8 rad, they exhibit minor parametric variations immediately following radiation. They have low

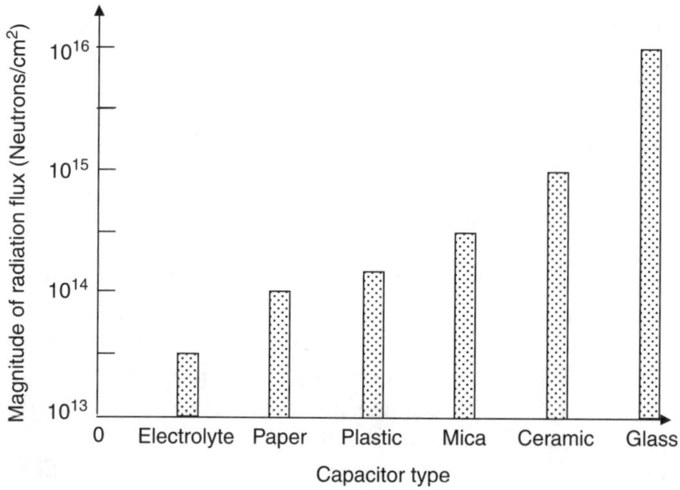

FIGURE 5.3 Radiation hardness of various capacitors under integrated fast neutron flux.

dielectric absorption, zero aging rate, and a temperature coefficient retraceability.

Outgassing of materials in a space vacuum can cause dielectric breakdown at voltages above the Paschen minimum. Proper material selection and venting is needed to avoid such breakdown, particularly in high-voltage systems. Any material can be used in space as long as the volatile content is less than 0.1% and the total mass loss in space is less than 1%. Permanent magnets do not degrade under radiation.

5.6 Mass Erosion under Atomic Oxygen

Atomic oxygen (AO) is a significant environmental constituent for spacecraft in a low Earth orbit. It is formed by photo-dissociation of diatomic oxygen by ultraviolet radiation from the sun, more in LEO, where recombination or the formation of ozone is improbable. It reacts with many commonly used spacecraft materials resulting in degradation and mass erosion of exposed spacecraft surfaces. The AO can oxidize and damage surfaces, especially in the ram direction with the spacecraft ram velocity simulating a high-energy beam. Volatile oxidation products may be lost, leaving a surface without its protective coverings and this can lead to electrostatic discharge to or from exposed conductors. Moreover, chemically active ions may be attracted by charged spacecraft surfaces, and the reaction rates increase with energy. Such considerations are important mainly for low planetary orbits, such as in LEO, and also in low Mars's orbit.

The AO has significant impact on the design and durability of the materials. The following are significant characteristics of AO in LEOs:

- Most abundant species in LEO atmosphere (80% AO, 20% nitrogen)
- Average velocity is approximately 8 km/s
- Average energy is about 5 eV
- Fluence (atoms/cm^2) is dependent on solar activity, velocity vector, and altitude
- Mechanism of material reaction is not well understood

The reaction rate on a material depends on the AO density, velocity vector, and the material reactivity. Materials in order of high to low degradation resistance are metal oxides, metals, inorganic polymers, and organic plastics. Thus, shielding with metal oxides, such as oxidized aluminum, can offer adequate protection against AO damage, provided that the shielding is strong enough to resist micrometeoroid damage.

Sputtering may also be a concern for long duration LEO missions carrying surfaces at high negative voltage. A complicating factor is that little information is available on the nature of chemical effects on sputtering rates under AO. However, some work has been done for developing a sputtering model to help evaluate the surface material loss rates in LEO. Measurements of low-energy sputtering by atomic oxygen support a theoretical model. One distinguishing characteristic of the model is that there is no absolute threshold for sputtering, but the rate at low energy is determined by the thermal tail of the surface material atom velocity distribution. The sputtering rate is a strong function of the surface voltage. Because the spacecraft voltage is usually well under 1 kV, sputtering is only important for mission durations in years in high-density plasma, such as that in LEO or low Mars orbit. Sputtering may also be a long-term problem in the lunar environment if the locally produced environmental plasma has a high enough density. For long life missions in a high-density plasma environment, surfaces should be insulated to prevent direct contact of high potential conductors with the plasma. Earlier in the ISS designs, engineers showed that the sputtering rate greatly increases by ions focusing onto the pinhole where micrometeoroids, debris or manufacturing defects produce small pinholes in insulators. Underlying conductors must be of sufficient thickness to withstand puncture at accelerated sputtering rates. Materials of low sputtering rate also may be used, provided that they satisfy thermal, atomic oxygen, and other requirements. Optical, thermal or other specialized coatings in the line of sight to the sputtering pinhole will become coated with the sputtered material over a long period of time. It is desirable to place surfaces such that they have no direct line of sight to surfaces that might undergo sputtering.

Sputtering by ions in LEO is also possible. However, such a sputtering rate per incident particle may be 0.1 and the flux on the order of 10^{12} per square centimeter per second. In contrast, reaction rates for atomic oxygen may be 1.0 and the ram flux 10^{14} per square centimeter per second.

Thus, instead of a time scale of 10 years for sputtering, we might expect a degradation time scale of a few days under atomic oxygen. Indeed, only a few weeks of time in LEO is required to destroy 1-mil layer of a highly reactive material like Kapton. However, not all materials are reactive to atomic oxygen. For LEO missions longer than a week, surface materials that are non-reactive with high-energy atomic oxygen must be used.

The spacecraft in LEO runs into the atomic oxygen with an impact energy of about 5 eV. Both the impact energy and the high reactivity of the atomic oxygen result in oxidation and erosion of most hydrocarbon polymers. The erosion is commonly measured in terms of the volume of the material removed per incident oxygen atom. Such measurements have been performed on board many spacecraft and have been reported by many investigators.[3]

The erosion rates of most polymers, except silicones and materials containing metal atoms which produce nonvolatile oxides, vary widely from 0.337×10^{-24} to 6.3×10^{-25}. Such wide variations make it difficult for the design engineer to know more accurately the durability of candidate materials at the design stage. Moreover, new materials continuously emerge which do not have such information due to expensive in-space testing. Ground-based testing has been used with limited success in predicting the in-space erosion rate based on the content and structure of hydrocarbon polymers.

The erosion rate (e) is often stated in terms of the surface thickness loss per unit time (i.e., mm/year). That is,

$$e = \frac{\Delta M}{A\rho} \tag{5.1}$$

where $\triangle M =$ mass loss, $A =$ surface area, and $\rho =$ mass density. The erosion rate is also influenced by the direction of atomic oxygen impingement. Most spacecraft have surfaces that are exposed to atomic oxygen at various angles during each orbit. Therefore, it is important to measure the erosion rates under impingement from various directions. Banks et al.[3] have reported a study for some 40 materials commonly used in space.

The silicone insulation on wires and the solar array could be adversely impacted by atomic oxygen. When exposed to atomic oxygen, silicon tends to shrink, harden, and experience surface conversion of silicon to silica. This causes surface erosion, which is more on the front surface than on the tail surface. Long duration spacecraft, such as the Russian Mir space station, have indicated significant deposits of silica on nonsilica containing surfaces. Such depositions cause UV darkening, which changes the solar absorption properties and degrades the solar array output power. Silicones intensively investigated are DC-93-500 used to attach the coverglass to solar cells and CV-1147 used between solar cells and the substrate on the ISS. Teflon shows the worst radiation damage, whereas Kapton (polyamide) is much better.

Silver Teflon erosion rate has been recently reported to be 1 mil/year in the ISS orbit. This rate is 10 times higher than some old estimates. In coating other solids or films with Kapton, the coefficient of thermal expansion must match; otherwise it may eventually crack. For LEO missions, it is recommended that: (1) sulfuric acid anodization be used as an insulating surface in regions where thermal control is not an overriding concern, and (2) AO protected aluminized Kapton blankets (Kapton surface on the outside) of 1000-V dielectric strength be used in all other outside surfaces of a high-voltage spacecraft.

5.7 Micrometeoroid and Debris Impacts

The solar array gradually becomes damaged by micrometeoroids punching small holes or shattering large surface areas of the cell. In the case of small holes of a couple of millimeters diameter, the impacted cell would still work without losing the whole string, but with a degraded performance. In the case of a shatter, the cell fails open, losing the whole string. In the past a few spacecraft have incurred such damage under micrometeoroid showers. The damage is indicated by a sudden decrease in power output immediately following a wobbling of the spacecraft.

Micrometeoroid and debris impacts may puncture insulators, fluid lines, pressure vessels, and manned compartments. Such damage potential must be accounted for in the spacecraft design. Studies of the expected flux of debris and micrometeoroids of a certain size are uncertain. This is due to the uncertainties in the amount of debris produced by abandoned spacecraft, its collision and breakup into smaller fragments, and models of the atmospheric drag that eventually remove it from orbit. The SP-100 had enough built-in redundancy to survive collisions with 1 cm size debris particles. On the other hand, the ISS required a shield to protect its pressurized manned modules against debris particles up to 1 cm in size.

In late 1998, the Earth passed through a Leonid meteor storm that originated from the dusty passage of the parent comet Tempel-Tuttle. It lasted about 10 h, and was one of the most intense meteoroid showers in decades. To guard against possible damage, TDRSs and other satellites were rotated to have the solar panels face the storm edge-on to minimize the target areas to speeding meteoroids. The satellites were also powered down for at least a part of the 10-h passage through the Leonid.

5.8 Predicting Damage

Radiation damage to spacecraft components can be predicted fairly accurately now. For example, the damage found on the solar array of MIR, when examined in 2000, was as predicted for the effects of radiation, micrometeoroids, thermal absorptivity, and emissivity, for example. The

MIR solar array did not have internal bypass diodes, only string isolation diodes. PV cells were silicon with 14% efficiency and 76% fill factor. No arcing or soldering damage or open circuits were found. Some cells were shorted due to micrometeoroid damage through a ceria microsheet coverglass. The array showed a high level of contamination, which was believed to be there when built due to poor bake-out procedures.

The most damaging effects of the environment on the power system occur during solar storms. The damage can be minimized if the intensity of such storms can be predicted in advance. A new scale for rating the severity of solar storms has been developed by NOAA. The storms are rated on a scale of 1 to 5, with 5 being the most severe. The scale categories and their effects on power system are given in Table 5.5. Satellite owners can take protective measures as a storm approaches, including reducing power levels on certain components that are vulnerable. For example, the solar array can be rotated on the edge to reduce the surface area normal to the radiation flux. Such precautionary measures are taken when required in view of the predicted risk.

A new sensor under the trade name CEASE is available on the market to monitor local radiation in the environmental and to provide an autonomous real-time warning of the following:

- Total radiation dose and dose rate
- Surface charging
- Single-event effects
- Solar cell damage

The sensor can alert satellite operators to hazardous conditions before the mission is affected. Under such an alert the operators can, in turn, reprioritize the operation, inhibit any anomalous sensitive tasks, such as attitude control adjustments, or initiate other prudent actions as indicated by the warning flag.

Table 5.5 Scale categories of space environment based on effects on spacecraft

Category	Effects
S-5	Permanent damage to solar panels, loss of controls, serious EMI including common mode noise
S-4	Accelerated degradation of solar panels, orientation problems
S-3	Damage to exposed components, decrease in soar array current
S-2	Isolated problems possible
S-1	None

References

1. Crofton, M.W. and Francis, R.W., ''ESD measurements on solar cell coupons in a simulated GEO environment,'' *Proceedings of the 35th Intersociety Energy Conversion Engineering Conference*, IEEE, 1999, Paper No. 2634.
2. Attia, J.O. et al. ''Radiation effects on dc–dc converters,'' *Proceedings of the 35th Intersociety Energy Conversion Engineering Conference*, IEEE, 1999, Paper No. 2696.
3. Banks, B.A. et al. ''A space experiment to measure the atomic oxygen erosion of polymers,'' *Proceedings of the 35th Intersociety Energy Conversion Engineering Conference*, IEEE, 1999, Paper No. 2695.

Chapter 6

Power System Requirements

6.1 Introduction

The overall mission requirements are set by the customer and given to the satellite contractor in the form of mission specifications and statement of work. The customer performance requirement specifies only the launch, transfer orbit, and operational orbit constraints, and the payload power demand over the mission life with specified power margin. This leads to the spacecraft requirements. All systems of the bus derive their requirements from the spacecraft requirements, and all components of a system derive their requirements from the respective system requirements. The flow-down of the requirements along with the design, manufacturing, assembly, and testing is shown in Figure 6.1.

The spacecraft requirement specifications generally do not specify the power system requirements *per se*. It merely includes a general statement such as " the power system shall generate, condition, distribute and store electrical power to meet the requirements of the spacecraft payload and bus systems during all phases of the mission." To meet such a broad statement of work, almost all requirements of the power system are self-derived based on the design analyses and approach taken by the satellite manufacturer.

The basic satellite level parameters impacting the EPS design are the orbit altitude, orbit inclination, and the mission duration. These parameters are used to determine the orbit period, sunlight and eclipse durations, and the solar β angle between the orbit plane and the Earth–Sun line. The power system design is then driven by the load power requirement in all phases of the mission, which are as follows:

- **Launch and ascent.** This phase extends from the lift-off to the sun acquisition, during which the battery supplies the needed power.
- **Transfer orbit.** During this phase, the geosynchronous spacecraft is transferred from LEO to HEO and then to GEO with the solar array still stowed. Only the outer panels are exposed to the sun and generate a fraction of the rated power. The load power is, therefore, rationed. The energy balance in this phase is desirable, but is not a rigid requirement. A small energy deficit may be drawn from the battery up to the allowable DOD as long as it is fully recharged before commencing the full load operation.

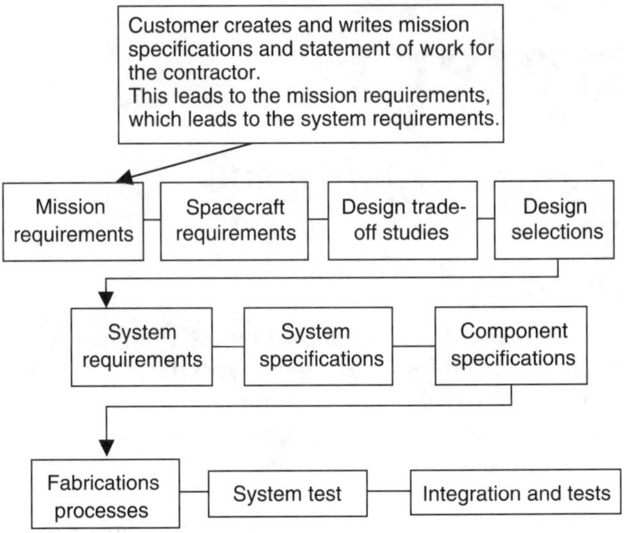

FIGURE 6.1 Satellite and power system requirements flow down.

- **Parking orbit:** Often this orbit is specified for storage until the satellite is placed in service in the operational orbit.
- **Operational orbit:** All solar panels are now deployed and generate full power. In this phase, all required loads are on. The battery charge and discharge energy balance must be maintained in each orbit.
- **Disposal orbit:** A disposal method at the end of the life may be specified, and the satellite may be required to function at a specified power level in the disposal orbit.

Thus, the power system is required to perform during various phases of the mission in a variety of orbits — LEO, HEO, and GEO. However, the orbit period, the maximum eclipse duration, and the load power requirement in the operational orbit determine the ratings of the solar array, battery, and other components. The power generation diminishes with increasing radiation dose, more so in low and highly elliptical Earth orbits which pass through the Van Allen radiation belts. For a given load power, therefore, the solar array area requirement increases as the mission duration becomes longer. The overall power system architecture is selected based on the prior experience on similar satellites that have met similar requirements in the most economical and mass effective manner.

6.2 Self-Derived Requirements

As mentioned earlier, the electrical power system requirements are seldom found in the customer specifications for the spacecraft under procurement.

They are derived from the spacecraft level requirements and in-house trade studies. The EPS self-derived requirements are based on various analyses performed on the power system under design. The final requirements generally come from the operational orbit analyses, but other orbits must also be analyzed to ascertain that the proposed requirement is met in the worst case. The power system design team commonly performs the following worst-case analyses:

- Power flow to determine the component ratings and heat dissipations
- Energy balance to determine the battery rating
- Voltages at the terminals of various equipment
- EOL and BOL solar array power generation capability
- Load switching and fault response, including major fuse blow event
- Bus stability under various feedback control loops
- Energy balance in transfer orbit
- dV/dt specifications, which come from three sources:
    - Voltage fall before the fuse blow
    - Voltage rise after the fuse blow
    - Sudden loss of power during integration and testing

The following are some major self-derived requirements of the power system:

- Solar array EOL power level
- Solar array pointing and rotation for sun orientation
- Battery capacity, in ampere-hours
- Battery DOD and charge control
- Bus voltage regulation
- EMI/EMC and ESD

6.3 System Specifications

The power system requirements vary over a wide range from low power (few hundred watts) to high power (several kilowatts) in LEO, MEO or GEO altitudes. The power system specification document is developed based on the performance analyses with the mission parameters as input. This is done well before the detail design starts. It is a well-drawn document containing a level of detail matching the complexity of the mission. A typical outline of such specifications is given in Table 6.1. The complete requirement specification is typically a 100 to 200-page document,

Table 6.1 Power systems requirement specifications — topical outline

1. Scope of the spacecraft
2. Applicable documents
3. Power system requirements
   a. Interface with other systems
      i. Functional interface
      ii. Logical interface
      iii. Physical interface
   b. EPS characteristics and performance
      i. Load capability
      ii. Operating phases and modes
      iii. Battery charge and discharge management
      iv. Main bus voltage, normal and transients
      v. Battery bus characteristics
      vi. Station-keeping power
      vii. External power input
      viii. Command and telemetry
      ix. Ground interface
      x. Reliability
      xi. Software
   c. Major component characteristics and performance
   d. Design and construction
4. Quality assurance
5. Documentation

which is generally unique to the particular spacecraft. This section lists some of the most basic requirements for a PV–battery power system for an Earth orbiting satellite. Such a mini-specification is generally used informally or with some formality at an early stage. The blanks below represent a number to be inserted, which is derived from the mission specific analysis performed by the system engineer. Some blanks are filled in *italics*, which are given merely as representative examples of a 50-V or 70-V GEO bus. They would change with the bus voltage and mission requirements.

The electrical power system shall provide for the generation, storage, control, protection and distribution of electrical power to the spacecraft payloads and housekeeping loads during the entire mission life of ______ years.

Operational orbit load: during sunlight _____ watts, during eclipse ____ watts.

Peak load of additional _____ watts for ____ minutes in every ________ minute period.

Pulse load of additional _____ watts for ____sec with duty ratio of 5 *percent*, with *1 ms* turn on time and *1 ms* turn off time.

Transfer orbit load: during sunlight_____ watts, during eclipse _____ watts, with solar array stowed and only the outer panel exposed.

The EPS shall provide sufficient capability to support the spacecraft transfer orbit loads with the solar array stowed for (a) 9 hours when the incident sun angle is in the range of 15 to 165° as measured from the spacecraft spin axis, or (b) 30 hours under incident sun angle in the range of 40 to 140°, or (c) indefinitely when the sun angle is in the range of 60 to 120°.

Load of ____ watts during launch, ____ watts during separation and spin, and ____ watts during spin and sun acquisition. Mission power profile up to entering the transfer orbit as per *Figure*____.

Bus voltage_____ Vdc ±_____V, regulated direct energy transfer bus.

Single fault tolerant system where faults shall not propagate to other components, subsystems or systems.

Autonomous redundancy changeover in case of a fault.

EPS Reliability of *0.95* over the entire mission life, including the ground and space storage, if required.

Telemetry and command to monitor the power system operation as per *Table*____.

Autonomous operation in all normal functions, unless overridden by ground command.

Power management software via *1773 data bus* to control the battery state of charge/discharge, temperature, pressure, V-T curves, charge/discharge balance between batteries and load shedding if and when necessary.

Each non-critical load fused for the EPS protection.

Bus transients to comply with ____________ document.

EMC to comply with ___________ document.

Orbit environment to comply with ____________ document.

Launch environment to comply with ____________ document.

Survivable to additional threat as per _______ document.

Piece parts derated as per ________________ document.

Umbilical cord for ________ watts of ground power to the EPS bus and for _______ amperes battery charging current.

Two batteries, each with ________ NiH_2 cells with ____ Ah ratings.

Maximum worst-case battery DOD *70 percent* during operational orbit and *80 percent* during transfer orbit.

Battery charge rates of *3 A* and *5 A*, and the trickle charge rate of *0.5 A*.

Battery temperature to be maintained between $-10\,^{\circ}C$ and $25\,^{\circ}C$, with $5\,^{\circ}C$ maximum cell-to-cell gradient and $5\,^{\circ}C$ maximum inter-cell gradient.

Ordnance controller powered from both batteries independently at voltages between 18 *and 34* volts.

The deployment ordnance controller shall contain relays (enable, arm and fire) required to switch power from the batteries via independent paths to the EEDs used for separation of deployable structures and ignition of the apogee kick motor.

Bus dynamic impedance $< 200\ m\Omega$ from 500 Hz to 20 kHz and $> 0.60\ m\Omega/Hz \times$ frequency for frequencies below 500 Hz.

EPS-induced ripples $< 50\,mV$ peak-to-peak from dc to 400 MHz, except $250\,mV$ p–p between 30 kHz and 45 kHz. Total RMS ripple $<120\,mV$.

Bus ripples at housekeeping loads due to effect of payload current ripples $< 500\,mV$ p–p for frequencies up to 50 MHz.

Harness wires no thinner than *24* AWG. Aluminum wires shall not be used.

Maximum voltage drop to users: *0.50* V for loads < 75 *watts, 0.75 V* for loads *>75 watts, 1.0 V* to heaters and mechanism loads.

All load power conditioners must supply power to their loads with *10 ms* power interruption from the bus under a fuse blow transient or other similar events.

The spacecraft power return bus to be hard grounded to the spacecraft chassis at the power regulator unit.

Testability shall be provided at the EPS level and the spacecraft level.

Chapter 7

Design Process and Trades

7.1 Introduction

The power system design process starts with analyzing the orbit parameters, the load power requirement, and heritage data on similar satellites built and flown earlier. The design that meets the mission requirements at minimum mass, volume, and cost at present generally incorporates new technologies as they become available and space qualified. The design process starts with a top-level trade analyses — formal or informal — in order to select the power system architecture that would lead to an optimized design. The detailed design then follows and continues in several iterations until it is fully optimized. The engineer would know that the design is fully optimized when a small change in any one component would result in a penalty of some sort at the power system level or at the spacecraft level.

7.2 Spacecraft Level Trades

As expected, the power system interfaces with other systems of the spacecraft as shown in Figure 7.1. The power system engineer works with the chief engineer in doing the spacecraft level design trades. The ultimate goal of the program is always the minimum mass and cost *at the spacecraft level*. For that reason, while focusing on the power system, the power system engineer keeps a peripheral vision on the overall spacecraft mass and cost with applicable input from the spacecraft system engineer and/or the chief engineer. Reducing 1 kg mass in one component that adds less than 1 kg in another component is a winning trade. Such trades are continuously revisited as new spacecraft technologies become available.

The following factors have a large influence on the architecture selection of the PV–battery power system. They are driven primarily by the top-level mission requirements and do not involve the power system level trades. However, understanding their influence on the power system can quickly lead to the selection of an optimum EPS architecture.

1. As the satellite altitude increases, the orbit period increases and the eclipse time, as a fraction of the sunlight time, decreases. The demand on the battery discharge and the charge power therefore decreases.

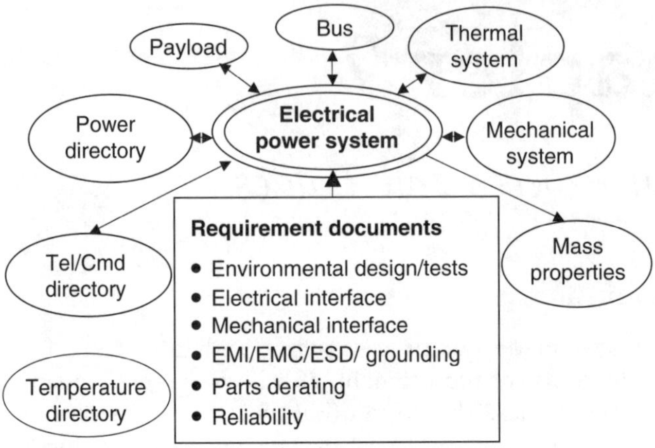

FIGURE 7.1 Power system interfaces with other systems of spacecraft.

2. The micrometeoroids and orbital debris are worse in the 50–55° inclination LEO orbits. The charged particle radiation environment is also worst in this orbit over the South Atlantic Ocean where the Van Allen belt dips to a low altitude. At higher inclinations in LEO, the Earth's infrared radiation decreases while the albedo (solar radiation reflected from the Earth) increases.
3. The β angle between the Earth–Sun line and the orbit plane varies seasonally. It reaches maximum values on the summer and winter solstice days, which is $\beta_{max} = i + i_o$, where i = orbit inclination, and $i_o = 23.45°$ (the angle between the equator and the ecliptic plane). As β increases, the eclipse time decreases, and the solar array temperature also decreases. At higher β, therefore, the EPS is capable of supplying greater load, both during eclipse and also during sunlight. The eclipse duration is the longest when $\beta = 0$ and the shortest or even zero when β is maximum.
4. The array pointed away from the normal sun-line reduces the solar flux, increases the coverglass reflectivity and changes the view factor to Earth affecting the array temperature. Typically, a 3° error in the gimbals pointing and 2° in the array flatness, thus a total of about 5° error, is allowed in the sun pointing. The cosine law up to 30° governs the power loss. In the 30–80° range, the loss is greater than that given by the cosine law, and the array generates no power beyond 80°. This has an impact on the transfer orbit maneuvering when the array could be way off-normal.
5. A planner solar array with dual-axis gimbals requires tracking of the sun-line to the array normal no more than $\beta°$. With single-axis gimbals, the array must have a cant angle equal to the annual average β angle. A concentrator array requires two-axis sun tracking to maintain the sun-line normal to the array surface.

6. Two-wing design is effective in maintaining the balance of solar pressure on the wings. One-wing design must compensate for the offset between the center of mass and the center of solar pressure. Otherwise, additional fuel is needed to maintain the spacecraft attitude.
7. Due to 1.5% eccentricity in the Earth's orbit around the sun, the solar flux in LEO and GEO varies seasonally about ±3% with an annual average of $1358 \pm 5\,\text{W/m}^2$, where $5\,\text{W/m}^2$ is an allowance for measurement errors. The flux is near the average value on both equinox days, maximum near the winter solstice and minimum near the summer solstice. Thus, the solar array receives about 6% more solar flux at the winter solstice than on the summer solstice day. However, it generates only about 4% more power because of the increased temperature under the increased flux. The 11-year solar cycle of ionized particles has no effect on the solar flux.

7.3 Power System Level Trades

Once the spacecraft level trades are settled, the power system engineer focuses on the internal EPS level trades that may reduce mass and/or cost. The power system mass as a percentage of the satellite dry mass can range from 25% in LEO satellites to 45% in GEO satellites. This mass must be supported during launch and then transferred to the final orbit at high cost. Saving of even a few percent of the power system mass can result in appreciable savings at the spacecraft level. For example, the impact of the electrical power system on the structural system, although not obvious and direct, is strong.

The first task of the power system design engineer is to select the optimum architecture. As established in Chapter 4, a direct energy transfer sun-regulated low-voltage bus is generally optimum for satellites having a load under a couple of kilowatts. For large high-power satellites, the fully regulated high-voltage bus would generally lead to the optimized design. For a shortl mission operating over wide orbit parameters, peak power tracking architecture may be suitable. For an interplanetary or deep space mission, nuclear power sources may be optimum. All these options must be considered for the mission under design in order to arrive at the optimum power system architecture.

Next, the bus voltage level and the power generation and energy storage technologies must be jointly selected to optimize the total power. The major driving factors are the payload power level, operating orbit parameters, mission life, and number of satellites in the program procurement. For the self-derived satellite level load requirement, the trade study is done to select various key component for the power system, such as the PV cell, coverglass thickness for radiation protection, array substrate, battery cell, power converter topologies, etc. An example of such trades is displayed in Figure 7.2. As the trade study proceeds from left to right, the selections made are shown by continuing lines. Since the solar array and the battery are two components that primarily contribute to the power system mass and cost, they receive more attention and see more rapid technological changes than other components. They impact not only on the EPS design, but also on other spacecraft systems.

7.3.1 PV Cell Trades

The fundamental design requirement is to generate the required electrical power at the end of life. The degradation of the solar array power output under charged particles is a trade within the power system design that often does not cross the system boundary. The PV cell selection depends on the mission environment and the design life. Some cells, such as low substrate resistivity cells, have high BOL power, but degrade fast, while others do the opposite. The orbit environment and the mission duration strongly

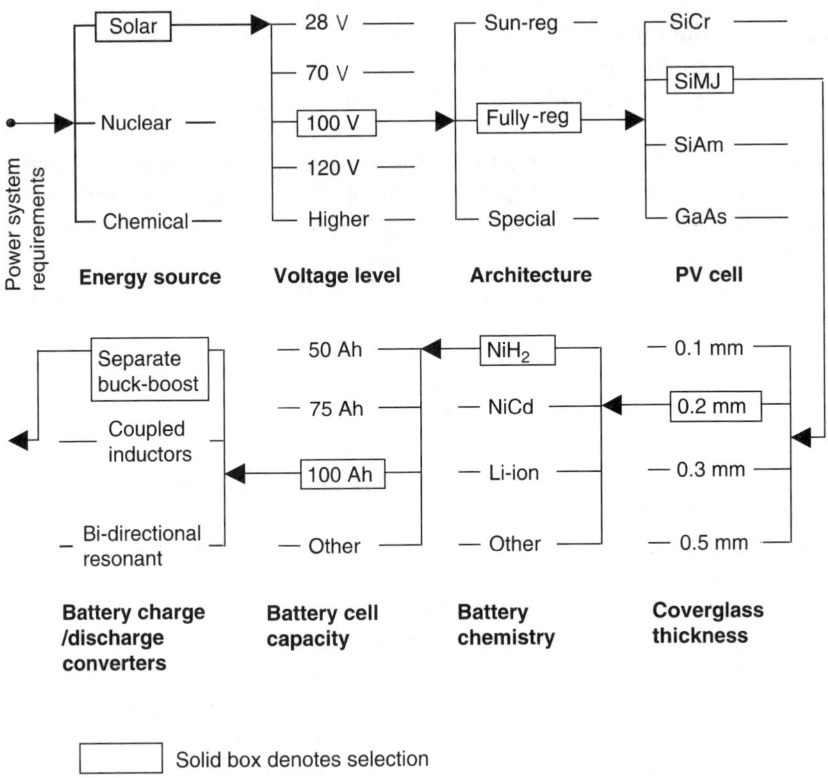

FIGURE 7.2 Power system design trades flow-chart.

influence the cell selection. A cell that gives high BOL power may be beneficial for a short life mission, whereas a cell that is radiation hard and gives high EOL power may be beneficial for a long mission. Figure 7.3 is an example of such a trade. The cell designated as A (10 Ω cm base resistivity) would be better for a mission which would accumulate more than 10×10^{14} MeV fluence over 15 years lifetime. It gives lower BOL power but offers high radiation resistance. For a shorter mission, the cell designated B (2 Ω cm base resistivity, such as used in EOS-AM spacecraft) would require smaller array area since it has high BOL power but is not as radiation hard. With the same PV technology, it is generally true that the cell that gives high BOL power degrades faster, giving less power at the EOL. The mission life thus plays a definitive role in cell selection.

For a LEO or GPS type MEO satellite, the radiation degradation in power generation over 10 years life is about 40–50% for silicon cells and 25–35% for GaAs/Ge cells. The additional cost of high radiation resistance cells may be more than recovered in long duration missions in such high radiation orbits.

For minimizing the mass of a long life satellite, attitude control reaction jets that use chemical propellant are often replaced with torque rods and reaction wheels. However, the torque rod performance is limited by the Earth's magnetic field and the reaction wheel performance is limited by the angular momentum storage capability. Subsequently, reaction control jets may be required to supplement the torque rod system in large disturbance torque environment. This is especially true for low park orbits where drag is high. In such situations, the cost of a high efficiency array with a small area may be more than compensated by not requiring reaction jets to supplement the torque rod.

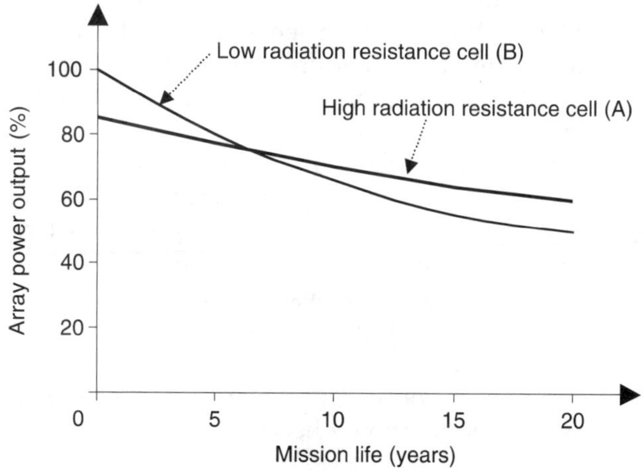

FIGURE 7.3 PV cell selection trade for radiation resistance.

The power system must also provide reliable power to ascertain a targeted disposal, especially if electric thrusters are used for disposal maneuvers. As for the end of life disposal of the satellite, international organizations may establish stricter regulation in terms of the timelines, disposal orbits, reliability, and liability for disposing of the satellite. The array area having high drag may directly impact the disposal time and reliability, and may cause the satellite to lose attitude control.

7.3.2 Solar Array Trades

A mass reduction in the solar array has multiple benefits at the satellite level, such as reduced drag, inertia, and propellant mass. Such win–win trades must be thoroughly investigated at an early design stage. The solar array area is linearly proportional to the required power generation. At the same time, it also has significant influence on the propulsion and attitude control systems. A large array increases the drag and moment on the satellite, more so in LEO satellites and in low Earth park orbits of geosynchronous satellites. A small array of high efficiency cells might be worth the added cost, considering that it would allow lower park orbits for the same drag. A lower park orbit allows the launch vehicle to insert more mass into the orbit, resulting in a reduced launch cost per kilogram of the payload. A small array may also result in a shorter transfer time from the park orbit to the geosynchronous operational orbit as a result of lower drag in the early part of the transfer orbit. This may yield greater benefit, considering that the transfer from LEO to GEO may take hundreds of days with large arrays. Aligning the center of pressure and center of gravity of the satellite body can minimize the drag.

If a narrow orbit control is required in LEO, a large array may add significant propellant mass in a long duration mission, costing probably more than the savings in using low efficiency PV cells. In GEO, a large array causes orbit drift due to solar pressure. Therefore, the satellite requires more propellant fuel for N–S station keeping in operational orbit during the entire mission life.

If a precision attitude control is required, it is necessary to have a low mechanical natural frequency of the solar array in the first mode of vibration. Small arrays inherently have high frequency. A large array, on the other hand, would require adding structural rigidity and decoupling at an additional cost. The torque compensation hardware must be sized for drag and solar pressure. The torque is generally canceled by changing the reaction wheel speed, and then by de-spinning the wheels using torque rods over the duration of the orbit. Thus, a large array can increase the size of the torque rods and the reaction wheels.

7.3.3 Battery Trades

LEO and GEO satellites differ significantly in their power system requirements. GEO satellites have a high launch cost and require fewer battery charge/discharge cycles. LEO satellites, on the other hand, have a low launch cost, high ionized radiation, and a much greater number of battery charge/discharge cycles. The battery chemistry selection may, therefore, differ significantly for these two classes of satellite. It would also influence the thermal control system. The NiCd battery was the workhorse of the space industry until the mid-1980s, when the NiH_2 battery replaced it for both LEO and GEO missions. The NiH_2 battery, however, requires a narrow operating temperature range of –10 °C to +10 °C. This adds some mass to the thermal control system. There are various derivatives in which the basic NiH_2 technology is available, commonly known as the IPV, CPV, SPV, and DPV cells as described in Chapter 9. There is only a small mass difference between each type. The lithium-ion battery has recently made inroads into the space industry for its significantly higher specific energy and is expected to replace the NiH_2 battery in the near future. However, it is sensitive to over-charge and over-discharge. This often adds some mass in the charge control electronics.

7.3.4 Bus Voltage Trades

For communications satellites, selecting one of the standard busses offered by prime contractors may be the most economical option. The bus voltages are standardized for various ranges of the load power. However, the voltage selection may be open for designing a special science or defense spacecraft. Even in such cases, the bus voltage is sometimes set by the heritage and the preferred database. Otherwise, a trade study is needed to select the optimum voltage for the mission on hand. Developing preliminary designs for various available voltages is the only way to arrive at the optimum. The mass of each component is estimated under various bus voltage options and added to the total power system mass and volume. As for the cost, it may be assumed that the cost is proportional to the mass, at least in the first approximation. Figure 7.4 is an example of such trade for a 2000-W TDRS-type GEO satellite with a 15-year mission life. Notice that the solar array mass is not shown in the chart, but is included in the total mass. As it is a series–parallel assembly of numerous PV cells, it is fairly independent of the bus voltage. In a high-voltage bus, more PV cells are connected in series and less in parallel, but still the same number of cells is needed to generate the required power. For this example mission, the figure indicates that a 50-V or 70-V bus is the optimum. The 100-V bus reduces the harness mass, but it also increases the battery mass and volume significantly. This is because the battery ampere-hour rating in a 100-V bus becomes so small that its energy storage per kilogram mass and liter of

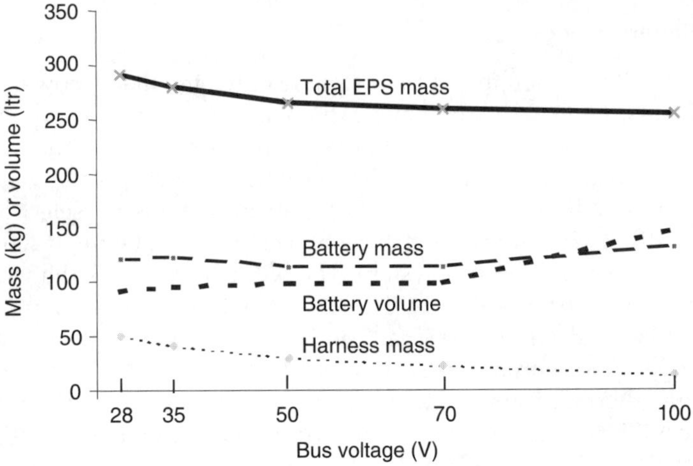

FIGURE 7.4 Bus voltage trade example for a 2000-W GEO satellite.

volume fall significantly. As a result, a 100-V bus does not reduce the total mass compared to a 50-V or 70-V fully regulated bus.

7.3.5 Pyro Power Trades

In the case where a fully regulated direct energy transfer bus has been selected, another top level trade is whether to power the EED pyro circuits directly from the regulated bus or from the battery, bypassing the battery discharge converter. This trade is depicted in Figure 7.5. It is not for cost or mass reduction, but for the EMI and reliability improvements as described hereafter. Both the primary and backup converters for each battery are normally active during operational orbit. During launch, the battery discharge converter can be off if the pyro power is supplied directly from the battery bus. However, it must be on if the pyro power is supplied from the main bus. In this case, the satellite is launched with the backup converter on and the primary converter off. The reason for launching in such backup mode is design related; that is, switching from the primary to backup converter is automatic, whereas switching from the backup to primary converter happens only on ground command.

The pyro firing during launch is generally a source of high EMI that may cause switching transients, which may disturb some critical components and sensors. The primary unit under severe transient glitches may get turned off, and the systems may automatically switch to the backup unit. If this happens, it is not possible to switch back to the primary unit. This amounts to losing the redundancy in this critical function. If the satellite is launched in the backup mode, an automatic switching to the primary mode under a transient voltage glitch does not amount to losing the redundancy.

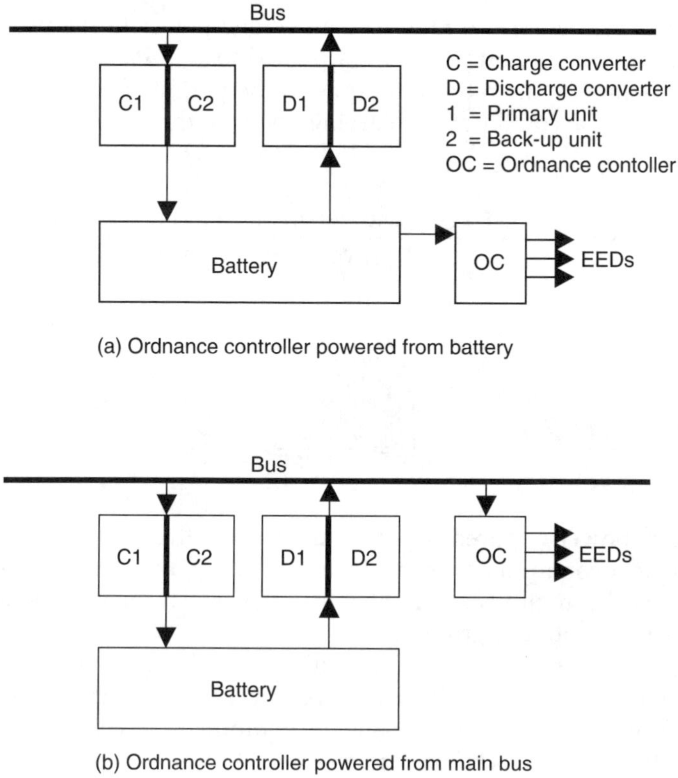

FIGURE 7.5 Powering EEDs by battery bus versus main bus.

If launched in the backup mode, it is switched to the primary mode by ground command shortly after the orbit acquisition, on the third or fourth revolution.

7.4 Load Power Profile

Once the satellite level trades are settled, and the EPS architecture has been selected, the next step in the design process is to compile a detailed load power profile of each component to be powered. It is also called the power budget or the power allocation in an early phase of the design. The input power for the on and off durations of each item of equipment 'peak power when on, quiescent power when off, and the thermal dissipation are compiled. The component dissipation may vary significantly during sunlight and eclipse, and during the beginning of life and end of life. Such variations are properly accounted for.

The power versus time curve for each item of load equipment may be condensed into two parameters: the peak power and the duty ratio. The duty ratio is defined as the ratio of the average power to the peak power over a certain time period. For a well-defined load, which is on or off, the duty ratio is merely the ratio of on-time to the period of repetition, that is $D = T_{on}/T$. For a continuously varying load at the other extreme, it is defined as

$$\text{duty ratio} = \frac{\text{energy consumed in one repetition period}}{\text{peak power} \times \text{repetition period}} \tag{7.1}$$

i.e.,

$$D = \frac{\int_{o}^{T} P(t)\,\mathrm{d}t}{P_{pk}T} \tag{7.2}$$

The average power required by any equipment is then $P_{avg} = DP_{pk}$. Figure 7.6 is a typical communication load profile for a geosynchronous satellite. The load reaches high peaks of short duration when the transponders are communicating with the ground.

The average power is used to determine the solar array and the battery ratings. On the other hand, a load averaged over five thermal time constants of the equipment is used to determine the equipment power rating, the wire gage, and the fuse size. The peak power and duty ratio of most equipment are significantly different during the launch, the transfer orbit, and the

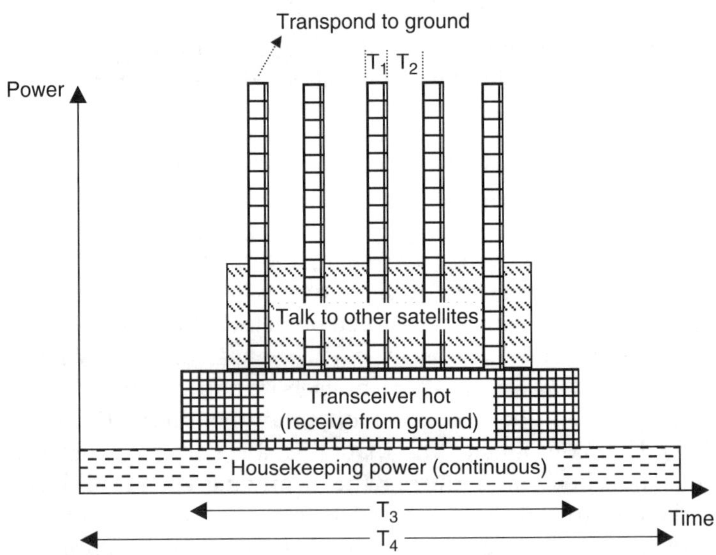

FIGURE 7.6 Typical communications satellite load profile.

Table 7.1 Seasonal variations in power generation and heater power requirement

Seasonal day	Solar flux[a]	Sun angle	Power generation[b]	Heater load
Vernal equinox	1.001	0°	1.001	High
Summer solstice	0.967	23.5°	0.887	Low
Autumn equinox	0.995	0°	0.995	Low
Winter solstice	1.034	23.5°	0.948	Medium

[a]Normalized with the annual average.
[b]Accounting for seasonal variations in both solar flux and sun angle.

operating orbit. They may also vary seasonally during the operating orbit. The worst-case seasonal power situation is sometimes known from a similar satellite designed and flown in the past. When there is no such heritage to draw from, considering 4 key days in the year, as listed in the first column of Table 7.1, may be adequate for a geosynchronous satellite. On these 4 days, both the power generation and the heater power requirements vary. Since the energy balance must be maintained all the time in the operating orbit, any one of these 4 days may most probably be the worst case for setting the power system capability. Because the heater power requirement is low when the power generation is low, it may not be obvious which day is the worst for the power system design. In the absence of some heritage data, it may be necessary to conduct a preliminary design for all 4 days to determine the worst case, and then to focus on the worst case for detail design. The summer solstice is usually the worst day for the load power capability of the GEO communications satellite. After establishing the worst day, the design process focuses on meeting the load power requirement on that day.

The load requirement of each of the bus systems and the payload components are then complied during the transfer orbit and the worst-case operating orbit. Table 7.2 and Table 7.3 are examples of loads and their characteristics in one satellite during the transfer orbit and operational orbit, respectively. The load power values in these tables are input powers to the components that must be supplied by the power system. In new

Table 7.2 Transfer orbit load power profile for System A

	Component 1	Component 2	Component n	System total
Quantity	X_1	X_2	X_n	X_{TO}
Peak power	Y_1	Y_2	Y_n	Y_{TO}
Peak heat dissipation	Z_1	Z_2	Z_n	Z_{TO}
Duty ratio	D_1	D_2	D_n	D_{TO}
Average power	P_1	P_2	P_n	P_{TO}
Average loss (dissipation)	L_1	L_2	L_n	L_{TO}

Table 7.3 Operation orbit load power profile for System A

	Component 1	Component 2	Component n	System total
Quantity	X_1	X_2	X_n	X_{OP}
Peak power	Y_1	Y_2	Y_n	Y_{OP}
Peak heat dissipation	Z_1	Z_2	Z_n	Z_{OP}
Duty ratio	D_1	D_2	D_n	D_{OP}
Average power	P_1	P_2	P_n	P_{OP}
Average loss (dissipation)	L_1	L_2	L_n	L_{OP}

designs when the loads are somewhat uncertain, appropriate margins are incorporated in the power estimates. For example, it may be acceptable to the design team to have no margin for heritage components requiring no change at all, 10% margin for components modified from a heritage design, and 30% margin for new component yet to be developed.

Various system loads obtained from Table 7.2 and Table 7.3 are then condensed into the total power system load budget as shown in Table 7.4. All bus systems are to be included in this table, including the payload and the housekeeping load for the electrical power system itself.

7.5 Solar Array Sizing

The load power profile developed above is then used to determine the solar array output power requirement. For example, for a communications satellite having a 2500-W communication payload power, the solar array power generation requirement is shown in Table 7.5 in two seasons and during an eclipse. The autumnal equinox day, having the longest eclipse, in this case determines the solar array output requirement of 3940 W, which includes a 5% power margin and a 5% reliability margin. Allowance must also be made for array shadowing and failed bypass diode, if applicable. The solar array must generate this power at the end of life. The beginning of

Table 7.4 Total average load and heat dissipation budget (allocation) for power system

System	Transfer orbit power	Operational orbit power	Transfer orbit heat dissipation	Operational orbit heat dissipation
System A	P_{TO}	P_{OP}	L_{TO}	L_{OP}
System B	..	..	..	..
System Z	..	..	..	..
Payload	..	..	..	..
Spacecraft total load	ΣP_{TO}	ΣP_{OP}	ΣL_{TO}	ΣL_{OP}

Table 7.5 Example of sizing solar array output power for a communications satellite (average watts)

	Autumn equinox[a]	Summer solstice[b]	During longest eclipse
Communication payload system	2500	2500	2500
Communication system load margin (10%)	250	250	250
Total communication system load	2750	2750	2750
Bus systems			
TT & C	50	50	50
Attitude control	50	50	50
Thermal control	150	100	50
Electrical power	20	20	20
$I^2 R$ conduction loss	50	50	60
Battery charging	500	100	0 (disch.)
Total EPS load	3570	3150	2980
Power margin (5%)	180	160	150
Reliability margin (5%)[c]	190	170	165
Total solar array power generation required at EOL	3940[d]	3480	3295

[a]Non-eclipse season near September 21.
[b]Eclipse season near June 22.
[c]Allows one string failure (in a typical 20 circuit solar array).
[d]Solar array output power required = 3940 W minimum.

life power would be greater depending on various degradation factors, which are:

- Environmental degradation due to ionized radiation
- Cell-to-cell current mismatch (assembly loss)
- Sun angle off-pointing angle (due to array flatness, structural misalignments, and sun-tracking errors)
- All other, such as micrometeoroids, space debris, contamination, plasma, ultraviolet radiation, yellowing of coverglass, and thermal cycling

The values of individual degradation factors come from the considerations detailed in Chapter 8, and the total degradation is obtained by multiplying all those factors. If the total degradation comes to 33%, as an example, then the solar array must generate a BOL power of 3940/(1 − 0.33) = 5880 W in order to meet the 3940 W load requirement at EOL. This is the P_{max} it must deliver at its knee point voltage V_{mp} with appropriate margins both in power and also in voltage. The P_{max} is temperature dependent, so we need to know the array operating temperature. The temperature is calculated by equating the heat input and output for the front and back of the array. All cells on an array are assumed to operate at the same temperature. The heat input is the solar radiation, albedo, Earth infrared radiation, and from the adjacent solar array. The heat output is the power converted into electricity and the power radiated back into space. If multiple wings connected in parallel operate at different temperatures, the operating voltage is selected to be a few volts less than the V_{mp} of the higher temperature wing to avoid the colder array operating above V_{mp} where the power output drops rather quickly.

Assuming that the solar array output power calculated above is at the shunt dissipater interface, where the calculated voltage is, say, 110 V in a 100-V bus, the BOL design power of 5880 W would require the solar array output current of 5880/110 = 53.5 A. Once the array output voltage and current are determined, the array design proceeds as described in Section 8.9. The number of strings in the array and number of cells in each string are then determined as follows:

- Number of series cells per string = array output voltage (110 V in the above example)/cell voltage at the operating point
- Number of strings in the array = array output current (53.5 A in the above example)/cell current at the operating point

7.6 Battery Sizing

The detailed battery design procedure is described in Chapter 9. At a preliminary design stage, the round trip energy efficiency approach can be used to derive the approximate rating of the battery. Both NiCd and NiH_2 batteries have a round trip energy efficiency of about 70%. For the life considerations, the maximum DOD must be below a certain value that depends on the number of charge/discharge cycles the battery is required to deliver during the mission life. With these considerations, the ampere-hour rating is derived from the expressions below:

- Battery cells in series = minimum permissible battery voltage for pyro or power electronics/minimum cell voltage at the end of discharge at EOL
- Wh discharged = eclipse load in watts × eclipse duration in hours
- Ah discharged = Wh discharged/1.25 average cell voltage × cells in series
- Ah capacity = Ah discharged/permissible DOD for required cycle life
- Wh charge = Wh discharged/0.70 round trip energy efficiency
- Charge power = Wh charge/charge time during sunlight (7.4)

7.7 Power Flow Analysis

The energy collected during sunlight must match the energy required by the loads, plus the energy used to charge the battery and the power losses in various components. Such a balance must be maintained over the orbit period with the worst-case eclipse duration. In the simplest case with constant load during the orbit (Figure 7.7), the two shaded areas A_1 and A_2 must be equal in each orbit period during the entire mission life. A small deviation from such a stringent requirement in an occasional orbit could be manageable for a short time, but must be thoroughly examined for acceptability from the battery DOD considerations.

The design optimization is done to maximize the power output to the load while maintaining the energy balance. The start point of the energy balance analysis is usually the beginning of the eclipse. The battery SOC is determined at the end of eclipse based on the load and the power losses. The design engineer then determines the required solar array output to bring the battery back to full SOC while meeting the load power during sunlight. During the entire orbit period, the engineer also determines the maximum power, voltage or current handled by each component, which then sets the component power, voltage, and current ratings. The best design is obviously the one that leaves minimum margins in all components, and also is acceptable to the program team for the level of uncertainty in the design or the system operation. At the minimum, the system should not be battery power limited or array power limited. Typically, a few percent margin in both the battery and the solar array is considered adequate.

The spreadsheet shown in Figure 7.8 illustrates the power flow necessary to maintain the energy balance for an EOS-AM satellite requiring 3000 W constant load during sunlight and also during an eclipse. It may use flywheel energy storage instead of a battery, but that would not change the analysis method. In addition to the power generation (solar array) and the energy storage (battery), it includes other components, such as the power

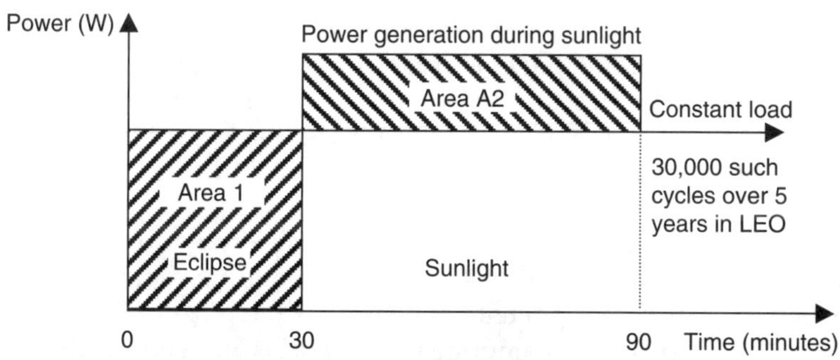

FIGURE 7.7 Energy balance over one orbit period with constant load.

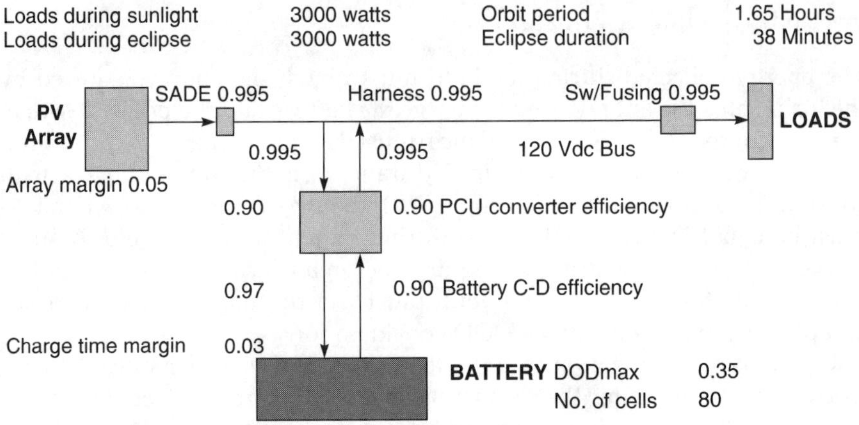

Energy balance analysis key results:

Energy discharged from EW during eclipse	2381.2 Wh
Charging power required from the bus	2779.8 watts
Load at PCU during sunlight (before charging power)	3030.2 watts
Load at PV array during sun (including charge+diode)	5839.2 watts

Required minimum steady state ratings of system components:

PV array EOL capability including margin	6131.2 watts
PCU charge converter rating	2489.3 watts
PCU discharge converter rating	3045.5 watts
Charge power in battery	2414.6 watts
Discharge power out of battery	3383.8 watts
Depth of discharge of energy during longest eclipse	0.350 pu
Energy storage rating of the battery	6803.5 Wh

Summary of Major Component Ratings:

Solar array power output rating	5839.2 watts
Battery Wh energy storage rating	6803.5 Wh
Battery Ah rating with 80 cells	68.90 Ah
PCU charge converter rating	2489.3 watts
PCU discharge converter rating	3045.5 watts

FIGURE 7.8 Power flow and energy balance analysis example for a 3000-W LEO satellite.

electronics and distribution harness. Various power electronic converters are used to change the voltage as needed at various points in the power system. Their efficiencies are accounted for in the systems model. The power consumed in the control electronics, such as the bus voltage controller; bus voltage error amplifier, are included in the efficiency numbers.

The power flow analysis determines the voltage and current throughout the power system, and establishes the power handling capability and heat dissipation of various components. It computes the power through various components, based on the component efficiencies marked on the spreadsheet. The converter efficiency, harness resistance and diode voltage drops are factored in the computations. The current, voltage, and power loss in the distribution harness are also determined in each segment of the system.

There is always an excess power generation, more so at the beginning of life. The shunts, therefore, are required to bypass the excess power to the ground, or leave on the solar array itself. In the design calculations at the end of life, the shunt power should be equal to the power margin plus the reliability margin.

The result of an energy balance analysis example for a 3000-W load LEO satellite on a spreadsheet is depicted in Figure 7.8. As seen in the lower half of the figure, the solar array must generate 5839 W and the battery must have an energy storage rating of 6804 Wh. An 80-cell NiH_2 battery would require a 69-Ah rating to support a 3000-W load during an eclipse. The assumptions used in this example are listed on the spreadsheet. The spreadsheet algorithm is derived from the design methods developed in various chapters of this book.

For a given EPS design, such a spreadsheet is also typically used to perform the following two types of analyses:

- Forward analysis to establish the EPS load capability, assuming that the battery must be fully charged at the end of the sunlight in each orbit.
- Backward analysis to verify whether a given load profile can be supported without exceeding the component ratings, assuming that the power can be drawn form the solar array and the batteries within the permissible DOD during that time period. This is often done to assess the peak power capability during a planned or contingency operation, or when the load profile is not flat during the loading period.

7.8 Design Analyses List

Various design analyses are conducted side by side as the design matures. They may confirm or change the design in order to meet the required specifications under all normal and abnormal conditions. Table 7.6 is a list of such analyses that are typically conducted as applicable for a given mission.

7.9 Worst-Case Error Margin

Various uncertainties in input parameters may introduce probable errors in the design estimates. Such errors are managed by providing appropriate margins. If no guidance is available from the management, a conservative engineer may stack all probable errors to derive the total error that is the sum of all individual errors. The probability of occurrence of such a total error is incredibly slim, of course. In situation where life safety in not

Table 7.6 Typical design analyses required for the power system design

Design analyses performed:
- Launch and ascent power
- Transfer orbit energy balance
- Operational orbit energy balance
- Component dissipations
- Stability and transient analysis
  - Transient due to step load changes (on and off)
  - Transient during fault and recovery (fuse blow)
  - Ripple analysis due to repetitive load pulses
  - Stability margin and sensitivity analyses
- Battery latch-up analysis for sunlight regulated bus

Each of the above analyses is performed at:

Beginning of life with
- Minimum load
- Maximum load
- Nominal and failure conditions

End of life with
- Minimum load
- Maximum loads
- Nominal and failure conditions

involved, it may be acceptable to use the root sum squared (RSS) error calculated as follows.

If errors in individual performance parameter are $E_1, E_2, \ldots$ with various input parameter uncertainties, then the total error stacked in series is $E_{\text{stacked}} = E_1 + E_2 + \ldots$. However, the *most probable error* is given by the RSS of the individual errors, i.e.,

$$E_{RSS} = \sqrt{\sum_{1}^{n} E_i^2} \tag{7.4}$$

As a simple example, consider a design having four probable sources of error, each ±2%. The total stacked error is ±8%, whereas the RSS error calculated as above is only ±4%, or one half as much. The difference between the stacked and the RSS errors widens with a greater number of error sources. In another example, say we wish to estimated the most probable variance from the expected performance of an integrated circuit under three damaging effects, namely the temperature, aging, and radiation of charged particle. The procedure of estimating the most probable RSS error is as follows:

- Estimate the component performance with the normally expected values of the operating stresses.

- Then determine the deviation in performance under possible sources of uncertainties at their worst-case levels individually.
- Now, determine the total worst-case deviation in performances due to all uncertainties combined using the RSS variance method, i.e.,

$$\Delta_{total} = \sqrt{\Delta^2_{radiation} + \Delta^2_{temp} + \Delta^2_{aging}} \tag{7.5}$$

If the most probable worst case deviation thus calculated is unacceptable with some margin, appropriate design changes must be made to bring it under the acceptable level by either decreasing the temperature, or increasing the device hardening, or both.

7.10 Design Process Phases

In today's competitive environment it is not enough to ensure that the system and the equipment are designed to be capable of meeting high standards of performance and reliability. A consideration of manufacturing processes and expected tolerances is part of good and complete design. A thorough and correct design the first time is of paramount importance for maintaining the program schedule and cost. The following *Rule of Ten* can be a strong motivator to the design engineer: *Not doing something right the first time costs ten times as much to find and fix each time it escapes to the subsequent stage from the design to the delivery.* This rule results in the following string of cost multipliers of fixing an error at various phases of the total design and delivery cycle:

Design	Production	Test	Launch pad
1	10	100	1000

The engineer must also detect and correct the design early enough for potential manufacturing and test problems. This would minimize, preferably reduce to zero as a goal, the design changes, test failures, manufacturing disruptions, trouble shooting, and re-work. For such cost-effective implementation of the design, all engineers — design, manufacturing, test, and field engineers — work in *the total team on the total satellite.* Only such a concurrent design team can ensure that the design of each system is compatible with all other systems, and also with the manufacturing and test procedures. For this purpose, the design process incorporates various reviews as it proceeds from the contract award to the delivery. A typical design phase sequence is listed in Table 7.7.

Table 7.7 Power system design, analyses and review phases

1. Requirements allocation and conceptual design phase
   - Requirements definition/allocations
   - Review of system requirements
   - Top level design trades and conceptual design
2. Preliminary design and analysis phase
   - Component specifications
   - Preliminary design and analysis
   - Preliminary design review
3. Detail design and analysis phase
   - Detail design and analysis
   - Critical design review
4. Support phase
   - Support component and system tests
   - Support outside suppliers
   - Support satellite integration and test
   - Prepare operations manuals
   - Support launch
5. Continuing tasks
   - Cost and schedule control through the spacecraft delivery

7.11 Factory-to-Orbit Events

In the entire design process, the engineer stays focused on what the power system is required to do not only in the operational orbit, but also during the factory tests, on the launch pad, and during the orbit acquisition. The following are some typical functions and requirements in various phases of the satellite, which the power system engineer must incorporate in the design:

Storage: Unforeseen circumstances may require a partially built spacecraft to be placed in storage, typically 3 years maximum in a controlled environment.

Ground tests: Solar array and battery may not be installed, requiring external power for testing via ground enable plug. Workhorse battery may be used during testing, which may not be an exact duplicate of the flight battery in performance.

Prelaunch: External or internal power is used, flight batteries may be installed, and solar array is stowed.

Launch: Battery power is used at high discharge rate, and the system operates through the launch environment of shocks and vibration.

Transfer orbit: Solar array generates limited power. The battery powers the pyro bus loads only. Redundancy management is enabled.

Deployment: Partial or full solar array power is available after deployment. Only the bus and the pyro loads are powered.

Pre-operational drift orbit: Full power generation, but still low loads.

Non-operational on-orbit: Full power generation, but less then full load. Supports eclipse and battery charging.

Normal operational orbit: Full power generation and full load. The eclipse and battery energy balance is maintained in every orbit.

Station keeping: Full power generation and full load. In addition, the battery supports arcjets during N–S station keeping.

Reacquisition: Loads shed by ground commands, arrays slewed back-to-back if this mode persists for more than 30 min or so.

Deactivation or de-orbiting: Power system supports this maneuver as the last function. Then the power system role is over.

7.12 Power System Functions Over Life

The power system engineer must also be cognizant of the major satellite level events from the pre-launch to the customer hand-over. A typical sequence used for DMSP satellites follows:

Prelaunch: The spacecraft is on the launch pad when the decision is made to launch or abort. The system is typically single-string, so potential single point failures are identified. A launch decision proceeds as follows:

- The power system is on and the satellite goes on the internal power about 10 min before lift-off, when the solar simulator is turned off.
- Both batteries are on to supply power during the final 10 min before lift-off and during ascent. No charging can occur until the satellite is deployed and faces the sun.
- The power converters are on and in discharge mode. The hardware detection and switches are enabled. Software-controlled redundancy is typically off.
- The array drive electronics is on, but the array is not driven. All components drawing power directly from the battery are on, except turned off by an internal switch.

Ascent mode: The power system is on, but has no change from the prelaunch phase.

Pre-handover: Two redundant cutters for each of the two bands are cut to deploy the solar array. Both bands must cut for the deployment to occur. Two cutters for the boom hinges and dampers are activated. Redundant pin pullers activate the solar array cant angle.

Handover: The spacecraft control computer initiates handover based on time as adjusted during the flight.

Orbit operation: Power system maintains the energy balance in each orbit. The satellite is fully operational.

Part B

PV–Battery System

Chapter 8

Solar Array

8.1 Introduction

The solar array is made of numerous photovoltaic (PV) cells connected in a series–parallel combination to obtain the required voltage and current. The fundamental characteristics of the PV cell — the building block of the array — is covered first in this chapter before moving on to the array construction, performance, and design.

8.2 Photovoltaic Cell

A photovoltaic cell converts sunlight into direct current electricity. The PV effect was discovered in 1839 by the French physicist Antoine Henri Becquerel. It remained in the laboratory until 1954, when Bell Laboratories produced the first silicon solar cell. It powered the first Russian satellite in 1958, followed by applications in the U.S. space programs. Since then, it has been an important source of power for satellites because of its high power output per unit mass. Having developed maturity in space applications, PV technology is now spreading into terrestrial applications ranging from powering remote villages around the world to feeding power grids. The installed capacity of terrestrial PV systems worldwide in 2004 was over 3000 MW. Almost 40% of the PV panels installed in the world are produced in the U. S. Unlike the dynamic systems using turbine or engine driven generators, the PV array is static, produces no vibration or noise, and needs no active cooling.

The photovoltaic effect is the electrical potential developed between two dissimilar materials when their common junction is illuminated with photons (Figure 8.1). When the junction absorbs light, the energy of the absorbed photons is transferred to the electron system of the material, resulting in the creation of charge carriers that are separated at the junction. The charge carriers may be electron–ion pairs in a liquid electrolyte, or electron–hole pairs in a solid semiconducting material. In either case, they create a potential gradient (voltage), are accelerated under the electric field, and circulate as electrical current if the external circuit is closed. The product of the voltage and current in the circuit is the photon power converted into electricity. The power that is not converted into electricity remains at the junction and elevates the cell temperature.

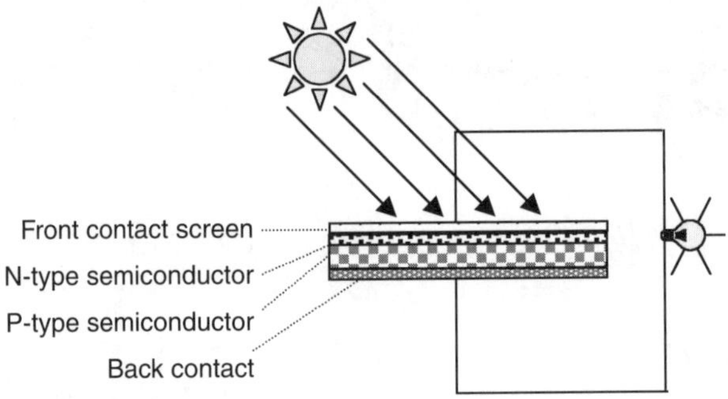

FIGURE 8.1 Photovoltaic effect converts sunlight into electricity.

The PV cell in practical use consists of a p–n or n–p junction of two semiconductors with a terminal for each. The physics of the PV cell is very similar to the classical junction diode. The origin of the photovoltaic potential is the difference in the chemical potential — called the Fermi level — of the electrons in the two isolated materials. When they are joined, the junction approaches a new thermodynamic equilibrium. Such equilibrium can be achieved only when the Fermi level is equal in the two materials. This occurs by the flow of electrons from one material to the other until a voltage difference is established between the two materials, which is just equal to the initial difference of the Fermi level. Absorption of photon energy in the silicon cell knocks out an electron from the atom, leaving a positively charged hole. Free electrons return through an external circuit under the potential difference delivering power to the load.

From the solid-state physics point of view, the cell is basically a large area diode with the junction positioned close to the top surface. Early cells were p–n type with p layer facing the sun. They degraded at a rapid rate in the Van Allen radiation belts. The cell design was changed to n-on-p to improve the radiation resistance. The new n–p cell extended the mission life over a decade.

The basic cell construction is shown in Figure 8.2. The metallic contacts are provided on both sides of the junction to collect electrical current induced by the impinging photons on one side. A conducting foil or solder contact is provided all over the bottom dark surface and on one edge of the top illuminated surface. Thin conducting mesh on the remaining top surface collects the current while allowing the light through. The spacing of the conducting fibers in the mesh is a matter of compromise between maximizing the electrical conductance and minimizing the blockage of the light. In addition to these basic features, several enhancements are also included in the construction. For example, the front face of the cell has an anti-reflective coating to absorb as much light as possible by minimizing the reflection. The protection against damage due to energetic particles hitting the surface is provided by the coverglass applied with a transparent adhesive.

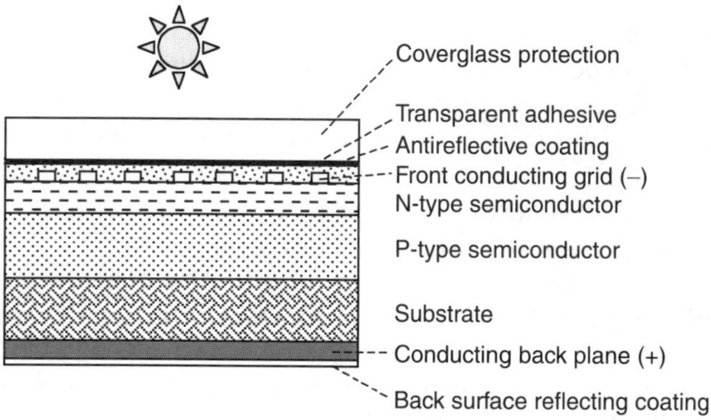

FIGURE 8.2 Basic construction of PV cell with performance enhancing features.

The series resistance is one of the limiting factors in determining the conversion efficiency of PV cell. It is influenced by the current collecting grid thickness (usually less than 1 μm), and the semiconductor structure. The cell manufacturer undertakes a multi-dimensional optimization to achieve the best design. The space qualified cell efficiency is quoted at air mass zero (AM0), which represents the condition in outer space. The voltage developed and the fraction of the incident energy converted into electricity depends on the characteristics of the junction material and the wavelength of the radiation.

8.3 PV Technologies

The most important measure of the PV cell performance is the photoelectric energy conversion efficiency and the cost per watt capacity. Together, these two parameters indicate the economic competitiveness of the photovoltaic with alternative power generation technologies. The conversion efficiency, η, of the photovoltaic cell is defined as

$$\eta = \frac{\text{electrical power output}}{\text{solar power impinging the cell}} \tag{8.1}$$

Theoretical maximum energy conversion efficiency for three common materials is 16% for germanium (Ge), 24% for silicon (Si) and 29% for gallium arsenide (GaAs). The energy conversion is different from the energy absorption. Not all energy that is absorbed by the cell is converted into electricity. The sunlight wavelength spectrum is very important. Certain semiconductors have a cut-off wavelength, such as 1.1 μm in the silicon cell. Photons with longer wavelengths are unable to generate hole–electron pairs. They pass through the p–n junction and are absorbed at the bottom of the cell in the form of heat. The effect of changing the wavelength of illumination is shown in Figure 8.3. It may be noted that about two thirds of the solar radiation energy lies between 0.4 and 1.1 μm wavelengths. A photon of blue light having 3 eV energy generates around 0.5 eV electricity and the remaining 2.5 eV is absorbed as heat. This heat must be dissipated back into the space; otherwise it would decrease the electrical conversion efficiency. A high-emissivity coating on the back surface helps to dissipate the heat effectively, thus improving the electrical conversion efficiency.

The continuing development efforts to produce more efficient low cost cells have resulted in various types of PV technologies available on the market today in terms of conversion efficiency and cost. The major types are discussed in the following sections.[1]

8.3.1 Single-Crystal Silicon

Single-crystal silicon is a widely available cell material that has been the workhorse of the space industry. In the most common method of production, known as the Czochralski process, the silicon raw material is first melted and purified in a crucible. A seed crystal is then placed in the liquid silicon and drawn at a slow constant rate. This results in a solid single-crystal cylindrical ingot (Figure 8.4).[2] The manufacturing process is slow and energy intensive, resulting in high raw material cost. The ingot is sliced using a diamond saw into 200 to 400 μm (5 to 10 mil) thick wafers. The wafers are further cut into rectangular cells to maximize the number of cells that can be mounted together on a rectangular panel. The cell is generally in several centimeter squares in size. Unfortunately, almost half of the expensive silicon ingot is wasted in slicing it and forming square cells. The material waste can be minimized by making the full size round cells from round ingots. Using round cells is uneconomical in satellites where the

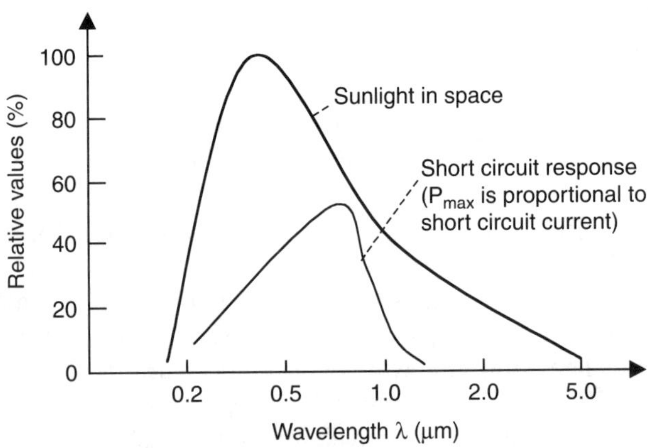

FIGURE 8.3 Power versus wavelength spectral response of silicon cell.

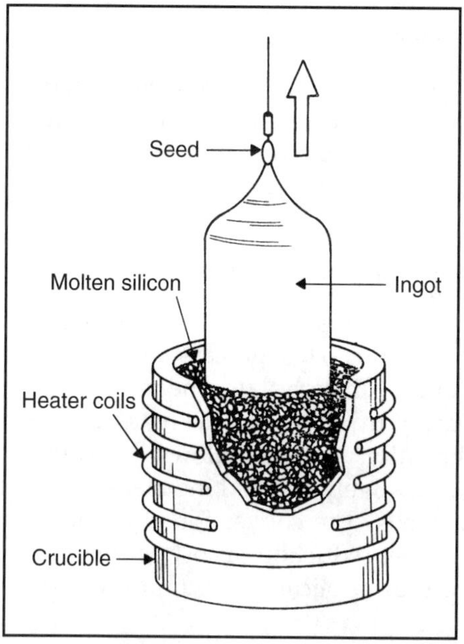

FIGURE 8.4 Single-crystal ingot-making by Czochralski process. (Source: G. Cook et al. DOE Report No. DE91015001, 1995.)

panel space is at high premium. Another way to minimize waste is to grow crystals on ribbons. Some U.S. companies have set up plants to draw PV ribbons, which are then cut by a laser to reduce waste.

The high level of cleanliness during the manufacturing process, followed by stringent screening for various possible defects, makes the space-qualified cells expensive. A flexure test eliminates failure-prone wafers at an early stage. The probability of cell cracking can be substantially reduced by mechanical proof testing of silicon wafers before further processing. This is based on a study[3] that demonstrated that weak wafers could be eliminated by subjecting all the wafers in a manufacturing batch to a biaxial-flexure test. The test is performed after wafers have been chemically polished but before junctions are formed in them. Wafers that are likely to crack during further processing or in service break as soon as they are flexed in a test jig. The wafers having no internal flaws that could eventually cause mechanical failure withstand the flexure test. The study also showed that the biaxial flexure test does not induce additional damage in otherwise good wafers. If a wafer is sound, the testing does not create flaws that would later weaken the specimen. The experimental results were analyzed statistically. The biaxial strength of silicon wafers that failed from surface flaws was found to correspond to a Weibull distribution. The critical flaw sizes were found to be 4.6, 3.1, and 2.1 μm in proof testing at stress levels of 215, 260, and 310 MPa, respectively.

8.3.2 Gallium Arsenide

GaAs has much higher conversion efficiency than silicon cells, but is expensive. It has a 1.43-eV direct band gap. It is relatively insensitive to wide temperature cycling, and has high radiation resistant. Modern GaAs is generally produced by growing GaAs film on a Ge substrate.

8.3.3 Semicrystalline and Polycrystalline

Instead of drawing single crystals using seeds, the molten silicon is poured into a rectangular crucible. This process under a controlled cooling rate forms partial and/or multiple crystals. The rectangular ingot thus produced eliminates the squaring-up process and the associated waste of materials. The remaining fabrication process follows the steps of the single crystal silicon cell. This is a relatively fast and inexpensive process for manufacturing thick crystalline cells. The conversion efficiency is low, but the cost is much lower, giving a net reduction in cost per watt of power. Such cells are developed for terrestrial applications, but are uneconomical at present for space applications.

8.3.4 Thin Film

Copper indium gallium diselenide (CIGS) and cadmium telluride (CdTe) are thin film materials, typically a few micrometers or less in thickness, directly deposited on glass, stainless steel, ceramic or other compatible substrate materials. While the basic copper indium diselenide cell has a band gap of 1.0 eV, the addition of gallium increases the band gap to 1.4 eV, resulting in more efficient collection of photons near the peak of the solar spectrum. The thin film technology uses much less material per square area of the cell, hence is less expensive per watt of power generated. These are new types of photovoltaic cells entering the market that has drawn some interest in the space industry.

8.3.5 Amorphous

In the amorphous technology, silicon vapor is deposited in a couple of μm thick amorphous (glassy) film on glass or stainless steel rolls, typically 2000 feet long and 13 inch wide. This process forms a random bond structure, as opposed to a crystalline structure. It significantly reduces the carrier mobility and results in poor conversion efficiency, which is about 10% at present. However, it uses only 1% of the material. Moreover, the sheet manufacturing, instead of single cells from ingots, is inexpensive. The cost per watt generated is significantly lower. With the stabilized efficiency about one half of that in the crystalline silicon (c-Si) cells, the amorphous silicon (a-Si) is not a serious candidate for space power at present. However, crystalline technology has been with us for several decades and has entered a plateau of a slow learning curve. On the other hand, amorphous technology is new and hence a steep learning curve is anticipated. The expected result is a rapid decline in the amorphous silicon price per watt by increasing the efficiency and decreasing the manufacturing cost. On this premise, in 1997, two plants in the U.S. started manufacturing a-Si sheets in large quantity for terrestrial applications. It has been tested in space and used in protoflights also.

8.3.6 Multi-Junction

Ever increasing demands on spacecraft power systems has produced double and triple junction cells with improved performance, radiation resistances, and lower cost. Recent advances have produced the tandem multi-junction GaAs cell similar to the a-Si tandem cell, where one cell consists of two tandem GaAs cells separated by thin tunnel junction of GaInP, followed by a third tandem GaInP cell, separated by an AlInP tunnel junction. The tunnel junctions mitigate voltage drop of the otherwise forward-biased p–n junction that would appear between any two tandem p–n junctions in opposition to the photon-induced cell voltage.

An n-on-p silicon junction converts only red and infrared light into electricity, not blue and ultraviolet. The GaInP/GaAs cell captures infrared photons as well, thus improving the efficiency. Efficiencies up to 35% in double junction and 40% in triple junction space-qualified cells have been measured by Spectrolab Inc. for DoE. The triple junction efficiency has a theoretical limit of 50%. That is like capturing one half the sun. Today, the GaInP/GaAs on Ge substrate multi-junction cells are being increasingly used by many spacecraft builders seeking high efficiency and good manufacturability.

8.4 Equivalent Electrical Circuit

The steady state equivalent electrical circuit of the cell is shown in Figure 8.5, which represents the complex physics of the PV cell. The cell acts as a

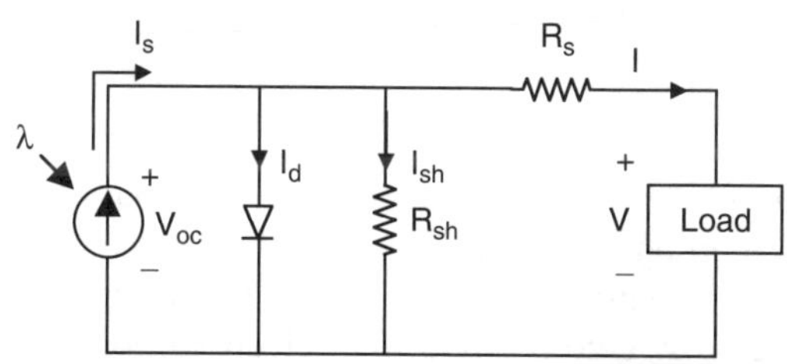

FIGURE 8.5 Equivalent electrical circuit of PV cell.

constant current source shunted by a perfect diode. The circuit parameters are as follows. The series resistance R_s represents the internal resistance to the current flow. It is primarily due to the resistivity of the material. The shunt resistance represents the leakage current across the junction. It depends on the n–p junction depth, the impurities, and the contact resistance. The value of R_{sh} is inversely related with the leakage current to ground. In an ideal PV cell, $R_s = 0$ (no series loss), and $R_{sh} = \infty$ (no leakage to ground). In a typical high quality 2.5 × 2.5-cm silicon cell, $R_s =$ 0.05 to 0.10 Ω and R_{sh} = 200 to 300 Ω. The PV conversion efficiency is sensitive to small variations in R_s, but is insensitive to variations in R_{sh}. A small increase in R_s can decrease the PV output significantly. Since magnitudes of the diode current and the resistances R_s and R_{sh} vary with temperature, the cell output and the conversion efficiency decrease with increasing temperature.

In the equivalent circuit, the current delivered to the external load equals the source current (photo current), I_s, proportional to the illumination, less the diode current, I_d, and the ground shunt current, I_{sh}. The open circuit voltage, V_{oc}, of the cell is obtained when the load current is zero, i.e., when $I = 0$, and is given by

$$V_{oc} = V + IR_s \tag{8.2}$$

The diode current is given by the classical diode current expression

$$I_d = I_o\left[e^{\frac{qV_{oc}}{AKT}} - 1\right] \tag{8.3}$$

where I_o = the diode saturation (dark) current

q = electron charge = 0.1592×10^{-18} coulomb

A = curve fitting constant

K = Boltzmann constant = 1.38×10^{-23} J/K

T = absolute temperature, in kelvin

The load current is therefore given by the expression

$$I = I_s - I_o\left[e^{\frac{qV_{oc}}{AKT}} - 1\right] - \frac{V_{oc}}{R_{sh}} \tag{8.4}$$

The last term in Equation 8.4 is the ground leakage current. In practical cells, it can be ignored since it is negligible compared to I_s and I_d. The terminal current versus voltage (I–V) characteristics under dark, partial light, and full light are shown in Figure 8.6. The voltage developed and the fraction of incident flux converted into electrical power depend on the junction material and wavelength of the radiation. The maximum power

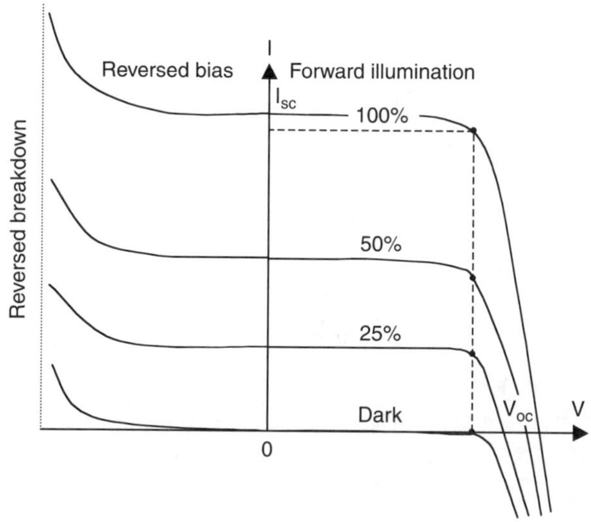

FIGURE 8.6 Terminal *I–V* characteristics under full, partial, and zero illuminations.

that can be extracted from the cell under full light is equal to the area of the dotted rectangle. The maximum power is approximately proportional to the short circuit current. In the forward direction, the power generation is maximum at the knee point of the curve, where the voltage is typically 0.50 V, the current about 40 mA/cm^2 and power about 20 mW/cm^2 of the cell area in crystalline silicon cells. For GaAs/Ge cells, these values are 1.0 V, 30 mA/cm^2, and 27 mW/cm^2, respectively.

The diode saturation current can be determined experimentally by applying voltage V_{oc} to the cell in the dark and measuring the current going into the cell. This current is often called the dark current or the reverse diode saturation current. If the voltage is externally applied in the reverse direction, say during a system fault transient, the current remains flat and power is absorbed by the cell. However, beyond a certain negative voltage, the junction breaks down as in diode, and the current rises to a high value. In the dark, the current is zero for voltage up to the breakdown voltage, which is the same as in the illuminated condition.

8.5 *I–V* and *P–V* Characteristics

The electrical characteristic of PV cell in sunlight in the useful working range is represented by the *I–V* curve in the first quadrant only, as shown in Figure 8.7(a). Two most important parameters widely used for describing the cell electrical performance are the open circuit voltage, V_{oc}, and the short circuit current, I_{sc}. The short circuit current is measured by shorting the output terminals, and measuring the terminal current under full

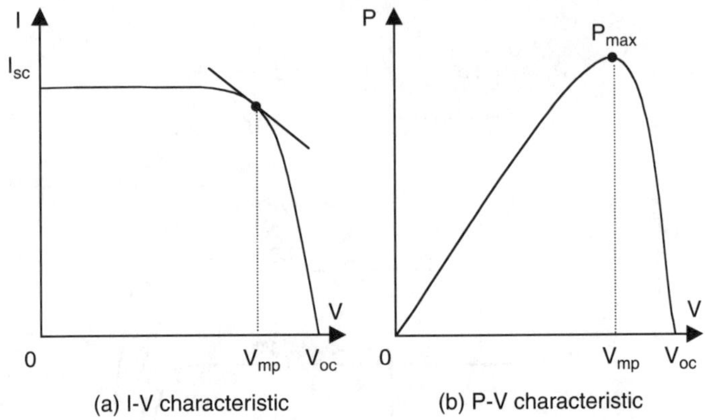

FIGURE 8.7 Cell characteristics in sunlight and the maximum power point.

illumination. Ignoring the small diode and the ground leakage currents under zero terminal voltage, I_{sc} is the photocurrent, I_s. The current under this condition is the maximum current the cell can deliver. The *I–V* characteristic is developed from the test data at various illuminations, temperature, and ionized radiation doses. The bottom right of the curve at zero current is the open circuit voltage measured with the output terminals open. The maximum photo voltage is produced under the open circuit voltage. Again, by ignoring the ground leakage current, Equation 8.4 with $I = 0$ gives the open circuit voltage as

$$V_{oc} = \frac{AKT}{q} \log_n \left(\frac{I_s}{I_o} + 1 \right) \tag{8.5}$$

The constant KT/q is the absolute temperature expressed in volts (300 K = 0.026 V). In practical photocells, the photocurrent is several orders of magnitude greater than the reverse saturation current. Therefore, the open circuit voltage is many times the KT/q value. Under constant illumination, I_s/I_o is a sufficiently strong function of the cell temperature, and the solar cell ordinarily shows a negative temperature coefficient of the open circuit voltage.

The output power of the panel is the product of the voltage and the current outputs. In Figure 8.7(b), the power is plotted against the voltage. Notice that the cell produces no power at zero voltage or zero current, and produces the maximum power at voltage corresponding to the knee point of the *I*–V curve. This is why the PV power circuits are designed such that the panel operates closed to the knee point, slightly on the left-hand side. Therefore, the PV panel is modeled approximately as constant current source in the electrical analysis of the system.

On the left hand side of the knee point, the cell works like a constant current source, generating voltage to match with the load resistance. On the right hand side, the current drops rapidly with small rise in voltage, and the cell works like a constant voltage source with an internal resistance.

Figure 8.8 and Figure 8.9 are typical data sheets of two representative space qualified cells. The basic requirement of the solar cell modules for space applications is generally described in MIL-STD-83576 for the specific

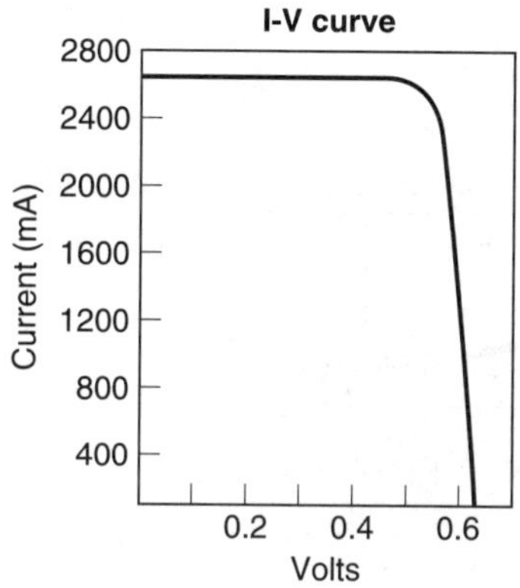

Temperature coefficients (28° to 60°C)

Parameter	%/°C
V_{oc}	−0.370
I_{sc}	0
V_L	−0.443
P_{max}	−0.460

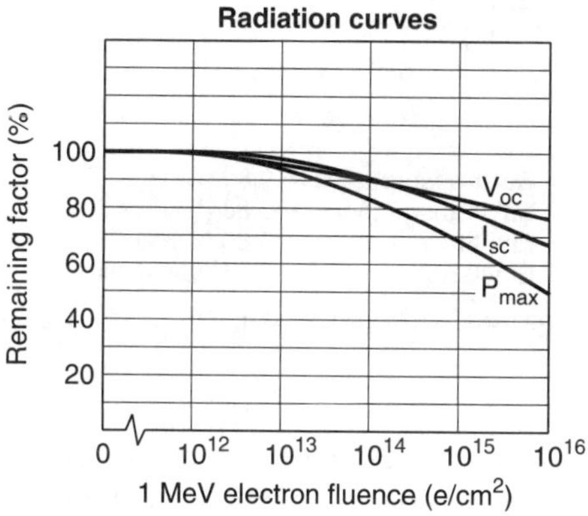

Test	Description	Results
Humidity	+45°C, 90% RH min., 30 days	<1.5%
Thermal cycle	+80°C to −180°C, 3000 cycles	<2.5%
Thermal shock	+140°C to −185°C, 5 cycles	<1.5%
Thermal soak	+140°C for 168 hrs., 5×10^{-5} torr	<1.5%
Radiation	Characterized thru 1×10^{16} 1 MeV e/cm^2	—
Pull test	90° pull, standard tab	>250 gms

FIGURE 8.8 Silicon PV cell performance data: Typical × 8-cm, 8-mil thick on 10-Ω·cm substrate with back surface field reflector, 14% efficiency in AM0.

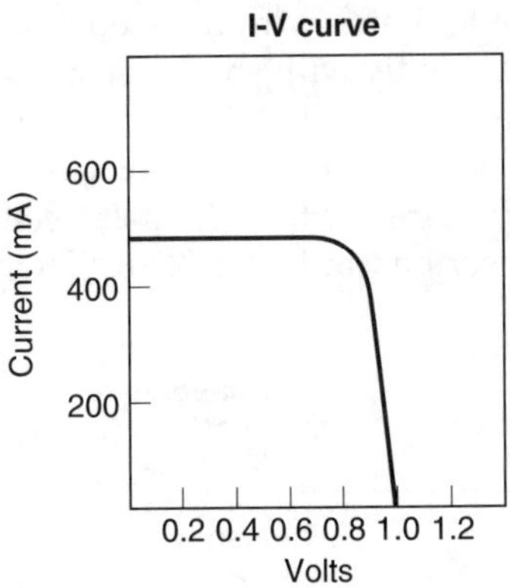

Temperature coefficients (28° to 60°C)

Parameter	%/°C
V_{oc}	−0.194
I_{sc}	+0.056
V_L	−0.232
P_{max}	−0.223

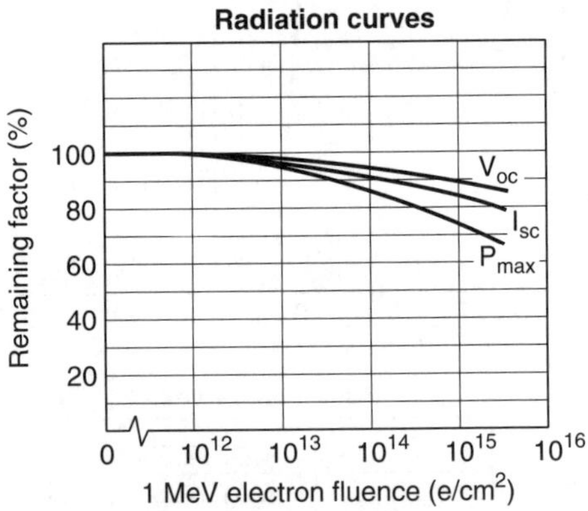

Solar absorptance = 0.89 (CMX)
Solar absorptance = 0.88 (fused silica)
Emittance (normal) = 0.85 (CMX)
Emittance (normal) = 0.81 (fused silica)

FIGURE 8.9 GaAs/Ge PV cell performance data: Typical 4 × 4-cm, 3.6-mil thick, 19% efficiency in AM0.

needs of the U.S. Air Force. However, the specifications of solar cells intended for general use in space are not well defined. Even though the PV cell performance standard is defined for specific procurement, the performance measurement standard also needs to be defined. The ISO Technical Committee C20 has issued one such standard ISO-15387 for the aircraft and space vehicles that is endorsed by ANSI. NASA/Jet Propulsion laboratory (JPL) performs such test in aircraft and balloons at 35 km height, and plans to offer such tests in the space shuttle at about 500 km altitude at a cost of around \$5000 for a 2 × 4 cm cell. Comparing such tests conducted by one organization with another is difficult. For example, in a round robin test by six organizations to ISO-15387, including tests conducted at JPL and in the PV engineering test bed at NASA GRC, the results showed deviation of up to 3% on the same cell. The cell temperature coefficient is extremely sensitive to the solar spectrum, and can vary by a few-fold, making the spectrum duplication in tests extremely important. For this reason, power engineers allow a 3% margin for standard cells, and perhaps more for multi-junction cells, for possible error in power generation estimates using test results.

8.6 Power from Albedo

Albedo is the ratio of the sunlight reflected by a planet to that received by it. The Earth's albedo impinging on the front and back sides of the solar array can generate power at low voltage, which can be used if the solar array is designed accordingly. NASA test data in LEO show that the PV conversion efficiency under the Earth's albedo on cells mounted on the back side is about 50% of that on the sunlit front side. This can increase the array current by 10 to 15% and about 5% increase in the voltage due to lower back side temperature. However, if cells were mounted on the back side of the array also, the cost per watt generated would be high, making it uneconomical in most cases. As for the front side albedo superimposed on the full sun, it does not significantly increase the power generation. Usually, the albedo effects are ignored for the sake of simplicity, particularly in GEO satellites.

8.7 Back-Surface Illumination

The PV cells are sometimes mounted on flexible cloth rather than on rigid panels. The flexible cloth array under the back side illumination would transmit significant energy to the front, which the front cell would convert into useful power. The power generation under back surface illumination on a coupon of 10 series connected cells representing the flexible cloth solar array of the ISS was tested under side and back illuminations. The cells were 8 cm × 8 cm × 200 μm, n-on-p, crystalline silicon cell with 125-μm borosilicate coverglass. They were mounted on 125-μm thick composite polyimide–glass screen cloth substrate with silicon adhesive. The cells were interconnected by 37-μm thick flat copper conductors welded to contacts on the back of the solar cells. The copper interconnects covered 31% of the solar cell area. The coupon was subjected to standard flash tests on the front side, sideways and on the back side. As the illumination angle of incidence varied, the cell *I–V* characteristic changed primarily due to the projected area loss up to about 60°. Above that, other mechanisms also become significant, such as Fresnel reflection, edge effect on light collection, scattering and refraction, shadowing and other effects covered under the Kelly cosine. The noteworthy conclusions of these tests were as follows:[4]

1. The back-illuminated cells produced 41% of the front-illuminated power at essentially the same voltage.
2. Under edge illumination, the cells produced about 70% of the voltage attained with normal front side illumination.

These data are valid for the BOL. However, it has been shown that under the equivalent radiation dose of 1.5×10^{13} MeV after 11.6 years of the ISS operating life, the back side response degraded by only 10%. This is a significant power generation from the back side illumination of the flexible cloth mounted cells. It can be factored in the analysis if needed in certain mission situations to avoid over-sizing the arrays on the extreme conservative side.

Analysis has shown that back side illumination from sources other than direct sun, such as albedo from the Earth, the moon, and the spacecraft itself, produce negligible power.

8.8 Array Construction

The solar array consists of numerous cells mounted on a base substrate. The construction can take various forms described below. With a variety of array constructions possible, the bulk of the arrays are still made with crystalline silicon cells on rigid flat panels.

8.8.1 Rigid Panels

In this traditional construction method, the cells are mounted on a rigid substrate made of an aluminum honeycomb with face sheets. Traditionally, the face sheets have been made from aluminum, but fiber composite sheets such as graphite, Nomex™ and Kevlar™ have also been successfully used for mass reduction. Honeycomb substrates are available in ¼, ½, ¾ and 1 inch thicknesses. Silicon adhesive is used to fix the PV cells onto the panels. Extreme care is exercised in applying the adhesive, since even a trace of silicon in the cell production facility can degrade the cell output voltage considerably.

The array is protected from the space environment primarily by a coverglass such as fused silica Microsheet™. An antireflective coating minimizes the light reflection. It enhances the light absorption by the cell and produces more power. A suitable coating is also applied to the back surface for temperature control. The ESD concern is addressed by ITO coating on the front surface. However, it reduces the power generation, which must be compensated by larger panel area or thicker coverglass.

Figure 8.10 shows a satellite with one wing of the solar array with key panel construction features. The series strings of cells on the panels are routed such that the magnetic moment is minimized. If the propellant plume is a concern, the strings are oriented perpendicular to the plume flow

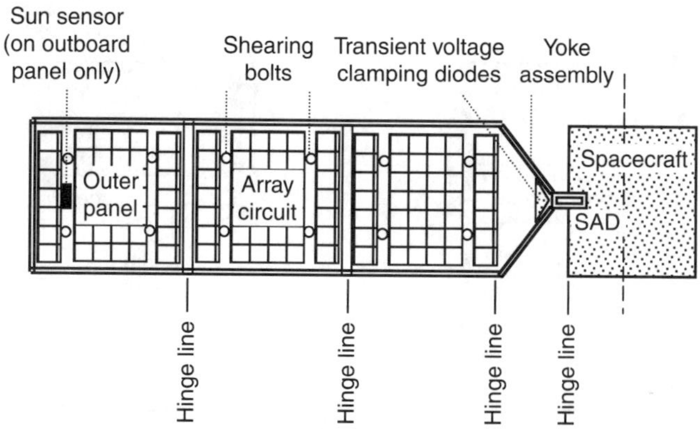

FIGURE 8.10 Satellite showing one wing of solar array fully deployed.

such that the plume affect only one whole strings versus some parts of all strings. This way, if one string were damaged due to a heavy plume, the required power would still be produced with the built-in redundancy in the number of string circuits. Exposing all strings to the plumes may pose relatively higher risk.

Figure 8.11 is a representative cross section of the panel depicting cell and various other layers with their typical thickness.

The rigid panels are initially stowed with the spacecraft body during launch. They are deployed by cutting various spring-loaded cables and bolts with electro-explosive devices. Figure 8.12 displays the sequence of one wing stowed, being deployed, and after full deployment.

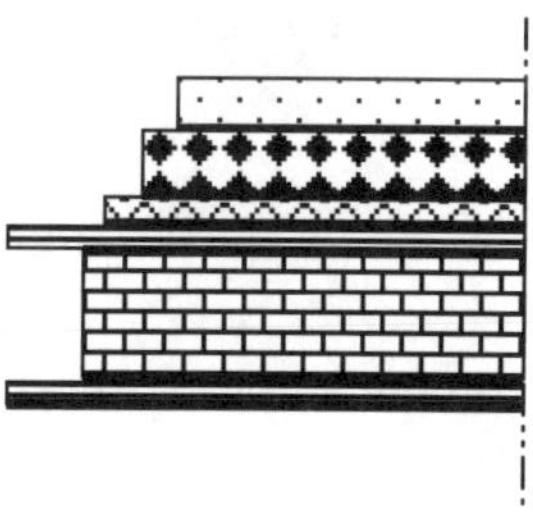

Components from top to bottom	Typical thickness, mm
Fused silica coverglass	0.15-0.50
Coverglass adhesive (DC93-500)	0.05-0.10
PV cell	0.1-0.2
Cell adhesive (RTV-566)	0.10
Kapton insulation	0.05
Adhesive (GRP-33A)	0.005
Face skin (aluminum)	0.10
Adhesive (FM-300)	0.10
Honeycomb (aluminum)	6.2-12.5
Adhesive (FM-300)	0.10
Face skin (aluminum)	0.10
Back paint (Z-306)	0.05

FIGURE 8.11 Panel cross section showing cells mounted on honeycomb substrate with face sheets.

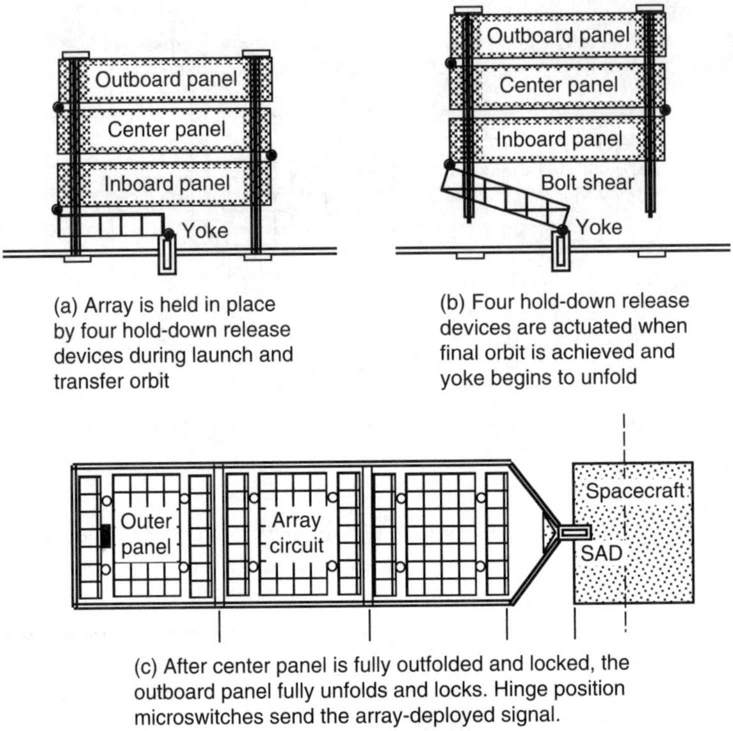

FIGURE 8.12 Solar array wing: (a) stowed, (b) being deployed, and (c) after full deployment.

8.8.2 Body Mounted Array

The body mounted array finds applications in small satellites for science missions, such as the one shown in Figure 8.13. The cells are mounted directly on the spacecraft body without using a honeycomb substrate. It saves substrate mass and also gimbals. The obvious disadvantage is that it needs of 57% more cells for the same power. However, it is generally found advantageous in the following situations:

- The spacecraft distance from the sun, the sun angle or the operating temperature vary over a wide range, or
- Flat deployed panels interfere with an instrument's line of view.

When the body mounted array rotates, the illuminated strings are at rated voltage and the dark strings are at zero voltage. This can cause ESD discharge along the circumference of the body. The remedy is to use bleed resistance across the isolation diode as shown in the figure. It equalizes the potential of the sunny side and the dark side through a bleed current.

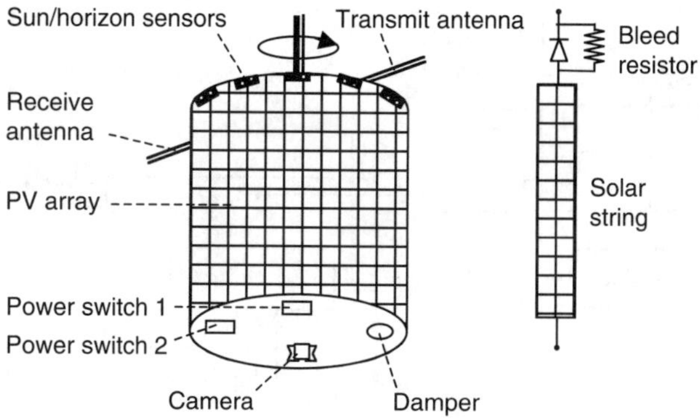

FIGURE 8.13 Body mounted array with bleed resistors in solar strings.

8.8.3 Three or More Wings

The three-wing array shown in Figure 8.14 may give essentially the same benefits as the body mounted array, provided that the deployed array wings do not interfere with the instruments. The wings can be canted at different angle as required for powering the mission. This architecture is most likely to be used for small science mission satellites with peak power tracking system design. With multiple wings, a separate PPT converter and control are required for each wing as shown in Figure 8.15. Otherwise, the PPT controls would conflict with each other.

8.8.4 Flexible Array

The flexible array made by installing PV cells on Kevlar cloth — open weave or close knit — can reduce the array mass. The array is folded when stowed,

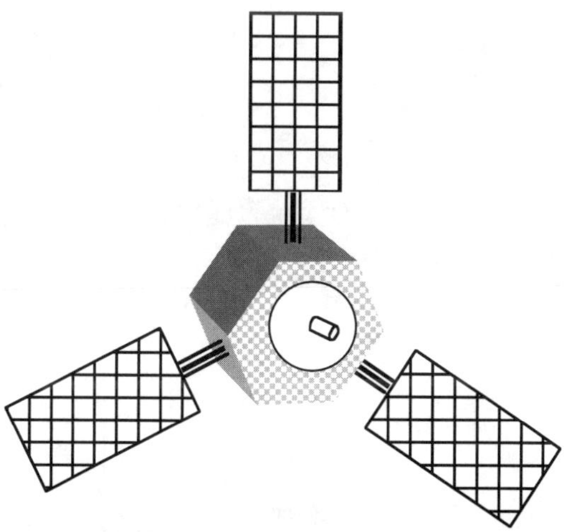

FIGURE 8.14 Three-wing solar array for small science mission satellites.

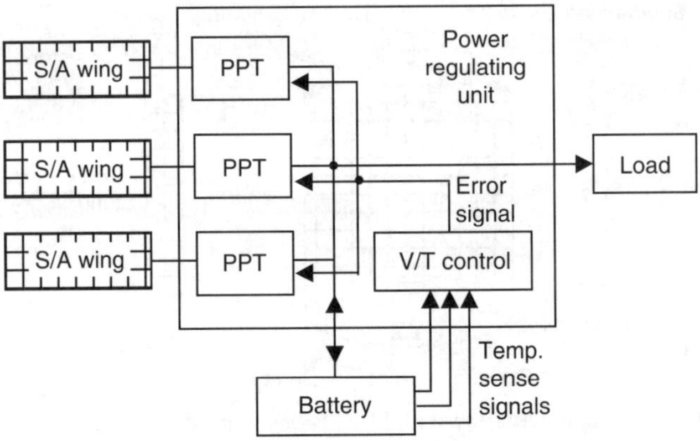

FIGURE 8.15 EPS architecture for 3-wing array with PPT for each wing separately.

and is deployed like an accordion panel or rolled out (Figure 8.16). Such arrays have been used on the first Hubble Space Telescope, EOS-AM, Olympus, and ISS. They must be designed to withstand the thermal snap on exiting the eclipse. The snap is caused by the cell side facing the sun heating up quickly, while the back side is still cold due to insignificant heat transfer from front to back. Hence, a large thermal gradient builds up across the cloth thickness. This and the mismatch in the coefficients of thermal expansion between the cell area and the metal frame jointly cause the front to bend

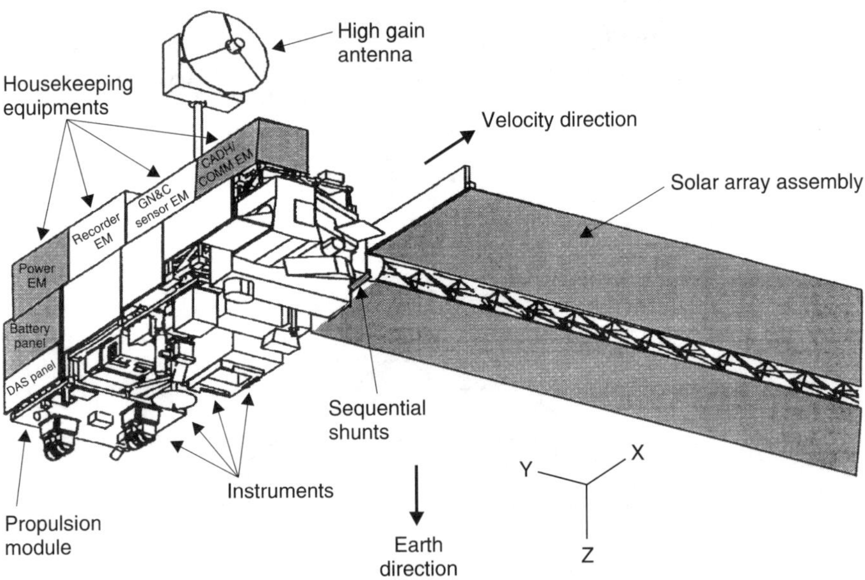

FIGURE 8.16 Accordion type flexible cloth solar array for EOS-AM satellite. (Source: NASA.)

(buckle) since the flexible cloth cannot support bending stress. This produces bowing in the array as shown in Figure 8.17. The electrical power is reduced, since the array is not completely flat now. It comes back to a flat shape in about 30 min when the temperature on the two sides is equalized across the thickness.

As an example of estimating the power reduction, consider a 2-m wide cloth panel. Such a width may bow by 20 cm or so. In that case, the angle θ in the figure is 11.3°, which is roughly the average off-angle across the width. Cos $\theta = 0.98$ gives 2% loss of power until the cloth regains its flat shape. This is about 30 min in every orbit. In LEO satellites with 1-h sunlight, this loss may need at least 1% larger array area.

The rigid array does not experience such bowing, since the aluminum honeycomb takes the front heat to the back quickly, and the front and the back face sheets spreads the heat evenly over the entire front and back surfaces.

8.8.5 Inflatable Arrays

Flexible arrays come in two types:

- The flat accordion type described above with conventional cells
- Inflatable, often in balloon shape, with a few micrometers thick amorphous PV sheets on 1 or 2 mils thick substrate.

The inflatable array is a new technology that has already been developed and flight tested. It significantly reduces the mass and can be packaged in small volume. Where usable, it reduces the overall program cost by reducing the launch vehicle size, or by allowing more satellites per launch. The inflatable array utilizes a flexible thin-film PV on a flexible composite laminate structure densely packed for launch, and deployed in space using inflation gas. It is finally cured and rigidized *in situ* in orbit without further need of inflation pressure to maintain structural rigidity.

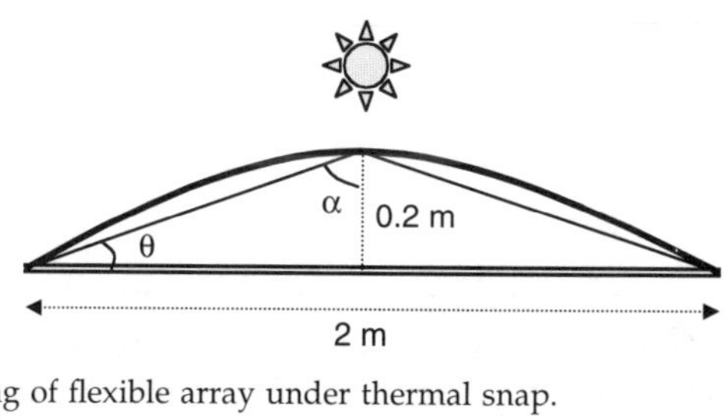

FIGURE 8.17 Bowing of flexible array under thermal snap.

8.9 Array Performance

The major factors influencing the electrical performance of the solar array are the sun intensity, sun angle, and the operating temperature. These factors are discussed below.

8.9.1 Sun Intensity

The *I–V* characteristic of PV array shifts down at lower sun intensity with a small reduction in voltage as shown in Figure 8.18. However, the photoconversion efficiency of the cell is insensitive to the solar radiation in the practical working range. Figure 8.19 shows that the efficiency is practically the same at full sun (1353 W/m^2) and ½ sun, and starts falling off rapidly only below ¼ sun (340 W/m^2). This means that the conversion efficiency is high in penumbra. We obtain a lower power output in penumbra only because of the less solar energy impinging on the cell. However, the power extractable during penumbra cannot be utilized in practical designs because of the low voltage. Hence, it is ignored.

8.9.2 Sun Angle

The cell output current is given by

$$I_s = I_o \cos\theta \tag{8.6}$$

where I_o is the photocurrent with normal sun ($\theta = 0$). The cosine law holds well for sun angles ranging from 0 to about 50°. Beyond 50°, the electrical output deviates significantly from the cosine value, and the cell generates no power beyond 85°, although the mathematical prediction would give

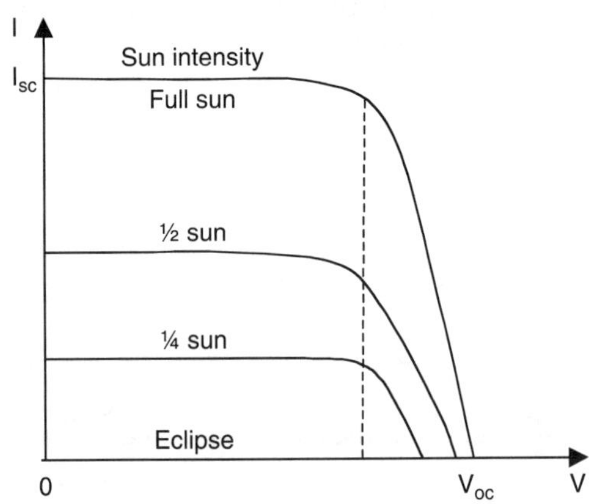

FIGURE 8.18 Current versus voltage characteristics at various illumination levels.

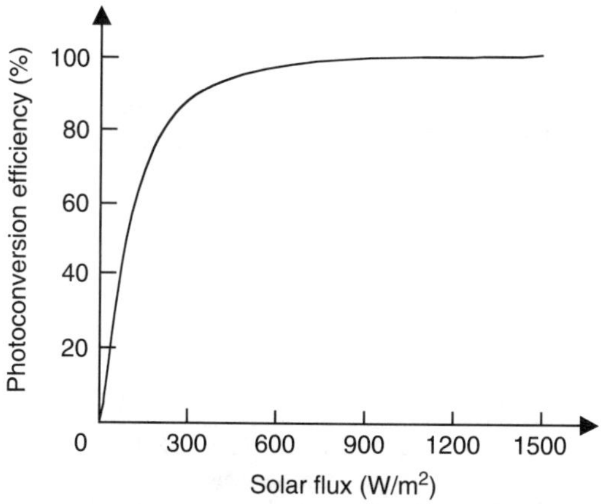

FIGURE 8.19 Photo-conversion efficiency over wide range of solar flux.

7.5% power generation. The actual power–angle curve of the PV cell is called Kelly cosine, and is given in Table 8.1. The Kelly cosine presented in the table is useful to assess accurately the power available from sun at low angles during the transfer orbit.

8.9.3 Temperature Effect

With increasing temperature, the short circuit current of the cell increases, whereas the open circuit voltage decreases as shown in Figure 8.20. Since the increase in current is much less than the decrease in voltage, the net effect is the decrease in power, which is quantitatively evaluated by examining the effects on current and voltage separately. Say I_o and V_o are the short circuit current and the open circuit voltage at reference temperature T, and α and β are their temperature coefficients in units of A/°C and V/°C, respectively. If the operating temperature is increased by $\triangle T$, then the new current and voltage are given by

Table 8.1 The Kelly cosine values of the photocurrent in silicon cells

Sun angle (degrees)	Mathematical cosine value	Kelly cosine value
30	0.866	0.866
50	0.643	0.635
60	0.500	0.450
80	0.174	0.100
85	0.087	0

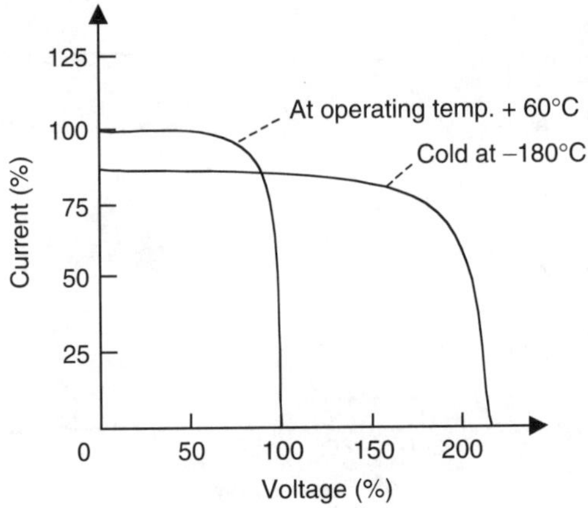

FIGURE 8.20 Temperature effect on *I–V* characteristic.

$$I_{sc} = I_o + \alpha\Delta T, \text{ and } V_{oc} = V_o - \beta\Delta T \tag{8.7}$$

Since the operating current and voltage change approximately in the same proportion as the short circuit current and open circuit voltage, respectively, the new power is

$$P = VI = (I_o + \alpha\Delta T)(V_o - \beta\Delta T)$$

This can be simplified in the following expression by ignoring the small term containing the products of α and β,

$$P = V_o I_o + \alpha\Delta T V_o - \beta I_o \Delta T,$$

which reduces to

$$P = P_o - [(\beta I_o - \alpha V_o)\Delta T] \tag{8.8}$$

For a typical 2 × 4 cm single crystal silicon cell, α is 250 μA/°C and β is 2.25 mV/°C. Therefore, power varies approximately as

$$P = P_o[1 - 0.005\Delta T] \tag{8.9}$$

This expression indicates that for every degree centigrade rise in the operating temperature, the silicon cell power output decreases by 0.50%.

Figure 8.21 depicts the power output versus voltage characteristics at two operating temperatures. It shows that the maximum power available at the lower temperature is higher than that at the higher temperature. Thus, a cold temperature is better for the PV cell, as it generates more power. However, the two P_{max} points are not at the same voltage. In order to extract

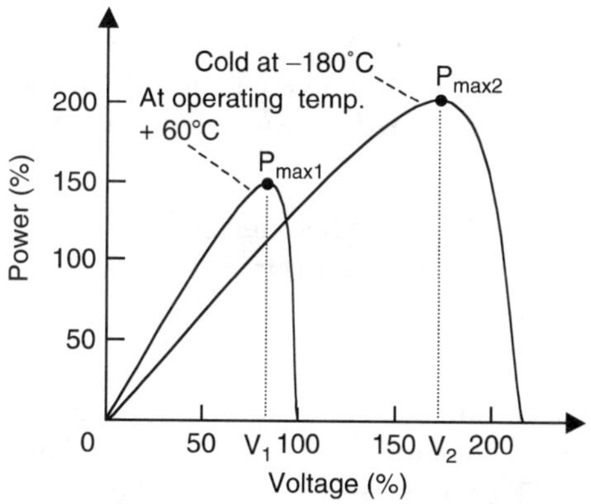

FIGURE 8.21 Temperature effects on *P–V* characteristic.

maximum power at all temperatures, the PV system must be designed such that the array output voltage can increase to V_2 for capturing P_{max2} at lower temperature, and can decrease to V_1 for capturing P_{max1} at higher temperature. If the array is operating at a fixed regulated voltage, the higher power generation capability at a cold temperature cannot be utilized by the loads, and the excess power generation by the cell must be wasted in the shunt circuits. One can capture and utilize the peak power all the time with varying temperature only by using the PPT system architecture. Otherwise, the power versus temperature variation would be as shown in Figure 8.22.

The array undergoes a wide temperature cycle in each orbit. During sunlight, it operates around 50 to 60°C on the front face, 40 to 50°C on the

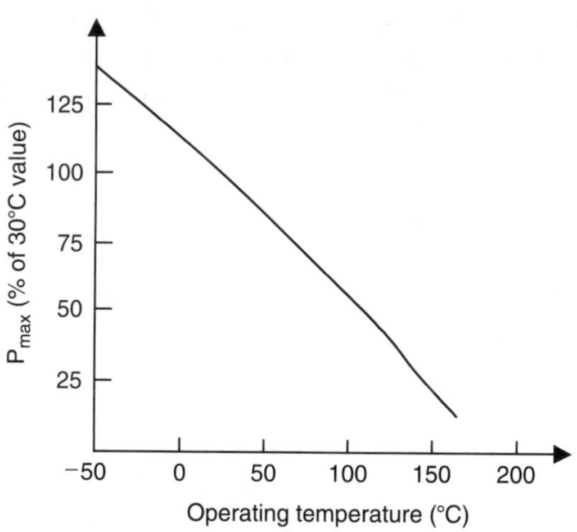

FIGURE 8.22 Power generation versus operating temperature of silicon cell.

back face, and with a 10°C gradient. The temperature is calculated from the thermal model with the heat received primarily from the sun. The solar array temperature is determined by the following thermal equation:

solar flux + Earth's albedo + Earth's thermal radiation + heat coming from adjacent components of the spacecraft = electrical power output + heat radiated back into space (8.10)

The incidence and radiated heat from the spacecraft body varies with location in the orbit as shown in Figure 8.23. The array temperature variations over one GEO and LEO orbits are shown in Figure 8.24. During an eclipse in GEO, the temperature drops exponentially to as low as −175°C. The time constant depends on the mass composition of the array component, and is typically in the 30 to 60 min range. The front to back face temperature gradient for rigid array with aluminum face sheet can be 5 to 10°C under steady sun and up to 20°C on sunlight snap after eclipse. Various techniques are used to control the temperatures of spacecraft parts. Figure 8.25 shows some passive techniques used in spacecraft thermal design.

8.9.4 Sun Acquisition

The array wings after deployment must be oriented to acquire the sun normal to the array surface. The sun sensor used for this purpose has two PV cells mounted on two 45° wedges and connected differentially in series opposition through an actuator motor as shown in Figure 8.26. When the sun is perfectly normal as in (a), the currents from both cells are equal to $I_o \cdot \cos\ 45°$. Since they are in series opposition, the net current in the motor is zero, and the array stays put. On the other hand, if the array is not normal to the sun, with an error angle δ as in (b), the sun angles on the two cells are different, giving two different currents, namely

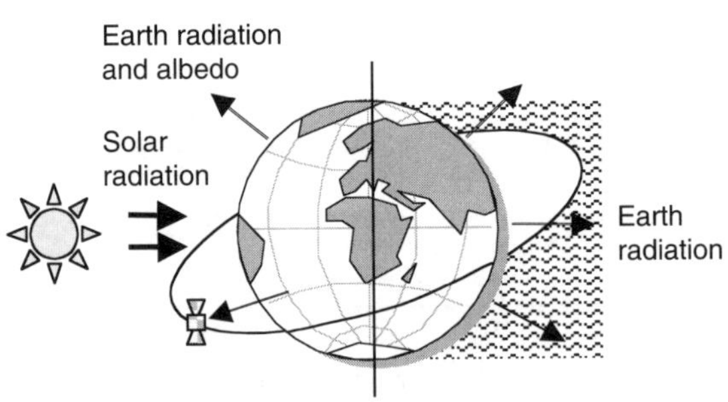

FIGURE 8.23 Heat input and output of satellite in Earth orbit.

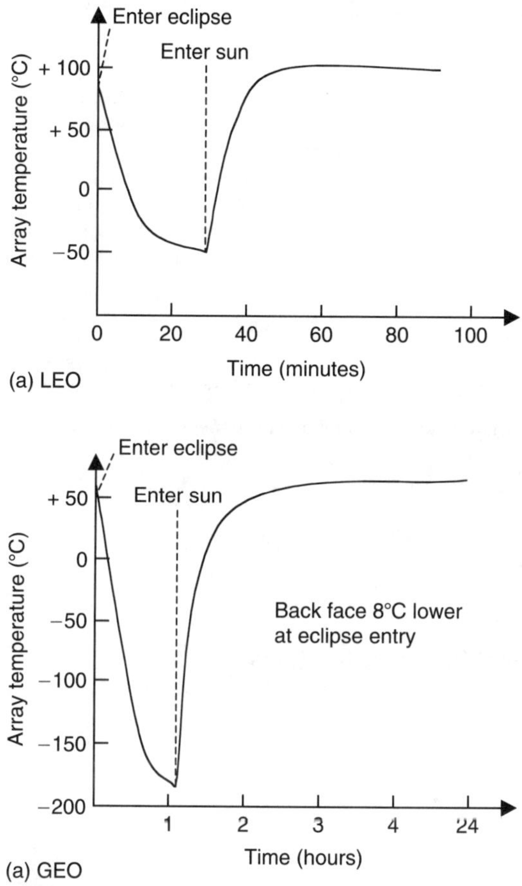

FIGURE 8.24 Solar array temperature variations over one LEO and GEO orbits.

$$I_1 = I_o \cos(45 + \delta), \text{ and } I_2 = I_o \cos(45 - \delta) \tag{8.11}$$

The motor current is therefore,

$$I_m = I_1 - I_2 = I_o \cos(45 + \delta) - I_o \cos(45 - \delta) \tag{8.12}$$

Using the Taylor series expansion

$$f(x + h) = f(x) + hf'(x) + \frac{h^2}{2!} f''(x) + \cdots$$

We can express the two currents as

$$I_1 = I_o \cos 45 - I_o \delta \sin 45,$$

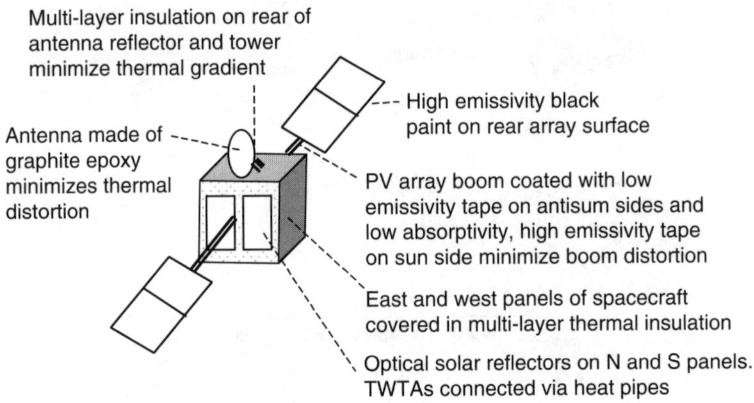

FIGURE 8.25 Thermal control on satellite components by passive techniques.

and

$$I_2 = I_o \cos 45 + I_o \delta \sin 45 \tag{8.13}$$

The motor current in Figure 8.26(c) is then

$$I_m = I_1 - I_2 = 2I_o\delta \sin 45^\circ = \sqrt{2}I_o\delta \tag{8.14}$$

if δ is in radians.

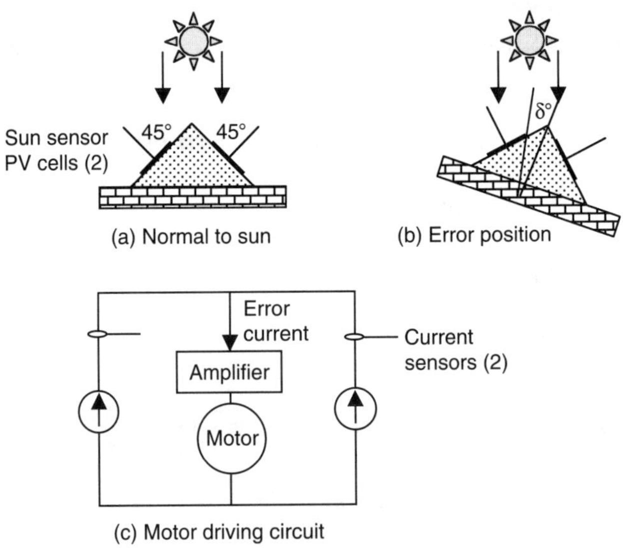

FIGURE 8.26 Sun pointing actuator principle.

The motor turns towards the sun until the current is zero, and then it remains there. At that position, the error angle, δ, is zero, i.e., the array is normal to the sun. The two current sensors shown in (c) work as the sunlight/eclipse binary detectors also.

As in the ISS, sun sensors are generally mounted on the main solar array body itself. They may indicate that the solar array is off pointed when in fact the array is shadowed from the sun, or the sun sensors themselves are shadowed, or when the solar array is feathered in order to avoid the rocket exhaust plumes during rendezvous with the space shuttle or any other vehicle.

8.9.5 Sun Tracking

Sun tracking is required for the solar array to face the sun continuously as the spacecraft orbits the Earth. This is done by an actuator that follows the sun like a sunflower. There are two types of sun trackers:

- One-axis gimbals, which follows the sun from east to west during the day.
- Two-axis gimbals which tracks the sun from east to west during the day, and from north to south during the seasons of the year. The dual-axis tracking is done by two linear actuator motors, which aim the sun within 1° of accuracy (Figure 8.27).

The EPS provides a means of independently orienting and rotating the deployed north and south solar wings about the pitch axis of the spacecraft. After the deployment, the ground control is involved in aligning the solar array to the sun. Once aligned, the array rotates by the clock time without ground intervention to maintain the sun orientation.

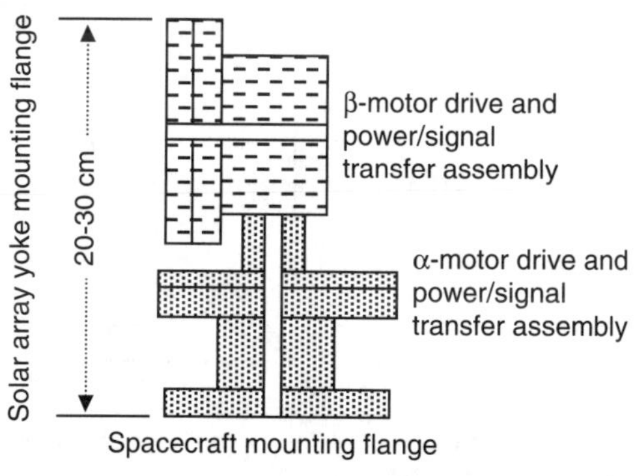

FIGURE 8.27 Dual-axis gimbals.

Some sun pointing error cannot be avoided even after acquiring the normal sun by the sun sensor and then tracking by the SAD. The error generally comes from the cell flatness error of about 2°, and the gimbal tolerance error of about 3°. The total 5° error must be accounted for in the array design. Cos 5° = 0.996, which means 5° sun pointing error will reduce the power generation by 0.4%.

8.9.6 Peak Power Extraction

The gimbal motor drives the module to face the sun for collecting the maximum solar flux. However, that in itself does not guarantee the maximum power output from the module. As was seen earlier, the module must electrically operate at the voltage that corresponds to the peak power point under the given operating conditions. First we examine the electrical principle of the peak power tracking operation.

Some loads such as heaters have constant resistance, with power varying with the voltage squared. On the other hand, most communications loads in satellites are constant power loads, drawing more current at lower voltage.

The operating point of any power system is the intersection of the source line and the load line. If the source having the *I–V* and *P–V* characteristics shown in Figure 8.28 is supplying power to the resistive load R_1, it will operate at point A_1. If the load resistance increases to R_2 or R_3, the operating point moves to A_2 or A_3, respectively. The maximum power is extracted from the module when the load resistance is R_2. Such load matching with the source is always necessary for maximum power extraction from the PV module.

Some loads such as heaters have constant resistance, with power varying with the voltage squared. On the other hand, most communications loads in satellites are constant power loads, drawing more current at lower voltage.

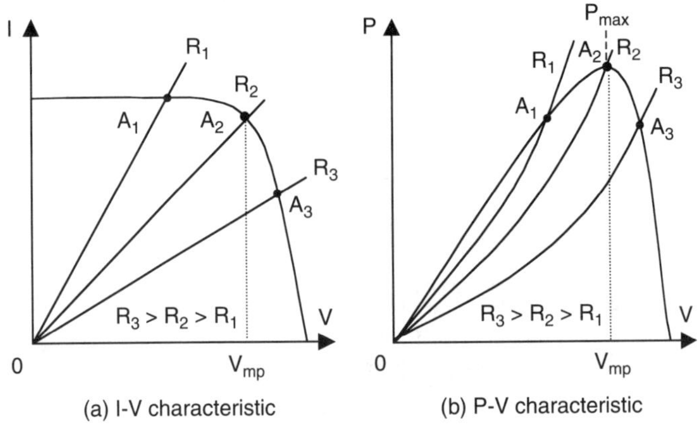

(a) I-V characteristic (b) P-V characteristic

FIGURE 8.28 Operation and load matching of PV source with constant resistance load.

The operation with constant power loads P_1 and P_2 is shown in Figure 8.29. The load line P_2 has two points of intersection with the source line, denoted by B_1 and B_2. Only point B_2 is stable, as any perturbation from it will generate a restoring power to take the operation back to B_2. Therefore, the system will operate at B_2.

The necessary condition for the electrically stable operation of the solar array is

$$\left[\frac{\mathrm{d}P}{\mathrm{d}V}\right]_{\text{load}} > \left[\frac{\mathrm{d}P}{\mathrm{d}V}\right]_{\text{source}} \tag{8.15}$$

If the array is operating at voltage V and current I on the I–V curve, the power generation is $P = VI$ watts. If the operation moves away from the above point, such that the current is now $I + \triangle I$, and the voltage is $V + \triangle V$, the new power is

$$P + \Delta P = (V + \Delta V)(I + \Delta I) \tag{8.16}$$

After ignoring a small term, Equation 8.16 simplifies to

$$\Delta P = \Delta VI + \Delta IV \tag{8.17}$$

The $\triangle P$ should be zero at the peak power point, which necessarily lies on a locally flat neighborhood. Therefore, at the peak power point, the above expression in the limit becomes

$$\frac{\mathrm{d}V}{\mathrm{d}I} = -\frac{V}{I} \tag{8.18}$$

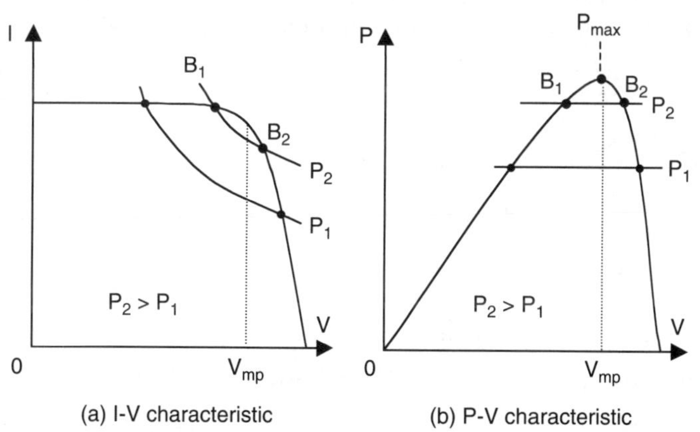

FIGURE 8.29 Operation and load matching of PV source with constant power load.

We take note here that dV/dI is the dynamic impedance of the source, and V/I is the static impedance. Various electrical methods of extracting the peak power from the module are now described below.

8.9.6.1 Impedance method

In the first method, a small signal current is periodically injected into the array bus and the dynamic bus impedance $Z_d = dV/dI$ and the static bus impedance $Z_s = V/I$ are measured. The operating voltage is then increased or decreased until $Z_d = -Z_s$.

8.9.6.2 Power gradient method

In another method, the operating voltage is increased as long as dP/dV is positive. That is, the voltage is increased as long as we get more power. If dP/dV is sensed negative, the operating voltage is decreased. The voltage is retained if the dP/dV is near zero within a preset dead band.

8.9.6.3 Voltage ratio method

The third method makes use of the fact that for most PV cells, the ratio of the voltage at the maximum power point to the open circuit voltage (i.e., V_{mp}/V_{oc}) is approximately constant, say K. For example, K equals 0.72 for high quality crystalline silicon cells. An unloaded cell is installed on the array and kept in the same environment as the power-producing module, and its open circuit voltage is measured. The operating voltage of the power-producing array is continuously adjusted at KV_{oc} to produce the maximum power.

8.9.6.4 Current regression method

Yet another method uses statistical least-squared error regression curve fitting of numerous test data on a given cell. For example, in extensive testing of GaAs cells under various illumination (50 to 2500 W/m^2) over a rather wide temperature range (−120°C to 140°C), JPL found that I_{sc} varies linearly with illumination I_o according to the following expression:

$$I_o = a_o + a_1 T + a_2 I_{sc} \tag{8.19}$$

where T is the temperature in degrees centigrade. Furthermore, V_{oc} varies linearly with log I_o.

8.9.6.5 Power regression method

The JPL tests also found that the PPT algorithm could also be set up as

$$P_{\max} = b_o + b_1 T + b_2 I_o + b_3 T I_o W/m^2 \tag{8.20}$$

This method can be used at least as a back-up for the software routine. Gluck[5] has published test results with values of all *a* and *b* constants in the above equations for numerous GaAs cells tested at JPL.

8.9.7 Shadow Effect

The array consists of many parallel strings of series connected cells. One such string is shown in Figure 8.30. A large array may be partially shadowed due to a structure interfering with the sun line. If a cell in a long string becomes completely shadowed, it loses the photovoltage but still must carry the current by virtue of its being in series with other fully operational cells in the string. Without internally generated voltage, it cannot produce power. Instead, it acts as a load, producing the local I^2R loss and heat. The remaining cells in the string must work at higher voltage to make up the loss of the shadowed cell voltage. Higher voltage in healthy cells means lower string current as per the *I–V* characteristic of the string. This is shown in the middle dotted *I–V* curve in Figure 8.30(b). The current loss is not proportional to the shadowed area, and may go unnoticed for mild shadow on a small area. However, if more cells are shadowed beyond the critical limit, the *I–V* curve is reduced below the operating voltage of the string, making the string current fall to zero, losing all power of the string.

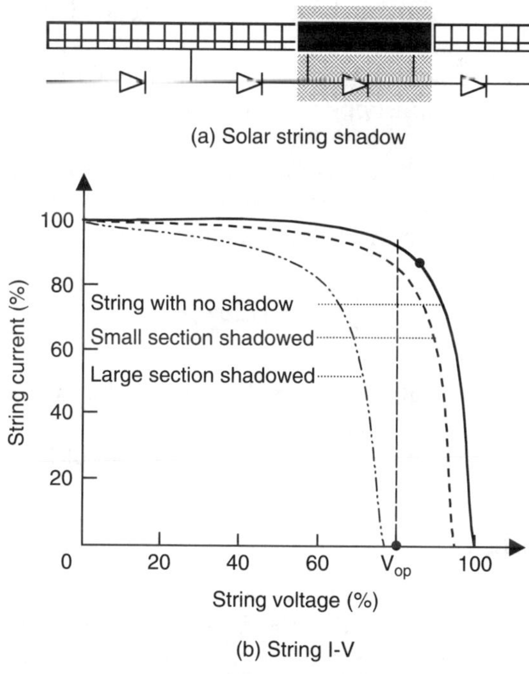

FIGURE 8.30 Shadow effect on a long PV string of solar array.

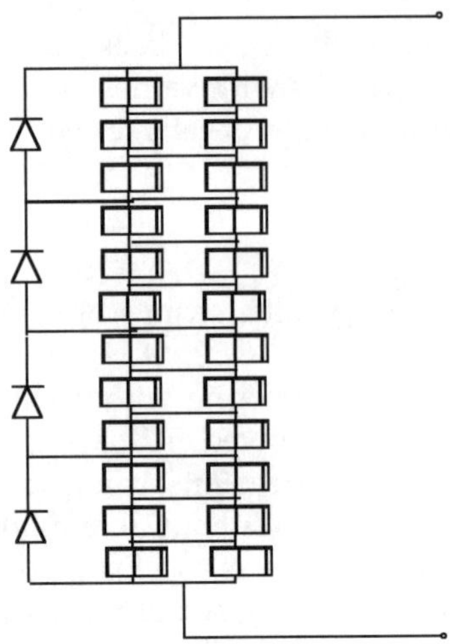

FIGURE 8.31 Bypass diodes in PV string to minimize power loss under heavy shadow.

The commonly used method of eliminating the loss of a string due to the shadow effect is to subdivide the string length into several segments with bypass diodes (Figure 8.31). The diode across the shadowed segment bypasses only that segment of the string. This causes a proportionate loss of the string voltage and current without losing the whole string power. Some modern PV cells come with such internally embedded bypass diodes to eliminate the need of separate diodes on the back of the panel. Cross-strapping cells in adjacent strings, also shown in the figure, is often used to improve the reliability against open cell, but is not related with the shadow effect mitigation.

If part of the array is shadowed for part of the orbit, the decrease in incident energy during the sunlight portion of the worst-case orbit may need to be accounted for in the energy balance. It is given by

$$E_{sun} = \frac{1}{T_{sun}} \sum_i (1 - \text{shadow}_i) \cdot \cos\theta_i \cdot \Delta t \tag{8.21}$$

where T_{sun} = sunlight duration, θ_i = angle between the solar vector and the array normal at time step i, shadow$_i$ = fraction of the solar array surface shadowed at time step i, and Δt = time step interval of the analysis.

8.10 Array Design

At the end of life, the array must generate required power equal to the sum of all load, the power losses and the battery charge power with some margin. The usual end of life solar power margin in the worst case with any one failure in the power system is typically 3 to 7% depending on the maturity of the load estimate. The solar array in LEO must recharge the battery in a short time during the sunlight. As a result, the ratio of the needed array power to the load power is approximately 2.0 in LEO compared to about 1.1 in GEO satellites.

The PV cells on the array are segmented in many strings connected in parallel as shown in Figure 8.32. The following relations approximately hold true with a suitable margin in the array segmentation. The margin varies as the design progresses from the conceptual stage to the final design.

$$N_{\text{series cells per string}} = \frac{\text{array voltage}}{V_{mp}} \tag{8.22}$$

$$N_{\text{strings per circuit}} = \frac{I_{\text{circuit}}}{I_{mp}} \tag{8.23}$$

$$N_{\text{parallel circuits}} = \frac{\text{array current}}{I_{\text{circuit}}} \tag{8.24}$$

where N denotes a number, and V_{mp} and I_{mp} are the cell voltage and cell current at the maximum power point, respectively. Each circuit delivers the output current to the bus via dedicated isolation diode and sliprings. One extra circuit is provided for single point failure redundancy. The number of power slip rings is thus equal to the number of parallel circuits. The segmentation is done such that the circuit current is usually below 5 A, which is the derated current capability of some standard SAD slip rings.

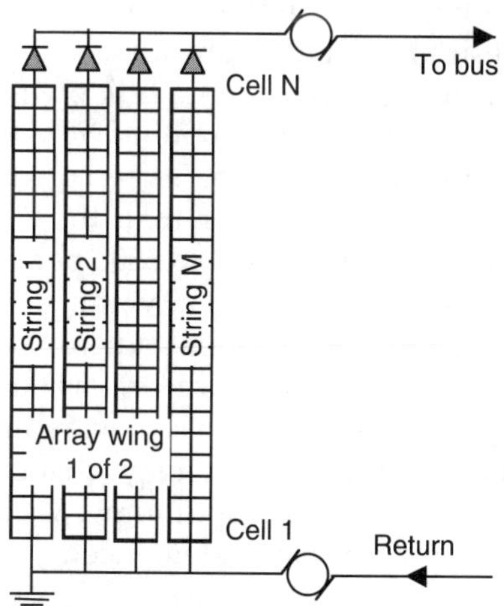

FIGURE 8.32 Array segmentation in strings with isolation diodes for each slip ring.

The solar array substrate is grounded using a separate slip ring and the signal slip rings are kept separate from the power slip rings.

Substrate materials are selected to meet the space environment over the mission life. Differential coefficient of thermal expansion between various array materials is important to avoid thermal cycling of the array under thermo-elastic stresses. Otherwise, the cell mounting on the substrate may delaminate, leading to premature failure. Mechanically, large solar array wings having output power of a few kilowatts are often required to have the natural frequency exceeding 20 Hz in the stowed configuration. In the deployed configuration, the typical requirement is 0.10 Hz in the bending mode and 0.5 Hz in the torsion mode. This may have some bearing on the cell and substrate selection.

The following factors directly or indirectly influence the total number of cells required in the array design. Some of them are discussed in this section, while others are covered elsewhere in the book:

- Environmental radiation
- Coverglass protection
- Degradation factors
- Operating temperature
- Temperature coefficients for I_{sc} and V_{oc}
- Cell size suitable for manufacturing
- Magnetic dipole moment minimization
- Power shunting method

As for the PV cell type, the silicon cell has met the industry need for decades. GaAs/Ge has been used in the past where needed for high power or high radiation resistance, although mainly in defense spacecraft. Compared to the silicon cell, the GaAs cell gives approximately 40% more power per square meter and 20% more power per kilogram with rigid panels on an aluminum honeycomb. The material cost of the GaAs cell is, however, an order of magnitude higher than the silicon cell. The space-worthy rigid array cost in 2002 was $400 to $500 per watt using Si cells and about three times as much using GaAs. In volume and/or weight limited power systems, GaAs may find applications where high cost can be justified from other considerations. Recently, however, various other PV cells, such as the double and triple junction cells, have made inroads into spacecraft power systems design. And the technology change will continue. The responsibility for selecting the most optimum PV cell at the time of the design remains with the power systems engineer.

8.10.1 Degradation Factors

Degradations in the solar array performance come from the following two distinct groups of factors: (a) initial degradations at BOL due to manufacturing imperfections and tolerances, and (b) accumulated degradation over time up to the EOL due to various environment effects.

(1) Initial degradations at BOL
- Cell mismatch in the assembly
- Cell-to-cell wiring loss
- Array wiring loss over panels and boom
- Plasma effect, causes leakage current from the array to space

(2) Accumulated degradation over time to EOL
- Ionized particles, cause damage due to cumulative radiation dose
- Ultraviolet radiation, changes the optical properties of the coverglass and the cell
- Thermo-elastic stress cycles, cause soldered joints to crack over time
- Micrometeoroids and debris impacts, reduce power generating area of cells and damage conductors, particularly at the solder joints
- Flue gases, change optical properties of coverglass and cell
- Bypass diode failure, loses a sub-string reducing the string current

Table 8.2 gives representative values of major degradation factors. The effect can be represented by the equation:

$$\text{total degradation} = \text{BOL degradation} + \text{in-service degradation to EOL} \tag{8.25}$$

8.10.2 Radiation Damage

Different charged particles in the space environment have different mass and velocity, resulting in significantly different energy levels. Heavy particles confine damage to the vicinity of the particle track, hence doing more damage for a given energy. Different particles having the same energy have different damaging effect on I_{sc} and V_{oc}. Therefore, their damaging effect is converted into the equivalent of 1 MeV of energy dose. The sum is expressed as, say, 2×10^{15} e/cm^2, where the letter *e* is called *electron fluence* but it really represents 1 MeV energy. Therefore, the fluence levels are different for I_{sc} and V_{oc}, and the higher of the two applies to the P_{max} point.

One way to approximately account for the P_{max} degradation is to take the product of the I_{sc} and V_{oc} degradation. However, such approximation in

Table 8.2 Degradation factors for a 900 W satellite in MEO 10,900 nautical miles, 69° inclination orbit, 15 years

I_{sc} factors	
• Natural radiation	Note a
• Assembly mismatch loss ^{bol}	0.98
• Coverglass charge particles	0.99
• Coverglass coating (ITO) ^{bol}	0.98
• Ultraviolet rays	0.97
• Propellant contamination	0.98
• Micrometeoroid damage	0.98
V_{oc} factors	
• Natural radiation	Note a
• Coverglass charge particles	0.99
P_{max} factors	
• Natural radiation	Note b
• Wiring loss (cell-to-cell) ^{bol}	0.98
Panel wiring voltage drop $0.05\sqrt{P^{bol}}$	1.50 V

a. To be determined from the radiation fluence over the mission life.
b. Approximate value equal to the product of the I_{sc} and V_{oc} degradation factors can be used in top-level calculations only. In detailed calculations, it is accounted through the I_{sc} and V_{oc} factors.

some situations may not be accurate. For the silicon cell, the equivalent fluence for V_{oc} and $P_{\max}$ is same, i.e., V_{oc} and $P_{\max}$ degradations are identical. But for GaAs cells, the equivalent fluence for V_{oc} and $P_{\max}$ are different, i.e., different particles degrade the GaAs cells differently with respect to V_{oc} and $P_{\max}$.

The degradation curves are established by testing numerous cells at various fluence levels. The most common source of degradation factors is NASA/JPL Handbooks, such as PUB 82-69 were originally published with data on silicon cells. The GaAs data were added in the early 1990s. The triple junction cell degradation curves have been published by organizations such as Aerospace Corporation. Some representative degradation curves for silicon cells are shown in Figures 8.33 through 8.37. The radiation not only affects the I_{sc} and V_{oc}, but also their temperature coefficients as shown in Figure 8.38 and Figure 8.39. The I_{sc} coefficient values in Figure 8.38 are for a 4 × 6 cell. Divide them by 3 for a 2 × 4 cell.

The coverglass is used to protect the cell from radiation damage. Its thickness is increased as long as the saving in the array mass is more than the added mass in the coverglass. This gives an optimum coverglass thickness for a minimum array mass including the coverglass. For example, estimated values for natural radiation in GPS orbit with various coverglass thicknesses on GaAs/Ge cells are given in Table 8.3. The parametric curves are generated at various coverglass thicknesses for candidate PV cells

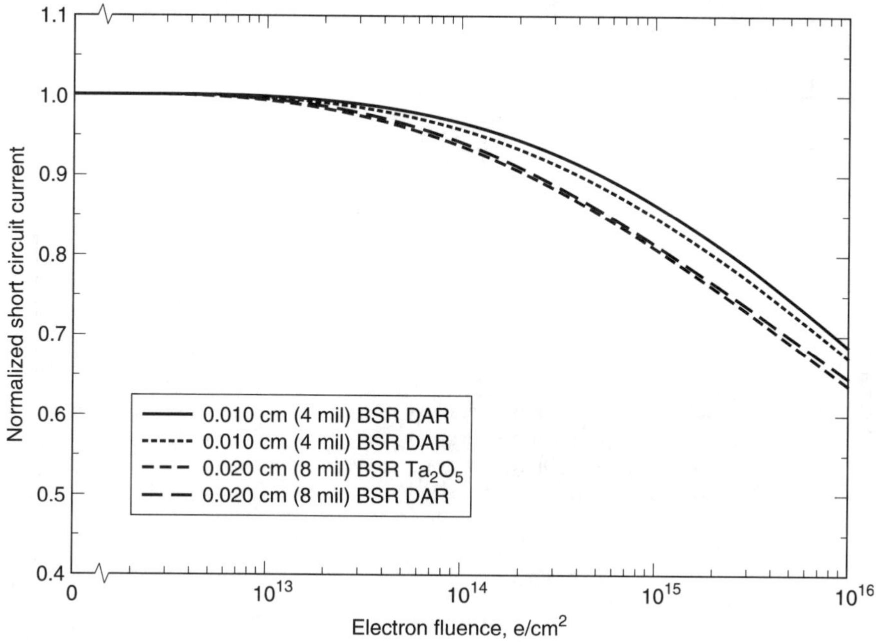

FIGURE 8.33 Normalized I_{sc} versus equivalent 1-MeV/cm^2 fluence for 2-Ω·cm silicon cell. (Source: NASA/JPL.)

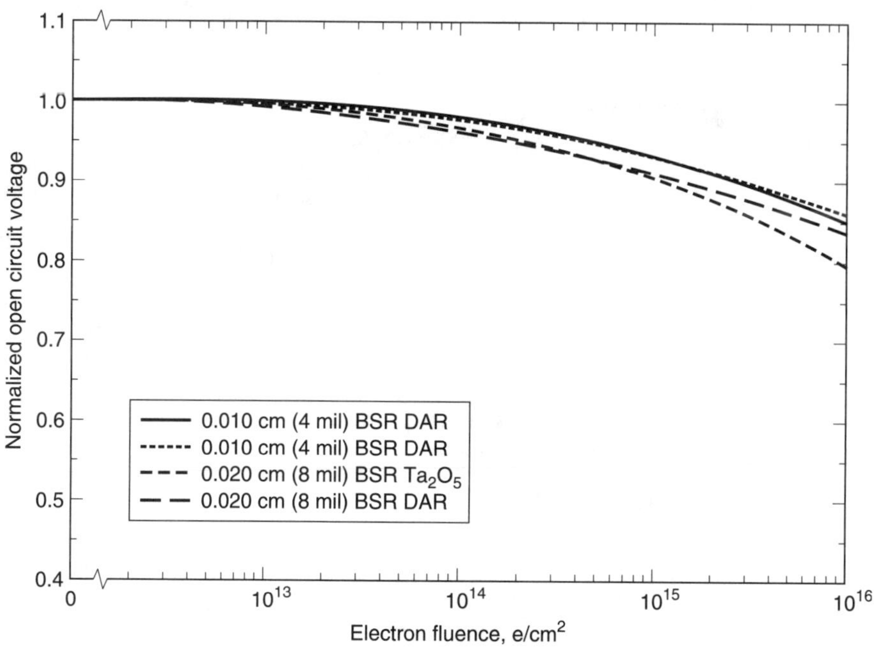

FIGURE 8.34 Normalized V_{oc} versus equivalent 1-MeV fluence for 2-Ω·cm silicon cell. (Source: NASA/JPL.)

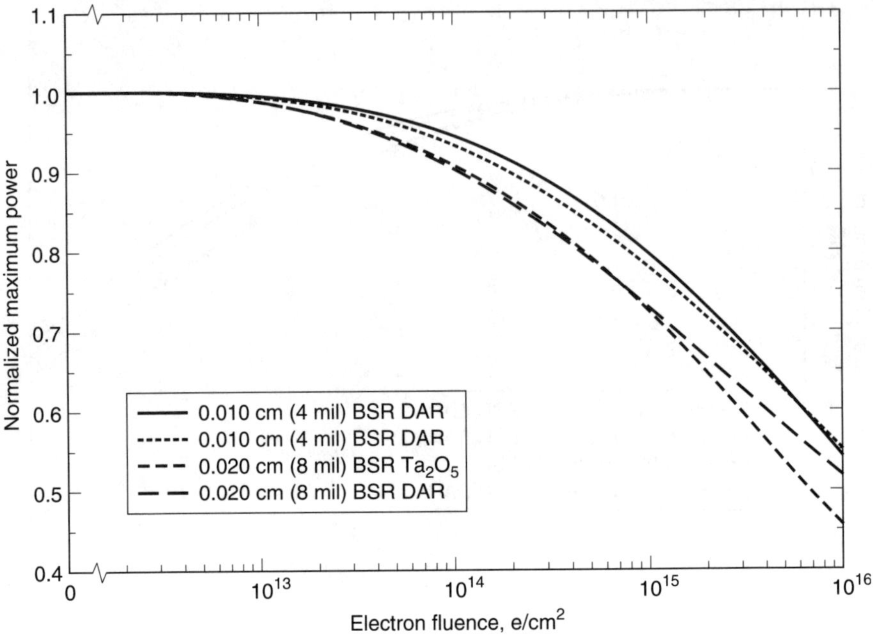

FIGURE 8.35 Normalized P_{max} versus equivalent 1-MeV fluence for 2-Ω·cm silicon cell. (Source: NASA/JPL.)

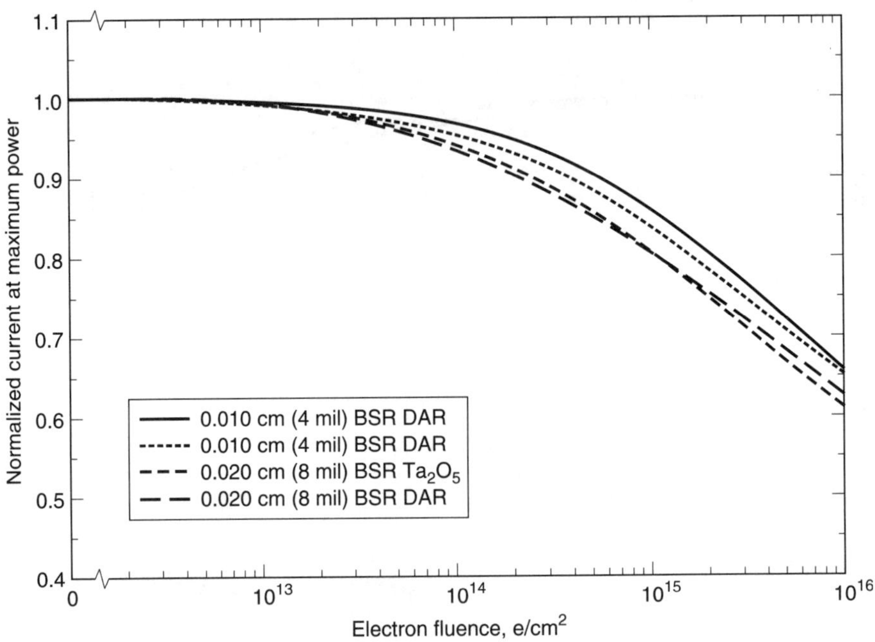

FIGURE 8.36 Normalized I_{mp} versus equivalent 1-MeV fluence for 2-Ω·cm silicon cell. (Source: NASA/JPL.)

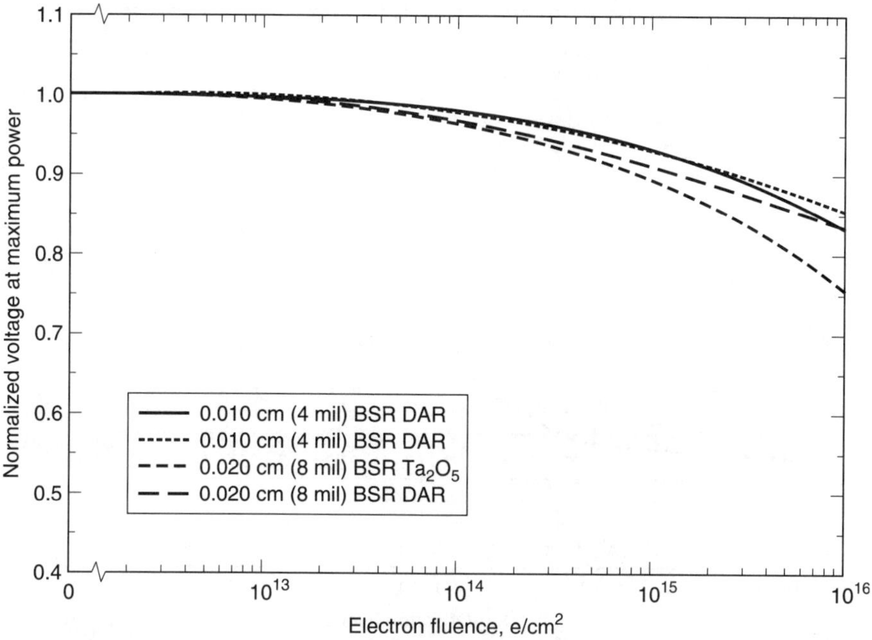

FIGURE 8.37 Normalized V_{mp} versus equivalent 1-MeV fluence for 2-Ω·cm silicon cell. (Source: NASA/JPL.)

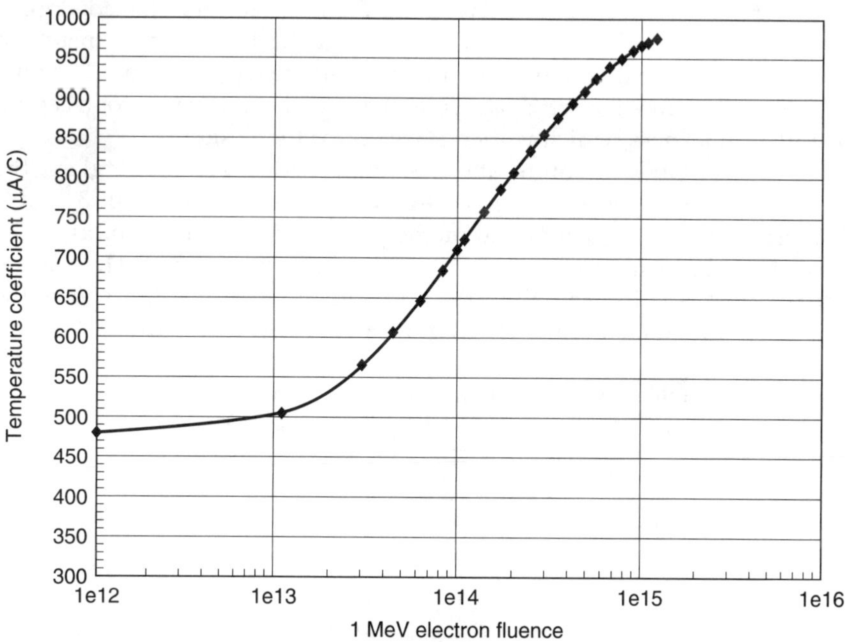

FIGURE 8.38 Temperature coefficient of I_{sc} versus equivalent 1-MeV fluence for 2-Ω·cm 4 × 6-cm silicon cell: divide by 3 for a 2 × 4 cell. (Source: NASA/JPL.)

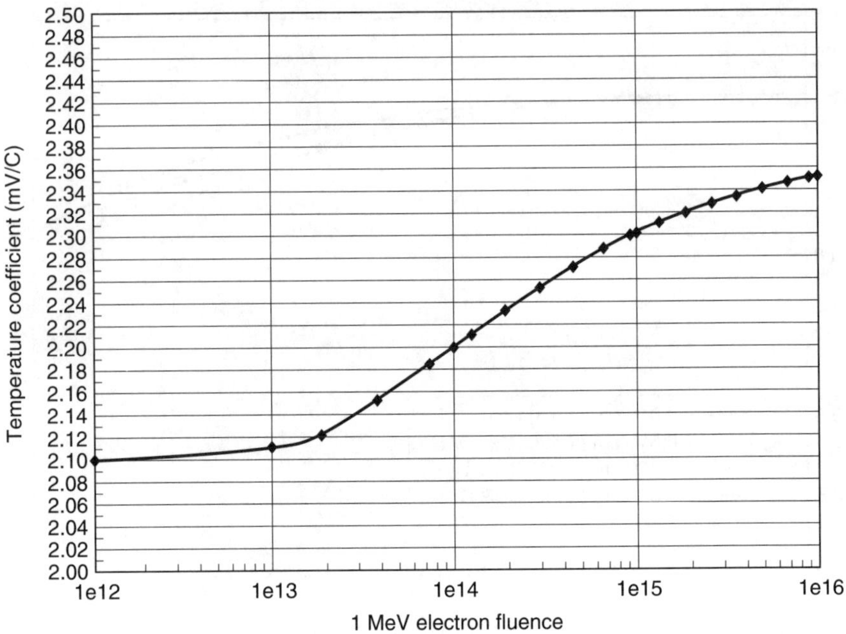

FIGURE 8.39 Temperature coefficient of V_{oc} versus equivalent 1-MeV fluence for 2-Ω·cm silicon cell. (Source: NASA/JPL.)

(silicon or any other) as displayed in Figure 8.40. As the coverglass thickness is increased, the coverglass mass increases but the array mass and coverglass decrease due to reduced fluence with better protection. The total array mass therefore takes the parabolic shape with a minimum for a certain coverglass thickness, which is then selected for the design.

In a typical GEO communications satellite with typical coverglass, the natural radiation fluence on silicon cells with coverglass is about 3×10^{13} MeV per year. Under such radiation, the power generation capability of the array degrades with the service life as shown in Figure 8.41. Natural radiation varies with orbit altitude and inclination. Figure 8.42 shows the variation in radiation and EOL/BOL power ratio required with two types of

Table 8.3 Natural radiation in GPS orbit[a] (1 MeV for GaAs/Ge cells of 0.075 g/cm^2 thickness)

Coverglass (mm)	I_{sc}	V_{oc}	P_{max}
0.10	9.0×10^{14}	9.7×10^{14}	9.7×10^{14}
0.20	7.2×10^{14}	7.3×10^{14}	7.3×10^{14}
0.30	6.2×10^{14}	6.2×10^{14}	6.2×10^{14}
0.50	4.8×10^{14}	4.8×10^{14}	4.8×10^{14}
1.00	3.0×10^{14}	3.0×10^{14}	3.0×10^{14}

[a]Nuclear radiation on I_{sc}, V_{oc} and P_{max} are to be added.

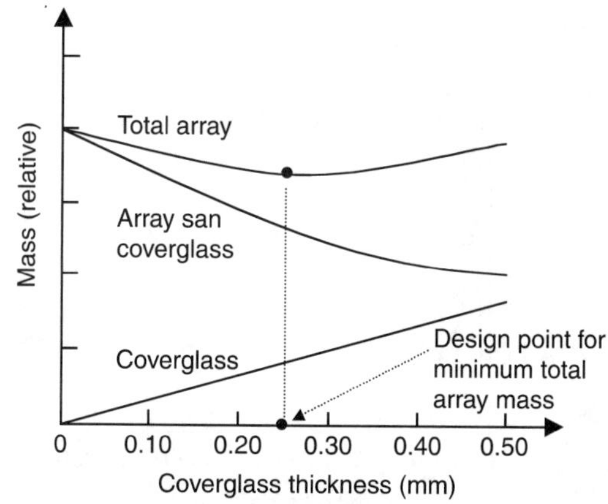

FIGURE 8.40 Optimum coverglass thickness for radiation protection results in minimum array mass.

traditional silicon cells designated by K4 and K6 in a 45° inclined orbit at various altitudes.

8.10.3 Fluence Calculation

The following parameters generally enter the total radiation fluence calculations:

- Orbit altitude, inclination, and duration of the mission
- Trapped electron model, such as NASA's AE7
- Trapped proton model, such as NASA AP8

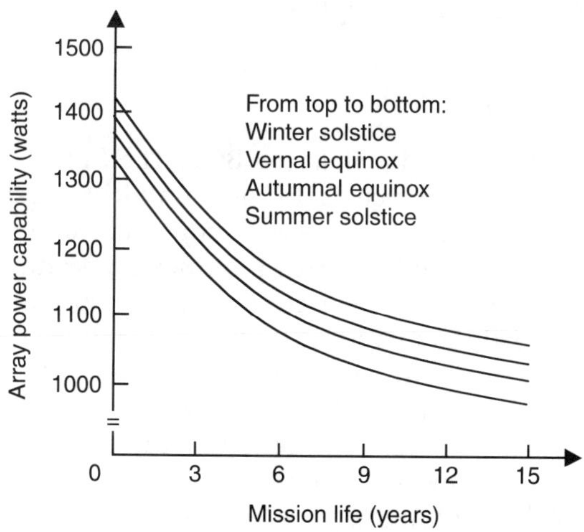

FIGURE 8.41 Solar array power capability degradation versus years in GEO service. (Si cell with 6 mil coverglass.)

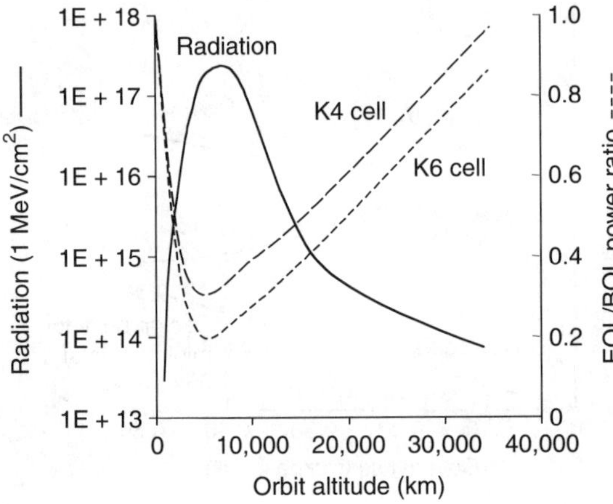

FIGURE 8.42 Radiation fluence and power degradation versus orbit altitude for 3-year mission in 45° inclination orbit.

- Solar flare model as specified by the customer
- Fission electrons and neutrons from nuclear threat, if specified
- Solar array cross section with material thickness and densities in the front and in the back of the cell junction. The surface mass density of all the front and back shielding is then converted to the total surface mass density in g/cm^2. The face sheets are included, but the honeycomb contribution is neglected.

The radiation damage equivalent to 1 MeV is then calculated for each design option, such as the PV cell type and the coverglass thickness, as shown in Table 8.4 for various options.

8.10.4 Offsetting from Sun

Excess power is always available at the BOL. The full load power is often not needed for a long time, such as during on-orbit storage. We may expect that offsetting the array from the normal sun could minimize the ionized radiation damage and extend the array life. Such expectation is based on the less than 90° angle of incidence of charge particle on the array. However, offsetting does not help in this regard, because the direction of the charged particles does not matter in their degradation effect. The fluence is like a cloud of homogeneous gas or liquid in which the array is immersed, particularly in GEO and MEO. The immersed area is what really matters. In LEO, however, there are some indications that the direction perhaps matters, but probably not much.

If the inboard panels are kept stowed for several years until they are needed later in the mission, the radiation damage on the stowed panels

Table 8.4 Radiation fluence on I_{sc} with one group of options[a]

Particle	Front thickness	Back thickness	Total fluence
Proton	7.08×10^{10}	0.00	7.08×10^{10}
Electrons	8.30×10^{14}	2.65×10^{14}	1.09×10^{15}
Flare proton	2.49×10^{14}	6.90×10^{13}	3.18×10^{14}
Flare alpha	1.22×10^{13}	8.30×10^{11}	1.30×10^{13}
Nuclear fission electron and neutron	TBD	TBD	TBD
TOTAL	1.09×10^{15}	3.35×10^{14}	1.42×10^{15}

TOTAL summary: equivalent fluence for $I_{sc} = 1.42 \times 10^{15}$ MeV/cm^2.
From similar table, equivalent fluence for $V_{oc} =$ .... MeV/cm^2.
Equivalent fluence for $P_{max} =$ .... MeV/cm^2.
Nuclear fluence of 3.2×10^{14} to be added, same for I_{sc}, V_{oc} and P_{max}.
[a]These values are representative of I_{sc} fluence on 10 mil silicon cells (0.061 g/cm^2 thickness), 8 mil ceria doped microsheet coverglass, 10 Ω·cm substrate, on 2 x 20 mil aluminum face sheets on 1 inch honeycomb, over 14 years in GPS II orbit.

would be somewhat less. However, this reduction in the damage is insignificant, may be 1% for every 2 years of preservation. As mentioned earlier, the radiation particles are like a cloud of gas in which the array is immersed. Deployed or not deployed, the charged particles enter the array and do the damage.

8.10.5 Magnetic Moment

The satellite's magnetic moment is primarily produced by the solar array current loops and the torque coils. It is a vector defined as $M =$ (current $\times$ area of the loop) A·m^2, regardless of the shape of the loop. The vector direction follows the right hand rule. For a loop current in one plane, the moment is perpendicular to the plane. In the Earth's magnetic field, B, it produces a torque, T (in N·m), of magnitude given by the cross-product

$$T = M \times B = M \times B \sin\theta \tag{8.26}$$

The potential energy, U (in joules), stored in the loop is given by the dot product

$$U = M \cdot B = M \times B \cos\theta \tag{8.27}$$

The moment increases with current. Since current is proportional to the array area and the array area is proportional to power, the moment increases roughly as square of the power, i.e.,

$$M = k \cdot P^2 \tag{8.28}$$

where k is the constant of proportionality derived from test values in similarly built designs.

The torque direction is such that the coil (dipole) tends to orient itself parallel to the B field such as to minimize the potential energy, which is always the stable position. In doing so, if the coil turns at an angular speed, ω, the power delivered by depleting the potential energy is $T\omega$. When changing the attitude (position and direction) to maintain the desired orbit, additional fuel must be spent by the attitude control system to overcome this torque. The amount of this additional fuel is proportional to the time integral of $T\omega$ over the mission life, i.e.,

$$\Delta\text{fuel} = \int_{\text{life}} T\omega \, dt \tag{8.29}$$

The magnetic moment of the solar array is minimized by laying the string circuits such as to minimize the loop area. This is done not only to minimize the fuel requirement, but also to minimize its influence on payloads, particularly the instruments that are sensitive to magnetic field.

8.11 Advances in PV Technology

Advances in PV cell technologies continue in the following areas:

- Manufacturing cost reduction
- Efficiency improvements
- Spectrum response improvement
- Using low resistivity base
- New semiconducting materials
- Radiation resistance enhancements
- This film and multi-junction (cascade) cells
- Concentration cells, parabolic and linear
- Amorphous cells (non-crystalline glass type)

Because much of the present PV technology uses crystalline semiconductors similar to integrated circuit chips, the production cost has been high. The PV cell manufacturing process is energy intensive. Every square centimeter cell area consumes a few kilowatt hours before it faces the sun and produces the first kilowatt hours of energy. However, the manufacturing energy consumption is steadily declining with continuous implementation of new production processes. Newer processes are being developed and continuously implemented to lower the energy use per square centimeter of the finished cell. In the current manufacturing process, the front and the back of the cells are diffused separately, taking 1 to 3 h of manufacturing time. Researchers at Georgia Institute of Technology are considering rapid thermal processing technology. This process simultaneously forms the front and the back of the cell in several minutes, and the metal contacts are applied by screen printing. It not only cuts the processing time, it also lowers the temperature and energy needed for fabrication.

Recently, double junction and triple junction $GaInP_2/GaAs/Ge$ cells have made inroads in the spacecraft power systems designs.[6] The n-on-p design provides enhanced radiation hardness. This is achieved simply by deposition on a p-type Ge substrate and forming an active junction in the substrate by *in situ* arsenic diffusion. Tunnel junctions are used to link the individual Ge bottom, GaAs middle and $GaInP_2$ top cells. The *I–V* characteristic of a typical large triple junction cell is shown in Figure 8.43. Triple junction cell efficiency up to 40% has been tested in controlled laboratory conditions, and 30% in production cells used in the spacecraft, with a long-term goal of 40%. However, it has been reported that the cells

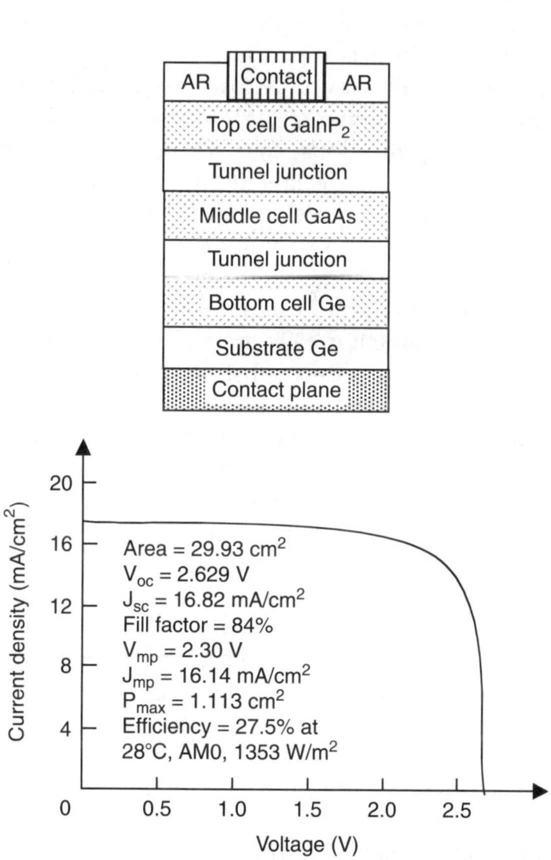

FIGURE 8.43 Cross section and *I–V* characteristic of typical large triple junction cell.

degrade faster than initially predicted. The present cost of 30% efficiency cells is 12 to 15 $/cm^2 (\$350 to \$450 per watt) at the cell level and \$800 to \$1200 per watt at the array level.

Developments are under way at JPL for a new 46% efficient cell using 40% efficiency triple junction cell in the front and 10% efficient thermo-electric cell in the back to use the waste heat. This technology, when available, may produce a new cell with a maximum theoretically possible efficiency of 47% in the laboratory and 37% in production cells.

The multi-junction c-Si cell significantly increases the efficiency, while a-Si cells drastically decrease the material cost. The disadvantage of the a-Si cell is that the cell efficiency is low to begin with, and degrades further by about 20% in the first several months under sunlight before stabilizing. However, afterwards, it suffers less radiation damage because of its very thin PV junction. Gamma rays do not cause damage while penetrating the super-thin junction. In fabricating the PV cell, better crystallivity means better efficiency, but also greater radiation damage. Even slight damage to the well-aligned crystalline structure degrades the power generation. The random atomic structure in the amorphous cell cannot become misaligned and there is no reduction in performance.

Amorphous silicon has an extremely low production cost, is inherently radiation hard and performs better at high temperature. Such cells in large volume for terrestrial applications cost \$3 per watt, with the annual production capacity in tens of megawatts at present. For space applications, their downside is lower efficiency compared to the c-Si cells. Multi-junction thin-film a-Si deposited on a thin flexible substrate is an emerging technology that may find space applications. The thin film cell is a few microns thick and is supplied on aluminum sheets each with a thickness of a few mils. The triple junction cells are produced by plasma-enhanced chemical vapor deposited on a thin substrate with a textured silver/zinc oxide back reflector. The top cell uses a-Si alloy of about 1.8 eV band gap to capture the blue photons, the middle cell uses an a-SiGe alloy of 1.6 eV band gap to capture the green photons, and the bottom cell uses a-SiGe alloy of 1.4 eV band gap to capture the red photons. The sunlight, which is not absorbed, is scattered back from the back surface at an oblique angle to facilitate multiple passes. Indium tin oxide is deposited onto the cells, which serves the dual purpose of ESD control and antireflective coating. The top silver grid collects the current. Assuming 9% efficiency, Guha[7] estimates that the specific power of the cell with 0.5-mil thick stainless steel substrate is about 770 W/kg, and about 2750 W/kg with 1 mil thick Kapton substrate. The triple-junction a-Si cell efficiency of 13% has been achieved in small cells under AM1.5 spectrum prevalent on the Earth. Since the AM0 spectrum in space is much richer in the blue wavelength region, the component thickness in the triple-junction cells are optimized differently. In space optimized cells under AM0 radiation tested in NASA GRC laboratory, 0.27 cm^2 cells have shown 12% efficiency, and large 11 cm^2 cells have shown 10.8% efficiency, all at BOL after the initial stabilization.

In space optimized a-Si cells, Guha also reports the following:

- Temperature degradation in power in the 20 to 70°C range is found to be between −0.2 and −0.3% per°C. This is about one half of that in the conventional crystalline silicon cells. For this reason, a-Si is expected to perform better at the high temperatures typically found in LEO and GEO.
- Radiation degradation is lower if the cells are annealed at moderate temperature. At 70°C, the power degradation was 5% after a dose of 1×10^{15} MeV electrons/cm^2 and 16% after a dose of 1×10^{16} MeV protons/cm^2 after annealing at 110°C. Crystalline Si cells would lose most power under such doses.
- No change in performance was found after more than 33,000 thermal cycles between +150°C and −110°C at a rate of 80°C/min, nor after exposure to atomic oxygen of 1.3×10^{20} atoms/cm^2.

Table 8.5 due to Guha compares the estimated EOL performance of the thin-film triple junction a-Si cells with the conventional c-Si cells. The most significant benefit of the a-Si cell is its low cost and low mass.

To drive the cost down further, scientists at the Swiss Federation of Technology have recently developed a titanium dioxide cell with 33% efficiency which, when fully developed, may cost about the same as a comparable thickness of glass.

A new cell on the horizon is the thin film copper–indium–selenide cells bonded on thin lightweight substrate of flexible sheets that are kept flat prior to deployment. The shape-memory springs would unfold the array

Table 8.5 Comparison between amorphous and crystalline silicon cells for space applications

Performance	Crystalline	Amorphous
BOL efficiency	14%	9%
Temperature effect	−20%	−8%
Radiation degradation	−25%	−3%
EOL efficiency	8%	8%
Cost	High	Low
Mass saving in 7 kW array	Reference	−50 kg
Interconnect	Reference	Much less
Stiffness	Rigid	Flexible
Technology status	Proven	Being tested

(Source: S. Guha et al. Proc. 34th IECEC, p. 2553, ASE, 1999.)

frame into panels. The specific power of such panels can be up to 100 W/kg at the array level.

Very large arrays generating 100 kW to 1 MW power for lunar or Mars colonies may become possible using thin film cells on a monolithic polymer substrate at a cost of about $100 per watt and specific power of 1000 W/kg. For conventional spacecraft applications, thin films must have at least 15% efficiency on monolithic polymer substrates to make them mass and cost competitive. A thin film means a few micrometers ($< 5\ \mu m$) thickness using any semiconducting material, but mostly in amorphous or polycrystalline single or multi-junction form. Their terrestrial versions use a glass substrate, but the cells for space may use aluminum, stainless steel, or polyimide substrates. The thin film cells are replacement cells (not monolithic) that come in a few × few inches rectangular sizes (e.g., 4 × 6 inches). At present, the specific power of the thin film cells is about twice that of silicon cells. The radiation resistance of these cells is inherently high because of both the thin film and also the randomly aligned non- or poly-crystalline structure, so the charged particle cannot damage the structure uniformly.

Table 8.6 summarizes major PV technologies on the market today.

The Naval Research Laboratory and NASA Glenn have incorporated a suitcase-size solar cell experiment on the ISS. It is designed to monitor and relay the voltage, current and temperature data on a variety of the present and emerging PV cells for 1 year. The primary emphasis of the experiments is on evaluating the long-term performance of current and future multi-junction solar cell technologies.

Table 8.6 Comparison of various PV cell technologies (typical representative values or relative values)

Cell	Efficiency BOL	W/m^2	\$/W[a] BOL	Radiation hardness	CIC[b] kg/m^2	Maturity
Silicon	12–14%	150	100	Low	1	Heritage
GaAs/Ge	18–19%	225	300	Med	1	Heritage
GaInP2/GaAs/Ge	24–26%	300	300	Med	1	New in use
Amorphous Si	8%	80	<< 100	High	0.1	Not ready
CuInSe	10%	135	<< 100	High	0.1	In Development
X 100 Suns	32–34%					GaAs tandem
X 150 Suns	30–32%					GaInP2/GaAs tandem
X 1000 Suns	36–38%					GaInP2/GaAs/Ge triple junction

[a]Rough order of magnitude costs in 2004.
[b]CIC = Cell + Interconnect + Coverglass.

8.12 Concentrator Array

In an attempt to improve the conversion efficiency, the sunlight is concentrated into tens or hundreds of times the normal sun intensity by focusing on a small area using low cost lenses. The primary advantage is that such a cell requires a small fraction of active area compared to the standard cell. This significantly reduces the PV material requirement, although the total energy collection area remains the same. Besides increasing the power and reducing the size or number of active cells, such cells have the additional advantage that the cell efficiency increases under concentrated light up to a point. For example, Krut reports that a GaInP/GaAs dual junction cell efficiency of 29% under one sun increased to 30% under two suns and 34% under 100 to 500 suns. The efficiency improves provided the cells are kept cool by radiating heat into space. Krut cites the following concerns about the introduction of concentrating solar panels in spacecraft:

- Need for major changes in conventional flat panel stowage and deployment methods
- Increases sun-tracking because of the optics dictating narrow acceptance angles
- Lack of space heritage of some concentrator material
- Higher operating temperature of cells, and the resulting drop in cell efficiency
- The technical problems, cost and schedule risk associated with any new technology.

Figure 8.44 is the technical data sheet for a GaAs concentrator cell. The concentrator silicon cell is already being used, whereas use of the GaAs cell is lagging behind, partly because a higher concentration is needed for peak efficiency. The conversion efficiency peaks at tens of suns in silicon cells, whereas it is at hundreds of suns in GaAs.[9] For solar power satellites discussed in Chapter 27, high concentrator GaAs cells would be needed with a concentrator factor over 1000, perhaps in the range of 5000 to 10,000, requiring extremely small area cells. It is easier to produce a high efficiency cell of small area than to produce large area cells with comparable efficiency. On the other hand, the major disadvantage of the concentrator cells is that they require focusing optics, which results in added cost.

Two options of concentration cells are depicted in Figure 8.45. In the dome shaped concentrator, the active cell lays in the center of an aluminum cup covered with typically a 15-mil transparent silicon dome. Since the dome lens produces high level of concentration, even slight misalignments due to a small pointing error can result in significant loss of power. The thin film inflatable array with dome concentrator is under development for

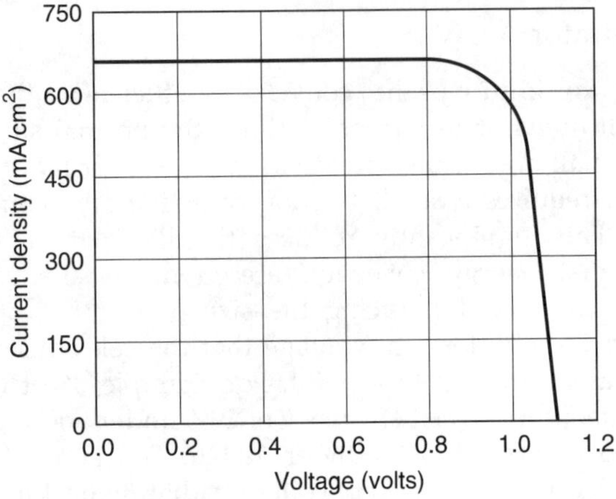

Temperature coefficients

I_{sc} = +24.0 microamperes/cm^2
V_{mp} = −1.9 millivolts/°C
V_{oc} = −1.8 millivolts/°C
P_{mp} = −3.9 mw/cm^2/°C

Radiation degradation*

Parameter	**1 × 10^{13}**	**1 × 10^{14}**	**1 × 10^{15}**	**3 × 10^{15}**
I_{sc}/I_{sc0}	0.99	0.95	0.84	0.69
I_{mp}/I_{mp0}	0.99	0.95	0.83	0.67
V_{mp}/V_{mp0}	0.98	0.94	0.90	0.87
V_{oc}/V_{oc0}	0.98	0.95	0.90	0.87
P/P_0	0.98	0.90	0.74	0.58

***Fluence e/cm^2 1 MeV electrons**

Thermal properties

Alpha (spherical) = 0.82 (CMX)
Emittance (normal) = System dependent

FIGURE 8.44 GaAs concentrator cell performance data: Typical 22% efficiency at 20 × Suns.

space antennas and is being considered for extending into power systems, particularly for high power electrical propulsion.[10] The line concentrator concept, on the other hand, produce relatively less concentration, but is less sensitive to the pointing error.

Concentrator solar cells can dramatically reduce the cell cost, which is 35 to 45% in conventional arrays. However, two key issues seriously affect the cell design at high concentrations. One is the minimization of the series resistance, and the other is more active cooling of the array to remove the energy not converted into electricity. The mounting structure can be a flexible cloth or rigid panel made of a honeycomb structure. The rigid panel approach takes away some mass benefit of the flexible blanket array, but

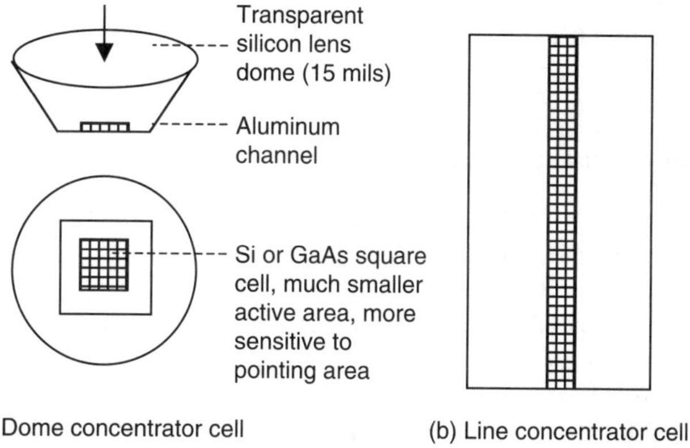

FIGURE 8.45 Solar concentrator cell construction.

offers well-proven deployment and ability of placing PV cells on the outer most panel for the transfer orbit power.

The primary advantage of the concentrator technology is low cost per watt, which is achieved by replacing the active solar cell area with less expensive optical surface. A secondary advantage come from the fact that the enclosed cell is naturally protected from radiation and space plasma. This makes it suitable for high voltage applications. Tests have shown the dielectric withstand capability of 1000 V with 400 V operational capability. A specific power of 200 W/kg at the cell level has been achieved. Concentration ratios of 3 to 10 have been considered for space and around 100 for ground-based applications. In designing such an array, some areas of concerns have been:

- Degradation of optics.
- Thermal control of the solar cells under concentrated sunlight, particularly on exiting the eclipse. However, the thermal gradients would be relatively small because of the thin structure.
- Error in pointing the array to the sun. The greater the concentration, the more power loss even with a small pointing error.

The concentration cells have limited benefits in traditional space applications for two reasons. The reduced active area saves some PV material cost, but the pointing error may offset the benefit. Moreover, the energy collection area remains the same, hence the same substrate mass. A concentration ratio above 10 results in a diminishing rate of return for space use. The potential space application is in a high voltage array operating around 500 V or higher. Its low leakage current at high voltage offers significant benefits, such as in electric propulsion.

The solar concentration technology falls into three main groups described below.

8.12.1 Stretched Lens Array

Stretched lens array (SLA) technology uses a stretched silicon membrane Fresnel lens to concentrate sunlight onto the PV cell, usually triple junction. The line-focused silicon lens provides 8- to 10-fold concentration onto a radiatively cooled PV cell. The lens is made of 200 μm silicon laminated on 75 μm ceria glass for support and UV radiation protection. The laminated lens is supported as a stretched membrane between end arches as shown in Figure 8.46. The PV cells are attached to 190 μm composite sheet supported as a stretched membrane between two edge elements. The composite has high thermal conductivity to keep the PV cells cool by radiation. The lens frames are foldable for launch in the same manner as in a planar solar wing. The entire solar array wing has a mass density of about 1.5 kg/m^2, which includes lenses, PV cells flex circuit, wiring, composite panels, hinges, yoke, and the deployment mechanisms. NASA Glenn has tested 27.4% power conversion efficiency based on the total panel area, yielding about 375 W/m^2 at 25°C. At GEO operating temperature of 80°C, the yield could be 300 W/m^2, or about 200 W/kg. The pointing error tolerance is approximately ±2° about the critical axis.

Among the advantages of SLA technology are:

- High efficiency, hence high specific power
- The line focus concentrator offers low cost mass-production of lens, as the active PV cell area is now 1/8th compared to the conventional solar array

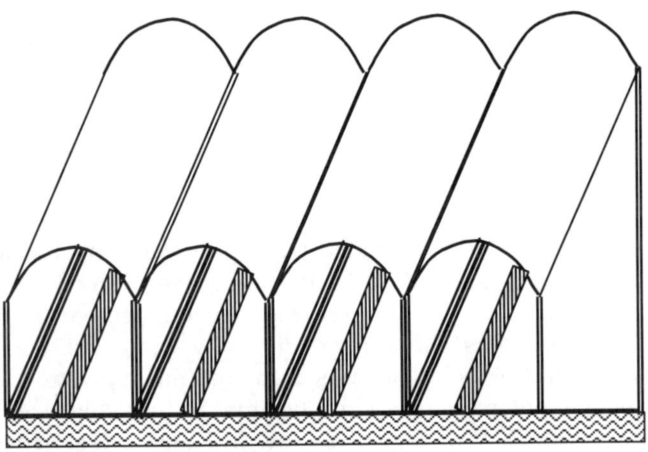

FIGURE 8.46 Stretched lens array prototype panel. (Source: NASA/GRC.)

- Relaxation of array alignment to only one axis
- It provides inherent protection of the active PV cells against charged particles, hence suitable for defense satellites in high radiation orbits or with nuclear threats
- Ability to withstand high level of space plasma. It is demonstrated to withstand cell voltage excursion of 500 V relative to the plasmas with minimal environmental interaction
- Minimized plasma interaction when biased to high voltage offers a potential use in high voltage electric propulsion around 500 V

The GaInP/GaAs on Ge substrate multi-junction cell is being increasingly used in SLA due to its high efficiency and manufacturability. The commercial dual and triple junction cells have 28°C AM0 conversion efficiency of about 30% at one sun, and even higher at concentration ratio of several suns. Figure 8.47 shows the *I–V* characteristic of an 8.5 × 24 cm lens aperture cell. The efficiency improvements come from the improved spectrum response. On the down side, keeping the operating temperature low for power output maximization requires effective heat spreading and thermal radiation of the concentrated energy into space. Another drawback is a potential loss of power generation capability under pointing error. However, it is noteworthy that such cells have soft angular dependence shown in Figure 8.48 and Figure 8.49 as reported by Krut and Karam.[6,8] The test module delivered over 97% of peak power at ±8° in the α plane (soft angle) and 97% at ±4° in the β plane (hard angle). Thus, the cell is capable of tolerating such pointing errors without significant loss in power generation.

The solar concentrator array with refractive linear element technology (SCARLET) is pursued by the Ballistic Defense Organization and NASA Glenn. The experiments flew first in Advanced PV and Electronics Experiment (APEX), and then again in 1995 on Comet/Meteor spacecraft

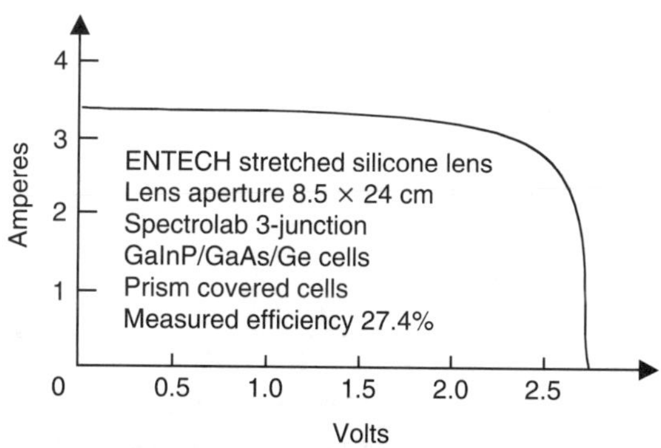

FIGURE 8.47 SLA module *I–V* test results at AM0. (Source: NASA/GRC.)

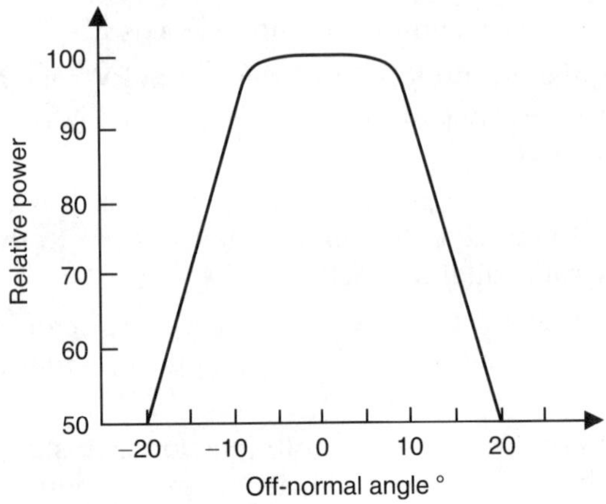

FIGURE 8.48 Off-normal performance of 10× crystalline silicon cell in α-plane (easy angle plane).

with both linear and parabolic concentration array with GaAs/Ge cells. The results are reported by Jones et al.[11] This technology is suitable particularly for high voltage arrays. It was also tested on NASA/JPL's Deep Space 1 spacecraft to provide 2500 W power to the spacecraft and the ion propulsion engine. It demonstrated 200 W/m^2 and 45 W/kg specific power[12] with a dome lens.

8.12.2 Light Concentrating Panel

In light concentrating panel (LCP) technology, a pair of optically flat panels in a channel shape reflect sunlight onto the PV cells as shown in Figure

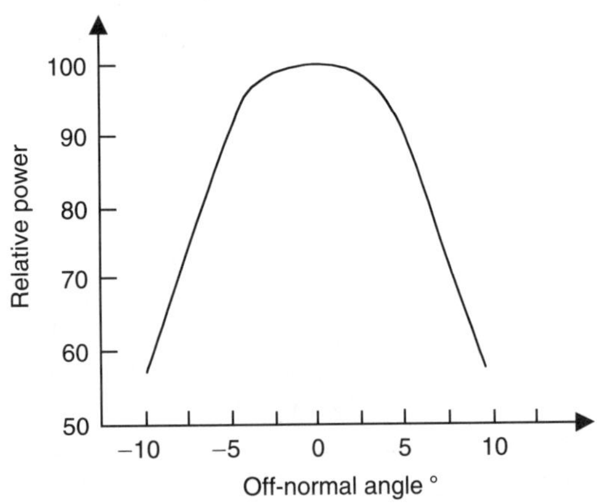

FIGURE 8.49 Off-normal performance of 10× crystalline silicon cell in β-plane (hard angle plane).

8.50.[13] The concentration ratio, which can be achieved using such reflectors, has an upper limit of 2.75. The reflector panels can be rigid aluminum or fluorinated polyamide film. In the latter, the reflective surface is achieved by vacuum sputtering aluminum onto a polyamide substrate. The latter is more advantageous for mass reduction. The reflectors roll up on themselves and are stowed besides the folded PV panels, leaving one panel exposed to generate power during the transfer orbit. A system of pulleys and cables synchronizes the panel deployment and maintains the support booms parallel to each other. Springs in the support booms keep the reflectors under tension after the PV panels are deployed. The booms at every other panel supports the reflectors and tensions them. The ends of the booms can rotate past the perpendicular to the PV panels, making the reflectors illuminate the entire length of the PV array even at extreme diurnal motion of the sun in geosynchronous satellites. There are no shadows on the cells even when the array β angle is not zero. A panel length of at least twice its width gives the maximum packing factor. The reflector flatness is critically

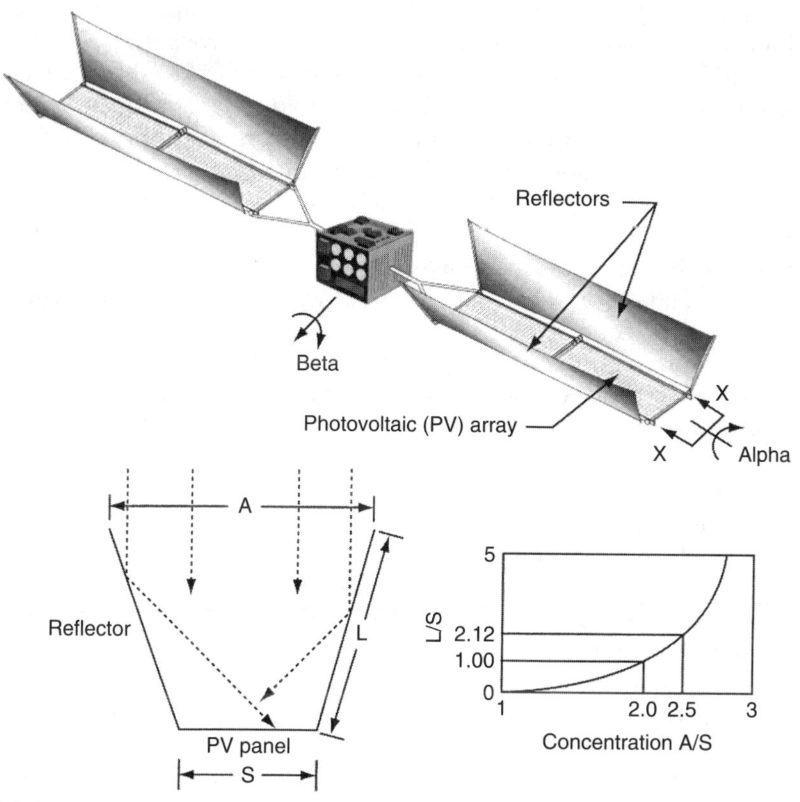

FIGURE 8.50 Light concentrating panel assembly.
(From M.A. Brown and J. Moore, *Proc. 36th IECEC*, ASME, p. AT-38, 2001. With permission.)

important to avoid local hot spots. It is estimated that such a channel concentrator design with polyamide film reflectors can yield specific power of 180 W/kg using 30% efficient PV cells.

LCP technology was demonstrated on *Mightysat II* launched in July 2000.[14] Its design concentrates sunlight using lightweight composite mirror strips forming an orthogrid of rectangular mirror troughs. Light with concentration ratio of about 2.75 to 1 is reflected onto standard solar cells. The cells are mounted on a high-conductivity flat composite laminate with an integrated Kapton film. The mirror and cells are mounted on a composite base sheet having high thermal conductivity to conduct the heat away. The PV cells interconnect to a flexible wiring layer integrated into Kapton film.

The box-shape array provides the needed structural rigidity, replacing the honeycomb structure in traditional planar panels. The active area with PV cells is now reduced to one third. Stern and Bonebright estimate that the solar panel cost can be reduced by 40% by using this technology.[14]

The original solar array design in Boeing's BSS-702 bus was the LCP type. It was used to generate maximum power from an array stowed in a given fairing. The design used angled solar reflector panels along both sides of the wing with GaInP2/GaAs/Ge PV cells mounted on a graphite face sheet on an aluminum honeycomb substrate. After all the tie-rods are released, the panels unfold like an accordion and the wings are deployed in a sequence shown in Figure 8.51. The reflectors then unfold in a coordinated manner. The BSS-702 concentrators were essentially mirrors fixed on both sides of the solar panels at an angle giving the panel a distinctive U-shape, reflecting the sunlight, and generating about 30% more power. However, Boeing detected premature power loss experienced by several BSS-702 satellites launched in 2000 and 2001. The average loss was 6% per year. The rate of decline depended on how the satellite was used. The problem was suspected in the concentrators' vulnerability to contamination by gaseous emissions from the spacecraft, gradually darkening the mirror surface. While the exact cause of the trouble remained unclear, it was determined that the xenon-ion propulsion system used on the spacecraft was not the cause. However, the LPC arrays were replaced with a traditional planar design. To boost the power without using concentrators, Boeing switched to solar panels with triple junction, instead of double junction cells, and adding panels to the arrays.

The root cause of the continuous linear degradation of power in the BSS-702 channel array was orally presented at the IECEC 2003 conference in Portsmouth, VA. It has been determined to be contamination with a volatile condensable material, coupled with the concentrator geometry. The two factors jointly reduced the reflectance, reducing the transmission of illumination, ultimately leading to mechanical failure due to additional shrinkage. The contamination was due to out-gassing *accumulated* on the channel panels. Planar arrays do not *accumulate* and also operate at lower temperature. Moreover, multi-junction cells are more sensitive to such

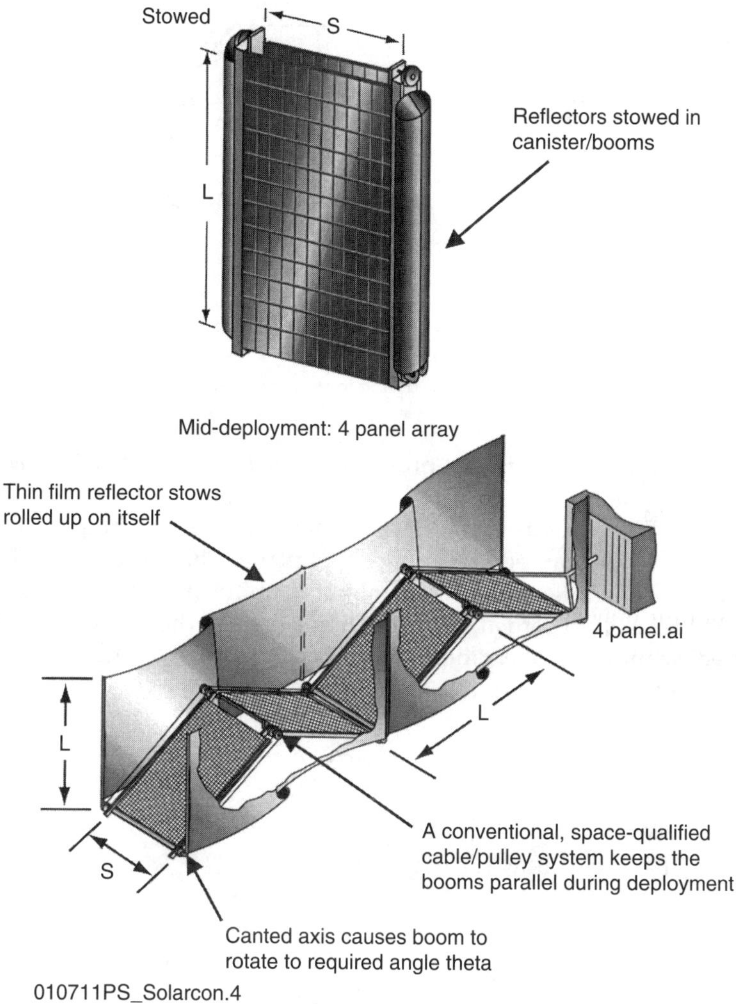

FIGURE 8.51 Light concentrating panel stowed and being deployment.
(From M.A. Brown and J. Moore, *Proc. 36th IECEC*, ASME, p. AT-38, 2001. With permission.)

contamination. Tests have eliminated possibilities of ion thruster and UV darkening, for example.

8.12.3 Parabolic Concentrator

In this technology, which is shown in Figure 8.52, the solar energy is concentrated onto a small area by using a reflective mirror in a parabola of revolution. Referring to the geometry shown in Figure 8.53, the solar ray bundle reflected from the center of the mirror form a circle on the focal

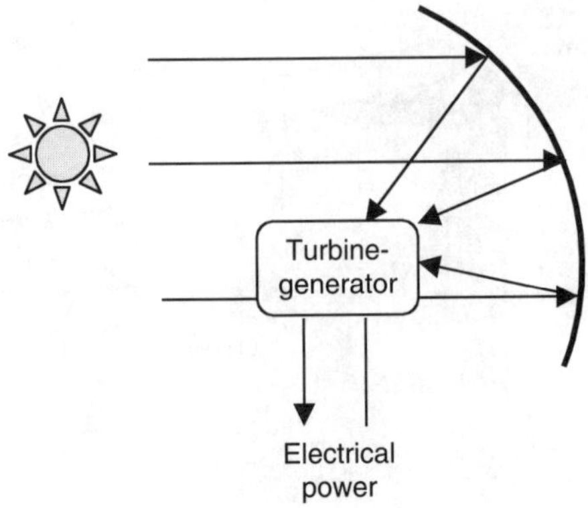

FIGURE 8.52 Parabolic concentrator for dynamic power system.

plane. The diameter of this first order image is $d_i = 2f \tan \phi$, where ϕ is the solid angle of the sun image, which is $0° \, 0' \, 32''$ from the Earth, and f is the focal length. This gives $d_i = 0.0093f$. A ray bundle reflected from the edge of the mirror will form an ellipse on the focal plane. This distortion is known as aberration. The intensity of the sunlight is uniform across the first order image, but falls off to zero at the aberration edge. All the energy collected is contained within the aberration image. The fraction of the total energy collected which is concentrated in the first order image is

$$\eta_i = \left(\frac{1 + \cos \theta}{2}\right)^2 \tag{8.30}$$

where θ is the angle shown in Figure 8.53.

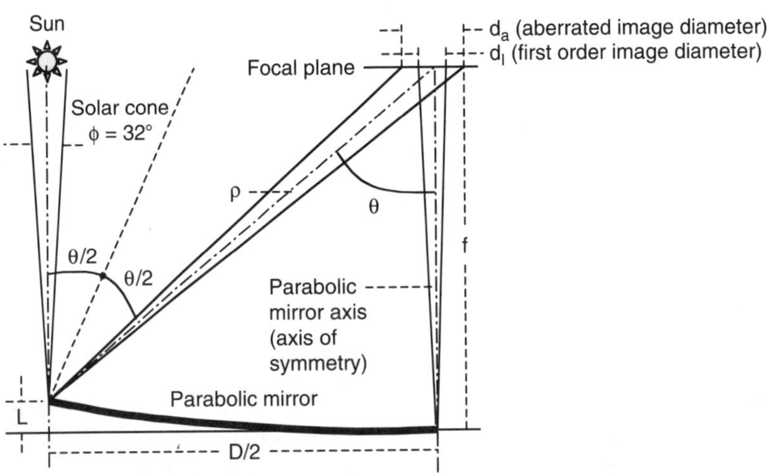

FIGURE 8.53 Parabolic concentrator geometry.

The concentration ratio, C_i, of the paraboloid is defined as

$$C_i = \frac{\text{energy density in the first order image}}{\text{energy density received from the sun}} \tag{8.31}$$

Obviously, there are two concentration ratios, one for the first order image and the other for the aberrated image. Their values can be derived as follows:

$$C_i = 46100 \ \sin^2\theta \tag{8.32}$$

$$C_o = 11550 \ (\sin 2\theta)^2 \tag{8.33}$$

where C_0 is the concentration ratio for the aberrated image.

Thus, very large concentration ratios of tens of thousand are possible in the first order image, creating the energy density of tens of megawatts per square meter in the Earth's orbit.

Due to tolerances of construction, practical concentrators depart appreciably from the above theoretical values. The deviations could be from imperfect parabola in the form of flat or over-curved spots, asymmetry or misalignment with the mirror axis. A practical mirror is often described by its index of perfection, defined as

$$\lambda_i = \frac{\text{actual } C_i}{\text{theoretical } C_i} \tag{8.34}$$

For optical quality mirrors, λ_i is near unity, whereas practical mirrors may have λ_i around 0.50 or less.

If a misalignment exists between the sun line and the mirror axis, the image moves with respect to the axis and also becomes distorted. The amount of the lateral displacement of the center of the image, or its edges, can be shown in terms of Figure 8.53 to be

$$\delta = \left(\frac{\Delta\varpi}{57.3}\right)\left(\frac{2\rho - f}{\cos\theta_m}\right) \tag{8.35}$$

where δ = displacement of the image from the zero error position and $\triangle$ = angular misalignment error in degrees.

References

1. Carlson, D.E., Recent advances in photovoltaics,in *Proceedings of the 30th Intersociety Energy Conversion Engineering Conference*, ASME, 1995, pp. 621–626.

2. Cook, G., Billman, L., and Adcock R., *Photovoltaic Fundamental,* DOE/Solar Energy Research Institute, February 1995, Report No. DE91015001.
3. Chern, P.C. and Leipold, M.H., *Stress Fiat and Proof Testing of Silicon Wafers,* Jet Propulsion Laboratory, 1987, Report No. NASA TSP-5.
4. Delleur, A.M., Kerslake T.W., and Scheiman D.A., Analysis of direct solar illumination on the backside of space station cells, in *Proceedings of the 34th Intersociety Energy Conversion Engineering Conference,* ASE, 1999, Paper No. 01-2431.
5. Gluck, P.R. and Bahrami, K.A., A simple algorithm to compute the peak power output of GaAs/Ge solar cells on the Martian surface, in *Proceedings of the 30th Intersociety Energy Conversion Engineering Conference,* ASME, 1995, pp 321–325.
6. Karam, N. H. et al. Development and characterization of high efficiency GaInP/GaAs/Ge dual and triple junction solar cells, *IEEE Transactions on Electron Devices,* July 1999, pp. 2116–2125.
7. Guha, S. et al. Low cost and light weight amorphous silicon alloy solar array for space applications, in *Proceedings of the 34th Intersociety Energy Conversion Engineering Conference,* ASE, 1999, Paper No. 2553.
8. Krut, D.D. et al. "High efficiency space concentrator with soft angular dependence and related cell technology, in *Proceedings of the 36th Intersociety Energy Conversion Engineering Conference,* ASME, 2001, Paper No. AT-90.
9. Algora, C. and Diaz, V., Influence of series resistance on guidelines for manufacturing of concentrator p-on-n GaAs solar cells,in *Progress in PV Research and Applications,* John Wiley and Sons, 2000, Vol. 8, pp. 211–225.
10. Laug, K.K. and Holmes, M.R., Paraboloidal thin-film concentrators and their use for power applications, in *Proceedings of the 34th Intersociety Energy Conversion Engineering Conference,* SAE, 1999, Paper NO. 2552.
11. Jones, P.A., Murphy, D.M., and Piszczor, M., A linear refractive PV concentrator solar array flight experiment, in *Proceedings of the 30th Intersociety Energy Conversion Engineering Conference,* ASME, 1995, Vol. I, pp. 309–314.
12. O'Neill, A. et al. The stretched lens array (SLA), an ultra light concentrator for space power, in *Proceedings of the 36th Intersociety Energy Conversion Engineering Conference,* ASME, 2001, pp. 79–84.
13. Brown, M.A. and Moore, J., Solarcon concentrator solar array, in *Proceedings of the 36th Intersociety Energy Conversion Engineering Conference,* ASME, 2001, Paper No. AT-38, pp. 75–78.
14. Stern, T.G. and Bonebright, P., Flight tests of a solar array concentrator on *Mightysat II,* in *Proceedings of the 35th Intersociety Energy Conversion Engineering Conference,* ASME, 2001, Paper AT-37, pp. 69–74.

Chapter 9

Battery

9.1 Introduction

The energy storage is required to meet the spacecraft load demand not only during eclipse, but also when the demand exceeds the power generation at any time. The most widely used energy storage technology is the battery, which stores energy in an electrochemical form. There are two basic types:

Primary battery, in which the electrochemical reaction is irreversible. It cannot be reused after full discharge, so it is discarded. It finds applications in very short duration missions as the primary source of power. It has a specific energy that is an order of magnitude higher compared to the secondary battery.

Secondary or rechargeable battery, in which the electrochemical reaction is reversible. After a discharge, it can be recharged with direct current from an external source. It finds applications in long duration missions where another independent power source is available for recharging the battery periodically. Most satellites use the secondary battery.

9.2 Electrochemical Cell

The battery is made of numerous electrochemical cells assembled in a series–parallel circuit combination to obtain the required voltage and current. The internal construction of a typical electrochemical cell is shown in Figure 9.1. It has positive and negative electrode plates with insulating separator and a chemical electrolyte in-between. The two electrode plates are connected to two external terminals mounted on the casing. The cell stores electrochemical energy at a low electrical potential. The cell voltage depends solely on the electrochemistry, and not on the physical size. Commonly used electrochemistries produces 1.5–3.5 V when fully charged. The cell capacity, denoted by C, depends on the physical size. It is defined as the ampere-hours (Ah) charge the cell can deliver at room temperature until it reaches a cut-off voltage of about $2/3^{rd}$ of the fully charged voltage. The battery can deliver C amperes for 1 h or C/n amperes for n hours. The capacity measures the Ah output at the terminals, not the input or that

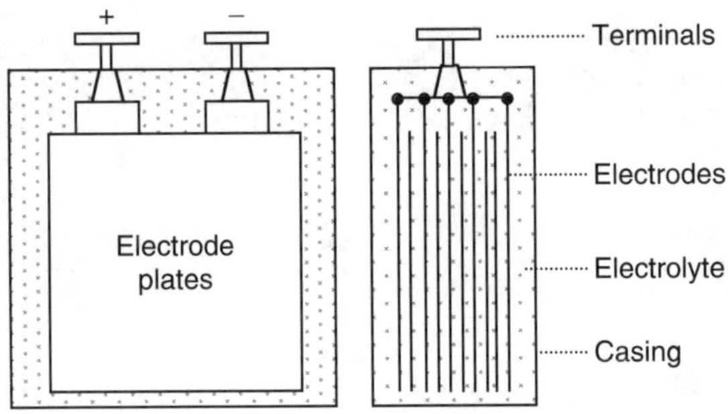

FIGURE 9.1 Typical electrochemical cell construction.

stored between the plates. A 1.5-V cell discharged to 1.0 V delivers practically full capacity of the cell. The Ah capacity to 0.1 V discharge is only a few percent greater than that at 1.0 V discharge.

The battery voltage rating is stated in terms of the average voltage during discharge. The higher the battery voltage, the higher the number of cells required in series. The product of the voltage and the Ah rating makes the energy rating in watt-hours (Wh) the battery can deliver to a load from the fully charged state.

The battery charge and discharge rates are stated in fraction of the capacity. For example, charging a 100-Ah battery at 10-A rate is said to be charging at $C/10$ rate. Discharging this battery at $C/2$ rate means drawing 50 A. At this rate the battery will be fully discharged in 2 h. The state of charge (SOC) of the battery at any instant is defined as

$$\text{SOC} = \frac{\text{Ah capacity remaining in the battery}}{\text{rated Ah capacity}} \tag{9.1}$$

The state of charge affects the cell voltage, specific gravity, and freezing point of the electrolyte. The electrolyte in a fully charged battery has high specific gravity and freezes at a much lower temperature. On the other hand, a fully discharged battery freezes at a higher temperature. This shows the importance of keeping the battery fully charged when exposed to low temperatures.

The battery depth of discharge (DOD) is defined as

$$\text{DOD} = \frac{\text{Ah capacity drained from fully charged battery}}{\text{rated Ah capacity}} \tag{9.2}$$

Obviously,

$$\text{DOD} = 1 - \text{SOC} \tag{9.3}$$

Major secondary batteries used in the spacecraft industry at present are nickel cadmium (NiCd), nickel hydrogen (NiH_2), and lithium-ion (Li-ion). New electrochemistry is being continuously researched by NASA for space applications,[1] and by the United States Advance Battery Consortium for a variety of ground based applications — consumer electronics, electric vehicles, utility load leveling, and renewable power systems. Lithium–polymer (Li-poly) and nickel metal hydride (NiMH) are two such examples in the commercial world.

The selection of electrochemistry for a given application is a matter of performance and cost optimization. The following figures of merit are often used in comparing relative performance of various electrochemistries:

Specific energy, defined as the gravimetric energy density, which is the energy stored per unit mass, Wh/kg.

Energy density, defined as the volumetric energy density, which is the energy stored per unit volume, Wh/l.

Specific power and power density, defined as the power the battery can practically deliver per kilogram of mass and liter of volume, respectively. It is sometimes necessary to think in terms of the power parameters also, since the internal resistances of the battery may limit the rate at which the energy can be discharged within practical design limits.

Cycle life, defined as the number of charge/discharge (C/D) cycles the battery can deliver while maintaining the cut-off voltage.

9.3 Types of Battery

The NiCd battery had been the workhorse of the spacecraft industry since the earliest missions, and is still used in some missions. However, since the mid-1980s, it has been replaced with the NiH_2 battery for general use. The NiH_2 provides deeper DOD for comparable cycle life, thus requiring lower Ah capacity, which translates into lighter weight. Today, the industry appears to be moving towards the lithium based batteries for potentially 2 to 5 times the specific energy compared to the NiH_2 chemistry. No single electrochemistry available on the market can meet the wide range of space mission requirements. All chemistries will perhaps continue in use where they fit the best for a minimum mass and cost design. However, the more commonly used electrochemistries and their unique construction and performance features are presented in this chapter.

9.3.1 Nickel Cadmium

The NiCd is a matured electrochemistry that powered almost all satellites until mid-1980s. It comes in a sealed vented prismatic cell with positive and negative electrode plates made of a sintered nickel structure. The positive plate is impregnated with nickel hydroxide active material and the negative plate with cadmium hydroxide active material. The plates are die-cut with corner radii and coined edges to prevent shorting. The negative plate design includes a thin Teflon™ impregnation on the surface. This feature extends the operational life by enhancing oxygen recombination and retarding cadmium migration. The positive and negative plates are separated by nonwoven nylon or asbestos, which has been replaced with Pellon 2536™. The electrolyte is potassium hydroxide and the casing is made of stainless steel.

The NiCd electrochemistry offers relatively lower specific energy, is temperature sensitive, and has shorter cycle life. Moreover, the cadmium has been placed under environmental regulatory scrutiny. For these reasons, NiH_2 has replaced the NiCd for most space applications.

9.3.1.1 Memory effect

A major disadvantage of the NiCd battery comes from its peculiar characteristic known as the *memory effect*. The NiCd battery remembers its DOD in most frequent use in the recent past, does not work well beyond that, and partially loses its unused capacity for subsequent use. For example, if a NiCd battery is repeatedly charged and discharged 25% of its capacity to point M in Figure 9.2, it *remembers* point M. If the battery is discharged beyond point M in a subsequent use, the cell voltage will drop much below its original normal value as shown by dotted line in the figure.

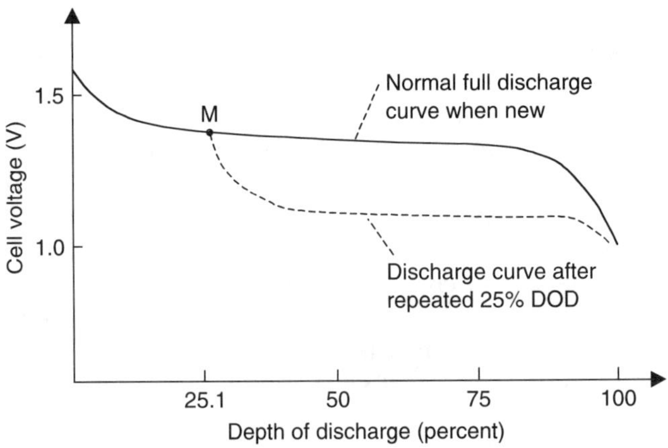

FIGURE 9.2 Memory effect in NiCd cell.

The end result is a partial loss of capacity after repeated shallow discharge cycles. The phenomenon is like losing a muscle in the human body due to lack of use over a long time.

9.3.1.2 Reconditioning

A remedy to counter the memory effect for restoring the battery to full capacity is *reconditioning*, in which the battery is fully discharged to almost zero voltage and then fully charged to about 1.55 V per cell. In the spacecraft power system, this is done twice a year at a convenient time in the orbit. In GEO, it is done outside the two eclipse seasons. In LEO, however, the spacecraft must carry an extra battery just for reconditioning purposes, adding significant mass and cost. Other types of batteries have negligible memory effect, offering a significant advantage over the NiCd battery.

A battery letdown unit is used to recondition the battery in orbit operation. It is capable of discharging the battery completely at more than one discharge rate through dedicated resistors, either at the cell level or at the battery level. In the latter case, each battery has its own battery letdown resistor. Generally the battery reconditioning starts at high rate around $C/50$. Upon the first cell reaching 0.5 V, the battery letdown unit is commanded to disconnect one of the discharge paths, and continue the battery discharge at $C/100$ rate. Upon any battery cell reaching 0.1 V, the remaining discharge path is disconnected to terminate the discharge. After the full discharge, the battery is then brought back to full charge in a normal manner. The battery cell voltage monitor provides information on each cell voltage during reconditioning and also during flight. The cell voltage data during flight is not directly useful, as little can be done if anomalies are found in flight. However, the data can be useful to the customer in the following ways:

- Predict the remaining life
- Plan replacement in serviceable spacecraft
- Reduce the load if necessary and possible
- Redesign the battery for future spacecraft

9.3.1.3 Advanced NiCd

The NiCd electrochemistry has been relatively fixed for several decades with some incremental improvements in performance in terms of life and specific energy. For example, reduction in electrode loading levels has been used in recent years to limit internal stresses that cause cell degradation. Despite a series of improvements, NiCd does not match other electrochemistries at present.

A major improvement program was undertaken under Department of Defense funding. Under this program, optimized electro-deposited nickel and cadmium electrodes, advanced separator system, and special charge control methods were developed to achieve longer life at higher DOD, and eliminate the memory effect. Such space-qualified cells are available from a few sources, one under the trade name SuperNiCd^{TM}. These cells compete well with NiH_2 at a power level of 1 to 2 kW, and can reduce the battery weight by a factor of 2 or more. Such an advanced NiCd battery gives about twice the life at a given DOD, or twice the DOD for a given life. The final result is the same — a significantly lighter battery. In the tests simulating LEO service, Sanyo's advanced NiCd battery of 35-Ah capacity survived more than 35,000 cycles at 20°C and 40% DOD.

The advanced NiCd competes with NiH_2 for some niche low power spacecraft, particularly in defense applications. It has negligible memory effect. Its cycle life for a given DOD is several-fold higher than the conventional NiCd and a few-fold higher than NiH_2.[2] It had been used in over 50 missions by the year 2000 in capacities ranging from 5 Ah to 50 Ah, typically in small satellites in LEO for NASA, the U.S. Navy, and the U.S. Air Force.[3] These batteries are not being reconditioned. The typical design DOD has been around 75% in 5- to 12-year GEO missions, and 10 to 20% for 1- to 8-year LEO missions. Advanced NiCd gives better performance than the old NiCd only below 50-Ah capacities, and matches with NiH_2 in higher capacities. The primary limitation for this capacity restriction is the heat removal difficulty across the plates. Although the charging is endothermic during initial charge, it becomes exothermic beyond 75 to 80% SOC.

9.3.2 Nickel Hydrogen

The NiH_2 has been the most widely used cell in space during the last 20 years, both in LEO and GEO satellites. Over 100 satellites carrying more than 10,000 cells have been flown. They have accumulated almost one billion cell-hours of highly reliable operation. The NiH_2 battery takes the best from two electrochemistries, the nickel oxide electrode from the NiCd and the hydrogen catalyst electrode from the fuel cell. The NiH_2 cells can withstand some abuse in over-charge and over-discharge. On the other hand, NiH_2 has the following disadvantages:

- Low energy density in Wh/l
- Pressure vessel rupture, handling, and safety concerns
- High self-discharge rate around 0.5%
- High loss of capacity on storage

In the most widely used configuration of NiH_2, each elementary cell developing 1.25 V is packed in an individual pressure vessel (IPV). Figure 9.3 depicts a simplified view of the IPV cell components. The positive electrode is made of aqueous impregnated slurry of sintered nickel. The

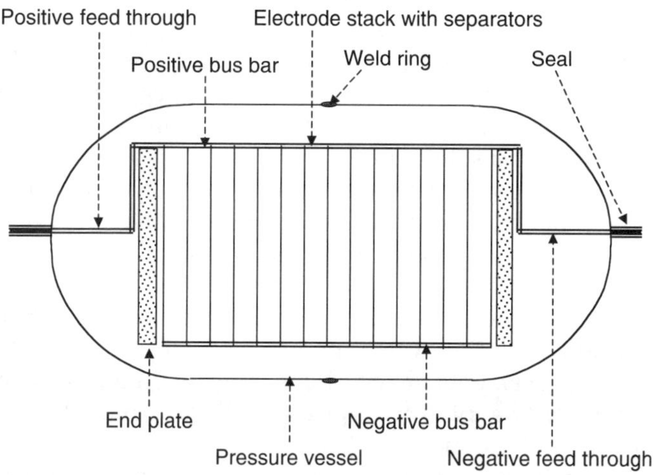

FIGURE 9.3 NiH_2 cell components — simplified view.

negative electrode is made of platinum. The electrodes are stacked in pairs with the leads extending through the center core. A separator is placed on each side of the positive electrode. The electrolyte is 26 or 31% potassium hydroxide (KOH). Other factors include ceramic to metal seals with integral terminals, double stack supports (no cantilever) for shock resistance, and zirconium separators that yield longer life. Life tests approaching 42,000 cycles have been demonstrated at 60% DOD.[4]

A plastic diffusion screen is placed on the back side of each negative electrode to diffuse gas to the back side. A number of such electrode pairs with separators are placed in one pressure vessel to obtain the required Ah capacity. The internal gas pressure rises to about 1000 psig when the cell is fully charged. To sustain such high pressure, the vessel is made of age hardened Inconel-718 with burst pressure of at least 2.5 times the maximum operating pressure at full SOC. It must comply with MIL-STD-1522 pressure vessel requirements. The vessel is made in two parts, each with an end-dome and one half of the cylindrical center body. The assembled vessel has the ring of electron beam weld in the center. Because of the high-pressure design requirement, the vessel contributes significant portion (about 40%) of the total cell mass.

Two types of positive electrode plates have been used. Both are based on a sintered nickel substrate, with one impregnated using aqueous procedure and the other using an alcohol procedure. Both behave almost similarly in all aspects. Two commonly used electrodes thicknesses are 30 mils for GEO and 35 mils for LEO applications. The 35-mil electrode gives longer wear life and slightly higher discharge voltage. However, it uses heavier electrodes and double separators, hence longer cell length and higher weight. Some manufacturers process the electrodes in aqueous solution, whereas some in alcoholic solution, although both are purged later on. The alcoholic solution gives a little better performance, but is believed to result in a capacity fading in wet storage in discharged condition. That is, when revived, the battery does not restore to the same capacity for hundreds of cycles. Hence, it needs some trickle charging in wet storage. The cell processed in aqueous solution does not pose such a problem. It comes to full life on the first recharge after wet storage for up to 3 years.

Battery technology is rapidly changing. Some of the recent improvements for increasing the specific energy in NiH_2 chemistry are described by Britton.[5] A few examples are the catalyzed wall-wicks incorporated in the KOH solution, and the replacement of traditional sintered substrate used in nickel electrodes with lightweight porous nickel felt. A positive plate is the key to all improvements in cycle life, voltage stability, and extended storage. The positive plate is typically 0.75 to 1 mm thick with active material loading of around 1.5 g/cm^3.

The cell capacity is a function of the loading level of the positive active material, KOH concentration, and temperature. The earlier NiH_2 construction used 21% KOH and a double layer zirconium separator, which was replaced by a single layer for mass reduction in GEO satellites. However, the cell for LEO applications still uses the double layer zirconium with thicker electrodes to meet the longer cycle life requirement.

The 3.5-inch diameter cell had been standard for a long time, but is inefficient for larger than 100 Ah capacity due to increased length. The 4.5-inch diameter cell is a direct scale-up of the widely used 3.5-inch diameter cell, while maintaining the stack length. The 3.5 and 4.5-inch diameter cells in 35- to 150-Ah capacity range is currently flying in many satellites. To extend the Ah range on both sides, 2.5- and 5.5-inch diameter cells have also been fully developed and are flying in a number of satellites. As seen in Figure 9.4, the specific energy varies with the ampere-hour rating and also

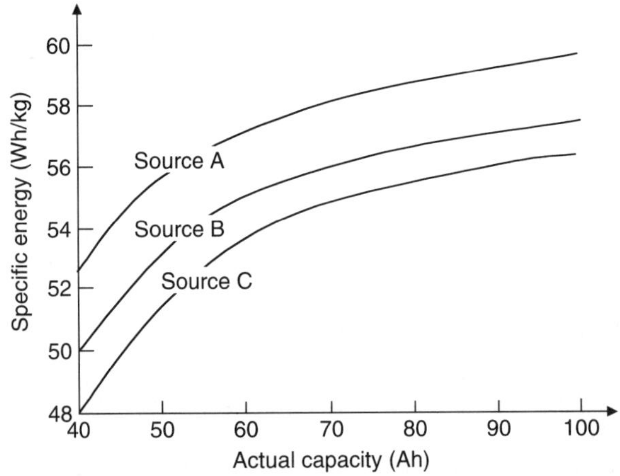

FIGURE 9.4 Specific energy of NiH_2 cells — industry range.

with design variations in the separator, electrolyte concentration, and gas management provisions such as a catalyzed wall-wick.

For the International Space Station (ISS), the NiH_2 was selected after evaluating many other options, including the regenerating fuel cell. The ISS NiH_2 cell is made of dry sintered Ni electrode, sintered Teflon-bonded platinum black H_2 electrode with Teflon backing and photo etched nickel grid, 31% KOH concentration electrolyte, and zirconium oxide clothe (Zircar^{TM}) separators. The positive electrode is electrochemically impregnated using an alcohol–water process. The cell contains two back-to-back stacks of many pairs of electrodes, placed in two halves of the cylindrical stainless steel case welded at the longitudinal center.

Compared to NiCd, the NiH_2 chemistry has superior charge/discharge cycle life and low internal resistances. Another attractive feature of the NiH_2 cell is the linear dependence of the SOC with the internal pressure that can be measured by strain gages mounted on the outside wall. Temperature compensated strain gages are accurately matched in bridge circuits, and processed by signal conditioning amplifiers. A two-stage regulator provides stable power for the circuit. Adhesive and conformal coating materials are used for mounting the gages and the flexible circuit board is made of space-qualified, low out-gassing materials. The strain gages are calibrated individually on each cell.

Although the NiH_2 cell does not exhibit a noticeable memory effect, a number of battery design engineers believe that reconditioning is required in both NiCd and NiH_2; in fact, in all batteries with a positive nickel electrode. As a result, reconditioning has been done in the past as a practice inherited from the NiCd experience. Cumulative flight data to date suggests no need of reconditioning. Most customers do not require it anymore, and most power system designs have eliminated this function. However, the battery letdown unit is often installed on NiH_2 battery to discharge before shipping, on the launch pad to measure the capacity, or before placing in storage. If the letdown unit is not made part of the battery assembly, a separate unit is needed for any one of these reasons.

The mass distribution in the cell is shown in Figure 9.5. It is dominated by the pressure vessel, which requires a thick wall to limit the hoop stress under high pressure. Moreover, the cylindrical shape makes packaging of a large battery in a 15 to 20 kW satellite difficult due to the large volume. Figure 9.6 depicts a NiH_2 battery assembly with cells mounted vertically.

Over the past decade, cell manufacturers have introduced the following design improvements to reduce the battery mass and volume.

9.3.2.1 Common and single pressure vessels

The vessel portion of the individual pressure vessel mass can be reduced by packing two cells in a common pressure vessel (CPV), or by packing all cells

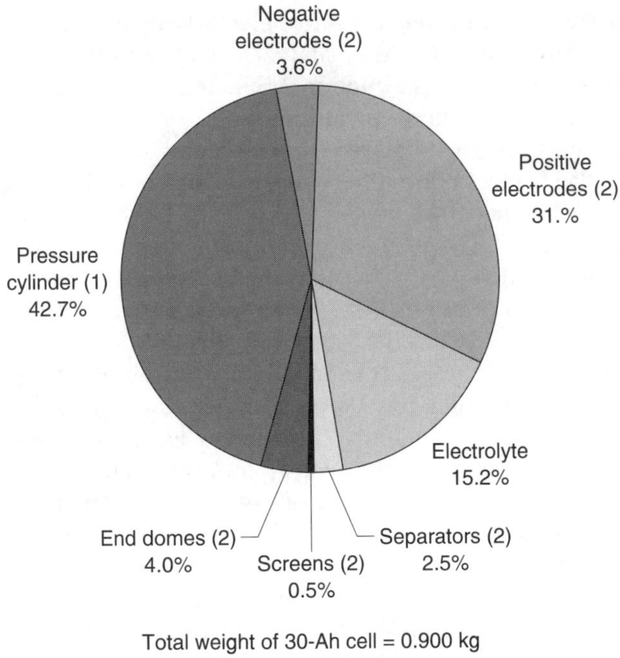

FIGURE 9.5 Mass composition of NiH_2 cell.

FIGURE 9.6 50-Ah 22-cell battery assembly using cylindrical NiH_2 cells mounted vertically. (Source: Dornier Satellitensysteme Gmbh, Daimler-Benz Aerospace. With permission.)

in a single pressure vessel (SPV). The CPV looks just like the IPV, except for the internal connections (Figure 9.7). Two cells internally connected in series in CPV doubles the terminal voltage to 2.5 V. The battery is then assembled as usual, using one half as many CPV cells. Handling and testing fewer cells save in the manufacturing cost. The CPV cells have been used extensively on NASA/JPL programs. However, venting one vessel results in the loss of two cells, partially offsetting the gain in mass and volume, or compromising the reliability.

The SPV design places all cells in a single vessel. About 100 SPVs have successfully flown in various NASA/JPL/Navy/Air Force programs. Iridium satellites are flying with 50-Ah SPV batteries. The single pressure vessel requires a thicker wall to limit the hoop stress, partially compensating the mass savings. Overall, the SPV does save mass and volume, but the reduced reliability is a significant concern. A pressure leak or an internal arcing would result in a total loss of the battery. Moreover, it does not have the means of monitoring the individual cell voltages. The failed-cell bypass circuit generally used in the IPV battery design cannot be used in this case.

The following are commonly available NiH_2 cells from various vendors:

- 2.5 inch diameter IPV for small satellites — 2 to 20 Ah
- 3.5 inch diameter IPV — 20 to 100 Ah
- 4.5 inch diameter IPV — 100 to 330 Ah
- 5.5 inch diameter IPV for large satellites — 200 to 400 Ah
- 4.5 inch diameter CPV 2.5 V — 60 to 100 Ah

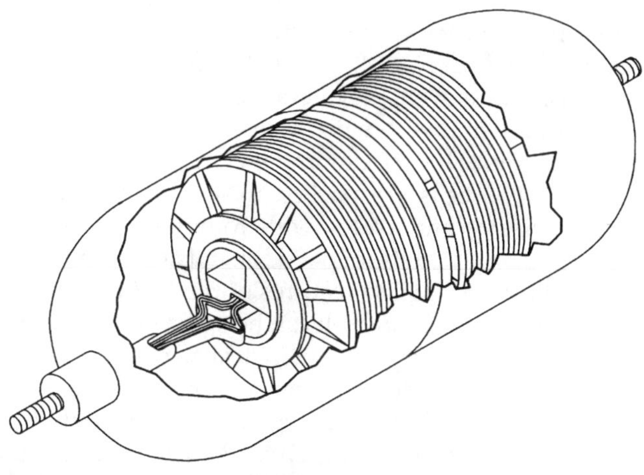

FIGURE 9.7 Common pressure vessel NiH_2 cell.
(From D.K. Coates et al. *Proc. IECEC*, ASME, p. AP-90, 1995. With permission.)

Table 9.1 NiH_2 cells for space applications

Rated 20°C Ah	Design type[a]	Diameter (inches)	Tip–tip length	Weight (kg)	BOL pressure[b]	10°C Cap. Ah	72 h OC Ah[c]
25	BB	3.45	9.2	0.830	500	33	24
50	BB	3.45	10.8	1.450	625	53	37
110	BB	4.67	11.4	2.760	600	120	88
160	BB	4.67	13.3	3.625	850	167	112
200	BB	4.67	14.4	4.700	1000	208	134
250	BB	4.67	17.5	5.900	1000	260	170
25	Re	3.45	9.30	1.000	500	25	18
55	Re	3.45	11.1	1.700	700	55	39
110	Re	4.67	12.1	2.975	600	120	88
160	Re	4.67	14.4	4.100	850	167	112

[a]Back to back has single separator suitable for GEO applications. Recirculating has double separator suitable for LEO applications.
[b]Internal cell pressure in pounds per square inch.
[c]Ampere-hour capacity at 20°C after 72 h of self-discharge at open circuit.
Typical for all ratings: Discharge rate C/2 to C, charge rate C/20 to C/2, trickle charge rate C/100 to C/200, 26–31% KOH, operating temperature −10°C to 30°C, Optimum operation at 5°C, Non-operating temperature −20°C to 40°C.

- 5.5 inch diameter CPV 2.5 V — 100 to 200 Ah
- 10 inch diameter SPV for small satellites — 15 to 50 Ah

Table 9.1 gives data on commonly available IPV cells.

9.3.2.2 Dependent pressure vessel

The dependent pressure vessel (DPV) uses hockey-puck shape cells.[6] Each cell has its own pressure vessel, but with reduced wall thickness and flat sides. The casing is made of SS-304, a laser closure weld and the usual nylon compression seals. Nonwoven polymeric separators, 26% KOH, a microfiber nickel electrode and a binary supported catalyst in a hydrogen electrode are common. The maximum operating pressure is about two thirds that in IPV. The cells are packed in a canteen-rack configuration, clamped between heavy end plates with four tie-rods (Figure 9.8). The pressurized fully charged cells depend on adjacent cells for structural support, offering all the reliability features of the IPV battery with reduced volume. An internal failure in one cell can be accommodated by having one spare cell as in the IPV battery. The electrical performance of the DPV is identical to that of the IPV under the same loading conditions. The DPV cell results in somewhat higher specific energy and significantly higher energy density. In conceptual designs using DPV, 30% mass reduction and 50% volume reduction have been estimated. However, some of the concerns of the DPV are:

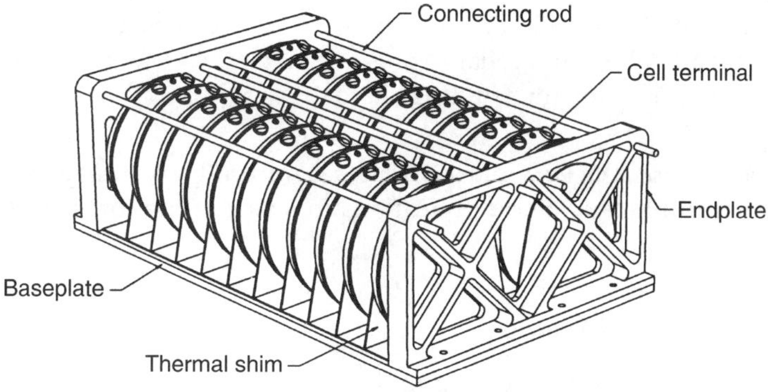

FIGURE 9.8 Dependent pressure vessel NiH_2 battery.
(From D.K. Coates et al. *Proc. IECEC*, ASME, p. AP-90, 1995. With permission.)

- A pressure leak in one cell may potentially crush all cells in a zipper mode. This concern is addresses by the cell designed with an internal bracket to support the failed cell and prevent crushing under the pressure of adjacent cells (Figure 9.9).
- Complex thermal control due to concentrated heat dissipation
- Must buy the whole battery from the vendor, which is uncommon at most satellite builders.

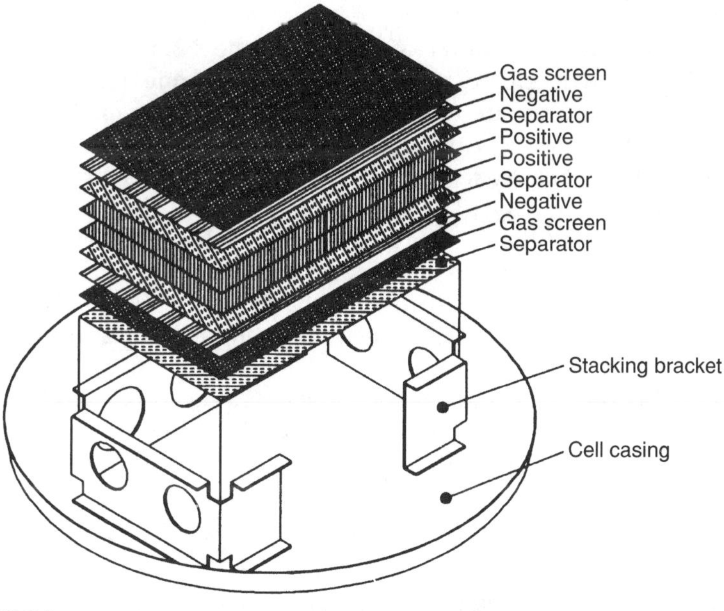

FIGURE 9.9 Dependent pressure vessel internal support bracket.
(From D.K. Coates et al. *Proc. IECEC*, ASME, p. AP-90, 1995. With permission.)

- Calibrating the strain gage reading with pressure on a hockey-puck cell is difficult. Without a reliable pressure data, the DPV has no advantage over the NiCd for interpreting the battery SOC.

For these reasons, the DPV does not offer significant advantage over the IPV cell, and is unlikely to find applications in space in the foreseeable future.

9.3.3 Nickel Metal Hydride

Concerns in working with cadmium in the NiCd battery and disposing of it helped to spark interest in using the NiMH battery for a variety of applications. In general, NiMH electrochemistry presents few health and safety risks, primarily because the electrode materials are nontoxic. Its anode is made of a metal hydride, which eliminates the environmental concerns of cadmium. The NiMH offers an improvement in specific energy over that in NiCd. Another performance improvement is that it has negligible memory effect. The NiMH, however, is less capable of delivering high peak power, has high self-discharge rate, is adversely sensitive to high temperature, and therefore is susceptible to damage due to over-charging. Compared to NiCd, NiMH is expensive at present, although the future price is expected to drop significantly. This expectation is based on current development programs targeted for its large-scale application in electric vehicles. For space applications, it has been flown on a few Japanese, European, and American flights.

Electrochemically, the NiMH is similar to the NiH_2, except that the working hydrogen is stored as solid metallic hydride rather than gas. Having no internal pressure to withstand, these cells are made prismatic. This significantly improves the energy density and safety. Having less volume to pack, the battery requires less support structure, hence improves the specific energy as well. The disadvantage is that the linear pressure versus SOC relation of the NiH_2 is no longer available. Hermetically sealed aerospace cells have been successfully tested for several thousand C/D cycles at 50% DOD. Sanyo's 37-Ah NiMH battery survived 15,243 cycles at 40% DOD. Their loss in capacity when operating at 0°C is much greater than that in a NiCd battery.

The NiMH is prismatic and has the same operating voltage as the NiCd. Therefore, the two are easily interchangeable. The NiMH, however, produces heat during charging as opposed to the endothermic charging of NiCd.

The key features of NiMH, as compared to NiH_2 cells, are as follows:

- It offers some (10 to 20%) improvement in Wh/kg
- It offers significant (~100%) improvement in Wh/l

- It is less sensitive to temperature, hence it has lower thermal control cost
- It has a much lower operating pressure, hence is much safer

The cell rupture due to internal gas build-up is one possible hazard. The cell design with vent to release the gas pressure is therefore needed. The risk of venting hydrogen is minimized by limiting its amount during overcharging. As a health risk, nickel and its compounds may be carcinogenic, as in other cells.

9.3.4 Lithium-Ion

Lithium-ion (Li-ion) technology is a new development that offers significantly higher specific energy and energy density over that of NiH_2. The prismatic shape (Figure 9.10) results in higher energy density. The large improvement in specific energy comes from lithium's low atomic weight of 6.9.

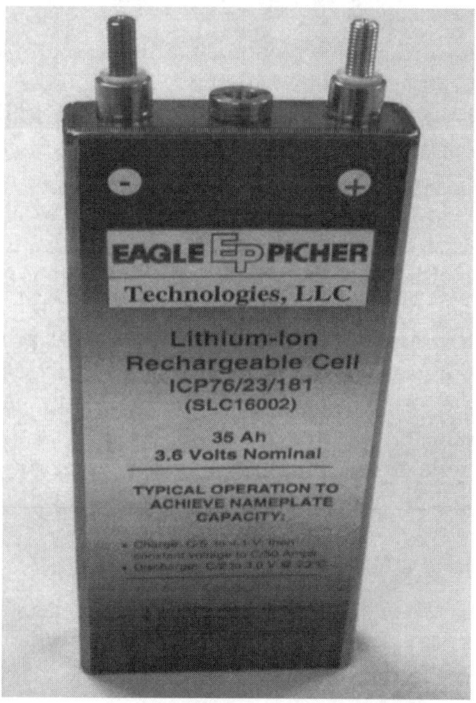

FIGURE 9.10 Lithium-ion cell of 35-Ah, 3.6-V ratings in prismatic casing. (From J.A. DeGruson, *Proc. 36th IECEC*, ASME, p. AT-57, 2001. With permission.)

Lithium-ion batteries are widely used in laptop computers, digital cameras, and other commercial applications. The positive electrode is $LiNiCoO_2$, the negative electrode is a mixture of two graphites with a non-PVDF binder, and the electrolyte is $LiPF_6$ salt. The cell construction uses a stainless steel prismatic case with a safety vent between the two terminals. The terminals are insulated from the case by polymeric seals as in the NiH_2 cell. Small cells are cylindrical and large cells are rectangular prismatic in shape. The prismatic shape gives high tolerance to shock and vibrations environments.

Since the Li-ion is non-aqueous chemistry, it is free from electrodes corrosion. On the negative side, lithium is highly reactive metal and must be stabilized by encapsulating the molecules in carbon using an expensive form of graphite. The lithium electrode reacts with the liquid electrolyte, creating a sort of passivation film. Each time the cell is discharged and then charged, the lithium is stripped away, a free metal surface is exposed to the electrolyte and a new film is formed. The cell uses thick electrodes to compensate for this. Otherwise the life would be short.

The Li-ion has the end of discharge cut-off voltage of 2.7 V, average discharge voltages of 3.5 V and end-of-charge voltage of 4.2 V. The average discharge voltage of 3.5 versus 1.25 for NiCd and NiH_2 requires about one third the number of cells in series for a given battery voltage, thus reducing the assembly and test costs.

In operation, lithium-ion electrochemistry has a high charge efficiency. The coulombic efficiency is 100% all the way to 100% charging. Other desired characteristics of a space-qualified Li-ion cell are high-rate capability and potentially long cycle life at high DOD. Its low temperature sensitivity supports operation over a wide temperature range. It has low internal impedance, less than 1 mΩ in a 50-Ah cell, which supports high charge and discharge rates. It can deliver high short time peak power without an adverse effect on life.

The Li-ion cell has a potential of achieving 150 Wh/kg and 400 Wh/l, along with abuse tolerance. However, it costs more at present, but can be cost competitive with NiH_2 in the same volume production. It is vulnerable to damage under over-charging or other shortcomings in the battery management. Therefore, it requires more elaborate charging circuitry with adequate protection against over-charging. Batteries need to be trickle charged, which requires extreme care in Li-ion. Each cell may need charge control, which requires additional electronics.

Several roadblocks for space applications that have been recently addressed are:

- Relatively short life demonstrated so far
- Sensitive to over-charge and over-discharge
- High internal resistance at low temperature

The long calendar life for GEO and long cycle life for LEO are two issues in Li-ion technology that still need to be proven with a sufficiently long flight history. To date, 6000 cycles of charge/discharge at 25% DOD has been demonstrated on 35-Ah cells with stable performance. A Li-ion battery requires trickle charge at a rate of $C/50$ to $C/100$. The self-discharge rate is a few percent of that in the NiH_2, but increases at high temperature. The internal impedance is low but increases at low temperature. The two must be traded in the design.

The present development funded by NASA and the U.S. Air Force has the following goals at the cell level:

- Cell capacity of 20 to 200 Ah
- Life of 30,000 cycle at 50% DOD
- Life of 40,000 cycles at 40% DOD
- Temperature range −40°C to +80°C
- Charge rate $C/5$ to C
- Discharge rate $C/2$ to $2C$

The 2001 test results on *Mars Lander* batteries showed specific energy of about 150 Wh/kg.[7] Nominally 5-Ah cells were tested for capacity when discharged at a 1-A rate at various temperatures. The cells were cycled from 2.4 V to 4.1 V. The tests showed significant degradation at low temperature, with only 60% capacity at −30°C. In internal heat generation tests on 25-Ah cells at 10°C during charge and discharge at 5-A rate, the heat generation spiked at the end of discharge due to increased polarization as the cell reached full discharge. The heat generation reduction at the end of charge was due to charge current tapered down to zero.

The tests on 10-Ah cells charged and discharged at the same temperature have shown[8] that 600 cycles at 100% DOD at −20°C resulted in 6% capacity loss. Fellner[9] reports test data on Li-ion cells cycled as follows for GEO and LEO applications:

- For LEO: 20-Ah cells, 40% DOD, charged to 4.1 V at $0.5C$ rate for 1 h, discharged at $0.8C$ rate for 30 min
- For GEO, 20-Ah cells, 100% DOD, charged at C rate to 4.1 V in 3 h, discharged at $C/5$ rate to 2.5 V.

The initial test data for a discharged cell show that the internal resistance grows exponentially with decreasing temperature for a given state-of-charge.

The Li-ion batteries can release lithium gas and flames when over-charged. This danger needs effective control of the over-charge. For this reason, the maximum charge voltage and minimum discharge voltage must be maintained for successful incorporation of Li-ion cells into the space power system design. In this respect, Li-ion chemistry is very different to that of the NiH_2 in that the Li-ion does not tolerate over-charge or over-discharge. As a battery of many cells in series and parallel is cycled, it often becomes unbalanced due to small cell-to-cell differences in the rate of capacity fade, rate of impedance increase, or rate of self-discharge that are accumulated over many cycles. For charge balancing, various techniques such as switched capacitor and resistance or inductor equalization have been explored.

Li-ion has a better low temperature performance in the −20°C to +40°C range. It can operate even at −40°C with some capacity loss. Poor cell performance at lower temperature is attributed to the electrolyte becoming viscous or solid, and poor lithium diffusivity in the electrodes. An 8-cell 25-Ah battery was space qualified in 2001 for 2000 cycles at 100% DOD for the *Mars Lander* mission requiring such a low temperature range. The battery was designed for a 3.0 to 4.1-V operation, 2.5-V cut-off voltage, and C/5 and C/10 charge rates. Tests reported by Bruce and Marcoux[10] at various temperatures cover design and fabrication features of large capacity cells up to 100 Ah. In cells for aircraft applications, the main concern is irreversible capacity loss at elevated temperature. The loss in capacity at high temperature comes primarily from the positive electrode degradation. In cells for space use, however, the shelf life and cycle life are critical.

9.3.5 Lithium Polymer

The electrochemical basis of the Li-ion solid polymer electrolyte (Li-poly) battery is similar to that of Li-ion, except that the electrolyte of the Li-ion is replaced by a solid polymer that encapsulates the electrodes. The cell is constructed with a film of metallic lithium bonded to a thin layer of solid polymer electrolyte. The solid polymer enhances the specific energy by acting as both the electrolyte and the separator. This construction also offers an improvement in safety compared to Li-ion by reducing flammability. Moreover, the metal in solid electrolyte reacts less than it does with liquid electrolyte. There are various types of Li-polymer cell, with markedly different characteristics. An advanced battery is under development by NASA with an industry team for aerospace and consumer applications. The goals of these programs are listed in Table 9.2. The 35,000 cycles life at 25% DOD addresses the LEO satellites requirements, while the 2000 cycles life at 100% DOD would meet the GEO satellite requirements.

On the safety side, the Li-ion system is relatively less reactive and more reliable than the present lithium metal systems, and may have performance advantage over some solid polymer electrolyte systems. However, the possibility of electrolyte spill is a disadvantage over a solid electrolyte. Some materials used in the Li-ion cells may be toxic and carcinogenic, such

Table 9.2 NASA goals for Li-ion research programs at cell level

Performance parameter	Li-poly (goals)	Present Li-ion	Present NiH_2
Specific energy, Wh/kg	200–250	125–150	50–65
Energy density Wh/l	350–400	300	–
Life at 100% DOD, cycles	2000	1000	–
Life at 25% DOD, cycles	35,000	–	–

as lithium, nickel and arsenic compounds, and possibly cobalt, copper and polypropylene.

9.3.6 Silver Zinc

The silver zinc (AgO-Zn) secondary battery has high specific energy but shorter cycle life. It is suitable in missions requiring a low number of C/D cycles. It is used in military aircraft and spacecraft, launch vehicles, and undersea applications. The specific energy ranges from 125 to 250 Wh/kg and energy density 200 to 600 Wh/l. Prismatic cells are available in a wide range of ampere-hours in low-rate (LR) and high-rate (HR) types. Because resistance of the silver monoxide electrodes vary little with the state of charge, a stable operating voltage is provided until nearly all the capacity is drawn.

The AgO-Zn cell comes in two varieties: dry charged and dry unformed. In the dry-charged cell, the positive and negative electrodes are charged at factory, but the electrolyte is added at the launch site. The dry unformed cell is charged and the electrolyte is added at the launch site. In both cases, the cell comes to a full charge voltage of 1.86 V open circuit after the soak time. In subsequent C/D cycles, the full cell charging is terminated at 2.05 V, and the open circuit voltage drops to 1.86 V. Full charge level is indicated by the voltage as well as the electrolyte level on a visible scale.

9.4 Electrical Circuit Model

The battery works as a voltage source with small internal resistance. Its electrical circuit model is shown in Figure 9.11, where E_i is the internal electrochemical voltage and R_i is the internal resistance. The E_i decreases and R_i increases linearly with the ampere-hour discharge. That is, the internal voltage is lower and the internal resistance is higher in a partially discharge battery as compared to their values — designated as E_o and R_o, respectively — in the fully charged state. Quantitatively,

$$\begin{aligned} E_i &= E_o - K_1 \cdot \text{DOD} \\ R_i &= R_o + K_2 \cdot \text{DOD} \end{aligned} \tag{9.4}$$

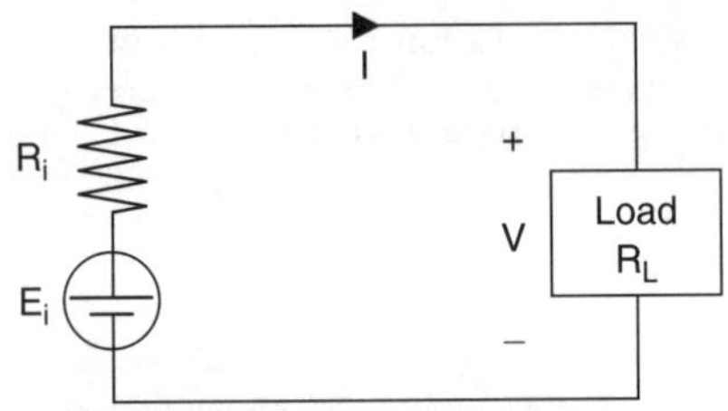

FIGURE 9.11 Equivalent electrical circuit of battery showing internal voltage and resistance.

where K_1 and K_2 are constants of the electrochemistry to be found by curve fitting the test data. Under load, the terminal voltage drops with increasing load, and the operating point *P* is the intersection of the battery line and the load line as shown in Figure 9.12.

9.5 Properties and Performance

The electrical performance of the battery depends on the electrochemistry and on many other parameters in a highly nonlinear manner. This makes the battery design one of the most difficult tasks for the power system engineer. Major performance characteristics of the battery in general are discussed in this section, mostly for NiH_2, the widely used electrochemistry at present.

9.5.1 Charge/Discharge

The cell voltage varies during a typical charge/discharge (C/D) cycle as was shown earlier in Figure 4.3. The charge/discharge characteristic also depends on the charge rate, discharged rate, temperature, and age of the cell. Figure 9.13 shows such variations for a LEO satellite.

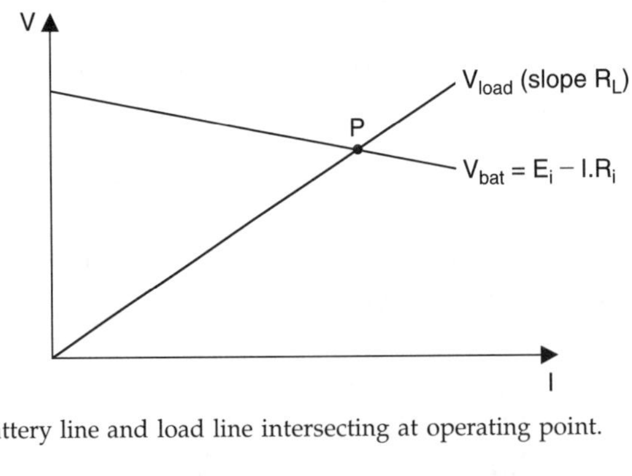

FIGURE 9.12 Battery line and load line intersecting at operating point.

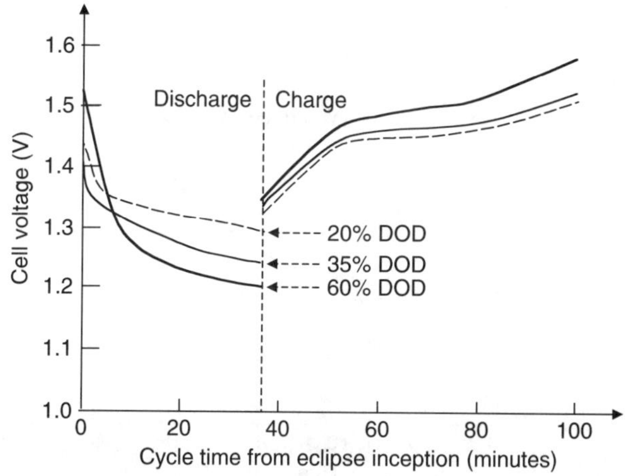

FIGURE 9.13 NiH_2 charge/discharge cycle at different rates in LEO.

Typical average NiH_2 and NiCd battery cell voltages are as follows:

- 1.55 V when fully charged
- 1.45 V average during charge
- 1.25 V average during discharge
- 1.10 V at 80% DOD
- 1.00 V when fully discharged

After discharging certain ampere-hours, the battery requires more ampere-hours of charge to restore the full state of charge. The charge/discharge ratio is defined as the ampere-hours input over the ampere-hours output for full state of charge. This ratio depends on the charge and discharge rates and also on temperature as shown in Figure 9.14. For example, a C/D ratio of 1.1 means the battery requires 10% more ampere-hours for restoring to fully charged state than what was discharged. Charging in LEO is usually at a much faster rate, which is more efficient than the slow charging rate typical in GEO. Therefore, a low C/D ratio of 1.05 to 1.1 is needed in LEO missions, while a higher ratio of 1.1 to 1.2 is needed in GEO missions. A high C/D ratio causes excessive corrosion of the nickel electrodes, and large amount of oxygen leading to heating and popping during recombination on the hydrogen electrode. On the other hand, low C/D ratio causes reduced usable capacity, low discharge voltage, and capacity degradation.

A new cell stores over 20% higher capacity when charged at −20°C compared to that at +20°C (Figure 9.15), but only about 10% higher after about 100 cycles of charge/discharge cycles (1 year in GEO). The maximum charge voltage also varies with temperature (Figure 9.16). The end of charge

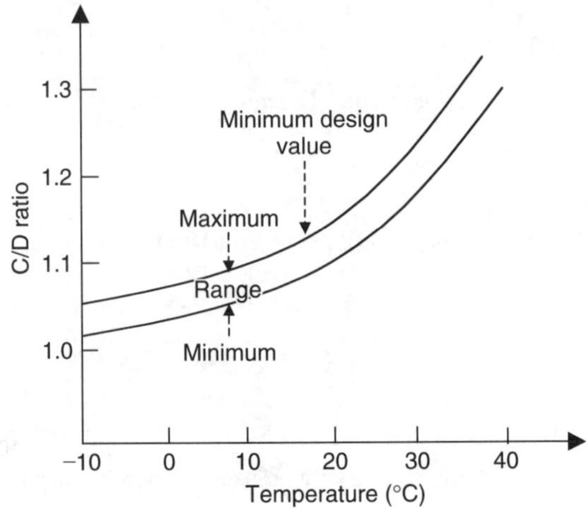

FIGURE 9.14 NiH_2 charge/discharge ratio versus temperature.

and end of discharge voltages in NiH_2 remain fairly flat with the number of C/D cycles even for several years in LEO missions, but not in NiCd.

The battery performs better under slow charge and discharge rates. A fast discharge rate results in fast voltage degradation and low ampere-hour capacity to the load. For these reasons, high charge and discharge rate applications require different design considerations than the low-rate applications.

After the battery is fully charged, the charge efficiency drops to zero. Any additional charge will be converted into heat. If over-charged at higher rate than the self-discharge rate for long time, the battery would overheat posing

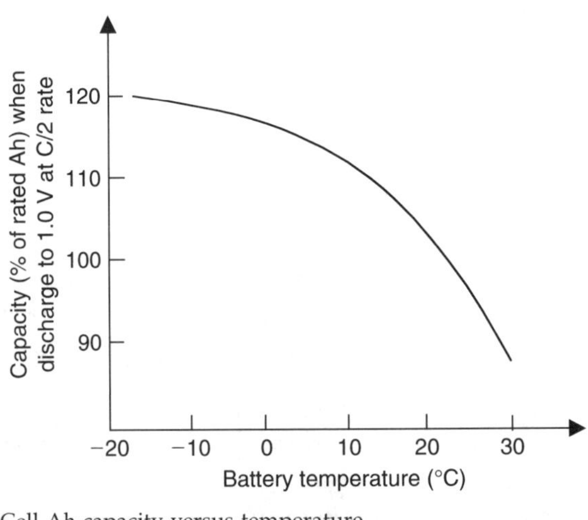

FIGURE 9.15 Cell Ah capacity versus temperature.

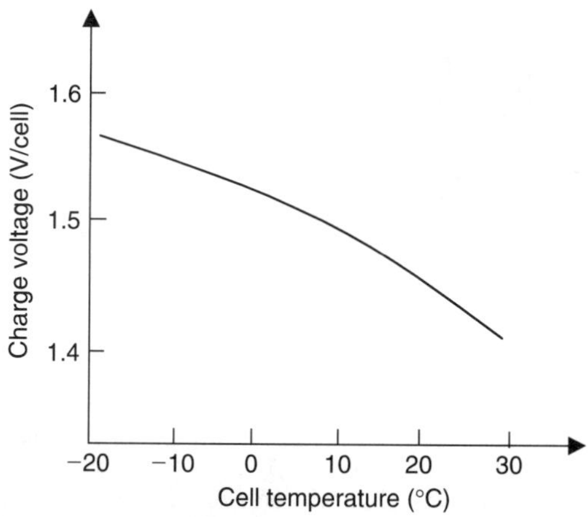

FIGURE 9.16 Full charge voltage at $C/10$ for 16 h.

a potential explosion hazard. Any excessive over-charging produces excessive gassing, which scrubs the electrode plates. Continuous scrubbing at high rate produces excessive heat, wears out electrodes, and shortens the life. For this reason, the battery charge circuit requires a regulator to cut back to the trickle charge rate once the battery is fully charged. The trickle charging produces a controlled amount of internal gassing. It causes mixing action of the battery electrolyte, keeping it ready to deliver full charge.

9.5.2 Internal Impedance

The internal impedance of the cell is made of the resistance, inductance, and capacitance. The resistance comes primarily from the electrodes and electrolyte, the inductance from the leads in and out of the cell, and the capacitance from the two electrode plates. The internal power loss during discharge or charge can be seen as having an internal resistance R_i. The value of R_i depends on the battery capacity, operating temperature, and the state of charge. The higher the cell capacity, the larger the electrodes and lower the internal resistance. The R_i varies with the state of charge as per Equation 9.4. It also varies with temperature as shown in Figure 9.17. The internal resistance and inductance of 50-Ah and 100-Ah NiH_2 cells at 10°C are given in Table 9.3.

9.5.3 Charge Efficiency

The charge efficiency, also known as coulombic efficiency, is defined as the ratio of the ampere-hours being deposited internally between the plates over that injected from the external terminals during charging. It is different from the energy efficiency. The coulombic efficiency varies as a function of

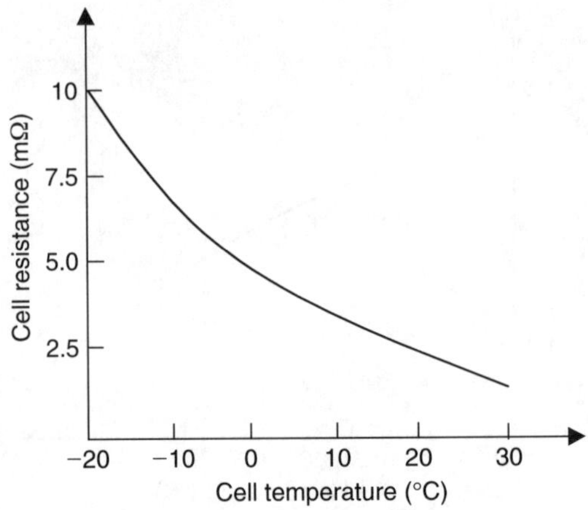

FIGURE 9.17 Internal resistance of 50-Ah NiH_2 cell at 100% SOC.

the SOC, charge rate, and the operating temperature, and is generally higher than 97%. The charge efficiency is almost 100% when the cell is empty of charge; the condition in which all charge received is deposited between the plates. As the SOC approaches 1, the charge efficiency tapers down to zero. The knee point where the charge efficiency starts tapering off depends on the charge rate and temperature (Figure 9.18). For example, at $C/2$ charge rate, the charge efficiency is almost 100% up to about 75% SOC. At a fast charge rate of $C/40$, on the other hand, the charge efficiency at 70% SOC is only about 50%.

9.5.4 Energy Efficiency

In both the charge and the discharge modes, a small fraction of energy is converted into heat. The energy efficiency over a round trip of full charge and discharge cycle is defined as the ratio of the energy output over the energy input at the electrical terminals of the battery. For a typical battery of capacity C at room temperature with an average discharge voltage of 1.25 V,

Table 9.3 Approximate impedance for IPV NiH_2 cells

Parameter	50-Ah cell	100-Ah cell
Cell resistances	4.5 mΩ at 1 kHz	2.25 mΩ at 1 kHz
	20 mΩ at 10 kHz	10 mΩ at 10 kHz
Cell inductance	0.70 μH at 1 kHz	0.60 μH at 1 kHz
	0.35 μH at 10 kHz	0.30 μH at 10 kHz

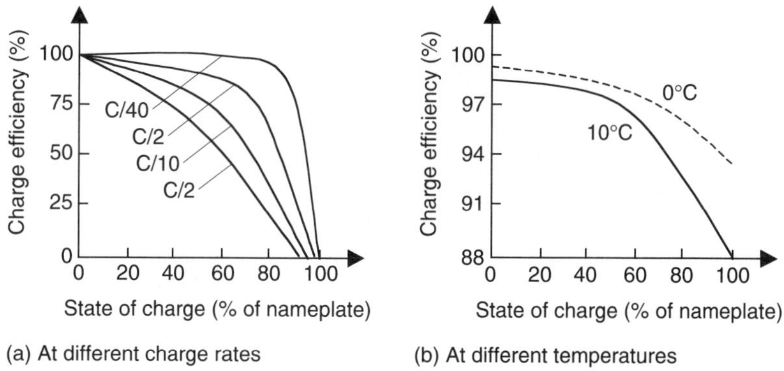

FIGURE 9.18 Charge efficiency versus state of charge for NiH_2 cell.

average charge voltage of 1.45 V, and the charge/discharge ratio of 1.10, the efficiency is calculated as follows:

- The energy output over the full discharge = $1.25 \times C$
- The energy input required to restore full charge = $1.45 \times 1.10C$
- Therefore, the round trip energy efficiency is

$$\eta_{energy} = \frac{1.25 \times C}{1.45 \times 1.10\,C} = 0.78 \text{ or } 78\% \tag{9.5}$$

These calculations indicate that about 22% energy is lost per charge/discharge cycle, which is converted into heat. Most of this energy loss occurs during discharge. The round trip energy efficiency varies with various factors, but is typically 75 to 80% for NiH_2 and 80 to 85% for NiCd.

9.5.5 *V–T* Characteristics

For most electrochemistries, the full charge voltage is a function of temperature and age of the battery a shown in Figure 9.19, where the limit numbers represent equally divided periods of the mission life. A similar relation for the C/D ratio is shown in Figure 9.20. Such voltage versus temperature (*V–T*) curves are used for on-orbit energy management as discussed in Chapter 13. With a NiH_2 battery, an alternative is to use the pressure–temperature (*P–T*) curves, which exhibit similar relations with age. The energy management software shifts the operation on-board as the battery ages in service.

9.5.6 Self-Discharge

The battery slowly self-discharges even with no load under open circuit. It must be continuously trickle charged to counter the self-discharge rate in

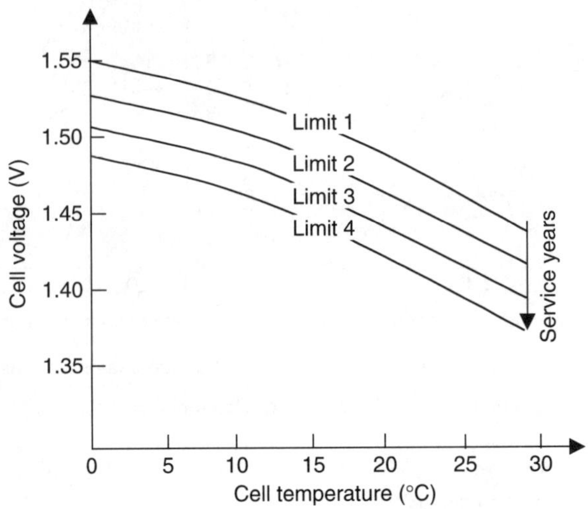

FIGURE 9.19 Temperature compensated *V–T* levels for NiH_2 cell.

order to maintain the full state of charge. This rate is usually less than 1% per day for most electrochemistry in normal working conditions. In NiH_2 cells, the self-discharge rate shows as a proportionate loss in pressure as seen in Figure 9.21. After the battery is fully charged, the charge efficiency drops to zero. Any additional charge will be converted into heat. If overcharged at higher rate then the self-discharge rate for long time, the battery would overheat posing a potential explosion hazard.

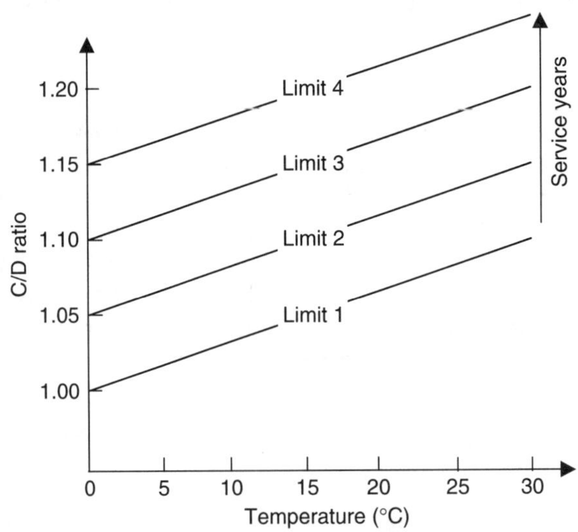

FIGURE 9.20 Temperature compensated charge/discharge ratios for NiH_2 cell.

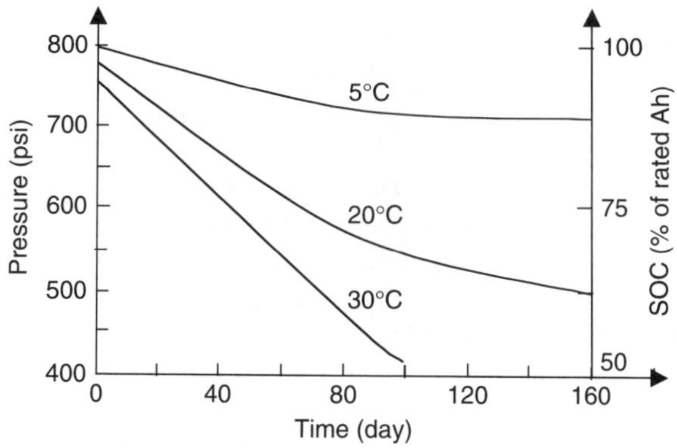

FIGURE 9.21 Self-discharge rates versus time in NiH_2 cell under open circuit.

9.5.7 Self-Heating

When the battery is initially put to charge, the internal heat generation rate is negative for a while, meaning that the electrochemical reaction during the initial charging period is endothermic (absorbing heat). The reaction changes to exothermic and generates heat as the battery approaches full charge. The temperature rise during this phase depends on the cooling method used to dissipate the heat by conduction, convection, and radiation.

The internal power loss (in watts) causes self-heating in the battery. It has a major impact on the battery temperature. During charge, it is equal to the sum of heat generated by charging of active materials (am) and the heat generated by over-charge reaction (or). That is,

$$H_c = H_{c,am} + H_{c,or} = -I_c\eta_c\{1.45 - E_c\} + I_c(1 - \eta_c)E_c \tag{9.6}$$

where I_c = charge current, η_c= instantaneous charge efficiency, E_c = charge voltage, and 1.45 is the ratio of the enthalpy in joules equivalent to the Faraday's constant for NiCd and NiH_2. The above equation simplifies to

$$H_c = -I_c\{1.45\eta_c - E_c\} \tag{9.7}$$

At the beginning of the charging process when η_c is near unity and E_c is low, $1.45\eta_c > E_c$ and H_c is negative, meaning the battery absorbs heat and self-cools. Near the end of charging, as η_c approaches zero and E_c rises, the battery generates heat and self-heats.

The heat generation rate (in watts) during discharge is given by

$$H_d = I_d\{1.50 - E_d\} \tag{9.8}$$

where I_d = discharge current and E_d = discharge voltage.

As seen in the above two equations, the required information for determining the internal heat generation rate is the current, voltage, and the instantaneous charge efficiency. During discharge, approximately 15% of the total energy released is heat, 5% is pressure–volume work and side reactions, and 80% is useful electricity. In this exothermic mode, it releases heat in addition to electrical power. Therefore, the battery temperature rises. During charge, on the other hand, the battery temperature falls as the chemical reaction requites heat. This is true until the full state of charge is attained as indicated by the battery voltage of 1.55 V or so. Beyond this point, the chemical reaction stops, and all energy supplied are turned into heat. The battery may over-heat quickly and may explode.

During over-charge and trickle charge, the heat generated is equal to the electrical power input. Thus, the efficiency depends upon the achievable trickle charge rate, which in turn depends upon the charge rate, temperature, charge regulator operation, etc. Typically, the battery is recharged before the eclipse starts with ample margin in time. Then it is placed into taper charge and over-charge if the *V–T* limit charge approach is used. During this period, the entire over-charge will convert into heat inside the battery. The thermal control is designed to maintain the battery within its desired operating temperature limits.

Figure 9.22 shows the power loss in a 40-Ah 22-cell NiH_2 battery delivering a 1200-W load during eclipse in GEO with 65% DOD and C/D ratio of 1.2. Figure 9.23 depicts the power loss in the NiH_2 battery on a 3000-W LEO platform with a maximum DOD of 35% and C/D ratio of 1.1 (EOS-AM batteries, each 54-cells 50-Ah capacity, 1500-W average load, 80-V battery voltage). For a lower C/D ratio, the power loss between 90 and 98 minutes would be lower. Note that the heat generation increases as the

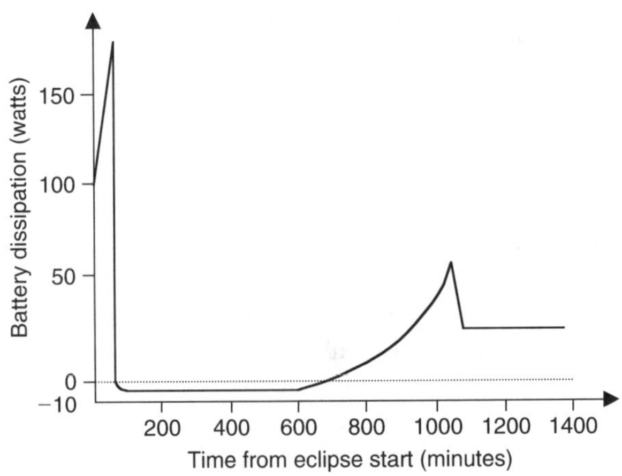

FIGURE 9.22 Battery power loss versus time starting with eclipse for 1200-W load and 1.2 C/D ratio in GEO orbit.

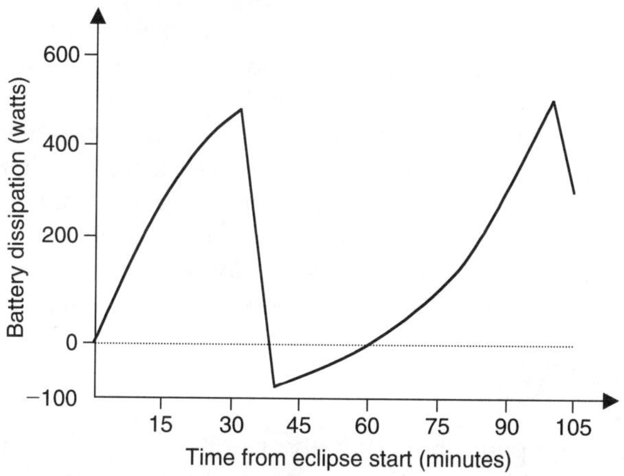

FIGURE 9.23 Battery power loss versus time starting with eclipse for 3000-W load and 1.1 C/D ratio in LEO orbit.

depth of discharge increases because of the increased internal resistance. In order to overcome possible cell voltage divergence, the C/D ratio can be increased. However, that would increase the orbit average dissipation. For NiH_2 cells, increasing the C/D ratio from 1.05 to 1.1 can increase the orbit average dissipation in LEO by 10% at 35% DOD and 15% at 70% DOD.

Different electrochemistry, however, generates internal heat at different rates. The heat generation between various batteries can be meaningfully compared in terms of the adiabatic temperature rise during discharge, which is given by the following relation:

$$\Delta T = \frac{\mathrm{WH_d}}{M \cdot C_p}\left[1 - \eta_v + \frac{E_d}{E_o}\right] \tag{9.9}$$

where

ΔT = adiabatic temperature rise of the battery, in degress centigrade

$\mathrm{WH_d}$ = watt-hour energy discharged

M = mass of the battery

C_p = battery specific heat, Wh/kg·°C

η_v = voltage efficiency factor on discharge

E_d = average cell entropy energy per coulomb during discharge, i.e., average power loss per ampere of discharge, W/A

E_o = average cell open circuit voltage, volts

For full discharge, the $\mathrm{WH_d}/M$ ratio in the above expression becomes the specific energy. This indicates that higher specific energy cells would also tend to have higher temperature rise during discharge, requiring enhanced cooling design. Figure 9.24 depicts the adiabatic temperature rise $\triangle T$ for various electrochemistries after a full discharge in short burst.

The thermal properties of the cell, such as the heat generation rate, thermal conductivity, and specific heat are useful for the thermal analysis and design. The specific heat of the NiCd cell is about 1000 J/kg°C, which is about that of pure aluminum. For the NiH_2 cell, it is about 1200 J/kg°C. In terms of the ampere-hour capacity, the following equation gives a reasonable first order approximation to the specific heat (in Ah·J/°C) of the cell:

$$C_p = 200 + 25 \tag{9.10}$$

The battery temperature generally means the wall temperature. A temperature gradient across the hydrogen gap between the wall and the internal stack in NiH_2 cell is typically 2 to 3°C.

9.6 Cycle Life

The battery fails when at least one cell fails. The cell failure is theoretically defined as the condition in which the cell voltage drops below certain low value before discharging the rated capacity at room temperature. The low value is generally taken as 1.0 V in cells with nominal voltage of 1.25 V. The remaining ampere-hours the cell can deliver from 1 V to 0 V is only a few percent. However, this is a conservative definition of battery failure. The battery does not fail like a light bulb, but gradually degrades in performance. In a degraded state with one cell dropping below 1.0 V, the rest of the cells providing voltage above 1.0 V may make up for the failed cell, and the battery as a whole may still provide the minimum required voltage. If the battery voltage falls below its required level at full load, it can

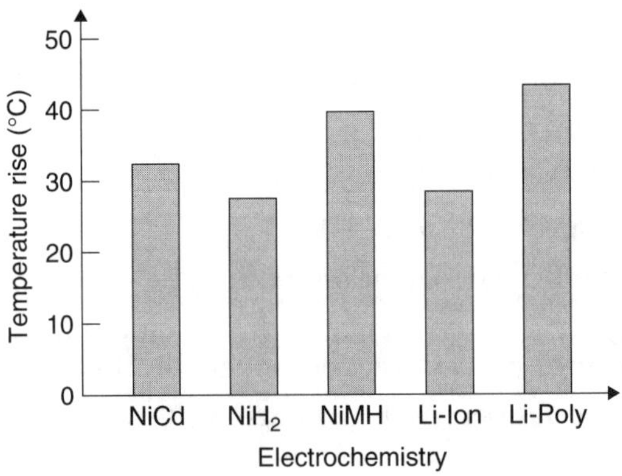

FIGURE 9.24 Adiabatic temperature rise under full discharge of capacity in various electrochemistries.

still deliver the required voltage for a short time if the load can be shed to lower the DOD.

Battery end of life is defined as one or more cell voltages dropping below 1.0 V at any point during the C/D cycle, usually at the end of discharge. The cell can fail either randomly or due to wear. It can fail open, short, or somewhere between (a soft short). A short that starts soft eventually develops into a hard short. In a low voltage battery, any attempt to charge with a shorted cell may result in physical damage to the battery and/or the charge regulator. On the other hand, the shorted cell in a high voltage battery with numerous series connected cells may work for a long time. However, it loses the voltage and ampere-hour capacity, and hence would work as a load on the healthy cells. An open cell, on the other hand, disables the entire string of series connected cells.

A spacecraft battery rarely fails randomly. Its primary failure mode is associated with electrode wear due to repeated charge/discharge cycles. The battery life is measured as the number of times it can be discharged and recharged before the electrodes wear out. The life depends strongly on the electrochemistry, the depth of discharge, and the temperature. Figure 9.25 shows the cycle life characteristic of NiCd and NiH_2 batteries. The life also depends to a lesser degree on electrolyte concentration, electrode porosity, and charge and discharge rates. The electrolyte concentration makes a significant difference to the cycle life of the NiH_2 cell. The cell with 26% concentration gives a greater cycle life compared to one with 31% concentration. The first set of factors is application related, whereas the others are construction related.

The number of charge/discharge cycles in a satellite equals the number of eclipses during the mission life. It is at least an order of magnitude greater in LEO satellites than in GEO. Such a long cycle life requirement in LEO can be achieved only by limiting the battery design to a low depth of discharge, typically 30% compared to 80% in GEO satellites. Obviously, such a design requires a proportionately much larger battery. A long cycle life at a given DOD and operating temperature reduces the battery mass in direct proportion.

It is noteworthy from Figure 9.25 that the life at a given temperature is an inverse function of the depth of discharge. If the life is 100 units at 50% DOD, then it would be about 200 units at 25% DOD. This makes the product of the cycles to failure and the DOD roughly constant in the first approximation. The cycle life multiplied by the DOD product decreases with increasing temperature. Such is true for most electrochemistries. This means that the battery at a given temperature can deliver the same number of equivalent full charges regardless of the depth of discharge. Phrased differently, the total watt-hour energy the battery can deliver over its life is roughly constant in the first approximation. The battery lasts proportionately longer if less energy is used per cycle. Such an observation is useful

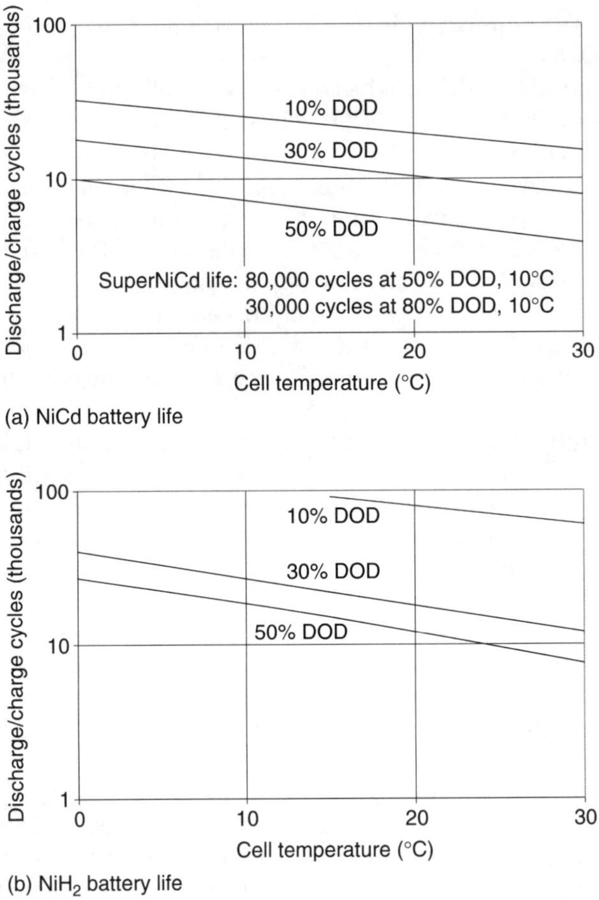

FIGURE 9.25 Cycle life versus temperature and DOD for NiCd and NiH_2 cells.

in comparing the costs of various batteries for a given application at the conceptual design stage.

The battery does not work fully well all the way until it dies. Its capacity remains fairly constant until it rapidly fades at the end as shown in Figure 9.26.

Since the battery cycle life is influenced primarily by the end of discharge DOD and temperature, the cell life prediction model is generally developed by fitting exponential curves to the test data on cycles to failure versus DOD at temperatures the battery would be subjected in orbit. The factors such as the charge and discharge rate and the electrolyte concentration are not included in such a model, but are factored into the cell life data by actual test or from previous experience. A reliability model then incorporates the predicated cell wear-out life and the random failure rates to estimate the mean time before battery failure by taking into account the number of cells

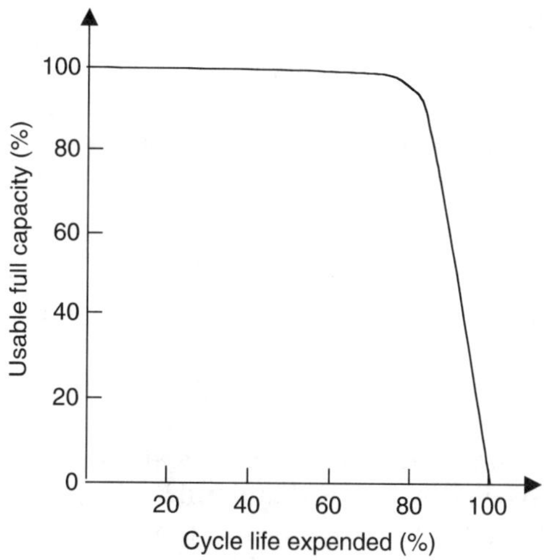

FIGURE 9.26 Capacity fade versus cycle life expended.

in series, the bypass diodes, and other components in the battery as a whole.

The battery life testing is generally done under accelerated cycling. For example, the GEO cycle is 24 h, but the battery testing is done in 8-h cycles. Smith and Brill[11] report such test data in NiH_2 cells. Many spacecraft design engineers believe that some electrochemical differences in actual cycling and accelerated cycling are still not fully acceptable.

9.7 Burst Power Capability

In fast discharge applications, such as for meeting a pulse power demand, the battery may be required to deliver the maximum possible power for a very short time. The peak power it can deliver can be derived using the maximum power transfer theorem in electrical circuits. It states that the maximum power can be transferred from the source to the load when the internal impedance of the source equals the conjugate of the load impedance. The battery can deliver maximum power to a d.c. load when $R_L = R_i$. This gives

$$P_{\max} = \frac{E_i^2}{4R_i} \tag{9.11}$$

Since E_i and R_i vary with the state of charge, the $P_{\max}$ also varies accordingly. The internal loss is I^2R_i, which is also equal to $P_{\max}$ delivered to the load. The power transfer efficiency while delivering theoretical possible peak power is therefore only 50%. Such poor efficiency may not matter for a short-duration peak load. Both NiCd and NiH_2 batteries are capable of following pulse discharge and charge lasting a few milliseconds. The battery voltage follows the load variations as expected even under pulse loading. The power peaks shorter than a few milliseconds durations primarily comes from the capacitance between the electrodes plates.

9.8 Performance Comparison

Figure 9.27 compares the specific energy and energy density of various electrochemical cells. The voltage varies with DOD. For 60% DOD, it varies from 1.22 to 1.26 V at moderate discharge rate. The average charge and discharge voltages are listed in Table 9.4. Other performance characteristics and properties of different types of cell are summarized and compared in Table 9.5 and Table 9.6. With NiCd and NiH_2 cells, the battery mass taking into considerations all design factors and the packaging hardware is given in Figure 9.28 for various load levels, service life, and DOD; and their qualitative differences are listed in Table 9.7. The battery mass required to meet typical power system performance with various cells is shown in Figure 9.29 in percentage of the NiCd battery mass. The life and cost are compared in Table 9.8.

9.9 Battery Design

The battery design starts with selecting the electrochemistry and the number of batteries based on the mission requirement and the heritage database. The following factors have strong influence the chemistry selection:

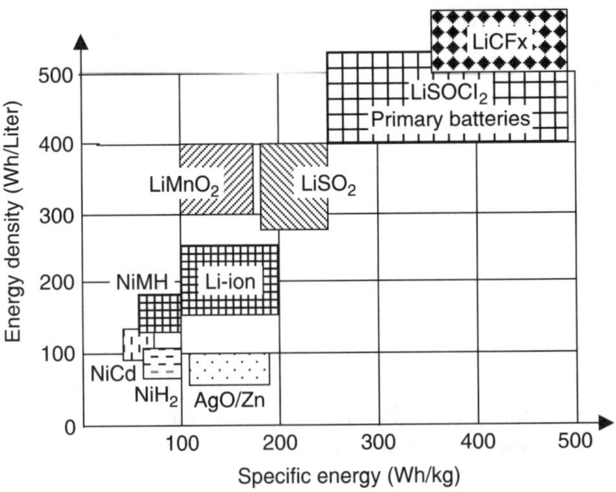

FIGURE 9.27 Specific energy and energy density of various electrochemistries.

Table 9.4 Voltage characteristics of various batteries for space applications

Electro-chemistry	Cut-off voltage	Discharge voltage	Charge terminate voltage	Remarks
NiCd	1.0	1.25	1.55	Exhibits memory effect
NiH_2	1.0	1.25	1.55	No memory effect
NiMH	1.0	1.25	1.55	Temperature sensitive
Li-ion	2.7	3.50	4.20	Safe, contains no metallic lithium
Lithium-polymer	2.7	3.50	4.20	Contains metallic lithium
Silver-zinc	1.3	1.50	2.05	Needs good air management to limit self-discharge rate

Table 9.5 Energy and power densities of various cells

Electro-chemistry	Specific energy (Wh/kg)	Energy density (Wh/l)	Specific power (W/kg)	Power density (W/l)
NiCd	40–5	50–100	150–200	300–500
NiH_2	45–65	35–50	150–200	200–300
NiMH	50–70	140–180	150–200	300–500
Li-ion	90–150	150–250	200–220	400–500
Lithium-polymer	100–200	150–300	>200	>400

Table 9.6 Battery characteristics affecting thermal design

Electro-chemistry	Operating range (°C)	Over-charge tolerance	Heat capacity (Wh/kg·K)	Mass density (kg/l)	Entropic heating on discharge W/A·cell
NiCd	−20 to 50	Medium	0.30	1.7	0.12
NiH_2	−10 to 50	High	0.35	1.0	0.10
NiMH	−10 to 50	Low	0.35	2.3	0.07
Li-ion	10 to 45	Very low	0.38	1.35	0
Lithium-polymer	50 to 70	Very low	0.40	1.3	0

- Specific energy and energy density
- Cycle life, stability of capacity and voltage
- Round trip energy efficiency
- Mass and volume constrains
- Temperature effects on the performance
- Ampere-hour capacity ratings available

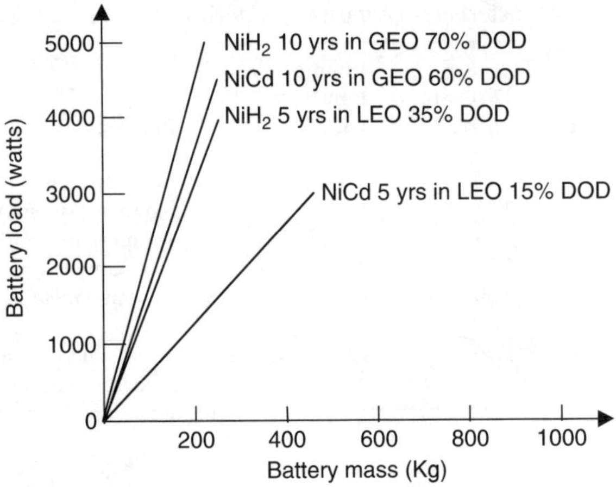

FIGURE 9.28 NiCd and NiH_2 battery mass versus load and missions duration.

Table 9.7 Qualitative differences in NiCd and NiH_2 batteries

Battery	Advantages	Disadvantages
NiCd	Prismatic cell requires less supporting Low operating temperature, hence simpler thermal management	Cell performance not easily scalable to high Ah ratings. Sensitive to over-charge at high temperature Must be maintained within narrow temperature range of –5 to 10°C. Memory effect requires periodic reconditioning, making the fully autonomous operation difficult.
NiH_2	Less sensitive to over-charge Longer life	Cylindrical cells results in low energy density, heavy supporting structure, and poor cooling. High self-discharge rate

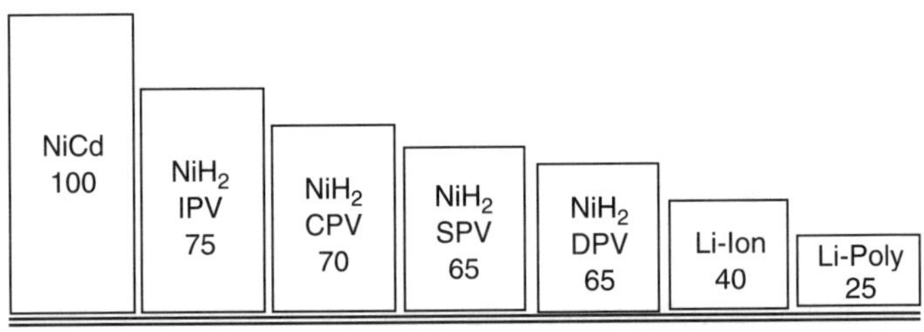

FIGURE 9.29 Battery mass with various electrochemistries relative to NiCd cell.

- Ease and speed of recharging
- Self-discharge rate
- Safety issues

Table 9.8 Life and cost comparison of various batteries

Electo-chemistry	Cycle life in full discharge cycles at 25°C	Calendar life, in years	Self–discharge (percent/month at 25°C)	Relative cost ($/kWh)
NiCd	1000–2000	10–15	20–30	1500
NiH_2	2000–4000	10–15	20–30	1500
NiMH	1000–2000	8–10	20–30	2000
Li-ion	500–1000[a]	a	5–10	3000
Lithium-polymer	500–1000[a]	a	1–2	>3000

[a]Under development.

Although NiH_2 offers higher energy storage capability per kilogram, both NiCd and NiH_2 continue to find applications. One can be more advantageous over the other depending on the mission-specific details. For example, NiCd is a more likely candidate for low power missions, because it is available in a wide range of capacity starting from a fraction of 1 Ah. The NiH_2 is generally better suited for high power satellites, and is not mass and cost effective below 25-Ah capacity.

Regarding the number of batteries, all GEO satellites have traditionally used two batteries, each with one half of the total Ampere-hours capacity needed to meet the energy demand during the worst case eclipse. For small LEO satellites, such as *Iridium*, one battery can be advantageous. The mass and cost of one large battery is always less than two small batteries of one half the capacity. On the business side, the customer generally has a strong preference, sometimes for documented or undocumented reasons, which is taken into account.

Once the chemistry and the number of batteries are settled, the battery design depends on the following system parameters:

- Bus voltage and load current
- Charge and discharge rates and duration
- Operating temperature during charge and discharge
- Life in number of charge and discharge cycles

The life consideration is the dominant design driver in setting the battery Ampere-hour ratings. Even when the load may be met with a smaller capacity, the battery is oversized to meet the cycle life requirement. For example, with the same watt-hour load, the battery that must deliver twice as many C/D cycles needs approximately double the capacity.

The battery design proceeds in the following steps:

- Determine the number of series cells required to meet the voltage requirement.
- Determine the ampere-hour discharge required to meet the load current demand.
- For the required number of charge/discharge cycles, determine the maximum allowable depth of discharge.
- The total ampere-hour capacity of the battery is then determined by dividing the ampere-hour discharge required by the allowable depth of discharge.
- Determine the number of battery packs required in parallel for the total ampere-hour capacity from packaging and other considerations.
- Determine the temperature rise and the thermal control requirement.
- Determine the required charge and discharge rate controls.

The actual ampere-hour capacity of each battery at operating temperature is given by the following expression:

$$Ah_b = \frac{P_e \cdot T_e}{N_b \cdot \eta_{dis} \cdot \left\{ (N_c - 1)\, V_{cdis} - V_d - V_{hdis} \right\} \cdot \mathrm{DOD}} \tag{9.12}$$

where
Ah_b = capacity of each battery
P_e = load power demand during eclipse
T_e = eclipse duration, in hours
N_b = number of batteries in parallel
η_{dis} = discharge converter efficiency
N_c = number of series cells per battery
V_d = voltage drop in bypass diode in case a cell failed
V_{cdis} = voltage per cell, average during discharge
V_{hdis} = voltage drop in harness from battery to PRU
DOD = maximum allowable DOD in the worst case eclipse

Equation 9.12 allows one cell failure in each battery for single fault tolerant design. The failed cell is bypassed using diodes as shown in Figure 9.30. There is generally one bypass diode in the charge path and three diodes in the discharge path to limit the leakage current. Without such bypass diodes, if one battery in a two-battery system fails, the power available to the payload is significantly reduced, plus one arcjet would not work. The satellite must carry more fuel to compensate for this. Some

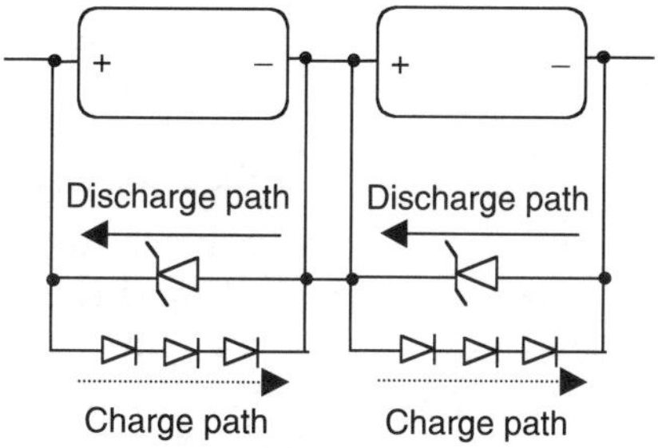

FIGURE 9.30 Cell bypass diodes in charge and discharge paths for reliability.

customers may prefer relay switches to the bypass diodes to minimize the power loss.

The battery round trip energy efficiency under the worst case with one cell failed is determined using the following expression:

$$\eta_b = \left[\frac{(N_c - 1)\,V_{cdis} - V_d - V_{hdis}}{(N_c - 1)\,V_{cchg} + 3\,V_d + V_{hchg}}\right] \cdot \left[\frac{\mathrm{AH}_{dis}}{\mathrm{AH}_{chg}}\right] \tag{9.13}$$

where

N_c = number of series cell, V_d = voltage drop in one diode

V_{cdis} and V_{cchg} = cell voltage during discharge and charge, respectively

V_{hdis} and V_{hchg} = voltage drop in harness during discharge and charge

AH_{dis} and AH_{chg} = ampere-hour discharged and charged, respectively.

There is a trade in selecting the cell ampere-hour rating in fully regulated DET bus. It is between using more small capacity cells giving closer to the bus voltage, versus fewer large capacity cells giving lower than bus voltage. The latter option requires greater buck and boost ratios in the battery regulators, but saves in mass and cost. Fewer larger cells have lower mass and volume and higher reliability. The battery voltage in any case is regulated by the duty ratio, and the regulator mass and efficiency do not change much with the battery voltage. A general guideline is to place cells in series to make the nominal battery voltage during discharge equal to about 80%, and during charge about 93% of the bus voltage. Even fewer cells may yield more savings as long as the buck and boost converter design is not adversely impacted.

Other factors also enter the battery design, such as the end of charge and end of discharge voltages. They remain fairly constant in NiH_2, but degrade in NiCd over the service life.[4] The ampere-hour rating thus determined must be adjusted upwards if the battery voltage does not meet the bus voltage requirement, cannot be fully recharged after the deepest DOD, or the nominal eclipse starts from a partially drained battery in some orbits.

All series cells must have equal Ampere-hour ratings within a few percent to avoid a cell voltage reversal leading to a total battery failure. For this reason, the cell selection is a critical acceptance test conduced at the vendor site. The cells are discharged to 1.0 V and the ampere-hour output measured at room temperature to match the nameplate capacity. All cells within ±3 to 5% capacity variations around the average are selected for assembly in the flight battery. Since the actual battery capacity is the capacity of the weakest link, it is taken as the average cell capacity less 3 to 5%. Battery cells are selected also by the cell voltage variance. One may consider the magnetic signature method to gage the internal details. Only the stabilized cells are included in the selection. The full SOC pressure in NiH_2 cells is significantly low when new, but stabilizes after a number of C/D cycles.

Typical specifications of the battery for 5- to 10-kW 100-V GEO communication satellite include the following:

- For 800 psi full SOC pressure, the battery should withstand 1350 psi without permanent deformation, 5000 cycles of pressure test cycles from fully charge to fully discharge state, and 2400 psi proof test without rupture.
- d.c. isolation resistance of 50 MΩ at 1000 V d.c.
- Reliability of 0.985 over 10 years of life based on 30 FITS per cell
- The radiation dose of 5×10^5 rads over a 10-year mission life
- Ability to withstand the required dynamic environment of the launch

9.10 Launch and Ascent Power

Generally, the battery is designed to meet the on-orbit load and energy balance. The same battery is also used for the spacecraft power needs during launch and ascent phases. Several possibilities that may arise during this time are dealt in the following manner.

Case A: The battery fully charged at the time of launch is sufficient for the launch/ascent energy requirement. This is usually the case in most satellites, and does not require any extra step for the power system engineer.

Case B: The battery fully charged at the time of launch is slightly short of the launch/ascent energy need. In this case, a widely practiced solution is to pre-cool the battery by cold air or nitrogen before the launch to increase the battery capacity by several percent or more as needed.

Case C: The launch/ascent power demand exceeds the battery capacity beyond the case B remedy. Increasing the on-board battery capacity, even in such a rare case, is almost never done, because it is very expensive. Instead, a primary battery having high specific energy is used to meet the shortage. This battery is jettisoned after the ascent. A primary battery that is widely used in space vehicles is lithium carbon monofluroid (LiCFx) having a specific energy of 350 to 500 Wh/kg, an order of magnitude higher than the commonly used secondary batteries.

9.11 Thermal Design

The battery is sometimes mounted on the exterior of the spacecraft body on a honeycomb panel on the anti-sun side. It is conductively isolated from the spacecraft honeycomb panel by G-10 fiberglass standoffs, radiatively isolated by multi-layer insulation blankets, and enclosed in an aluminum honeycomb micrometeoroid shield. A part of the shield front serves as the radiator, which is coated with silver Teflon. The remainder of the shield is insulated with multi-layer thermal blankets. The battery radiates heat to the radiator, which, in turn, radiates to space. Adequate thermal control for the battery, for both cooling and heating, is extremely important. A high temperature in a battery can lead to failure caused by electrolyte evaporation and decreased efficiency. A high cell-to-cell thermal gradient can cause uneven current sharing. The operating temperature significantly influences many performance parameters in the following ways:

- The capacity drops at a temperature above or below a certain range, and drops sharply as the temperature approaches the freezing point. The capacity significantly increases when charged at −20°C. Additional capacity obtained at low temperature is more stable than that obtained at 10°C when cycled.
- The battery voltage and charge efficiency decrease with increasing temperature.
- The self-discharge rate increase with temperature.
- The internal resistance increases as temperature decreases.

Table 9.9 shows the temperature sensitivity of NiCd cell. The process of determining the optimum operating temperature is illustrated in Table 9.10. It indicates that different attributes have different desirable operating temperature ranges shown by the vertical lines. With all attributes jointly considered, the most optimum operating temperature is the intersection of

Table 9.9 NiCd temperature sensitivity

Temperature	Charge efficiency (%)	Discharge efficiency (%)	Self-discharge rate, percent of *C*/day
−40°C	0	72	0.1
10°C	94	100	0.2
60°C	75	100	8.0

all the desirable ranges. For example, if we wish to limit the self-discharge rate below 1.0% and the charge efficiency above 90%, the optimum working temperature range is between −10°C and 25°C as shown by thin line in the first column. The actual temperature is then maintained between −5°C and +10°C shown by the heavy vertical line in order to allow some design and operating margins.

It is necessary to perform detailed thermal analyses for every mission, because it is mission unique. Controlling the battery temperature is a critical requirement of the thermal control system. This often requires dedicated radiators, conduction paths, heaters, blankets, temperature sensors, and

Table 9.10 Determination of optimum working temperature for a nickel cadmium battery

Operating temperature °C	Charge efficiency(%)	Discharge efficiency(%)	Self-discharge rate, percent capacity/day
−40	0	72	0.1
−35	0	80	0.1
−30	15	85	0.1
−25	40	90	0.2
−20	75	95	0.2
−15	85	97	0.2
−10	90	100	0.2
−5	92	100	0.2
0	93	100	0.2
5	94	100	0.2
10	94	100	0.2
15	94	100	0.3
20	93	100	0.4
25	92	100	0.6
30	91	100	1.0
35	90	100	1.4
40	88	100	2.0
45	85	100	2.7
50	82	100	3.6
55	79	100	5.1
60	75	100	8.0
65	70	100	12
70	60	100	20

control electronics. Multilayer insulating blankets are used to keep the heat or cold away as required for maintaining the orbit-average battery temperature. If the battery is running cold, blanketing a part of the external radiator surface can reduce the heater power requirement. The battery electrochemistry selection has a significant impact on the thermal system mass and cost. The chemistry that costs more but is less sensitive to the operating temperature can pay back at the satellite level.

Each cell in the battery pack is electrically insulated from each other and from the ground. The electrical insulation must be good conductor of heat to maintain low temperature gradient between the cells and also to the ground. The maximum temperature must remain below a set value to avoid damage to the battery or shortening the life. The minimum temperature must not be too low to reduce the battery efficiency in the operating mode. In the survival mode, the temperature must remain above the freezing point of the electrolyte to avoid the battery from being inoperable.

The cylindrical shape of IPV and CPV NiH_2 cells makes the thermal cooling difficult. Moreover, it increases the volume significantly and makes the heat conduction form the cell to the radiator plate inefficient compared to the battery with prismatic cells. Figure 9.31 shows a battery pack with prismatic NiCd or Li-ion cells. The cylindrical NiH_2 cells are packed in one of the following three configurations.

9.11.1 Flat Pack

The cells are laid flat on the radiator plate, offering the maximum cooling surface. Such design has been flown in communication satellites in several

FIGURE 9.31 Battery assembly using prismatic NiCd cells.
(Source: Dornier Satellitensysteme Gmbh, Daimler-Benz Aerospace. With permission.)

kilowatts power range. However, it requires much greater footprint and is vulnerable to shocks and vibration during launch.

9.11.2 Vertical Stack

This is the most common packaging method for NiH_2 cells at present, which was seen in Figure 9.6. The cells standing up and making contact with the radiator plate only at the dome point results in poor heat conduction. The conduction area is increased by providing conduction shims shown in Figure 9.32. The wedge shape minimizes the mass for the required heat transfer. The shims can be aluminum or graphite fiber composite for even lower mass.

9.11.3 Wine Rack

The cylindrical IPVs are sometimes packaged in a wine-rack configuration. In such design, the cell temperature can reach 15°C and maximum cell-to-cell gradient can reach 7°C in the worst case EOL. The cell-to-cell and cell-to-battery frame temperature can be improved by using metallic fiber, which could reduce the cell-to-cell temperature gradient to below 2°C, and maximum cell temperature to about 8°C.

The three NiH_2 cell stacking configuration options are compared in Table 9.11. The typical temperature requirements for large communication satellite battery design are listed in Table 9.12.

9.12 Safety Considerations

The battery design and operation require certain safety considerations. The most important is not to over-charge the battery. Any over-charge above the trickle charge rate is converted into heat, which can cause the battery to explode if allowed to build up beyond limit. This is particularly critical when the battery is charged directly from a dedicated photovoltaic module without a charge regulator in small science missions with short or infrequent eclipses. In such cases, the array is sized below certain safe

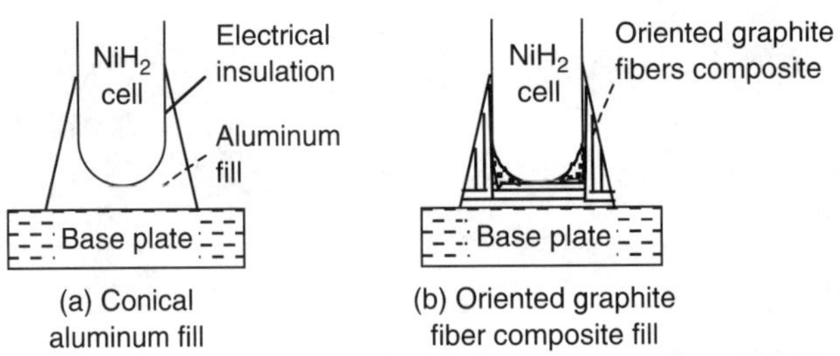

FIGURE 9.32 Mass-efficient conduction shims for vertical mounting of NiH_2 cell.

Table 9.11 NiH_2 cell stacking options and their relative performance

Configuration	Wine-rack	Flat pack	Vertical mount
Advantages	Low footprint, volume, and weight.	Lowest weight Largest foot print Better cooling	Best cooling Moderate footprint and volume
Disadvantages	Poor cooling	Large foot print	Heavy with metal cooling sleeve

limit. As a rule of thumb, such photovoltaic array rating is kept below the continuous trickle charge current that can be tolerated by the battery.

The battery design must consider a single failure mode of the battery without creating any safety hazard of contamination, corrosion, explosion, fire, injury or illness. This is required during all mission phases and ground operations. Major failure modes are listed below. Any one or more combinations could possibly cause venting of toxic, corrosive, and/or flammable materials with associated fire or explosion risk:

- Over-temperature from battery self-heat or ambient heat
- Short circuit, internal or external to the battery
- Reverse current due to inadvertent charging
- Cell voltage reversal due to over discharge
- Leakage of electrolyte or gases

Table 9.12 Typical thermal design requirement for NiH_2 battery in large communication satellite

Maximum gradient or any cell temperature	
Gradient along cell stack as measured at cell wall	5°C
Gradient between any two points of a cell wall	7°C
Gradient between any two cell stacks within a battery during charge	7°C
Gradient between average of the cell stacks between any two batteries or packs during charge	8°C
In transfer orbit during charge	30°C
During recharge following eclipse	15°C
During trickle charge	30°C
During EHT or arcjet firing	35°C
During reconditioning letdown	30°C
During reconditioning recharge	15°C
Minimum temperature in all scenarios	−10°C
Life average temperature of cell stack in a battery	15°C
Acceptance testing (excluding EHT/arcjet firing)	−10/+30°C
Qualification testing (including EHT/arcjet firing)	−15/+45°C

- Over-pressure due to over-charging or over-temperature

The design approach should preclude any of the above occurrences. If failure does occur, venting should be prevented. If that is not possible in the case of a major failure, it must be demonstrated that venting will occur in a manner that is not hazardous to the mission. Safe shipping requirements depend primarily on whether the battery is activated or not. If activated, the applicable shipping regulations depend on the chemical reactivity of the materials and the state of charge, both can be controlled by the battery manufacturer to meet the shipping regulations.

Govar and Squires[12] report safety tests on commercial off-the-shelf 9-V lithium-ion battery for navy applications. Battery packs consisting of two 9-V batteries in parallel were found relatively safe and experienced no violent behavior. On the other hand, packs with thirty 9-V batteries experienced venting, smoke, and flames under certain circumstances.

9.13 Charge Regulation

For safety reasons, it is extremely important that excessive charging of the battery is avoided at all times. Over-charging causes internal gassing, which causes loss of electrolyte in vented cell and premature aging. The charge regulator allows the maximum rate of charging until the gassing starts. Then the charge current is tapered off to trickle charge rate so that the full charge is approached gently.

During battery charging, the energy management software monitors the state of charge, the overall health and the safe termination criteria. The operating voltage, the current, and the temperature are monitored. The charging timer is started after all initial checks are successfully completed. Charging may be suspended (but not reset) if the software detects violation of critical safety criteria. The timer stops charging if the defect persists beyond a certain time limit.

The normal charging has the following three phases:

- Bulk (fast) charge, which deposits 80 to 90% of the drained capacity
- Taper (cut-back) charge in which the charge rate is gradually cut back to top off the remaining capacity
- Trickle (float) charge to counter the self-discharge rate after the battery is fully charged

The bulk charge and the taper charge termination criteria are pre-loaded in the battery management software to match with the battery electrochemistry and the system design parameters. For example, the NiCd and NiMH batteries are generally charged at constant current until about 95% SOC in 75 to 85% of the available sun time. Then onward, the charge current

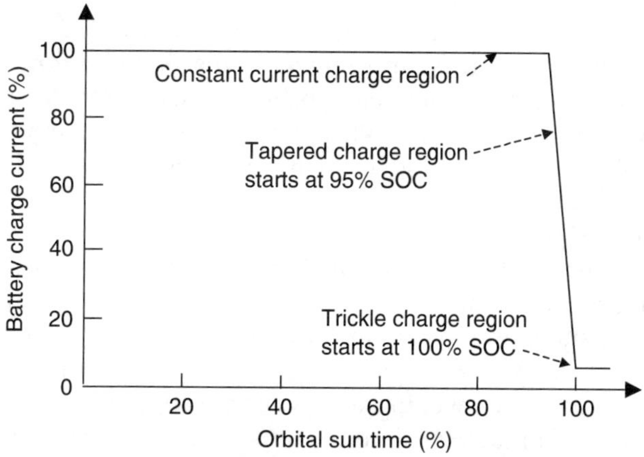

FIGURE 9.33 Battery charge profile versus orbital sun time.

is tapered off — gradually or in a few steps — to the trickle rate as shown in Figure 9.33. The decision is made to terminate the main charging and switch to the trickle charge when the continuously monitored $\triangle V$ is detected negative. On the other hand, Li-ion batteries, being sensitive to overcharging, can be charged at constant voltage, and then tapering off the charge current as needed.

Both NiCd and NiH_2 batteries show voltage roll-off and sudden temperature rise, i.e., negative dV/dt and positive dT/dt, soon after the full charge (Figure 9.34). Both of these characteristics are used in detecting the end of charge. In NiCd and NiH_2, the voltage roll-off is 20 to 30 mV per cell. In NiMH, on the other hand, the $\triangle V$ is much smaller (3 to 5 mV per cell) and the temperature rises steadily to the final value, making the

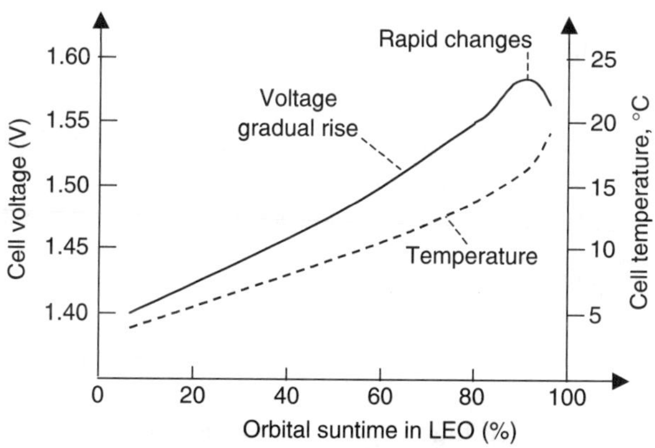

FIGURE 9.34 Battery voltage and temperature during charge showing rapid changes at the end of charge.

end of charge detection difficult. The rollover is attributed to the voltage decrease subsequent to temperature increase that results from the oxygen reaction with hydrogen. In some electrochemistries, the voltage rollover is distinct; in some it is rather flat. When the cells is over-charged at −20°C, the voltage rollover shows waviness with a smaller hump followed by a larger hump.

Rapid reaction of oxygen with hydrogen during over-charge is an undesirable feature, which causes an audible popping noise in certain cases. Popping results in melting of the gas screen pinholes in the negative plates, and heat stains in the core and the positive plate. In extreme cases, popping blows away the separator, resulting in shorts. Cells with significant positive plate swelling are prone to popping, which results in damage to the components. Swelling also causes extrusion of active material and reduction in the gap between the two positive plates. The final result is a decrease in the end of discharge voltage. Positive plates that show excessive swelling absorb more electrolytes. This is accompanied by a reduced absorption of electrolyte by the separator.

The batteries are charged in one of the following three ways using the voltage-limited tapered-current method or the constant-current ampere-hour return method.

9.13.1 Multiple Charge Rates

This is the best method, in which the battery is charged gently in multiple steps. First the battery is charged at full charge rate until 80 to 90% of the capacity is achieved. The charge current is then cut back, in steps, until the battery is fully charged. At this time, the charge current is further reduced to trickle-charge rate, keeping it fully charged until the next load demand comes on the battery. This method therefore needs at least three charge rates in the charge regulator design.

9.13.2 Single Charge Rate

This method uses simple low cost regulator, which is either on or off. The regulator is designed for only one charge rate. When the battery is fully charged, as measured by its terminal voltage, the charger is turned off by a relay. When the battery voltage drops below a preset value, the charger is again connected in full force. Since the charging is not gentle in this method, full charge is difficult to achieve and maintain. An alternate version of this charging method is the multiple pulse charging. Full current charges the battery up to the high preset voltage just below the gassing threshold. At this time, the charger is shut off for a short time to allow the battery chemicals to mix and voltage to fall. When the voltage falls below the low preset threshold, the charger is reconnected, again passing full current to the battery.

9.13.3 Unregulated Charging

This is the least cost method which uses no charge regulator. The battery is charged directly from a solar array module dedicated just for charging. The charging module is properly designed for safe operation for the given number of cells in the battery. When the battery is fully charged, the array is fully shunted to ground by a shorting switch (transistor). At this time, the battery voltage must still be slightly lower than the available solar array voltage so that the trickle charge can continue, but not much lower to avoid over charging. The shunt transistor switch is open when the battery voltage drops below certain value. An isolation diode would block the battery powering the array or the shunt at night.

If one cell in one of the two batteries working in parallel fails short, the two batteries would have different terminal characteristics. Charging or discharging such batteries in parallel can result in highly uneven current sharing, subsequently overheating one of the batteries. This can be avoided in two ways. One is to charge and discharge both batteries with individual current controls such that they both draw their rated share of the load. The other is to replace the failed cell immediately, which is not possible in space. In general, the individual charge/discharge control for each battery is the best strategy. It may also allow replacement of any one battery with different electrochemistry or different age, which would have different load sharing characteristics. The ISS batteries are designed for replacements several times during the life of the station.

The Li-ion battery requires additional circuitry for balancing individual cells to obtain maximum performance from the battery. The balancing method can vary from design to design. One method is to use d.c.–d.c. converters with multi-tap transformer to provide energy equilibrium in each cell. Another method is to use capacitor-switching technique called flying capacitor that shifts small packets of energy from higher voltage cells to lower voltage cells. The Saft Corporation has often used a resistive dissipation method that lowers the SOC of the highest cells down to match with the lowest common denominator cell. Sack, Croydon and Reynolds[13] report yet another method using segmented charger. In this concept, the battery charger is broken up into separate cell chargers. The concept was tested on a 26-V 45-Ah Li-ion battery under U.S. Army funding.

9.14 Battery Management

Battery management electronics include the battery monitor and controller. The main function is to monitor the battery status and optimize its performance by adjusting the charging process. Electronic circuits monitor the battery voltage, current, state of charge, and temperature. If not managed properly, the battery performance could suffer and the life shortened. The following are some potential performance problems:

- Low charge efficiency resulting in low state of charge
- Loss of capacity to hold the rated ampere-hour charge
- Premature failure leading to loss of load availability
- Excessive gassing and heating, leading to short life
- Positive plate corrosion shortening the life
- Stratification and sulfation degrading the performance

The following features incorporated in the battery management can avoid the above problems:

- Temperature compensated charging, in that the charge termination occurs earlier if the battery temperature is higher than the reference temperature
- Individual charge control if two or more batteries are charged in parallel
- Accurate set points to start and to stop the charge and discharge modes

The batteries in modern power systems are managed by dedicated computer software, which monitors and controls the following performance parameters:

- Voltage and current
- Temperature and pressure
- Ampere-hour in and out of the battery
- State of charge and discharge
- Rate of charge and discharge
- Depth of discharge
- Number of charge and discharge cycles

The above parameters are determined by a number of measurements and limits, which include the following:

- Minimum charging current threshold
- Terminal voltage threshold
- Absolute temperature cut-off
- Rate of temperature change $\triangle T/\triangle t$
- Drop of the cell voltage after the peak $\triangle V$
- Rate of voltage change ($\triangle V/\triangle t$), including the peak and the inflection
- Actual charging time

The ampere-hour integrating meter is commercially available, which keeps track of the ampere-hours in and out of the battery and send required signals to the mode controller.

With age, NiCd and NiH_2 batteries experience walk-downs in capacity, voltage, and end-of-charge pressure. These are accounted by shifting the software on board. The cell voltage divergence is another concern, more so after reconditioning. One method of avoiding it is to over-charge, so that all cells saturate to full voltage, provided the thermal system can handle the associated heating.

The temperature compensation on the maximum battery voltage and the state of charge can improve the battery management, particularly when cold. It can allow additional charging during cold periods when the battery can accept more charge. The low voltage alert is a good feature, as excessive discharging below the threshold low voltage can cause cell voltage reversal, leading to battery failure. The alert can be used to shed noncritical loads from the battery to avoid battery damage. The battery cell voltage monitor allows a comprehensive health check of the total battery. This is important just before launching, when a problem can be fixed. A 22-cell voltage monitor adds less than 200 to 300 g, but adds much more in numerous telemetry wires.

9.15 Dynamic Model

The term impedance refers to the ratio of the cell voltage change to corresponding change in the alternating current through the cell. It is given by

$$Z = \sqrt{R^2 + X^2} \tag{9.14}$$

where R and X are the battery resistance and reactance, respectively. The dynamic impedance characteristics discussed in this section are applicable to all electrochemistries. They apply at a given SOC, i.e., when net d.c. current flow or energy conversion is not involved. The internal impedance determines the battery response to ripple, noise, and transients during charging and discharging. It varies inversely with the capacity, as large capacity cells have large plate area. The impedance versus frequency of a 50-Ah NiH_2 cell is shown in Figure 9.35. It has a relatively constant value up to 10 Hz, but rises rapidly at higher frequency indicating a strong inductive component. The battery R and X are obtained by multiplying the cell values with number of series cells in the battery. A simple dynamic model of the battery for stability analysis with the dynamic impedance $Z(s)$ in the Laplace domain is shown in Figure 9.36.

Complex battery models developed by electrochemists are not expressed in terms of the electromotive force, resistance, inductance, and capacitance.

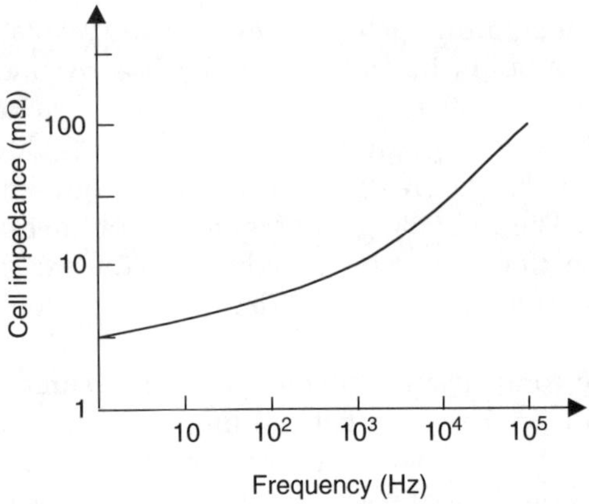

FIGURE 9.35 50-Ah NiH_2 cell impedance versus frequency.

Therefore, they are unsuitable for everyday use by the electrical engineer, who needs a model that represents the electrochemistry in the electrical network terms. Ceraola[14] has developed a model to represent the battery's nonlinear behavior using two electrical branches shown in Figure 9.37. The main branch is designated with subscript m and the parasitic branch with subscript p. The E_m and Z_m are the internal electromotive force and the internal impedance, respectively. Both are functions of the Laplace parameters, the electrolyte temperature, θ, and the SOC. The parasitic branch models the nonreversible reactions that draw some current, but does not contribute to the main reversible reaction. The energy that is absorbed by E_p ceases to be electrical and is converted into other forms, such as gas or water electrolysis that occurs at the end of the charge process. The power dissipated in the real part of Z_m and Z_p is converted into heat that contributes to the self-heating of the battery. The R_o is common to both branches and represents resistance from the battery terminals to the plates.

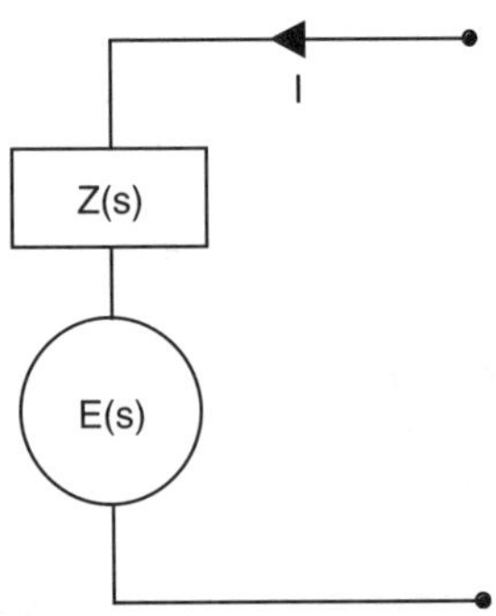

FIGURE 9.36 Simple dynamic electrical circuit model of the battery.

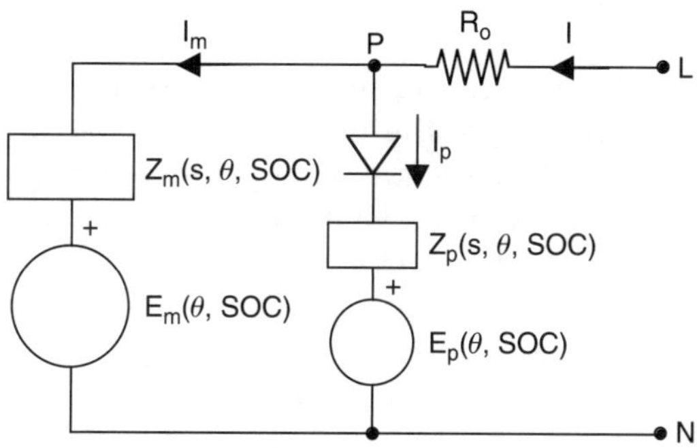

FIGURE 9.37 Dynamic circuit model of battery with main and parasitic reaction branches.

To make such a model useful, one need to determine the functional dependence of E and Z on s, θ, and SOC, and the battery thermal model from which the electrolyte temperature can be determined. The dependence on s can be stated by expressing each Z in terms of the equivalent R, L, and C values. The model can be simplified for a specific operating condition, if known. For example, the parasitic branch can be ignored in the initial charging phase when the charge efficiency is near unity. At the end of charge, however, it takes almost all power entering the battery terminals. Since the inductance is small and can be ignored, the battery model that is valid during both the charge and discharge is shown in Figure 9.38.

The battery capacity is maximum at certain optimum temperature and falls off on both sides of the optimum temperature. If the battery is maintained on the cooler side of the optimum temperature, the capacity at a constant discharge current and a constant electrolyte temperature is higher with higher temperature and lower discharge current. At a fixed discharge current, I, and temperature, θ, the capacity can be expressed as[14]

$$Ah(I, \theta) = Ah_o(I)\left\{1 - \frac{\theta}{\theta_f}\right\}^{\alpha} \tag{9.15}$$

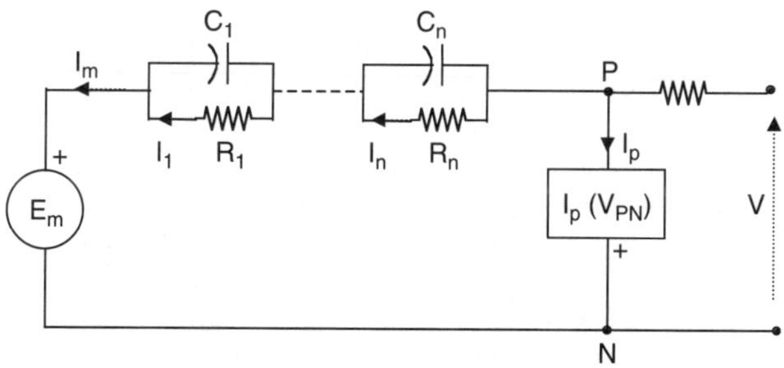

FIGURE 9.38 Dynamic circuit model of the battery during charge and discharge.

where Ah_o = battery capacity at 0°C at discharge current I, α = empirical constant, and $\theta > \theta_f$, the freezing temperature of the electrolyte (such as −40°C). Equation (9.15) gives the expected value of $Ah = 0$ at freezing temperature when the battery becomes inoperative. It is possible that $Ah_o(I)$ can be expressed in terms of its value at a reference temperature I_r and empirical constants K and β as follows:

$$Ah(I,\theta) = \frac{KAh_o(I_r)\left\{1 - \dfrac{\theta}{\theta_f}\right\}^{\alpha}}{1 + (K-1)\left\{\dfrac{I}{I_r}\right\}^{\beta}} \tag{9.16}$$

Equation (9.16) suggests that numerical values of the elements of the circuit model have to be established for each value of electrolyte temperature and SOC. In practice they can be established only for a few discrete values of these two quantities, and the values in-between can be obtained by interpolation. Determination of such parameters for the first time would take extensive analysis and test efforts. However, having made that investment, the model for a new battery can be established by using previous data on similar battery suitably modified for the new battery.

9.16 Cycle Life Model

As previously stated, the battery cycle life is related primarily to the DOD and the operating temperature. The life relationship can be expressed in terms of C/D cycles as the dependent variable, with the DOD and temperature as the independent variables. In developing a statistical model for the battery life, we must first select a mathematical expression that defines the relationship adequately, and then select a statistical distribution that best fit the test data. The selections can be guided by consideration of the physical behavior of the battery. For example, the number of cycles to failure decreases with increasing DOD and temperature. Various types of relationships commonly used for analyzing life test data can be considered to express such relationships. Among the alternatives, such as exponential, inverse power, and Arrhenius, more than one can give a reasonably good fit to the test data. However, the relationship that generally gives the best fit in similar life test data is the Arrhenius equation, which accounts for the chemical degradation limiting the life.[15,16] That is

$$L = C_1 T^{-C_2} \cdot e^{-C_3 D} \tag{9.17}$$

where L = mean number of cycles to the first failure, T = temperature in degress centrigrade, D = DOD in each cycle, and C_1, C_2, and C_3 are the relationship constants to be determined by multiple regression curve fitting techniques.

Based on the physical considerations of the wear-out mode of failure being studied, the statistical distribution selected to represent the spread of life around the mean value derived from Equation (9.17) at a given temperature and DOD should have a continuously increasing hazard function. The hazard function is a conditional failure rate, which expresses the rate of failure for the surviving population at a given number of cycles L. Also, the distribution should have a good shape flexibility to fit the data. Two distributions, which can satisfy these requirements, are lognormal and Weibull. The Weibull distribution is the most commonly used for presentation of such life test data. It is given by

$$F(L) = 1 - e^{\left(\frac{L}{\alpha}\right)^{\beta}} \tag{9.18}$$

where $F(L)$ is the fraction of the population failing by L cycles, and positive parameters α and β are scale and shape parameters, respectively. The parameter β can be assumed to be constant, while α is a function of temperature and DOD. The values of both α and β are derived from test data. The curve fitting can be a simple procedure using least squares regression techniques if the failure times of all samples are known. However, in typical situations, the battery test data includes samples: (a) which are known to have failed during an interval, rather than at a specific time, and (b) many samples not yet failed at the end of the test period. Such censored data can only be analyzed by statistical techniques developed for dealing with incomplete life test data.[17,18] They use the method of maximum likelihood, which involves selecting as an estimate for each unknown parameter the value that appears to be most probable on the basis of the given test data. The computations are based on the sample log likelihood which is a function of the unknown model coefficients (C_1, C_2, C_3). The estimates of the coefficients are obtained numerically by iterative methods on a computer, and the variances of these estimates are used to obtain approximate confidence intervals.

When the indicated analysis is performed on a set of test data, the expected life relationships is derived with its coefficients C_1, C_2, and C_3 established. Estimating the life from such considerations can be relatively less complex if the battery were used in an identical manner in each cycle over the mission life. It becomes more complex when it is used in one way sometimes and another way at other times, with significant variation in the usage patterns. In such cases, the life may be estimates using the *Milner's cumulative damage theory*[19,20] often used in fracture mechanics and fatigue studies on metals and other material. The theory utilizes the cycle ratio as its basic measure of damage. For the battery life with two independent variables as shown in Figure 9.39, it can be modified as follows. If a battery temperature is T_i at depth of discharge D_i for C_i number of cycles, and if the life relationship estimates L_i cycles as its life at T_i and D_i, then the proportional damage, PD_i, caused during that time is

$$PD_i = \left\{ \frac{C_i(T_i, D_i)}{L_i(T_i, D_i)} \right\}^{\alpha} \tag{9.19}$$

where α is a positive constant to be evaluated from correlating available life data and experience. The battery can be expected to fail when the cumulative damage equals unity (i.e., when $\sum PD_i = 1.0$). The evaluation of the exponent α requires extensive data from batteries, which are much closer in scale factor to actual battery to be modeled.

The theory keeps incremental account of the fraction of life depleted during each segment of time depending on the life degradation factors prevailing over that duration of use. It can be used to estimate the remaining life, i.e.,

$$\text{remaining life} = 1 - \sum PD_i \tag{9.20}$$

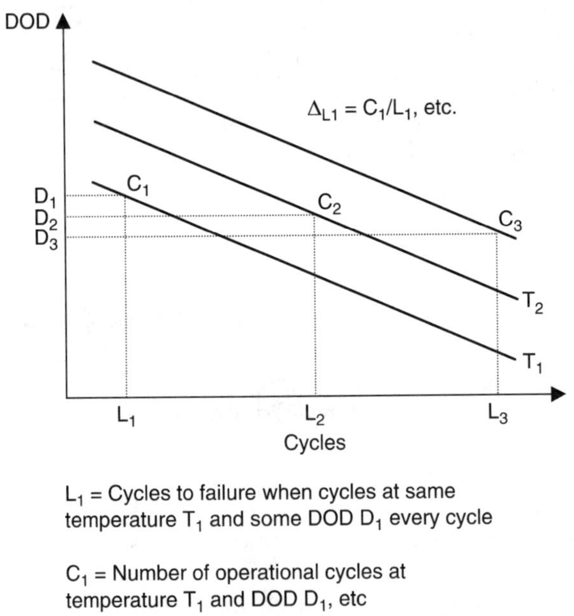

FIGURE 9.39 Life model for cycles to failure versus DOD and temperature.

Milner's cumulative damage theory suggests the following for use with some judgment and experience:

- The battery life ends when sum total of the accumulated proportional damage reaches its limiting value of 1.0.
- A short time at high wear depletes the same fraction of life as a long time at low wear.

Although the life model presented in this section could not justify direct use of the life relationships in this fashion at present, it can be regarded as a potential future application. It is hoped that the theoretical consideration proposed here may serve as a guide to define, plan and analyze future research in this field.

9.17 Primary Battery

Among primary batteries, lithium/polycarbon monofluoride (LiCFx) has the highest specific energy. It has been in used in manned space vehicles, missiles, terrestrial and marine applications since the early 1970s, and has accumulated a safe and reliable database. LiCFx cells are produced in many ratings, but only a few are space qualified for NASA and Air Force applications. Hundreds of LiCFx batteries have flown on space shuttle flights and in many spacecraft. The reliability estimated as per the MIL-HDBK-217 range from 1.6 to 10×10^{-6} FIT. They are insensitive to shock, vibration, forced charging, and forced reversal. Many such cells can survive external hard short circuit without opening the cell vent. A high-rate battery with low heat rejection rate could potentially vent following a sudden internal short circuit. However, sudden shorts are estimated to be highly improbable, yielding the reliability of the cell better than 0.9999 with 90% confidence. The space-qualified battery design using 40 Ah LiCFx cells increases the assembly mass by 60 to 70% over that of the cells.

9.18 Advances in Battery Technologies

Incremental advances in the cell designs continue. The heavy sintered nickel powder plaque has been the substrate of choice for many years to support the electrochemically active material in all nickel-based batteries. The nickel electrode is the most critical and the heaviest component in any nickel-based battery, such as NiH_2, NiCd, and NiMH. The nickel powder plaque amounts to about one half of the electrode weight. Highly porous light weight nickel fiber plaques are being considered to reduce the electrode weight by 50% and the same time increase the capacity of the active material by increasing the surface area available for deposition of the active material. This may lead to higher specific energy for the battery.

Table 9.13 Replacement candidates for NiH_2 cell (50 Wh/kg)

Candidate	Li-ion	Li-Poly	Fly-wheel	Super NiCd	NiMH	Sodium-sulfur	H_2/O_2 RFC
Wh/kg	130	150	a	50	65	120	75
Wh/l	200	300	a	115	200	250	50
Cycle life	> 10 k	> 10 k	High	> 45 k	>10 k	> 3 k	> 2 k
Used in space since	Since 2000	Not yet used	Not yet used	Since 1990	Quald in 1992	Experimental flights	Not yet used
Dev. Stage	Active	Active	Planned to fly on ISS	Used in 30+ spacecraft	Used in 2 spacecraft	Inactive	Was proposed for ISS

[a]Under development.

Another approach that may result in higher specific energy is a thicker nickel electrode. Fewer thick electrodes reduce the number of other components, such as hydrogen electrodes and separators, thus reducing the weight of the cell.

Entirely new or derivatives of time proven electrochemistry are continuously evaluated in search of a better battery. The following figures of merit ultimately determine the battery mass and cost, and should be of interest to the power system design engineer:

- Specific energy and energy density
- Round trip energy efficiency
- Cycle life at the operating temperature and DOD
- Wider operating temperature range
- Better packaging

In recent years, Li-ion has emerged as the battery of choice for high performance commercial applications — cell phones, laptop computers, etc. It offers significant advantages over alkaline-based NiCd and NiMH batteries. The space applications often require the battery to operate over a wide temperature range (−40°C to +65°C), long cycle life (> 30,000 cycles in LEO), and long calendar life (>10 years in GEO). NASA has qualified commercial off-the-shelf Li-ion technology to power electronic peripherals on the space shuttle and the ISS. The *Spirit* and *Opportunity* rovers that landed on Mars in 2004 were powered during eclipse by Li ion batteries capable of 200 cycles at −20°C.

At present, Li-ion appears to be a candidate that may eventually replace the NiH_2 for general use in space.[21] There are other candidates as well, such as the regenerative fuel cell and the flywheel. The flywheel is under significant research funding with a schedule to fly to the ISS in 2006. The NiH_2 replacement candidates as seen at present are compared in Table 9.13.

References

1. Parez, M. E. et al. Energy storage for space applications, in *Proceedings of the 36th Intersociety Energy Conversion Engineering Conference*, ASME, 2001, pp. 85–89.
2. Rao, G.M., Ahmad, A., and Chetty, P.K.R., Super nickel cadmium battery operation and performance on-board the Sampex spacecraft, in *Proceedings of the 30th Intersociety Energy Conversion Engineering Conference, Vol. I*, ASME, 1995, pp. 111–116.
3. Pickett, D.F., Hayden, J.W., Lucero, D., and Johnson, Z., Comparison of advanced NiCd space cell technology with other technologies in consideration for LEO and GEO orbit and planetary missions, in *Proceedings of*

the 35th Intersociety Energy Conversion Engineering Conference, AIAA, 2000, Paper No. 2989.

4. Brown, R., Nickel hydrogen life modeling, in *Proceedings of the 35th Intersociety Energy Conversion Engineering Conference*, AIAA, 2000, Paper No. 2991.
5. Britton, D.L., Progress in the development of lightweight nickel electrode for NiH_2 cell, in *Proceedings of the 34th Intersociety Energy Conversion Engineering Conference*, SAE, 1999, Paper No. 2537.
6. Garner, J.C., Braun, W.R., Loo, D.V., and Bowers, D., 90-Ah dependent pressure vessel NiH_2 battery, in *Proceedings of the 34th Intersociety Energy Conversion Engineering Conference*, SAE, 1999, Paper No. 2590.
7. Puglia, F., Gitzendaner, R., Ehrlich, G.M., and Marsh, C., Advancing Li-ion technology for aerospace applications, in *Proceedings of the 34th Intersociety Energy Conversion Engineering Conference*, SAE, 1999, Paper No. 2637.
8. Croft, H., Staniewicz, R., Smart, M.C., and Ratnakumar, B.V., Cycling and low temperature performance operation of Li-ion cells, in *Proceedings of the 35th Intersociety Energy Conversion Engineering Conference*, AIAA, 2000, Paper No. 27-AP-B1.
9. Fellner, J.P. and Loeber, G.J., Li-ion performance testing and ac impedance characterization, in *Proceedings of the 34th Intersociety Energy Conversion Engineering Conference*, SAE, 1999, Paper No. 2591.
10. Bruce, G. and Marcoux, L., Large lithium-ion batteries for aerospace and aircraft applications, *IEEE AESS Systems Magazine*, 16(9), 24–28, 2001.
11. Smith, R. and Brill, J, "Updated life cycle test results for flight qualified NiH_2 cell design, in *Proceedings of the 36th Intersociety Energy Conversion Engineering Conference*, ASME, 2001, pp. 55–59.
12. Govar, C.J. and Squires, T.L, "Safety tests of lithium 9-volts batteries for navy applications, *IEEE AESS Systems Magazine*, 16(9), 34–37, 2001.
13. Sack, T.T., Croydon, T., and Reynold, R., Segmented battery charger for high density 28-V lithium ion battery, *IEEE AESS Systems Magazine*, 16(9), 15–18, 2001.
14. Ceraolo, P., Dynamic model of the electrochemical battery, in *Proceedings of the 36th Intersociety Energy Conversion Engineering Conference*, ASME, 2001, Paper No. X001.
15. Hahn, G.J. and Shapiro, S.S., *Statistical Models in Engineering*, John Wiley & Sons, New York, 1967.
16. Kendall, M.G. and Stuart, A., *Advanced Theory of Statistics, Vol. 2*, Hafner Publishing Co, New York, 1961.
17. Hahn, G.J. and Nelson, H.B., A comparison of methods for analyzing censored life data to estimate relationship between stress and product life, *IEEE Transactions on Reliability, R-23*, April 1974, pp. 2–10.
18. Nelson, W. A survey of methods for planning and analyzing accelerated tests, *IEEE Transactions on Electrical Insulation. El-9*, March 1974, pp. 12–18.
19. Miner, M.A., Cumulative damage in fatigue, *Journal of Applied Mechanics*, 12(Sept). 1945, A159–164, 1945.
20. Madayag, A.F., *Metal Fatigue — Theory and Design*, John Wiley and Sons, New York, 1969, pp. 170–203.
21. Gross, O., Fox, C, Roller, D., and Shimanek, L., Lithium-ion polymer batteries for space applications, *Proceedings of the 36th Intersociety Energy Conversion Engineering Conference*, ASME, 2001, pp. 61–68.

Chapter 10

Power Electronics and Magnetics

10.1 Introduction

The spacecraft power electronics basically control the bus voltage and convert voltage levels to match the operating voltages of various components. These functions are performed by solid-state semiconductor devices used as controlled switches which are turned on and off at high frequency. Capacitors and inductors are used to store energy when the switch is connected to the power source. The stored energy is then discharged to continue powering the load when the switch is off. Transformers are used where needed. Major power electronic components routinely used in the spacecraft are:

- Shunt regulator for bus voltage control during sunlight
- Battery charge converter (buck converter)
- Battery discharge converter (boost converter)

10.2 Switching Devices

A variety of solid-state devices are available for use as controlled switches. However, the devices commonly used in space are:

- Metal-oxide semiconducting field effect transistor (MOSFET)
- Bipolar junction transistor (BJT)
- Insulated gate bipolar transistor (IGBT)

The device selection depends on the required voltage, current and the switching frequency. A common feature among these devices is that all are three-terminal devices. Their generally used circuit symbols are shown in Figure 10.1. The two power terminals 1 and 0 are connected in the main power circuit. The control gate terminal G is connected to the auxiliary control circuit. In normal conducting operation, terminal 1 is generally at higher voltage than terminal 0. Since the device is primarily used for switching power on and off as required, it is functionally represented by a gate-controlled switch. In absence of the gate control signal, the device

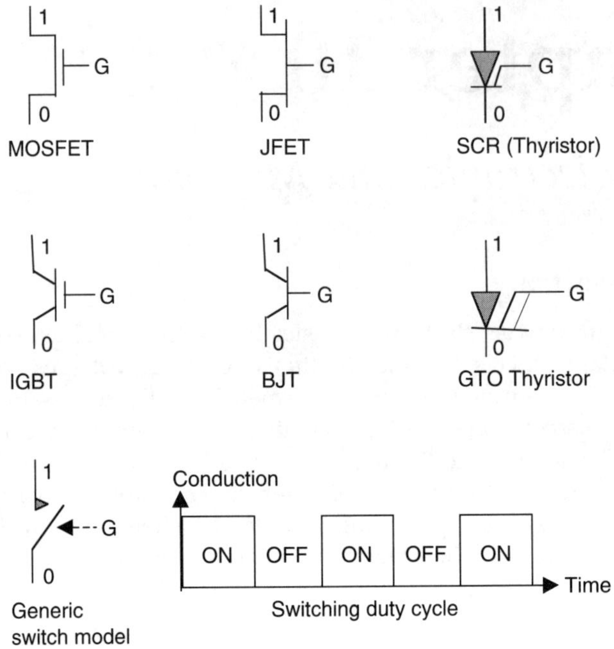

FIGURE 10.1 Semiconductor switching devices used in space.

resistance between the power terminals is large — the functional equivalence of an open switch. When the control signal is applied at the gate, the device resistance approaches zero, making the device function like a closed switch.

The switch is triggered periodically on and off by a train of gate signal of suitable frequency. A sharp gate signal of rectangular, triangular, or saw tooth wave shape without overshoot and ringing is generated by a separate triggering (firing) circuit. The gate circuit for the MOSFET is simpler than that for the bipolar transistor. Although the control circuit has a distinct identity and much different design features, it is often incorporated in the main power electronic component assembly. The transistor switch is turned on and off at high frequency, typically at 50 to 200 kHz, and sometimes higher. The duty ratio, D, of the switch is defined as

$$D = \frac{\text{Time on}}{\text{Period}} = \frac{T_{on}}{T} = T_{on} \cdot \text{ switching frequency} \tag{10.1}$$

The available voltage and current ratings of the switching devices and their gate triggering requirements vary with the device type. The presently available ratings are listed in Table 10.1, not all of which are space qualified. Various power electronic components built using such high-frequency switching devices are discussed next.

Table 10.1 Maximum voltage and current ratings of semiconductor switching devices

Device	Voltage rating Volts	Current rating Amps	Remark
MOSFET	1000	100	Offers higher switching speed, simpler firing circuit.
BJT	1500	200	Requires larger current signal to turn on.
IGBT	1200	100	Combines the advantages of BJT and MOSFET

10.3 Shunt Regulator

The power output of the solar array during sunlight normally exceeds the load plus the battery charge requirements, more so in the beginning of life. The excess power must be diverted (shunted) from feeding into the bus in order to control the bus voltage. The shunt load can be a dump resistor, which would convert the solar array power into heat. Such heat dissipation in the spacecraft body would pose a burden on the thermal system in providing adequate cooling. An alternative commonly used in spacecraft is to shunt some of the solar array strings to the ground. This forces the string to operate under short circuit condition, delivering I_{sc} at zero voltage. In this mode, i.e., in the shunt mode, no power is delivered to the load or to the ground. The photon energy remains on the array, raising the array temperature and ultimately dissipating the excess power to space. The solar array is essentially used here as the thermal dissipator.

Another application of the shunt regulator is in small dedicated solar array module used to directly charge the battery without a battery charge regulator. When the battery is fully charged, the solar array module is shunted to ground by shorting the switch. This way, the battery is protected from overcharging.

Figure 10.2 depicts a typical shunt regulator where a transistor is used as the switch. When the excess power is available, the bus voltage rises above the rated value. This is taken as a signal to turn on the shunt switch across the required number of solar array strings. Thus, the shunt is turned on or off by a transistor controlled by the bus voltage reference. For array with many strings in parallel, the basic configuration shown in Figure 10.2 is used for each string separately. The same gate signal is supplied to all modules simultaneously in small power applications. For shunting large power, multiple shunt circuits are switched on and off in sequence to minimize the switching transients and the resulting electromagnetic interference to the neighboring equipment. For a fine voltage control, the last shunt to turn on is operated in the pulse width modulation (PWM)

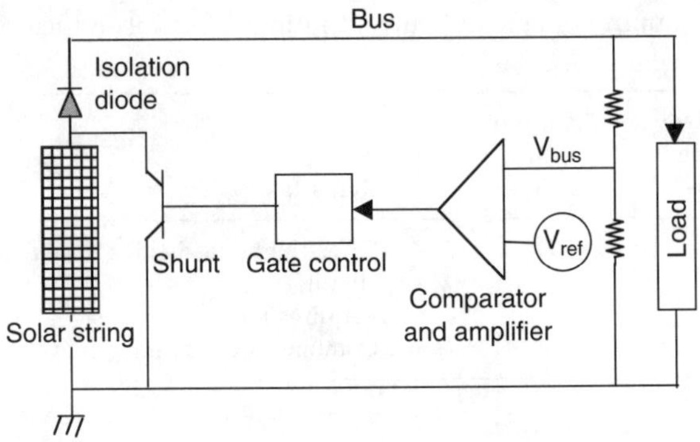

FIGURE 10.2 Full shunt regulator for bus voltage control.

mode, while all others are fully on or off. The constructional and performances features of various shunt circuits are described below.

10.3.1 Full and Partial Shunts

Figure 10.2 shows the schematic of a full shunt, where the whole string of the solar array is controlled *on, off, or in-between.* Figure 10.3(a) shows a partial shunt, where the shunt switch controls a part of the solar string. Only the lower part of the string is shunted when required. When the shunt is on, its operation is depicted in Figure 10.3(b) on the *I–V* characteristic of the string. The lower section thus works like a current source, whereas the upper part works like a voltage source. The unshunted and shunted sections of the string are mismatched. The mismatch thus created reduces the power fed to the bus by operating away from the maximum power point, leaving some of the solar power on the array where it appears as heat and is radiated back into space. A small percentage of power is dissipated as heat in the shunt. Since the shunting occurs only when excess power is available, the increased solar array temperature and the resulting decrease in photovoltaic conversion efficiency has no impact on the power system performance. In the partial shunt scheme, the shunt transistor can be located on the back of the solar array, which enables easy heat dissipation into space. However, this requires more signal sliprings than power sliprings, and the shunt transistor has to withstand wide temperature excursion cycles of the eclipse and sunlight.

The tap voltage of the lower segment $V_L=$ full string voltage − upper string voltage $= V_{sa} - V_U$. The value of V_L is typically 15 V in a 22- to 35-V sun-regulated bus. The shaded area in Figure 10.3(b) represents the shunted power. Quantitatively, it is equal to the product of the voltage and current in the lower shunted segments of the string, i.e.,

$$P_{shunt} = V_L I_L \quad (10.2)$$

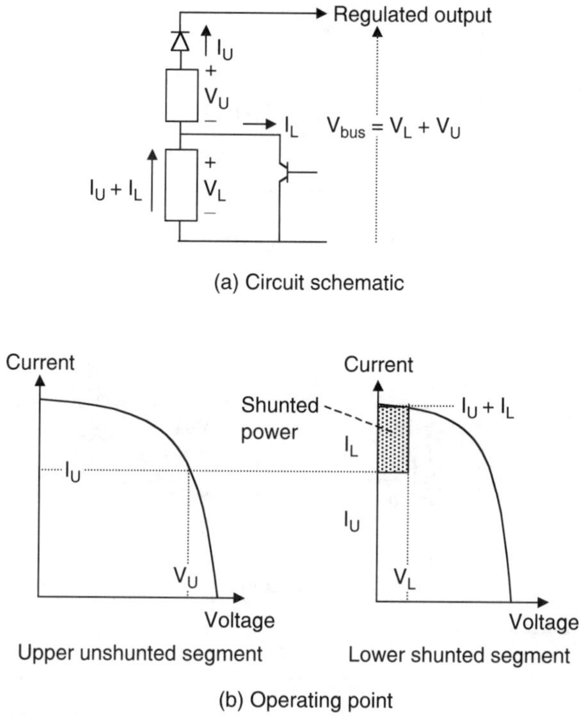

FIGURE 10.3 Partial shunt regulator with upper segment unshunted and lower segment shunted.

Partial shunts have been used in many satellites in the past, but have fallen out of favor in modern high power satellites, which generally use full shunts. Both the full and the partial shunts can be any one of the following types.

10.3.2 Linear and PWM Shunts

Figure 10.4 shows the schematic of a linear shunt. In configuration (a) without a series resistor, the gate current drives the shunt switch in unsaturated region when the bus voltage exceeds the reference value. In this region of operation, the switch offers a variable resistance depending on the gate current. The string power is partially shunted to the ground, and the remaining power feeds the bus. If needed, an additional fixed resistance can be added as shown in (b). A disadvantage of the linear shunt with resistor is that a part of the solar power is converted into heat in the resistor, which must be dissipated into space by the thermal system. For this reason, the resistor is seldom used in large satellites. The linear shunt without resistor keeps the unwanted power on the array for dissipation to

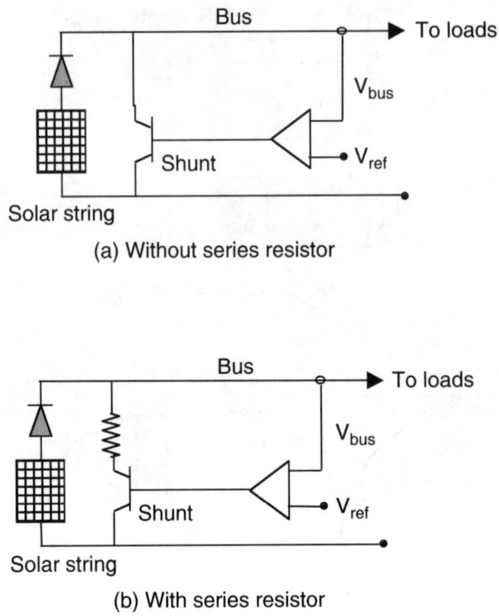

FIGURE 10.4 Linear shunt regulator.

the outer space, thus minimizing the sliprings, thermal management, and EMI related costs.

Another alternative is to use the switch-mode pulse width modulated (PWM) shunt shown in Figure 10.5. This scheme is used on most modern high power spacecraft. Instead of operating the linear shunt in the unsaturated region, the PWM shunt is operated in the off or all the way on (saturated) state when the bus voltage exceeds the reference value. The on and off duration are repeated at high switching frequency with a controlled duty ratio. This eliminates the power dissipation from liner

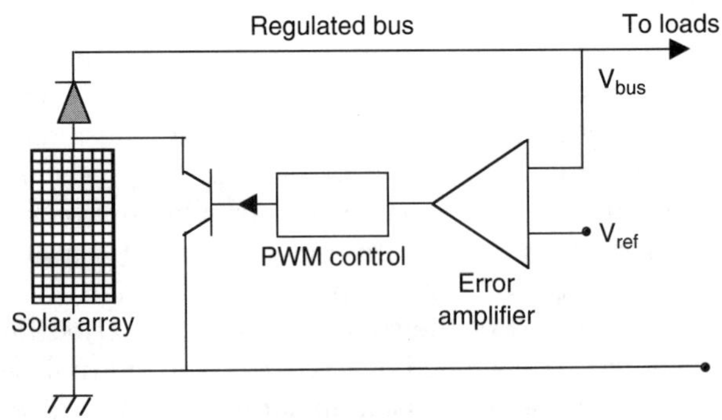

FIGURE 10.5 Pulse width modulated shunt regulator.

shunts. The string power fed to the bus equals the whole string power times the duty ratio. The duty ratio is adjusted to keep the bus voltage in a narrow range around the reference value. A disadvantage of the PWM shunt is that it switches full current at high frequency, causing EMI on the platform. However, the PWM shunt is widely used for its many benefits. On the other hand, the linear shunt is still used where EMI cannot be tolerated, or where heat is needed to maintain temperature inside the spacecraft body.

10.3.3 Sequential Linear Shunt

When more than one string is required to be shunted, the one-string linear shunt shown in Figure 10.4(a) is used as the building block to build the array shunt circuit as shown in Figure 10.6. When the solar output power exceeds requirements, the shunt drive controller first raises the gate current of shunt 1 from zero to the value that can maintain the bus voltage at the reference value. In doing so, if shunt-1 current reaches the saturation level, shunt-2 is gradually turned on. If more shunting is still needed, the controller may reach the nth shunt until the bus voltage is maintained at the set level. Turning on the shunt switches is done in sequence.

10.3.4 Multistage PWM Shunt

Multiple solar strings are generally needed to shunt the excess power, particularly at the beginning of life. The shunt circuit using multiple PWM shunts is shown in Figure 10.7. The bus voltage is controlled by first placing one string in the PWM. The duty ratio controls the power shunted. If more power needs to be shunted, the duty ratio of the first string is raised. After reaching the duty ratio limit of the first string, the second string is placed under the PWM, and so on. At the end, all shunts are in PWM mode, switching the power on and off simultaneously. This may create an EMI problem around the shunts. The EMI is minimized by designing the PWM

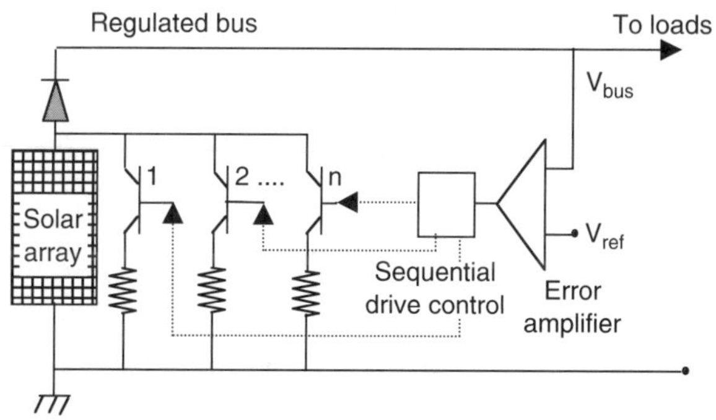

FIGURE 10.6 Sequential linear shunt regulator.

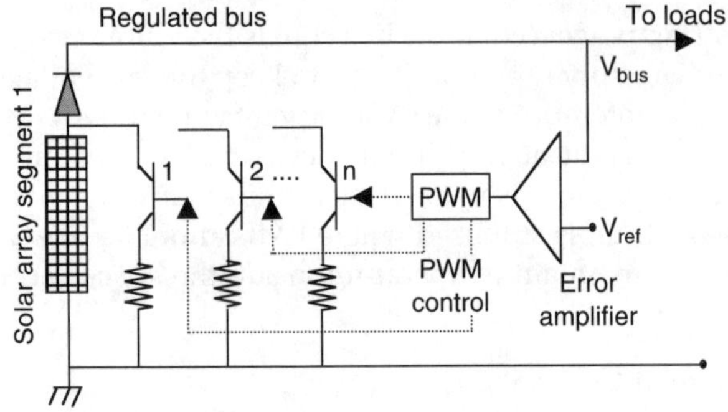

FIGURE 10.7 Multistage PWM shunt regulator.

control circuit to have the maximum duty ratio of 1. That way, all but one shunts are operated continuously in the saturated state, and only the last one is operated in PWM mode to shunt a fraction of power on or off at a time.

10.3.5 Polyphase PWM Shunt

In this scheme, the multistage PWM shunt schematic shown in Figure 10.7 remains the same, but the PWM modulation control is modified to operate simultaneously as many strings as needed for the bus voltage control. The difference here is that the gate signals for all strings under PWM are symmetrically out of phase with each other. For example, the gate signals for an 8-string array are shown in Figure 10.8. This way, all strings are not simultaneously switched on or off, but one after another in a rotation. Major advantages of this scheme are significant reductions in ripples and EMI.

The polyphase PWM shunt regulator reduces the ripple generation, hence lower the filter mass. Figure 10.9 depicts approximate mass saving with number of phases used. The minimum filter mass is achieved with four to six phases depending on the power level of the bus. The saving can be up to 20% compared to the multistage PWM shunts.

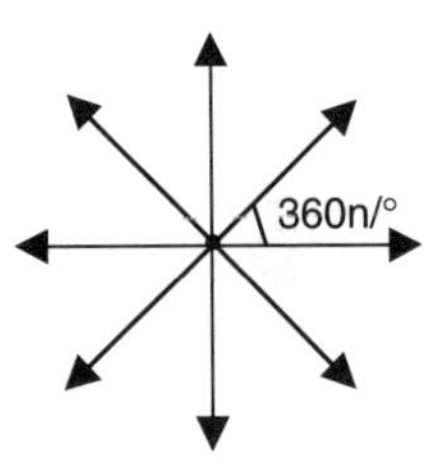

FIGURE 10.8 Polyphase PWM gate signals with $360/n^{\circ}$ phase difference.

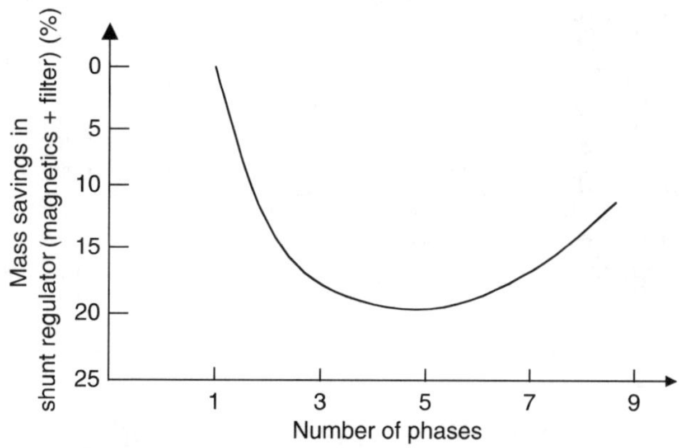

FIGURE 10.9 Shunt regulator mass versus number of phases.

10.3.6 Calibrated Step Shunt

The scheme of Figure 10.10 shows multiple shunts, each having power capability in calibrated steps such that great many values of power can be shunted by appropriate selection of the shunts and placing them in the saturation region. It provides a good resolution on the bus voltage control, but not as fine as other schemes.

10.4 Shunt Circuit Design

In designing the shunt circuit, the following trade options exist for the design engineer.

10.4.1 Full Versus Partial Shunts

The full shunt has the following features:

- It requires fewer slip rings and wires
- More array power is lost if one circuit fails
- Gives more power dissipation per circuit

The partial shunt, on the other hand, requires twice as many slip rings, one set from the upper section and one set from the lower section.

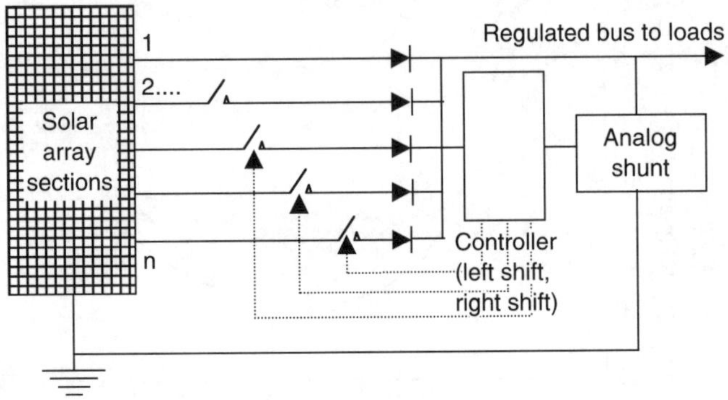

FIGURE 10.10 Calibrated step digital shunt regulator.

10.4.2 Number of Shunt Circuits

In selecting the number of solar circuits, too few circuits leads to:

- Higher power per circuit, hence more power margin is needed to allow for one circuit failure for reliability
- Higher local dissipation, hence hot spots
- Fewer slip rings in the solar array

Too many circuits, on the other hand, leads to:

- Smaller voltage control steps, i.e., better bus voltage resolution
- Small steps require higher shunt control amplifier gain. This may be harder to work with due to excessive amplification of the small error signal band.

10.4.3 Typical Shunt Circuit Design

Typically, the solar array is divided into as many string circuits as required to keep the BOL current below 5 A in order to use standard 5-A slip rings in each circuit. For a 22 to 35 sun-regulated bus with full shunt as an example, the maximum shunt dissipation would then be $35 \times 5 = 175$ W in the shunt just entering the linear range. This is 87.5 W per transistor with two in parallel for redundancy. Dissipation in each shunt when saturated, on the other hand, is $(\frac{1}{2}I)^2 \times$ transistor on-resistance $= (5/2)^2 \times 0.100 = 0.75$ W in each of the two switches in parallel for each solar string, or 1.50 W total in both switches combined. Thus, the total power loss per shunt circuit in the saturated region is less than 1% of the maximum before entering the saturation.

The solar array shunt control should limit the bus voltage dV/dt to less than 1% of the bus voltage per microsecond, and conduction voltage drop below 0.1% of the bus voltage per ampere of the shunt circuit current. For example, a 5-A circuit on a 100-V bus should keep the dV/dt less than 1 V/μs and the conduction drop below 0.50 V. The control signals to switch shunt circuits are degraded, buffered, shielded, and double insulated for reliability.

10.5 Bus Ripple Filter Design

The ripple produced by the PWM shunt operation is kept below the minimum required limit by using a bus filter capacitor shown in Figure 10.11. The following design analysis establishes the required capacitor size. The capacitor becomes charged when the shunt switch is off, and is discharged when the switch is on. The ripple comes from two sources as shown in the figure. As the capacitor charges and discharges, the equivalent series resistance (ESR) of the capacitor causes the voltage drop of opposite polarity, resulting in the capacitor voltage rise and fall. The peak-to-peak ripple voltage V_{pp} on the bus can be derived in to the following expression:

$$V_{pp} = I_s R_c + D(1 - D)\frac{I_s T}{C} \tag{10.3}$$

where

I_s = current switched per shunt circuit

R_c = bus capacitor ESR

D = duty ratio of the PWM switching

T = switching period

C = bus capacitor value in farads

The ripple voltage varies with the duty ratio. It has the maximum value at $D = 0.5$, which is equal to

$$V_{pp} = I_s R_c + 0.25\frac{I_s T}{C} \tag{10.4}$$

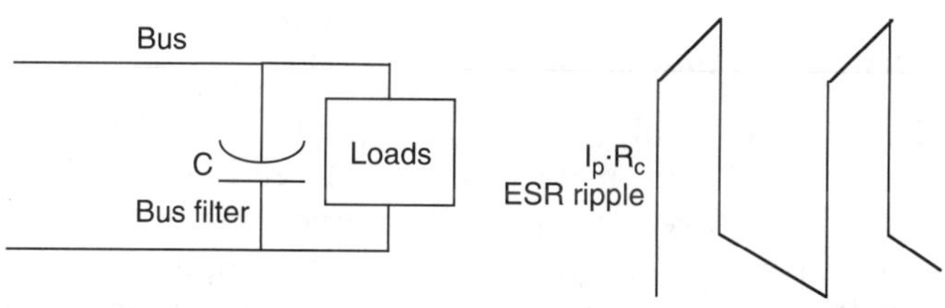

(a) Bus filter capacitor (b) Switching frequency ripple

FIGURE 10.11 Bus voltage ripple at shunt switching frequency with bus filter capacitor.

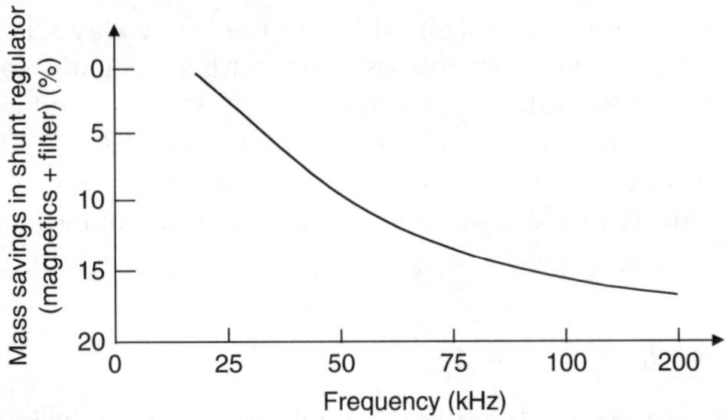

FIGURE 10.12 Shunt regulator mass versus switching frequency.

For a known ESR and switching frequency, the minimum bus capacitor required to meet a given ripple specification is then found from

$$C_{\min} = \frac{0.25\, I_s T}{V_{pp} - I_s R_c} \tag{10.5}$$

A large bus filter capacitor is desired not only to get low ripple, but also to reduce the bus impedance at high frequency. On the other hand, fast transient response requires a low value of the bus capacitor for a given loop gain crossover frequency. The engineer trades these two performance features.

Use of high frequency, i.e., low switching period, reduces the filter capacitor size as indicated by Equation 10.5. Approximate mass saving is shown in Figure 10.12 for the frequency range of 20 to 200 kHz.

10.6 Power Converters

The converter basically converts one voltage level into another at the interface of two power components. The following sections describe the topology and operation of various types of power converter.

10.6.1 Battery Charge (Buck) Converter

During sunlight, the bus voltage is maintained by the solar array output. The battery is charged only if the battery voltage is lower than the bus voltage. A d.c.–d.c. voltage converter is required to buck (step down) the bus voltage to the battery voltage during charging. Figure 10.13 is the most widely used battery charge converter circuit topology. Its operation during one on and off cycle of the triggering signal is shown in Figure 10.14. During the on time, the switch is closed and the circuit operates as in (a).

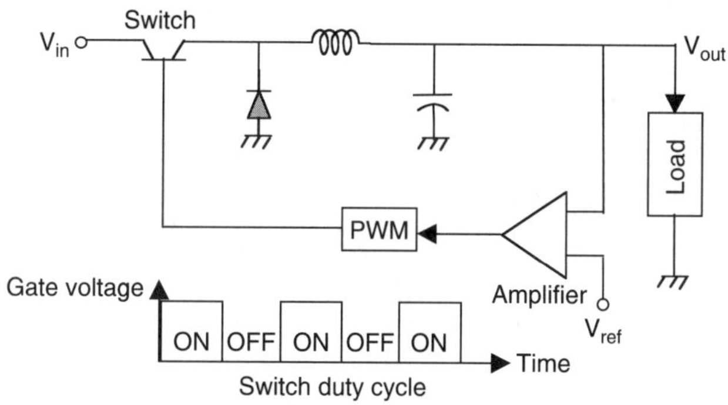

FIGURE 10.13 Battery charge (buck) converter.

The d.c. source charges the capacitor and supplies power to the load via the inductor. During the off time, the switch is open and the circuit operates as in (b). The power drawn from the d.c. source is zero. However, full load power is supplied by the energy stored in the inductor and the capacitor, with the diode carrying the return current. Thus, the inductor and the capacitor provide short-time energy storage to ride through the off period of the switch. The load current during this period is known as the free-wheeling current, and the diode is known as a free-wheeling diode. The voltage and current waveforms over one complete cycle are displayed in Figure 10.15. A suitable bleeding resistor at the load terminals is sometimes incorporated in the design to keep the converter working without the load.

The analysis of the power train of the buck converter follows. The analytical principles that follow are applicable also to other d.c.–d.c. converters. The power train analysis is based on the energy balance over one switching period. Since the inductor stores energy during the on-time and discharge during the off time:

- Energy supplied to the load over total period T = energy drawn from the source during the on time, which powers the load and charges the inductor and capacitor, and
- Energy supplied to the load during off time = energy discharged from the inductor and the capacitor during off time.

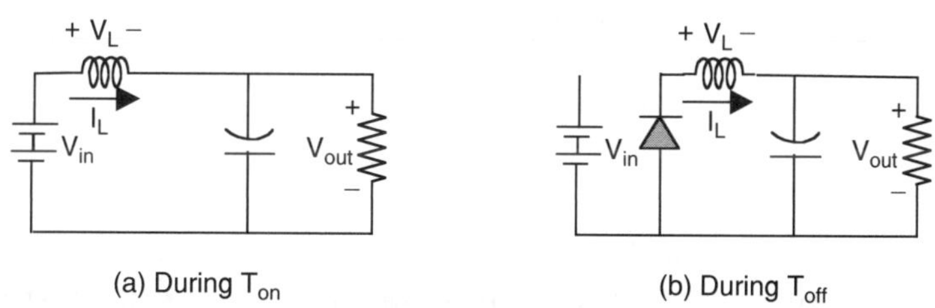

FIGURE 10.14 Buck converter operation during on and off switch.

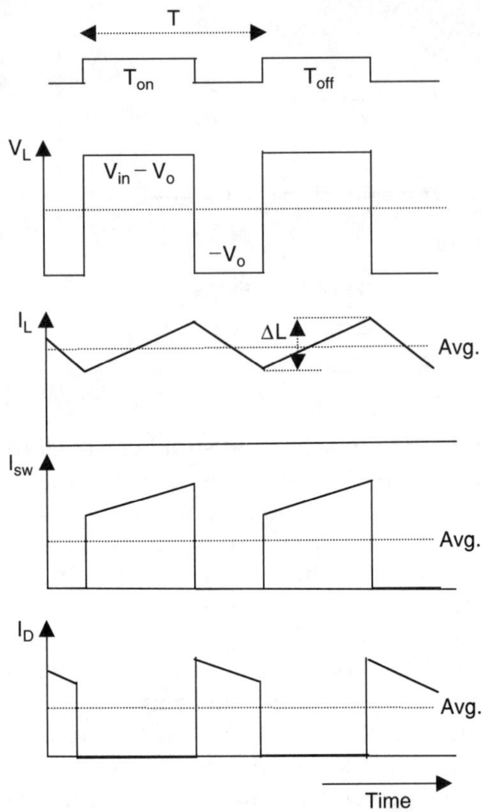

FIGURE 10.15 Current and voltage waveforms in the buck converter.

The inductor generally uses magnetic core with lumped or distributed air gap. The flux density in the core has to be kept below the magnetic saturation level. For that reason, the net change in flux over one cycle of switching has to be zero. Otherwise, the core would eventually *walk away* to saturation on the high side, or be depleted of energy on the low side. The magnitude of voltage drop across the inductor is given by Faraday's law, $V_L = N\,\mathrm{d}\varphi/\mathrm{d}t$, where N = number of turns, φ = flux in the core and t = time. The equation can be written as $V_L{\cdot}\mathrm{d}t = N{\cdot}\mathrm{d}\varphi$. As $\mathrm{d}\varphi$ must be zero over one switching cycle, we must have

$$\int_0^T V_L \cdot \mathrm{d}t = 0 \tag{10.6}$$

i.e., the change in the volt–second product over one cycle must be zero. Thus, in the steady state condition, the inductor volt–second balance during the on and off periods must be maintained. This in essence gives the energy balance.

We now apply the volt–second balance in the voltage and current waveforms shown in Figure 11.15. In terms of the inductance, the voltage drop across the inductor is $V_L = L\, dI_L/dt$. Therefore, during on-time,

$$\Delta I_L L = (V_{in} - V_{out})T_{on} \tag{10.7}$$

and during off-time,

$$\Delta I_L L = V_{out} T_{off} \tag{10.8}$$

If the inductor is large enough, as is usually the case in practical designs, the change in the inductor current is small, and the peak value of the inductor current is given by

$$I_{peak} = I_o + 0.5\, \Delta I_L \tag{10.9}$$

where the load current $I_o = V_{out}/R_{load}$ = average value of the inductor current.

The algebraic manipulation of the above equations leads to

$$V_{out} = V_{in} \cdot D \tag{10.10}$$

Since the duty ratio D is always less than unity, the output voltage of the buck converter is always less than the input voltage. Thus, the buck converter can only step down the voltage, hence the name.

It is seen from Equation 10.10 that varying the duty ratio controls the output voltage. This is done in a feedback control loop with the required battery charge current as the reference. Modulating the pulse width of T_{on} controls the duty ratio. Such converter is therefore known as the pulse width modulated (PWM) converter.

The efficiency is calculated as follows. During the on time, the input voltage, less the voltage drop in the transistor switch, supply power to the circuit. The energy transfer efficiency is therefore $(V_i - V_s)/V_i$, where V_i = input voltage and V_s = voltage drop in the switch. During the off period, the inductor voltage, less the diode drop, is the output voltage. The efficiency during this period is therefore $V_o/(V_o + V_d)$, where V_o = output voltage and V_d = voltage drop in the diode. In addition, there is some loss in the inductor, the efficiency of which we denote by η_i. The losses in the capacitor and the wire are relatively small and can be ignored for simplicity. The overall power train efficiency of the buck converter is, therefore, the product of the above three efficiencies, namely

$$\eta_c = \left(\frac{V_i - V_s}{V_i}\right)\left(\frac{V_o}{V_o + V_d}\right)\eta_i \tag{10.11}$$

With commonly used devices, the transistor switch and diode voltage drops are typically 0.6 V each. If we denote both by the device voltage drop V_d, then a sufficiently accurate simplification of the above equation is

$$\eta_c = \left(\frac{V_o}{V_o + V_d}\right)^2 \eta_i \tag{10.12}$$

The inductor efficiency can approach 0.99. The power train efficiency of 28-V output converters is therefore around 95%. The losses in the wires, capacitors, and magnetic core may further reduce the efficiency by a couple of percent, making it in the 92 to 94% range. The converter efficiency is a strong function of the output voltage. It decreases with decreasing output voltage, as the switch and diode voltage drops remain constant. For example, a 5-V output converter would have efficiency in the 75 to 80% range.

Since the charge converter in LEO satellites handles more power and works longer hours, it must be designed for high efficiency and better cooling. The LEO charge converter design, therefore, gets more complex than its GEO counterpart.

10.6.2 Battery Discharge (Boost) Converter

In a fully regulated bus, the bus voltage is maintained within a narrow band during the entire orbit. Since the battery voltage under discharge sags during eclipse, a voltage converter is required to boost (step up) the battery voltage to the bus level. The boost converter topology is shown in Figure 10.16. When the transistor switch is on, the inductor is connected to the d.c. source. When the switch is off, the inductor current is forced to flow through the diode and the load. The output voltage of the boost converter is derived again from the volt–second balance in the inductor. With duty ratio D of the switch, it can be shown, in a manner similar to that in the buck

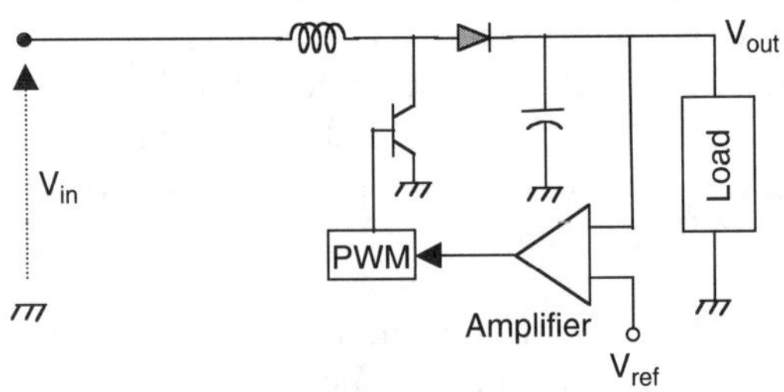

FIGURE 10.16 Battery discharge (boost) converter.

converter analysis, that the output voltage is given by the following expression,

$$V_{out} = \frac{V_{in}}{1 - D} \tag{10.13}$$

Since the duty ratio is always less than unity, the output voltage of the boost converter is always greater than the input voltage. Therefore, the boost converter can only step up the voltage, hence the name. The efficiency of the boost converter can be shown to be

$$\eta_c = \left(\frac{V_i - V_d}{V_i}\right)^2 \eta_i \tag{10.14}$$

10.6.3 Buck–Boost Converter

Cascading of the buck and the boost converters as shown in Figure 10.17(a) results in a buck–boost converter, which can step down or step up the input voltage. A modified direct buck–boost converter that uses fewer parts is shown Figure 10.17(b). The output voltage is obtained by cascading the buck and boost converter voltage relations. That is,

$$V_{out} = \frac{V_{in} \cdot D}{1 - D} \tag{10.15}$$

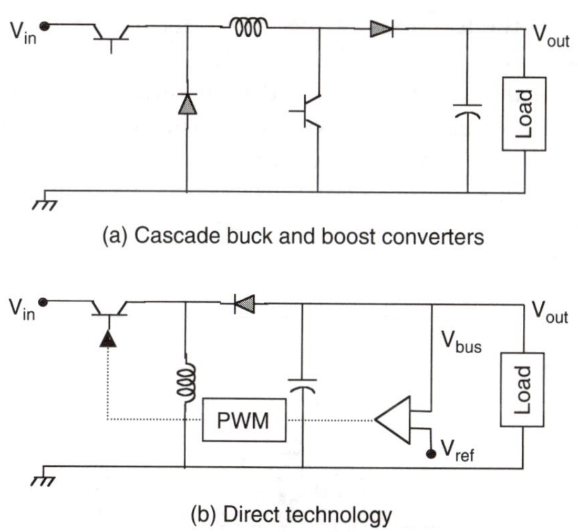

FIGURE 10.17 Two configurations of buck-boost converter.

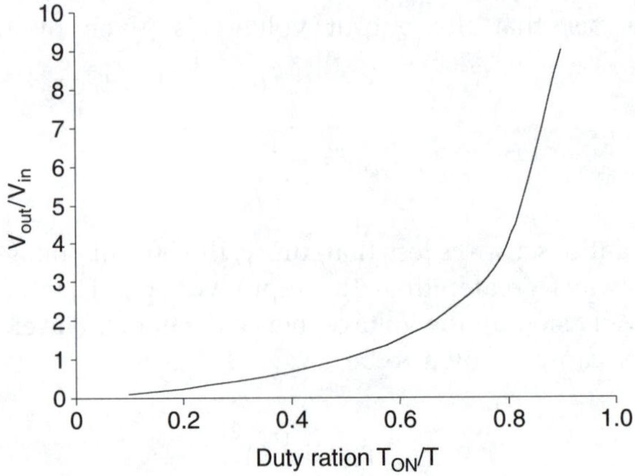

FIGURE 10.18 Buck-boost converter output and input voltage ratio versus duty ratio.

This equation shows that the output voltage of the buck–boost converter can be higher or lower than the input voltage depending on the duty ratio (Figure 10.18). In addition to its possible use to charge and discharge the battery with one converter, the buck–boost converter is capable of the four-quadrant operation with a d.c. machine in the variable speed applications and with the flywheel energy storage. In the flywheel system, the converter would step-up voltage during generating mode, and step-down during motoring mode.

10.6.4 Flyback Converter (Buck or Boost)

Figure 10.19 shows the topology of this converter, which can buck or boost the voltage depending on the turn ratio of the coupled inductor. The energy is stored in the primary side of the inductor during the on time. It then becomes inductively transformed (kicked) to the secondary side when the switch is turned off, and is delivered to the load via diode. The polarity marks of the two coils are of essence in this circuit. The inductor core is generally made of potted magnetic metal power. The energy is stored in distributed air gap of the core. The voltage ratio and efficiency are given by

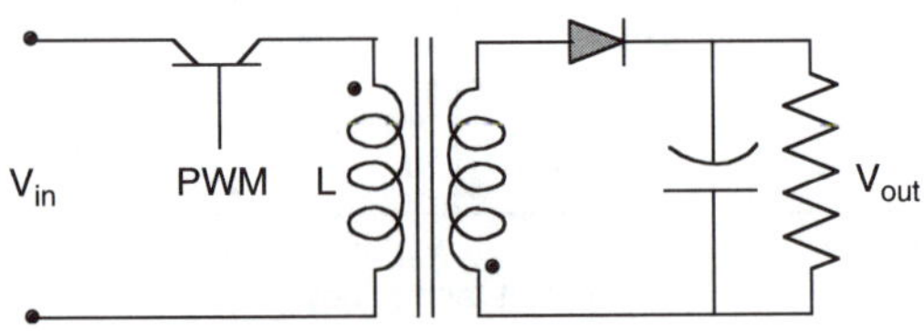

FIGURE 10.19 Flyback converter (buck or boost).

$$\frac{V_o}{V_i} = D \cdot \left(\frac{N_s}{N_p}\right) \quad \text{and} \quad \eta_c = \left(\frac{V_i - V_s}{V_i}\right)\left(\frac{V_o}{V_o + V_d}\right)\eta_i \tag{10.16}$$

where η_i = efficiency of the coupled inductors, which is typically 95 to 98%.

The voltage ratio is a mere multiple of the duty ratio with the turn ratio. This gives great flexibility in the output voltage without the limitation of the duty ratio. It can buck or boost depending on the turn ratio less than or greater than one.

An advantage of the flyback converter over the classical buck and boost converter is that it electrically isolates the two sides, thus minimizing the conducted EMI and enhancing the safety. Moreover, it stores the inductive energy, and simultaneously operates like a classical forward buck converter with a transformer, stepping up or down the input voltage as needed.

10.6.5 Transformer Coupled Forward Converter

The converter circuit shown in Figure 10.20 employs a transformer with suitable turn ratio in conventional converter such that the output voltage can be below or above the input voltage. The polarity marks are of essence. The bleeding load resistor is required to keep the converter working without load. The free wheeling diode is not required. Its voltage relation is identical to that for the flyback buck–boost converter described above. That is,

$$\frac{V_o}{V_i} = D \cdot \left(\frac{N_s}{N_p}\right) \tag{10.17}$$

and the efficiency

$$\eta_c = \left(\frac{V_i - V_s}{V_i}\right)\left(\frac{V_o}{V_o + V_d}\right)\eta_t \tag{10.18}$$

where η_t = efficiency of the transformer, which is typically 95 to 98%.

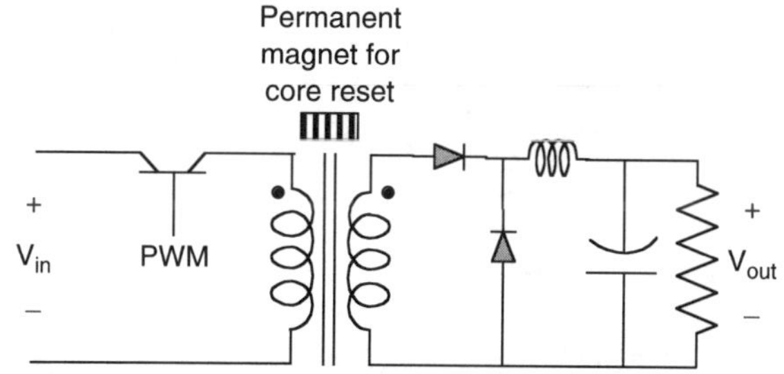

FIGURE 10.20 Transformer coupled forward converter.

10.6.6 Push–Pull Converter

This is again a buck–boost converter and is shown in Figure 10.21. The converter topology uses a center-tapped transformer, which receives square wave excitation. It is seldom used at low power levels, but finds applications in systems dealing with several tens of kilowatts and higher. It can provide desired output voltage by setting the required transformer turn ratio. The voltage ratio and the efficiency are the same as those given by Equations 10.17 and 10.18.

10.6.7 Inductor Coupled Buck Converter

All converters presented above need an L–C filter to control the ripple in output voltage. The buck converter needs heavier filter than the boost converter. Since ripples are present in both the input and output sides, it is possible to cancel them by coupling the ripple current slopes of two sides with matched inductors[1,2] as shown in Figure 10.22. The ripple magnitudes are matched by using required turn ratio, and the polarities are inverted by winding in magnetically opposite directions. The net ripple on the output side is zero at certain gap, below which it changes the polarity. A gap adjusting screw is needed to tune the gap for a precise match. The output capacitor C_o is not really needed in a perfectly coupled design, but a small value of C_o improves the performance. The coupling capacitor C is needed for the performance. The total value of the two capacitors is about the same as that needed in the classical buck converter. On the output side, some ripple may be left due to the winding resistances. The input side ripple is the same as that without the coupled inductor. The load current changes do not affect the ripple cancellation.

10.6.8 Resonant Converter

Another type of d.c.–d.c. converter now used in many high power applications is the zero-voltage switching converter due to its higher

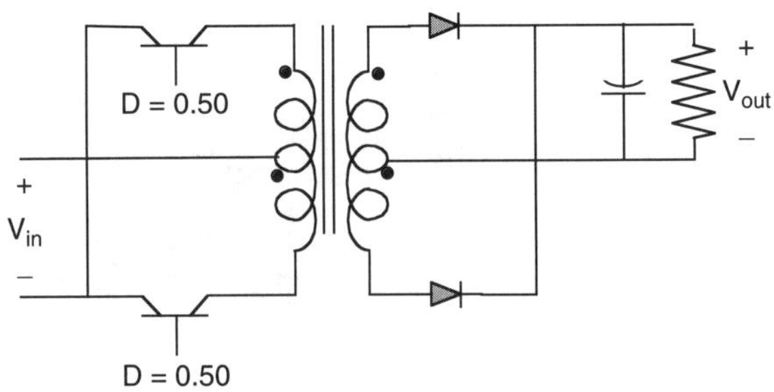

FIGURE 10.21 Push-pull converter with center-tap transformer.

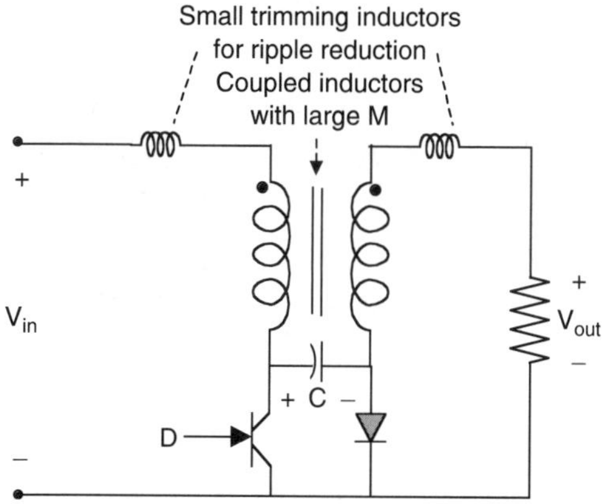

FIGURE 10.22 Inductor coupled buck converter with integrated magnetics (Cuk converter.)

switching frequency, higher efficiency, smaller size, lighter weight, and better dynamic response. It is a full bridge converter that includes inductors and capacitors as resonant elements. The switching is done when the voltage is passing through its natural zero, thus reducing the switching losses.

10.6.9 Multiple Output Converter

Multiple voltage output from the same converter is obtained by using two or more inductors or transformer coils on the secondary side as shown in Figure 10.23(a). In such converters, the regulation in one output voltage is influence by the load changes in the other via magnetic coupling. Analytically predicting the load regulation of a new design can save the prototyping cost and time, as opposed to a trial and error process. The three-winding transformer equivalent circuit shown in (b) can facilitate the voltage regulation analysis. It is based on the premises that between any two coils, there is a leakage flux and the corresponding leakage reactance X. The equivalent circuit shows the total impedance — made of the winding resistance and the leakage reactance — denoted by Z_{12} between coils 1 and 2, Z_{23} between coils 2 and 3, and Z_{31} between coils 3 and 1. The value of Z_{12} is determined by a short circuit test between winding 1 and 2, and so on. The analysis is significantly simplified by representing the three coil-to-coil impedances in an equivalent Y-circuit shown in (c). Through a circuit analysis, their values can be derived as follows:

$$
\begin{aligned}
Z_1 &= \tfrac{1}{2}(Z_{12} + Z_{13} - Z_{23}) \\
Z_2 &= \tfrac{1}{2}(Z_{12} + Z_{23} - Z_{13}) \\
Z_3 &= \tfrac{1}{2}(Z_{13} + Z_{23} - Z_{12})
\end{aligned}
\qquad (10.19)
$$

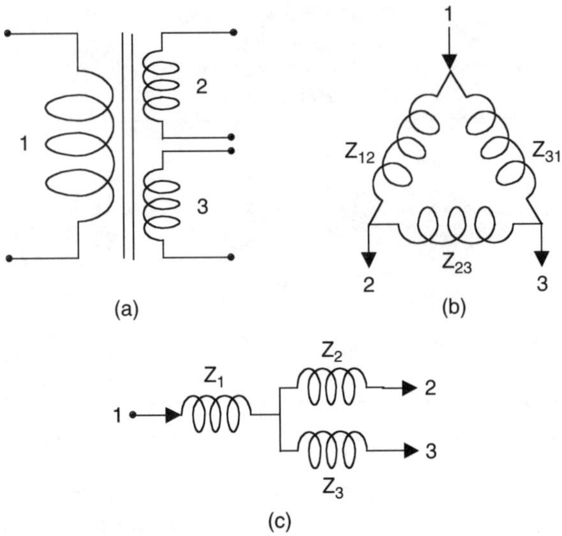

FIGURE 10.23 Three winding transformer equivalent circuit for multiple output converter.

The calculations become easier by expressing the impedances in percentage of the rated impedance of the transformer. It is necessary to emphasize that these equivalent impedances have no real meaning. Any one can be zero or negative, as long as the sum of impedances of any two branches is equal to the short circuit impedance between those two windings. By using the Y-circuit, it is easy to see that the terminal voltage V_2 depends not only on the secondary load, but also on the tertiary load because of the voltage drop in Z_1. It is also easy to see that when load I_2 is shed, the voltage of the other load will rise by I_2Z_1. The converter with more than two outputs can be similarly analyzed using the same general principle.

10.6.10 Load Power Converter

Many loads, particularly payloads, require input voltage within certain regulation and quality. For example, many communications payloads are constant power loads, which require higher current at low voltage and vice versa. In most cases, the payload has a load power converter at its interface with the power system bus (Figure 10.24). The LPC input voltage varies over a wide range in sun-regulated bus, but in a narrow range in a fully regulated bus. Loads having a stringent ripple requirement may require a prohibitively large bus filter capacitor. Alternatively, the L–C filter shown in Figure 10.25(a) used at the bus–LPC interface would be more mass effective. The inductor stores much greater energy per unit mass, and therefore significantly reduced the filter mass.

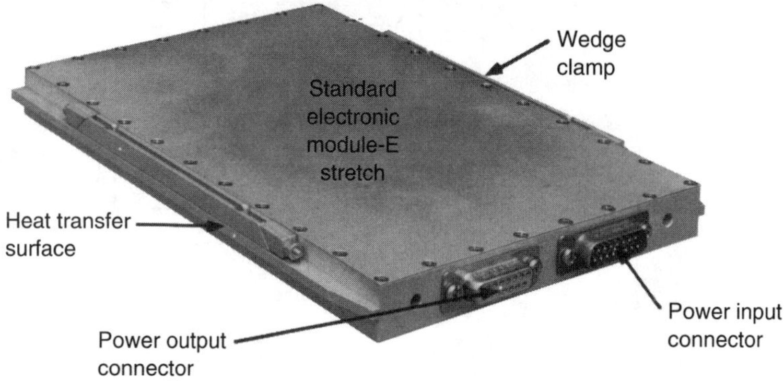

3.3 Volt Power Converter

Output voltage	+3.3 V
Load current	30 A
Regulation	1% static 2% dynamic

5 Volt Power Converter

Output voltage	+5 V
Load current	30 A
Regulation	1% static 2% dynamic

Three Output Power Converter

Output voltage	+5 V	+15 V	−15 V
Load current	12 A	2 A	2 A

Five Output Power Converter

Output voltage	+5 V	−5 V	+15 V	−15 V	+28 V
Load current	4 A	4 A	1 A	1 A	1 A

FIGURE 10.24 Load power converter with single or multiple outputs. (Source: Ratheon Corporation E-Systems Division. With permission.)

In the converter using an L–C filter, transient analysis becomes important on switch-on, as both currents and voltage may overshoot to damaging levels. Large negative voltage transients could occur at the power system and user interface when the relay contact is opened. This can result in component damage, which can be avoided by adding a free-wheeling diode as shown in Figure 10.25(b).

10.6.11 Voltage and Current Regulators

When a constant voltage output is required from a variable voltage input, a voltage regulator is used at the interface. If the input voltage variation is small, a dissipative series voltage regulator can be used, otherwise one of the PWM converters covered in this section is used. The two types are compared as shown in Table 10.2. The linear and buck converters are simple and dynamically stable. All other converters are relatively complex and may become unstable, as they do have poles in the right hand side of the Laplace plane. The linear converter is more reliable than the switching

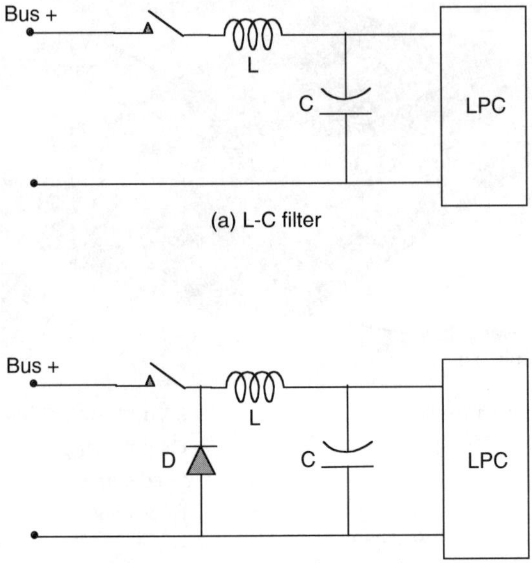

FIGURE 10.25 L–C filter at converter interface and free-wheeling diode protection.

converter, because of fewer parts and no high-voltage switching stress, although it runs hotter.

When a current limiting feature is desired, the converter control circuit regulates and limits the current. In such a design, the current is limited to some high value during overloads, or even under a load fault during which the bus voltage is not driven to zero. It just folds back to some low value.

Thus, the converter control loop can be designed to maintain constant voltage, constant current or constant power. It can also be designed for a combination of two or more performance parameters. For example, the converter may maintain constant voltage up to certain current limit, and then maintains constant current at the set limit, or even folds back a little in

Table 10.2 Linear versus switch mode PWM voltage regulators

Parameter	Linear–dissipative		Switch mode–PWM	
	Series	**Shunt**	**Buck**	**Boost**
Dissipation	High	High	Low	Low
Efficiency	50–90%	50–90%	85–90%	90–95%
Ripple output	Negligible	Negligible	Moderate	Moderate
Load regulation	0.1–0.2%	0.1–0.2%	0.5–1.0%	0.5–1.0%
Output impedance	~ 1 mΩ	~ 1 mΩ	~ 20 mΩ	~ 20 mΩ

current. It is all in the control system, rather than in the basic power train design.

In all types of converter, there is a trend to increase the switching frequency to reduce the mass of the power train inductor and the output capacitor filter. Frequencies up to a few hundred kilohertz are common at present. However, the design challenges increase as the frequency approaches megahertz. At these frequencies, the lead wire inductance and capacitance become significant due to stray and parasitic effects, which are difficult to analyze.

10.7 Magnetics

Inductor and transformer are two magnetic components used in the spacecraft power electronics. The inductor stores energy in the form of magnetic flux. The transformer changes voltage level from one coil to another via mutual flux in its magnetic core. Figure 10.26 is the simplest magnetic circuit showing one coil on a magnetic core. The coil current produces magnetic flux in the core. The magnetic resistance of the core, called the reluctance, $\Re$, is given by the follow expression:

$$\Re = \frac{L_m}{\mu_r \mu_o} \tag{10.20}$$

where

L_m = mean length of the magnetic path, cm

A_c = cross sectional area of the core, cm^2

μ_r = relative permeability of the core material

μ_o = permeability of free space (air or vacuum).

The energy storage or power transfer capability of the core is approximately equal to $A_p{}^{\alpha}$, where $\alpha = 0.85 - 0.95$ and A_p = area product defined as,

$$A_p = A_c A_w \tag{10.21}$$

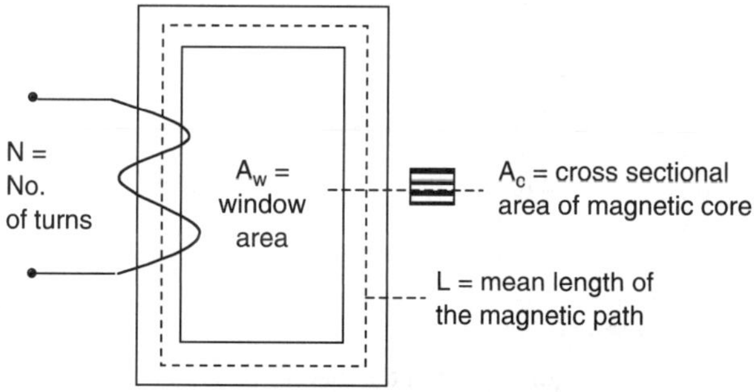

FIGURE 10.26 Coil on a magnetic core produces flux.

where A_w = window area of the core available for winding a coil through it. The inductor generally has one coil on the core, whereas the transformer has at least two coils, and sometimes more. The core can take various configurations shown in Figure 10.27, all of which are used in space power electronics. The core selection depends on the design drivers, such as efficiency, voltage regulation, EMI, and manufacturing cost. The toroidal core contains the flux well internally, and therefore is widely used in high frequency designs where EMI may be of concern.

The flux density, B, in the core is proportional to the magnetizing intensity H (ampere-turns per meter length of the flux path). The B varies linearly with H up to certain value, and then saturates at high H. Under alternating H, the B also alternates, but does not follow the same path while rising and falling. This is due to magnetic resistance (hysteresis) in the atomic structure of the core material. As a result, one complete cycle of H

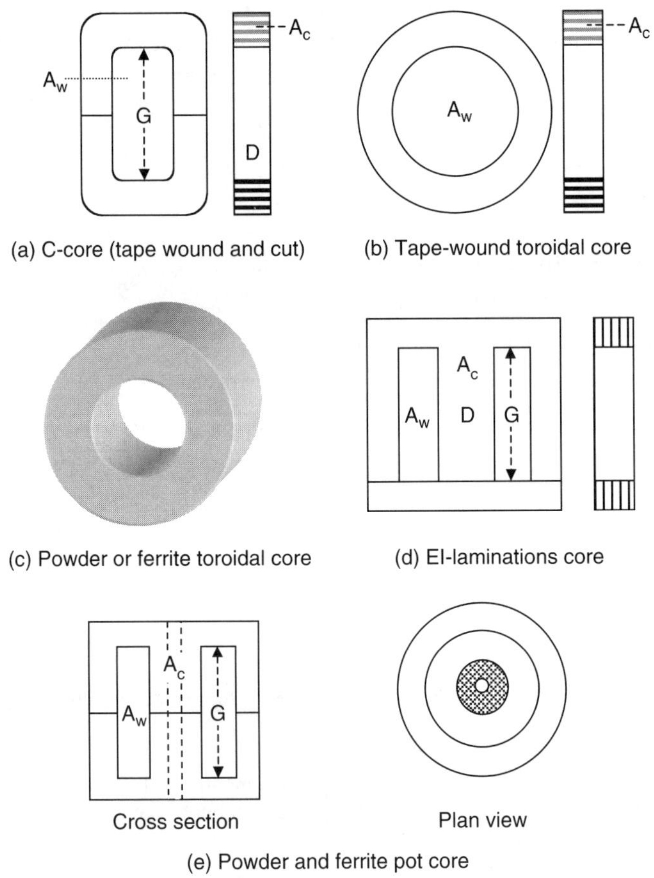

FIGURE 10.27 Common types of magnetic core.

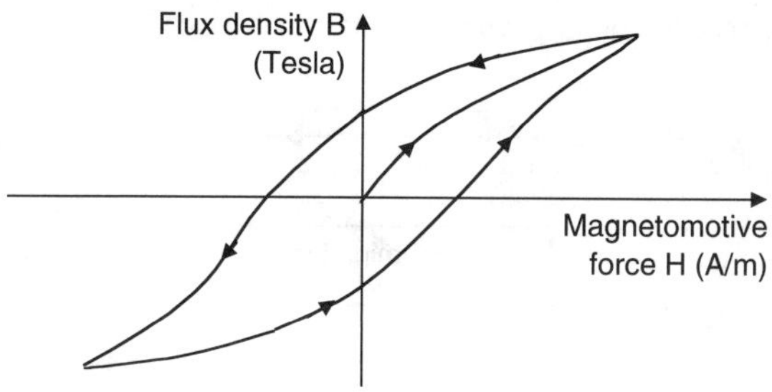

FIGURE 10.28 *B–H* hysteresis loop of the magnetic core.

traces a loop shown in Figure 10.28, known as the *B–H* or hysteresis loop. The area of the *B–H* loop represents energy loss per magnetic cycle. The power loss in the magnetic core is, therefore, the *B–H* loop area times the frequency. It is proportional to $B^{\alpha} f$, where α is known as Steinmetz exponent (or Epstein constant), which depends on the material. A great variety of magnetic core materials having significantly different *B–H* loop are available for various applications. Figure 10.29 is a collection of data on several commonly used materials.

Under alternating flux, the core metal induces voltage as per Faraday's law of electromagnetic induction. The resulting eddy currents circulate internally to the core metal and produce power loss. The eddy current power loss is proportional to the core lamination thickness squared. For this reason, the core is always made of insulated thin laminations or potted powder to reduce eddy loss, as shown in Figure 10.30. The total loss (in

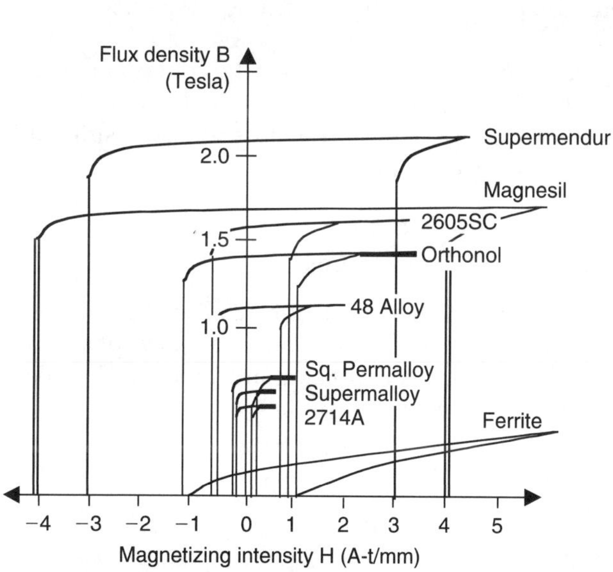

FIGURE 10.29 *B–H* loops of common core materials.

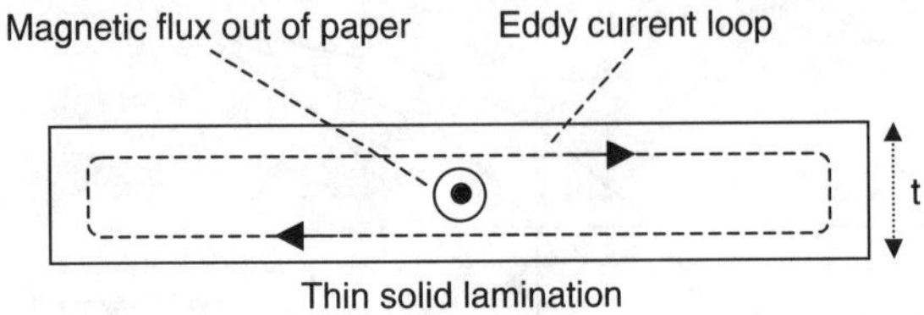

FIGURE 10.30 Eddy current loss in magnetic core.

watts) in the core is made of the hysteresis loss and the eddy loss, and is given by the following expression:

$$P_c = K_h B_m^\alpha f + K_e B_m^2 f^2 t^2 \tag{10.22}$$

where K_h and K_e = proportionality constants for the hysteresis and eddy losses, respectively, and α = Steinmetz constant equal to 1.4 to 1.8 for magnetic steel laminations.

Technical data sheet available from core vendors plot the total loss per kilogram of the core material as functions of the flux density and frequency. Figure 10.31(a) and (b) are examples of such plots useful to the magnetic design engineer.

Two major properties define the merit of the material for a given application:

- Maximum flux density it can carry before saturation
- Total core loss per gram at the switching frequency

The magnetic saturation values of some of the common materials are listed in Table 10.3. The higher flux density comes with a cost. It results in wider hysteresis loop, hence higher hysteresis loss. It also results in higher eddy current loss in the square relation.

10.7.1 Power Losses in Core and Coil

The total power loss in the core and coil under alternating flux is proportional to the flux density, current, frequency, physical dimensions, and the material. The loss components are broadly classified as follows based on the source of the loss:

Losses in the core:

- Hysteresis
- Eddy current
- Fringing flux

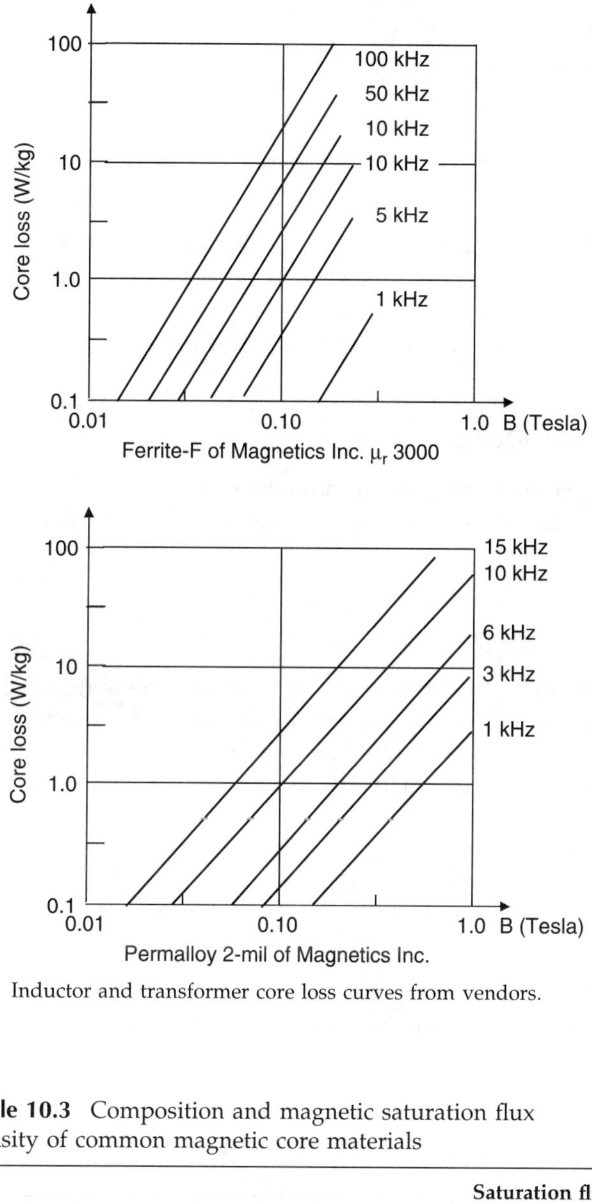

FIGURE 10.31 Inductor and transformer core loss curves from vendors.

Table 10.3 Composition and magnetic saturation flux density of common magnetic core materials

Core material	Composition	Saturation flux density (tesla)
Supermendur	Proprietary	2.1
Magnesil	97% Fe, 3% Si	1.6
Orthonol	50% Fe, 50% Ni	1.4
Alloy 48	50% Fe, 48% Ni	1.1
Permalloy, Supermalloy	79% Ni, 17% Fe, 4% Mo	0.7
Ferrites	Various	0.3

Losses in the coil conductors:

- Ohmic resistance
- Eddy current
- Skin effect
- Proximity effect

All of the above loss components must be included in the efficiency and thermal calculations. Some losses, such as those due to fringing flux, skin effect and proximity effects, can be significant at high frequency. Details of their calculations can be found in books dedicated to magnetics.[3]

10.7.2 Inductor Design

The inductance (in henry) of an inductor is defined as the flux linkage per ampere. In terms of the notations of Figure 10.32, it is given by

$$L = \frac{N \cdot \phi}{I} \tag{10.23}$$

where N = number of turns in the coil, φ = magnetic flux linking (passing through) the coil, and I = current in the coil. The energy stored in the inductor is given by $U = ½LI^2$. The total reluctance of the core is made of two parts, one in the core materials itself and the other in the air gap, and is given by

$$\Re = \frac{(L_m - L_g)}{\mu_r \mu_o A_c} + \frac{L_g}{\mu_o A_c} \tag{10.24}$$

where the total air gap L_g is lumped in a discrete gapped core or is distributed in powder cores. Calculating the flux and substituting in

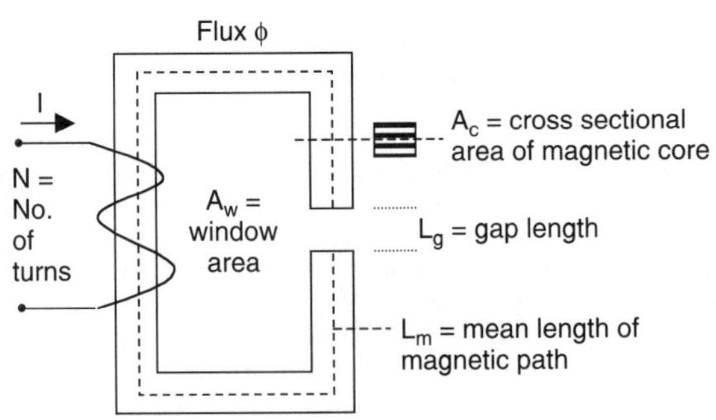

FIGURE 10.32 Energy storage inductor.

Equation 10.23 give the following expression simplified for most practical designs having $L_m/\mu_r << L_g$:

$$L = \frac{0.0125N^2 A_c}{L_g} \; \mu h \tag{10.25}$$

In discretely gapped cores, the fringing flux around the air gap increases the inductance by a factor

$$\xi = 1 + \frac{L_g}{\sqrt{A_c}} \ln\left(\frac{2G}{L_g}\right) \tag{10.26}$$

where G = window height (leg height) in which the gap is located (see Figure 10.27). A good inductor design maximizes the energy storage and minimizes the losses. The design is driven by the following major factors:

- Required inductance
- Direct current it must carry
- Alternating current (ripples or otherwise)
- Power loss and temperature rise limitations
- Mass, manufacturability, and cost

The following is an approximate expression often useful in estimating the energy storage capability (in joules) of an inductor based on its core and coil dimensions:

$$U = \tfrac{1}{2} K A_p^{0.9} B_m J_m K_u \times 10^{-4} \tag{10.27}$$

where

K = constant of proportionality derived from similarly built designs

A_p = area product of the core

$B_m = B_{dc} + ½B_{ac}$ = maximum flux density in core before saturation

J_m = maximum current density in coil before overheating

K_u = window utilization factor, i.e., fraction of the window occupied by the conductor cross section. It is about 0.5 for pot cores, 0.7 for laminated cores and 0.8 for tape wound cores.

It is noteworthy that the magnetic energy cannot be stored in a perfectly magnetic core with infinitely large permeability. This is because it would produce flux in the core with no current in the coil. The energy can be stored only in nonmagnetic materials such as air or free space. For this reason, the inductor core must have some air gap in the flux path, either discrete as shown in Figure 10.32, or distributed as in potted metal powder

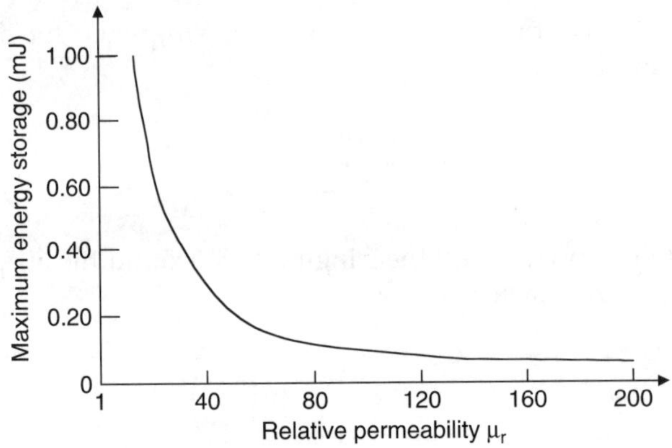

FIGURE 10.33 Energy storage capability versus magnetic permeability.

cores. The energy storage is mostly in the air gap, and not in the magnetic material. As a matter of fact, the energy stored in an ideal magnetic core is always zero.

The above point is illustrated by a practical design of an inductor with potted core 1.35 cm OD, 0.7 cm ID × 0.155 cm high with 3.1 g mass. The coil conductor mass varies with the conductor size around 9.1 g, making the total mass approximately 12.2 g. The energy storage capability of this inductor depends inversely with the relative permeability of the core material as seen in Figure 10.33. The higher the permeability, the lower the energy storage.

For a circuit simulation, the coil inductance, ohmic resistance and capacitance to ground can be represented by the equivalent circuit shown in Figure 10.34. The calculated parameters often need to be calibrated by tests due to various factors that are not well understood or are complex to analyze.

10.7.3 Transformer Design

The transformer is a magnetic device, which steps up or down the a.c. voltage. It has at least two coils mutually coupled via a magnetic core. Its power transfer rating, P, is expressed as the volt–amp product and not

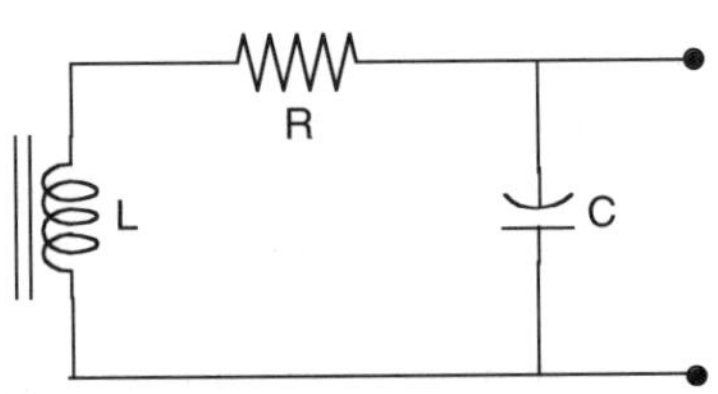

FIGURE 10.34 Inductor equivalent circuit.

watts. This is because an a.c. device can deliver zero watts when non-zero voltage and current outputs are out of phase by 90°.

In steady state performance, the voltage ratio of the two coils is the same as their turn ratio, the current ratio is equal to the inverse of the turn ratio, and the volt–amp product on both sides remains the same. A good transformer design maximizes the power rating while minimizing the losses and the temperature rise. The following approximate relations are useful in sizing a new transformer based on a similar unit built earlier. The transformer power is the product of V and I. Since V is proportional to the core cross section, and I is proportional to the window area available to wind coils, the power rating is proportional to the area product of the core $A_p = A_c A_w$. That is, $P = KA_p^{0.9}$, which has the unit of cm^4. Alternatively, it can be expressed (in volt–amp) as

$$P = A_p^{0.9} K_u K_j B_m J_m f \times 10^{-4} \tag{10.28}$$

where
f = frequency
K_u = window utilization factor
K_j = current density factor
B_m = maximum flux density the core can carry before saturation
J_m = maximum current density the coil can carry before over heating

Assuming that the total mass is proportional to the volume, which has the unit of cm^3, the transformer mass varies as the ¾th power of the area product, i.e.,

$$M = K_u A_p^{0.75} \tag{10.29}$$

where values of the window utilization factor K_u are the same as given for the inductor design, i.e., 0.5 for pot cores, 0.7 for laminated cores and 0.8 for tape wound cores. The total surface area A_{th} available for thermal dissipation of the power losses, having the unit of cm^2, varies with the ½ power of A_p, i.e.,

$$A_{th} = K_s A_p^{0.5} \tag{10.30}$$

where K_s is a surface constant, which is about 30 for pot cores, 40 for lamination cores and 50 for tape wound cores.

We note here that higher frequency gives more power from the same core, or it requires lower A_p for the same power. This is true for the energy storage inductor also. That is why high frequency is preferred for reducing the mass of the magnetics, which is relatively heavy in the power electronics assembly. However, the skin effect at high frequency requires

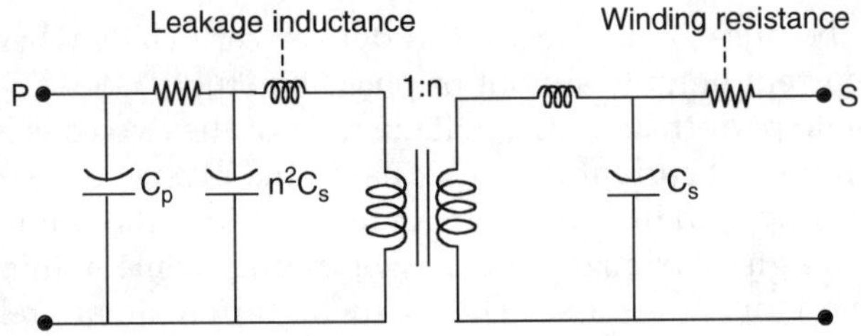

FIGURE 10.35 Transformer equivalent circuit.

thin insulated wires, which results in poor space factor. The 20 kHz and higher frequency coils are often wound with copper foil. Off course, there is a diminishing rate of return after a certain frequency. Beyond that, the losses, heating, and parasitic parameters may start impacting the design negatively.

As in the inductor, the transformer can be represented in the circuit simulation model by the winding resistance, leakage reactance, and capacitance to the ground for each coil as shown in Figure 10.35. Looking from the primary sides, the secondary side impedance becomes transformed by a^2, where $a = N_p/N_s$, the turn ratio.

10.7.4 Manufacturing and Testing of Magnetics

In the inductor and transformer manufacture, each device is impregnated in a vacuum chamber to achieve void-free impregnation. This is done after drying long enough at high temperature to remove all moisture. The potting compound is degassed prior to impregnation and potting. The use of dry air or dry nitrogen may be required in the process. The insulation requirements that generally apply for spacecraft magnetics are listed in Table 10.4.

10.8 Design for Maximum Efficiency

Since the power system is as efficient as its components are, designing an efficient system means designing each component to operate at its maximum efficiency. The electrical and electronic components while

Table 10.4 Insulation test requirement for inductor and transformers

Test on assembled unit	Inductor	Transformer
Dielectric withstand voltage from 1 atm to 5 μtorr	500 V rms 60 Hz	300 V rms 60 Hz
Insulation resistance by Megger	10,000 MΩ at 500 V d.c.	10,000 MΩ at 500 V d.c.

transferring power from the input side to the output side lose some power internally in the form of heat. In practical designs, the efficiency of 95 to 98% is typical in large power equipment in tens of kilowatt rating, and 80 to 90% in small equipment in less than a kilowatt rating. Figure 10.36 shows the efficiency versus load power for a battery converter boosting the voltage to 28 V from three different input voltages corresponding to the battery state of charge. Although less boosting means higher efficiency, the efficiency at each input voltage varies with load.[4] It increases with load up to a certain point beyond which it decreases. A good design maximizes the efficiency at the load that the equipment supplies most of the time. For example, if the equipment is loaded at 75% of its rated capacity most of the time, it is beneficial to have the maximum efficiency at 75% load. This is important for battery charge converter having output current that varies from the maximum to a trickle charge level over the sunlight duration. The battery discharge converter, on the other hand, works near the full load current all the time during eclipse. Such a design practice is also important in designing redundant battery converters, which may be working in active or dormant redundancy. The method of achieving the maximum efficiency at a desired load level follows.

The total loss in any power equipment has in general three components:

- Fixed loss representing the quiescent no-load power consumption. The fixed loss primarily includes the eddy and hysteresis losses in the magnetic core. It is always there as long as the unit is energized, and remains constant at all loads ranging from zero to full load or even overloads.

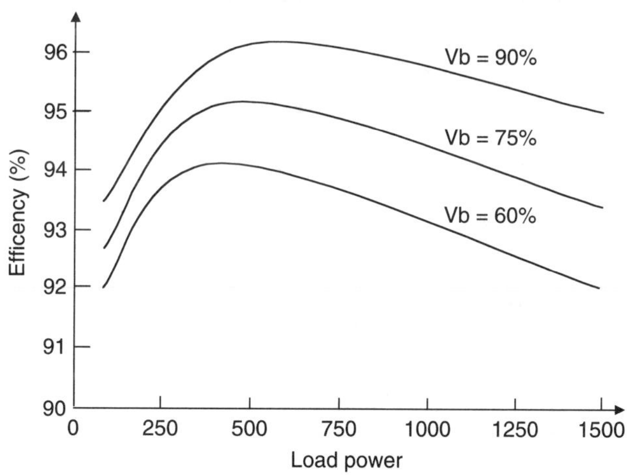

FIGURE 10.36 Typical buck–boost converter efficiency versus output load at various battery voltage levels.

- Diode and switching losses varying linearly with the load current, and
- Ohmic loss in conductors varying with square of the load current.

The total loss in a component operating under a constant voltage is therefore expressed as (Figure 10.37),

$$\text{loss} = L_o + K_1 P + K_2 P^2 \tag{10.31}$$

where P is the output power delivered to the load, L_o is the fixed loss and K_1 and K_2 are proportionality constant derived from curve-fitting the power loss data. The component efficiency is given by

$$\eta = \frac{\text{output}}{\text{input}} = \frac{\text{output}}{\text{output} + \text{loss}} = \frac{P}{P + L_o + K_1 P + K_2 P^2} \tag{10.32}$$

For efficiency to be the maximum at a given load, its derivative with respect to the load power must be zero at that load. That is,

$$\frac{d\eta}{dP} = \frac{P(1 + K_1 + 2K_2 P) - (P + L_o + K_1 P + K_2 P^2)}{(P + L_o + K_1 P + K_2 P^2)^2} = 0 \tag{10.33}$$

This equation reduces to $L_o = K_2 P^2$. Therefore, the component efficiency is the maximum at the load under which the fixed loss is equal to the square variable loss. The diode loss does not enter the efficiency optimization. This

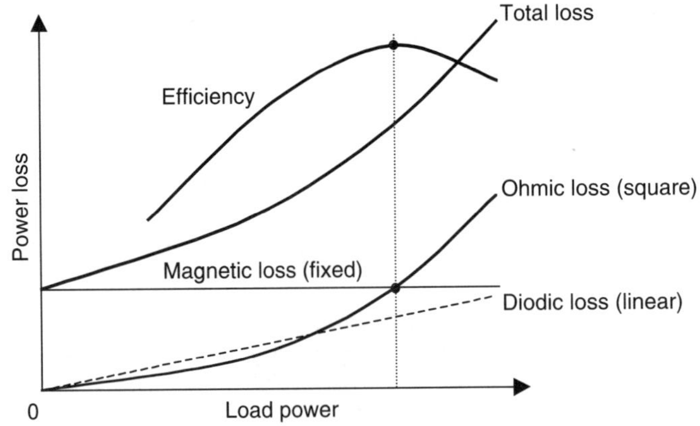

FIGURE 10.37 Power converter loss components versus load current — fixed, diodic and ohmic losses.

is an important design rule for all power components such as power converters and magnetics.

10.9 Trends in Power Electronics

In striving for mass reduction in converters and the magnetics, the switching frequency has increased steadily from early designs of 20 kHz to hundreds of kilohertz at present. Even 500 kHz and 1 MHz converters are under development for small power levels up to 3 kW at laboratories like the California Institute of Technology, Virginia Tech, and General Electric. High switching frequency basically reduces mass of the magnetic and capacitive components, resulting in light weight and high efficiency. The high frequency converter power density can approach 1 to $2\,W/cm^3$ (including heat sink), efficiency up to 85% at 5-V output and up to 90% at 15 to 24 V output. They can meet EMI MIL-STD-461 with common mode noise filter, typically 0.5 cm diameter × 3 mm toroid for a 100 W converter. Typical frequency response has a 20-kHz bandwidth.

The increased power loss at high frequency is addressed by developing new materials, semiconducting devices, and matrix transformers. The major cost driver is packaging. Planar magnetic provide compact packaging and better cooling. However, it has higher leakage inductance and higher mass per watt output. With presently available materials, the technology trend is to increase the switching frequencies to the 500 kHz to 1 MHz range using quasi-resonant topology with an efficiency goal of 90%. The low-cost, low-voltage, multiple output voltage converter designs for computer applications use magnetic amplifier post regulator for high efficiency. Similar approach may be considered for special applications in space. The tantalum and electrolytic capacitors are being replaced with film capacitors for further mass reduction.

The power electronics is a very wide and developing subject. The intent of this chapter has been to present an overview of the basic circuits used in the spacecraft power systems. Further design detail can be obtained from many excellent books on the subject.[5–7]

References

1. Cuk, S., *Switching dc-dc converter with zero input or output current ripple,* IEEE Industry Applications Society Conference Records, October 1978.
2. Cuk, S., *Analysis of integrated magnetics to eliminate current ripples in switching converters,* Power Conversion International Conference Proceedings, Intertec Publishers, Oxnard, CA, April 1983.
3. McLyman, W.T., *"Transformer and Inductor Design Handbook,"* Marcel Decker, New York, 1988.

4. Fox, D., Testing of the engineering model of electrical power control unit for the fluids and combustion facility, in *Proceedings of the 34th Intersociety Energy Conversion Engineering Conference,* SAE, 1999, Paper No. 2435.
5. Wu, K.C., *Pulse Width Modulated dc–dc Converters — Analysis and Design,* Kluwer Academic Publishers, Norwell, MA, 1997.
6. Wu, K. C., *Transistor Circuits for Spacecraft Power System,* Kluwer Academic Publishers, Norwell, MA, 2002.
7. Mohan, N., Undeland, T., and Robbins, W.P., *Power Electronics Converters, Applications and Design,* IEEE and John Wiley & Sons, 1995.

Chapter 11

Distribution Harness and Protection

11.1 Introduction

The harness includes the insulated conductors, connectors, and the shield. Its mass is determined from the detailed layout and routing of all the required wiring. For this reason, it is often considerably more than that estimated without actual layout at a preliminary design stage. A typical harness mass breakdown is given in Figure 11.1.

11.2 Ampacity of Wires

The maximum current carrying capacity (ampacity) of the wire in space is less than that on the ground due to lack of convective cooling. This requires derating of wire ampacity for space applications from the ground-based rating. The wire size is measured in American Wire Gage (AWG) or British Wire Gage (BWG). The gage number is an inverse measure of its bare diameter, and is set on a log scale:

$$\text{AWG} = 20 \log\left(\frac{0.325}{d}\right) \tag{11.1}$$

where d = diameter of bare conductor in inches. Thus for every one gage up, the diameter increases by a factor of 1.1225 and the area by 1.26. The diameter doubles every six gages and the area doubles every three gages. The ampacities of various gages of wires are listed in Table 11.1.

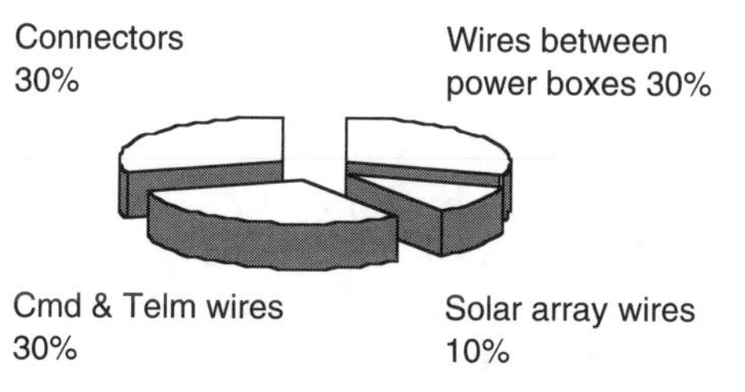

FIGURE 11.1 Harness mass breakdown in typical spacecraft.

Table 11.1 Maximum allowable current in amperes for wires and connector contact pins of the same gage

AWG	Diameter (inches)	Single wire in free air on ground (MIL-STD-5088)	Wires in space[a] in 70°C ambience (MIL-STD-975 and GSFC-PPL-19)	
			Single wire	Bundle or cable[b]
30	0.010 0	N/a	1.3	0.7
26	0.015 9	10.5	2.5	1.4
24	0.020 1	14	3.3	2.0
20	0.032 0	24	6.5	3.7
16	0.050 8	37	13.0	6.5
12	0.080 8	68	25.0	11.5
8	0.128 5	135	44.0	23.0
4	7 × 0.0772	260	81.0	40.0
0	19 × 0.0745	460	147.0	75.0

[a]For TFE teflon insulated wires rated for 200°C.
For 150°C rated insulation, use 80% of values shown.
For 135°C rated insulation, use 70% of values shown.
For 105°C rated insulation, use 50% of values shown.
[b]For cable bundles of 15 or more wires in 70°C ambience in hard vacuum. For smaller bundles, the allowable current may be proportionately increased as the bundle approaches a single conductor.
[c]Wires sizes 10, 14, and 18 are not used in aerospace for general wiring, and AWG 2 and 6 have no counterpart electrical connector contacts (pins).

11.3 *R-L-C* Parameters

The resistance (*R*), inductance (*L*), and capacitance (*C*) parameters of the harness are required for detailed simulation model of the power circuit. They are determined from the expressions given below, where the values are per meter length of cable (lead and return wires together). When wires are twisted, typically at a rate of one twist every few centimeters, the actual conductor is approximately 10 to 15% longer than the straight run. The resistance (in Ω/m) is given by

$$R = \frac{2\rho}{A} \tag{11.2}$$

where ρ = resistivity of the conductor and A = cross-sectional area of one-way conductor(s). At 20°C, $\rho = 1.724 \times 10^{-8}$ Ω·m for copper and 2.65×10^{-8} Ω·m for aluminum, both of electrolytic grades.

The resistance R_2 at temperature T_2 is derived from R_1 at temperature T_1 using the temperature coefficient of resistance α as follows:

$$R_2 = R_1[1 + \alpha(T_2 - T_1)] \tag{11.3}$$

where $\alpha = 0.00393$ 1/°C for electrolyte copper and 0.00390 1/°C for aluminum.

For a cable with two parallel round conductors of equal radii (Figure 11.2a), the cable inductance (in H/m) in free space is given by

$$L = \frac{2\mu_o}{2\pi}\left[\frac{1}{4} + \ln\left(\frac{s}{r}\right)\right]$$

or

$$L = \left(1 + 4\ln\frac{s}{r}\right) \times 10^{-7} \tag{11.4}$$

where μ_o = permeability of free space = $4\pi \cdot \times\ 10^{-7}$ H/m, s = spacing between conductor centers, and r = conductor radius.

The term ¼ is due to the magnetic flux internal to the conductor. It is small and approaches zero at high frequency due to skin effect. For twisted pairs touching each other, $s = 2(r +$ wire insulation thickness). Since the insulation thickness is proportional to the wire radius for mechanical reasons, the s/r ratio remains approximately constant, making L somewhat insensitive to the wire gage. For AWG 10 to 30 wires, the inductance remains in the range of 0.5 to 0.7 μH per meter length of twisted pair up to hundreds of kilohertz frequency. Twisting does not reduce the inductance per meter, but adds 10 to 15% in the length. However, it produces additive and subtractive B fields in adjacent spans (Figure 11.2b), canceling each other's emission at far distance. The inductance of open untwisted wire loops at the end of an equipment box is determined from Equation 11.4. It is about 5 μH/m for practical loops and 7.5 μH/m for infinitely large loops. It is noteworthy that the single point ground reduces the ground inductance by avoiding large ground current loops. The inductance of a ground strap is that of a single wire in free space with the ground as return. Its values are plotted in Figure 11.3.

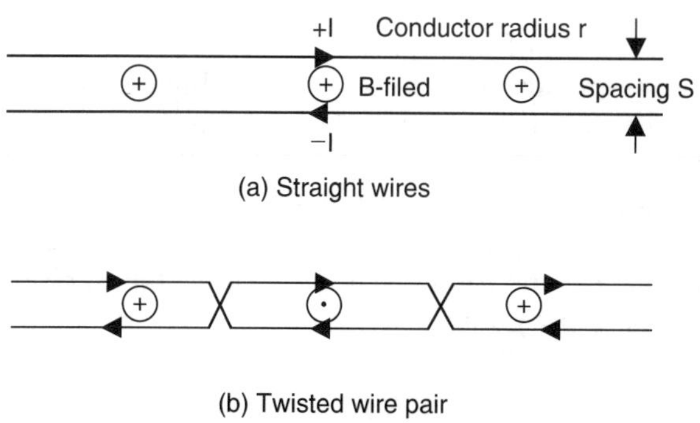

FIGURE 11.2 Parallel wires and resulting B field.

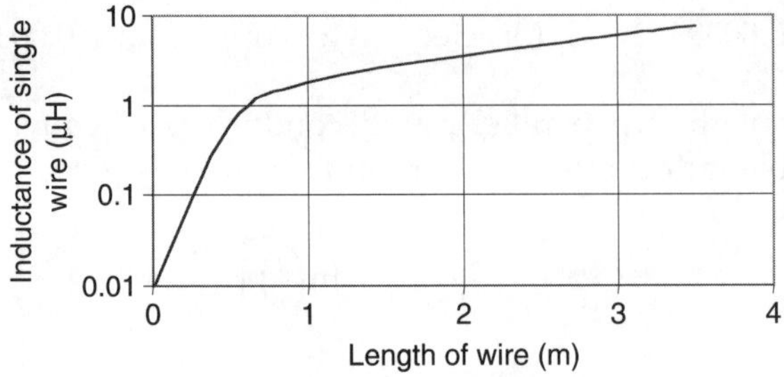

FIGURE 11.3 Inductance of single straight wire (ground strap).

The battery packaging influences the harness inductance and the magnetic moment. A one-pack battery has a small harness inductance but a two-pack battery can have a large harness inductance and the magnetic dipole moment depending on the spacing between them. Therefore, it is important to place all packs of the battery as close as possible to minimize the interconnecting harness loop.

The capacitance (in F/m) between two parallel round conductors in the cable is given by

$$C = \frac{\frac{1}{2}(2\pi\varepsilon_0\varepsilon_r)}{\ln\left(\frac{S}{r}\right)}$$

or

$$C = \frac{\pi\varepsilon_0\varepsilon_r}{\ln\left[\frac{S}{2r} + \sqrt{\left(\frac{S}{2r}\right)^2 - 1}\right]} \quad (11.5)$$

where ε_o = permittivity of free space = $8.85 \times \cdot 10^{(12}$ F/m, and ε_r = relative permittivity of the insulation between the conductors. The first expression is for open wires at the harness ends, and the second expression is for closely spaced conductors in the cable. Unlike the resistance, the *L* and *C* parameters do not change with temperature.

Bus bars of thin rectangular cross section shown in Figure 11.4 are often used in harness segments carrying high current. For such conductors, the *R*-*L*-*C* parameters after ignoring the end fringing are

$$R = \frac{2\rho}{ab} \quad \Omega/\mathrm{m} \quad (11.6)$$

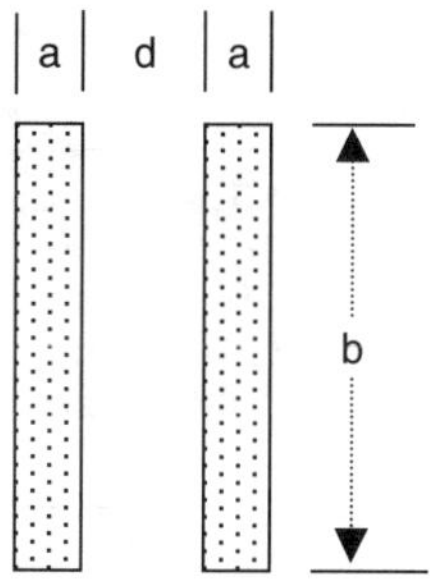

FIGURE 11.4 Rectangular bus bars for high current harness.

$$L = \frac{\mu_o}{b}\left(d + \frac{2a}{3}\right) \qquad \text{H/m} \tag{11.7}$$

$$C = \frac{\varepsilon_o \varepsilon_r b}{d} \qquad \text{F/m} \tag{11.8}$$

A typical communications satellite bus designed for 10 kW, 70 V power with copper-clad aluminum bus bars, 5 cm wide × 2 mm thick, would typically have $R_{dc} = 2$ mΩ at 70°C, $L = 50$ nH, and $C = 3$ nF per meter of the bus length. A cable using 20 AWG round copper wires would have $R_{dc} =$ 70 mΩ/m, $L = 0.375$ μh/m, and $C = 0.077$ nF/m.

The following approximate values can be used as a quick estimation guide for wire gages generally used in spacecraft. The AWG 10 conductor has 100 mils diameter, 40 g mass per meter and 3.3 mΩ resistance per meter at 70°C. Every three gages doubles the area and every six gages doubles the diameter. With 2.3 A/mm^2 typical current density, the harness mass (conductor + insulation + shield, but not the connectors) is about 7 g/A one way, and the watt loss is about 2.7 × conductor mass in kg. At 100 kHz, the skin depth is 0.22 mm and $R_{ac}/R_{dc} = 9$. At 1 MHz, the skin depth is 70 μm and $R_{ac}/R_{dc} = 27$.

11.4 Conductor Materials

The most commonly used electrical conductor is copper for its good performance and low cost. Annealed copper is often used for high conductivity, but it has low tensile strength. This makes it unsuitable for use in thin wires due to risk of breakage in handling. For this reason, wires thinner than AWG 20 are often required to use high strength copper alloy 135, which has 40% higher tensile strength, and 10% higher electrical resistance. Copper wire coated with tin, nickel or silver is used to resist corrosion and oxidation. The suitability of various coatings is compared in Table 11.2, where tin plating is included for comparison only. Tin plated

Table 11.2 Suitability of various coatings on copper conductor

Coating on copper	Silver	Nickel	Tin
Solderability	Good	Special process	Good
Crimp ability	Good	Special process	Good
Maximum operating temperature	200°C	260°C	150°C

wires are widely used on ground, but are forbidden in space due to growth of whiskers.

Aluminum is sometimes used in power equipment where lightweight and/or low cost is desired. It is used in overhead transmission lines and pole mounted power transformers on ground, and in some aircraft and commercial spacecraft harness. The performance of aluminum is compared with copper in Table 11.3.

For the same power loss or voltage drop, the copper can be replaced with aluminum of relative mass given by the product of the resistivity and the mass density ratios. An aluminum conductor, therefore, would theoretically have $(2.830/1.724) \times (2.70/8.89) = 0.50$, i.e., one-half of the copper mass. However, in practice, aluminum does not produce 50% mass saving owing to the following reasons:

- Aluminum has about 1000 times higher creep rate then copper. For long-term reliability of the terminal connectors, thicker wire is required to keep the crimp joint temperature below 130°C to minimize creep. Alternatively, soldered connections are used, which are expensive and often inconvenient.
- Aluminum oxidizes instantly when exposed to oxygen. This hinders the electrical contact between the wire and the connector. Crimp terminals that produce cold weld are often used to keep oxygen away from bare aluminum.

Table 11.3 Copper and aluminum conductor comparison

Characteristic	Copper	Aluminum
Resistivity Ω·m at 20°C	1.724×10^{-8}	2.830×10^{-8}
Mass density g/cm^3	8.89	2.70
Temperature coefficient of resistance α/°C	3.93×10^{-3}	3.90×10^{-3}
Melting point °C	1083	660
Flex life (relative)	1	0.5
Thermal coefficient of expansion (relative)	1	1.4
Creep rate at 65°C (relative)	1	1000

- The poor mechanical strength of aluminum wire often requires thicker wire for mechanical reason alone, even where it meets the ampacity requirement.
- The insulation and the EMI shield mass increases due to heavier gage aluminum wire required to carry the same current.

As a hybrid approach, copper-clad aluminum wires have been used in the past by some prime contractors in communications satellites with net mass savings of 10 to 15%. However, aluminum or copper-clad aluminum wires are generally not allowed on NASA or defense satellites.

11.5 Wire Insulation and Cable Shield

The insulation is designed to withstand the rated and abnormal transient voltages. The transients can be higher by several times the rated value. The insulation design must preclude corona and arcing at pressures below 10 torr, and withstand the radiation environment and atomic oxygen. In high radiation spacecraft, such as GPS, the system specifications required that the silicon-insulated wire be not used and the solar array wires and interconnects be welded to withstand the high radiation.

The cable shield is wrapped around the wire bundle to prevent electromagnetic interference from entering the cable or radiating out. The shield can weigh 15 to 40% of the cable weight. The material options are the braid versus tape and copper versus aluminum. The braid is used when extreme flexibility is required. Its mass as compared to the wire conductor is about 40% for flat cables and 20% for round cables. In tapes, 2-mil thick copper vapor deposited on Mylar or Kapton tapes are widely used. The tape is applied on the cable with insulation touching the cable, followed by another layer on the top. The shield mass with such tapes is roughly 80% of that with the braid. The EMI shielding is discussed further in Chapter 15.

11.6 Connectors

Both crimp and soldered connectors are used for connecting various wire segments. They come in Class A and B with maximum operating temperature of 125 and 200°C, respectively. They are further grouped in four types as listed in Table 11.4.

Series A rectangular connectors are limited to AWG 20 pins, whereas series B round connectors come in heavy pins, some of them in assorted pins to handle different power circuits. The wires are crimped with pins with a special tool. Wires thicker than pins are not permissible but can be up to two gages thinner. In both series A and B connectors, contact pins are further classified as Type A and B. The maximum voltage drop across type A and B contact pins of various gages are listed in Table 11.5.

Table 11.4 Connector classification

Type	Material	Application
A	Copper alloy	General purpose
B	Ferrous alloy	Hermetically sealed
C	Matching with thermocouples	Thermocouples
D	Copper alloy	Shielding (including coaxial, twin-axial and triaxial)

The general military standard isolation requirement on the connector is:

- High potential withstand capability between pin-to-pin and pin-to-shell at least 4 times the operating voltage
- Insulation resistance between pin-to-pin and pin-to-shell at least 5000 MΩ

The connector design practice as applied to the spacecraft power system generally includes the following:

- Pins in a connector and the SAD slip rings are assigned such that wires at different electrical potentials are separated and are not on adjacent pins or slip rings.
- Twisted wire pairs are terminated on adjacent pins in the connector.
- A minimum of 10% of spare pins is provided on the outer periphery of the connector.
- Each connector shell is grounded to the component case with impedance less than 2.5 mΩ.

Table 11.5 Permissible contact pin drop in millivolts at 200°C after conditioning as per Military Standard MIL-C-39029

Wire size AWG	Test current (amp)	Type A with silver-plated wires	Type A with nickel plated wires	Type B with silver-plated wires	Type B with nickel-plated wires
0	150	36	90	396	900
4	80	40	100	440	1000
8	46	45	113	495	1130
12	23	71	107	781	1177
16	13	84	126	924	1386
20	7.5	94	141	1034	1551
24	3	77	116	847	1276
28	1.5	92	138	1012	1518
32	0.5	74	111	814	1221

- In defense satellites, multiple contact connectors are not used for high voltages above 300 V d.c. or peak-to-peak a.c. voltage. Separate connectors rated for the specified high voltage are used for each connection.
- Hermetically sealed boxes must use hermetically sealed connectors with connector cases welded to the boxes.

11.7 Harness Mass Minimization

The harness mass is not small compared to other components. It is primarily driven by the voltage drop specifications between various harness segments. For example, the round trip voltage drop from the bus voltage control point in the power regulator unit (PRU) to any load is kept below 1 to 2% of the rated bus voltage. In a 28-V bus, the round trip drop from PRU to any load is typically kept below 250 mV for loads below 100 W, 500 mV for heavier loads, and 1 V for heater circuits drawing power only for a part of the orbit period. The voltage drop specifications, particularly in high voltage spacecraft, do not often result in an optimized design in the following sense. If a higher voltage drop in a thinner wire can be compensated with added mass in the source that is less than the mass saved in thinner conductor, the trade is beneficial for mass reduction at the power system level. Therefore, the harness power loss should match with the specific power of the source. For this reason, initially proposed voltage drop and derating guidelines in some spacecraft should be carefully examined before they become specifications.

The mass optimized harness design is derived from the considerations depicted in Figure 11.5(a). The power loss in the conductor is equal to the allowable voltage drop times the current. As the power loss increases, the harness mass decreases, but the mass of power source increases to make up the loss. The total of the two masses is a minimum at a certain optimum power loss in the harness, resulting in the minimum system mass. Figure 11.5(b) shows the harness power loss and the source power generation in watts per kilogram versus current density in the conductor. The harness with current density J_{opt} would result in the minimum systems mass. At the optimum current density, the loss per kilogram of harness equal the power generated per kilogram of the source.

Analytically, the power loss per pound of copper conductor at 75°C is given by $2.54\,J^2$, where J = current density in kiloamperes per square inch. For optimum conductor size, it must be equal to the specific power of the source P_{sp} W/lb. This leads to

$$J_{opt} = 627.5\sqrt{P_{sp}} \tag{11.9}$$

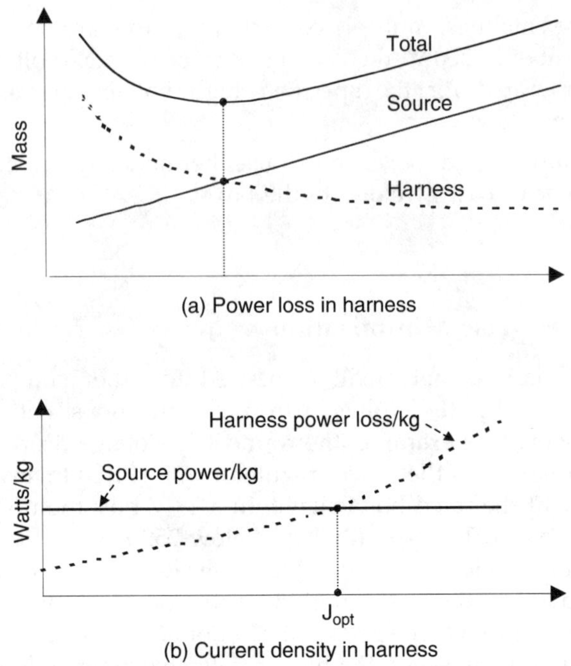

FIGURE 11.5 Harnessing mass optimization at the power system level.

The optimum conductor diameter (in inches) is then given by

$$d = \sqrt{\frac{4\, I_{rated}\, m}{\pi\, J_{opt}}} \tag{11.10}$$

where m = margin factor depending on the protection method used, typically 1.3 for fused circuits. The wire gage is then derived from Equation 11.1 rounded up. Such consideration leads to a simple expression for V_{opt}, the optimum voltage drop (in volts) per meter run of the harness (physical distance between the source and the load). It is given by

$$V_{opt} = k\sqrt{P_{sp}} \tag{11.11}$$

where P_{sp} = specific power of the source in watts/lb. The coefficient k is 0.03 for copper wires and 0.026 for aluminum wires. Harness segments having varying P_{sp} over the orbit period can be designed for optimum voltage drop using weighted average P_{sp}.

The harness mass optimization approach leads to the following observation for the conductor sizes in view of Figure 11.6.

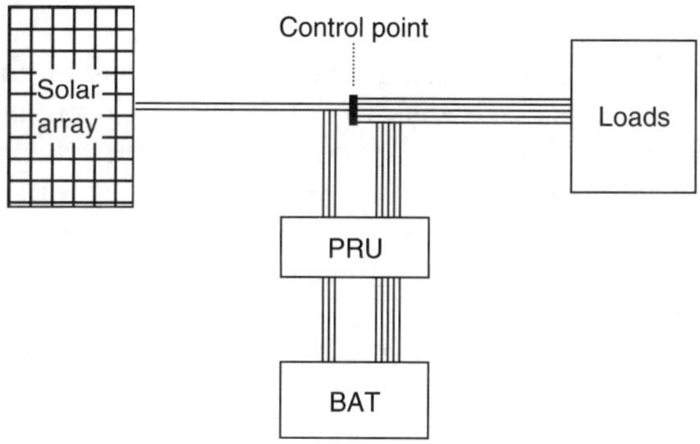

FIGURE 11.6 Progressive optimization of harness mass.

- The harness used with a source of high specific power, such as triple junction solar array, may operate at higher current density, allowing a higher voltage drop as percentage of the bus voltage.
- Thinner wire is justified from the solar array to PRU, because the power loss in that segment of the wire is supplied by the solar array alone.
- Thicker wire is required from the PRU to loads, because the power in that segment has been processed by the solar array, PRU, and the battery, all together during a part of the orbit. The whole system has much lower specific power compared to that of the solar array alone.
- For the same reason, the current density in the discharge path should be lower than that in the charge path. This is because the charge current comes from the solar array alone, whereas the discharge current involves the whole system.

So, as we move closer to the loads at the end of distribution lines, the current density should be progressively lower. Each watt becomes more precious at the load point compared to that at the generation point. In ground-based utilities, this is analogous to the cost of power to end users being more than five times the cost in the central power station yard. It must be noted that the above considerations lead to the minimum harness mass, and not the minimum cost at the power system level. The cost of adding one pound of source is always much greater than the cost of adding one pound of harness. However, the harness mass optimization may be warranted when the spacecraft mass is rigidly constrained to fit the launch vehicle or for some other reason.

11.7.1 Example

The harness mass optimization is more important in LEO satellites. The heavy battery in LEO makes the specific power of the entire power system about one-half of that in GEO satellites. For example, let us consider a fully regulated LEO power system where the battery mass is equal the solar array plus the PRU mass. We wish to determine the optimized wire gage for a solar array circuit carrying 5-A current from the silicon PV array to the load through slip rings, and also for a 5-A load circuit from the PRU to the user load. We are given that the specific power of the (array + PRU) is 10 W/lb, and the specific power of the (array + PRU + battery) is 5 W/lb. The optimum current density for the slip ring circuit is $\sqrt{\{10/(2.54 \times 10^{-6})} = 1984$ A/in^2. Allowing 30% margin, the required cross section area of the wire is (1.3 × 5/1984) = 0.0033 in^2, which requires AWG 14 wire. Using the GSFC-PPL-10, a bundle of 14 AWG wires has a derated current carrying capacity of 8.5 A, which is greater than (1.3 × 5) = 6.5 A. Thus, the mass optimized wire gage meets the preferred part list requirements. If the solar array were made of triple junction cells with specific power of 5 W/lb, then the optimum current density would have been $\sqrt{\{5/ \times 2.54 \times 10^{-6})\}} = 1192$ A/in^2. This results in conductor cross section 40% smaller, requiring AWG 12 wires. If such optimized wire gage results in a local hot spot somewhere in the harness run, spreading out the heat with a local aluminum foil should be considered first rather than increasing the wire gage for the entire harness run.

11.8 Harness Design Process

The harness design is influenced by the following factors:

- The rated current and the allowable voltage drop determine the conductor gage
- The bundle size and steady state temperature also determine the gage
- The rated voltage determines the insulation thickness and the temperature rise determines the insulation material
- The derating requirement increases the gage thickness
- Specific power of the source is factored in to determine the optimum wire gage
- Reliability goal may require redundant wires or double insulated wires

The harness design proceeds in the following steps:

1. Determine the maximum steady state d.c. or a.c. rms current for each harness segment on the spacecraft. For each segment, size the wire to carry the required current per applicable derating factors. The harness wires always run in large bundles, so the derating factor for the bundle of 15 or more wires generally apply.
2. Determine the voltage drop and compare with the requirement. If the requirement is not met with the wire selected in step 1 above, use more

wires in parallel or increase the wire size until the voltage drop requirement is met.

3. If an open wire in a current path constitutes a failure, i.e., if either the voltage drop or the derating criteria is not met with one wire open, and if there is no other redundant current path, then another wire should be added to guard against this mode of single point failure.
4. Ensure that the wire(s) in a particular current path will not be damaged to the point of propagating the damage to the surrounding wires when a smart short occurs. A smart short is one that draws current insufficient to blow the fuse but in excess of the wire derating. In such a case, determine the value of the steady state current that can cause damage to the wire insulation and could propagate to other wires. If that current is less than the expected steady state d.c. or rms fault current, then the harness must be increased in size to a point where the smart fault current cannot cause damage that could propagate through the harness.
5. The harness should be secured to withstand the mechanical repulsive forces, which can be significant during hard faults. The wires are generally tied with each other or all in a group, often attached to structural tubes as shown in Figure 11.7.

The mechanical repulsion force (in newtons per meter) between two parallel conductors in configuration of Figure 11.2(a), carrying currents I_1 and I_2 in opposite directions, is given by

$$F = 80\,K I_1 I_2 \frac{s}{d} \times 10^{-7} \tag{11.12}$$

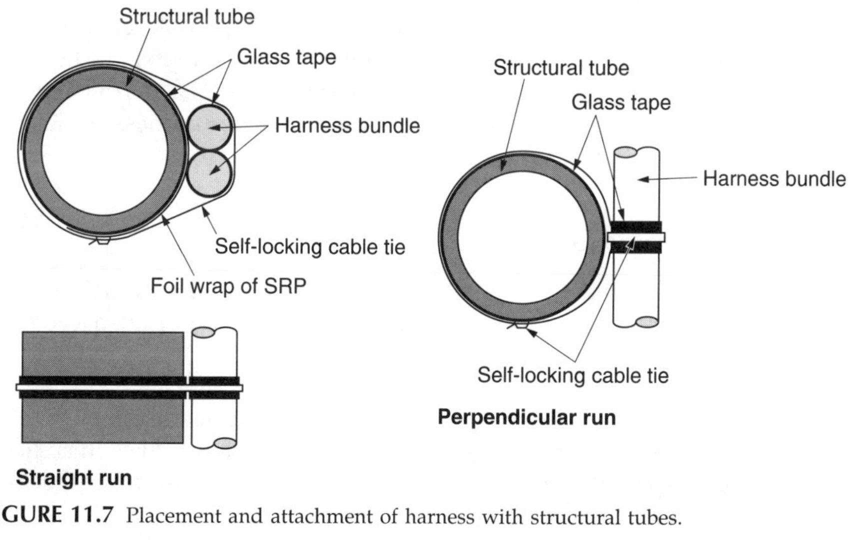

FIGURE 11.7 Placement and attachment of harness with structural tubes.

where K = conductor shape factor = 1.0 for round conductors. The force calculation on high power rectangular bus bars is covered in Chapter 22.

11.9 Flexible Harness

The slip rings can experience cold welding in high radiation satellites, as with a GPS. The flexible harness (coiled cable) that can accommodate the sun-tracking rotation of the solar array can eliminate the slip rings. The harness with $\pm 90^\circ$ flex has been considered for MEO satellites requiring 20,000 cycles of flexing over 15 years of mission life. Incorporating a flex harness in the design at present is within a development reach. Some key features of the design are:

- Flex harness is completely enclosed inside a Faraday cage (solar array boom), hence unaffected by the deployment behavior, environment shielding, EMI, etc.
- Harness flexing is required only between the SAD and the structural bulkhead.
- Harness outside the Faraday cage remains the same, so the cable interface with the bus is unaffected.

The cable needs to be ultra-flexible with very thin hair-like strands. This may impose a manufacturing problem in crimping and ascertaining that no whiskers are left projecting out. A fully developed space-qualified flex harness could eliminate the slip rings, which weigh several kilograms per spacecraft. The harness mass, however, would remain the same. A flexible fluoropolymer insulated cable with 20-gage wires for power and 24-gage wires for signals has been built and tested in the past.

The flexible harness has been used by ESA in 2002 for a Mercury mission. In this design, the solar panels rotate $\pm 180^\circ$ around one axis to keep the operating solar cell temperature below 150°C at AU < 0.375. The sun shielded solar array drive has no slip rings, but uses flex harness sufficiently flexible for $\pm 180^\circ$ rotation.

11.10 Fuse Protection

The fuses are provided to protect the power system from internal faults or from faults in load circuits. The general fusing practice in the spacecraft is as follows:

- Fuses are placed near the source for protection against faults in the wires, and also to reduce the local fuse blow voltage spikes in adjacent loads (cross-talks). Fuses are located as close as possible to the PRU (Figure 11.8). The fuse boards are often located inside the PRU.

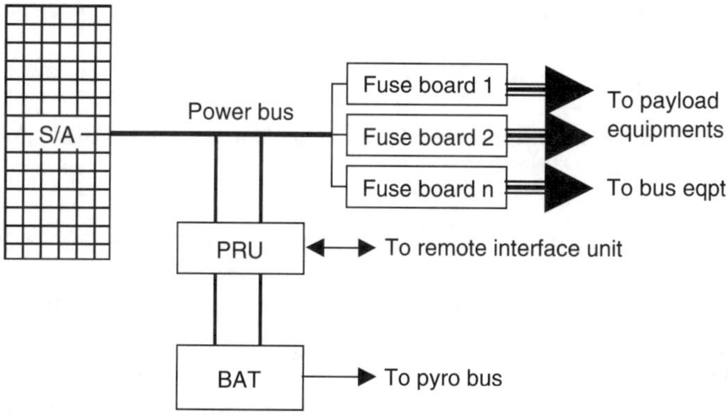

FIGURE 11.8 Fuse boards located in or near PRU for various load groups.

- All noncritical loads are individually fused to prevent a fault in any load from damaging the power system.
- In full-length redundant wires, fuses are placed in both lead wires (Figure 11.9).
- Critical loads are not fused to maximize the reliability, but they use double insulated wires. When a fuse is required in a critical load circuit, redundant fuses are used, as often required by NASA (Figure 11.10).
- Suitable load path redundancy is provided such that a single fault in a device or wiring does not permanently open or short the main bus itself or essential bus loads.

A typical construction of the fuse used for protecting semiconducting devices is shown in Figure 11.11. The fuse element may be silver, copper or nickel, with silver being more common for long-term stability. It has a reduced section for better predictability. The body is generally filled with sand-type filler to suppress sparks when the fuse element melts and interrupts the inductive energy of the load and the wires. The fuse life is determined primarily by the metal aging under thermo-elastic cycles.

Because of the lack of a sufficient market, only a handful of commercially available fuses have been space qualified in limited ratings. Among them are Piocofuse™ series 265 and Little™ fuses in small ratings; and Bussmann™ fuses in large ratings. They are basically commercial

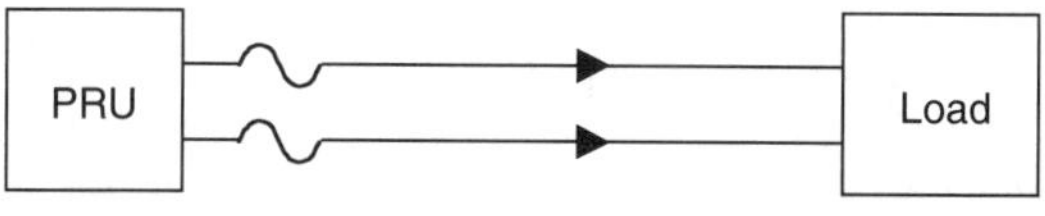

FIGURE 11.9 Full-length redundant wires with fuses in both lead wires.

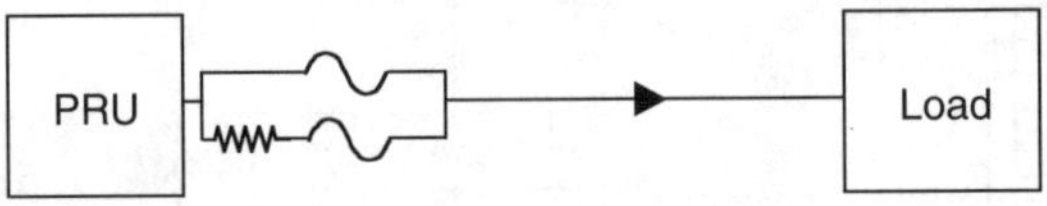

FIGURE 11.10 Redundant fuses in a single, nonredundant, double insulated wire.

nonsealed fuses, except that source control drawings are added for tractability to qualify for space.

Picofuses use nickel elements in less than 1 A ratings, 50% silver and 50% copper elements in 1 to 4 A ratings, and copper elements in greater than 5-A ratings. All element materials have high positive temperature coefficient of resistance. They rise in resistance with temperature, thus preventing an absolute short.

Nonsealed fuses, however, pose design difficulties when bus voltage exceeds 50 V d.c. This is due to arcing in partially evacuated construction. Hermetically sealed fuses address this problem. P600L™ fuses made by Mepcopal Corporation are such space-qualified fuses. These fuses are made using thick film material. The fuse element is primarily gold while the terminal pads are silver. These materials are deposited onto a 96% alumina substrate and coated with dielectric overglaze cured at high temperature. An arc suppressant glass of about 0.5 mm is then printed over the fuse element. When the fuse melts under a fault, the gold element melts and migrates readily into the arc suppressant glass coating, thereby effectively preventing subsequent vaporization and ionization. This suppresses arcs and restricts the circuit voltage after the fuse blow. For these reasons, Mepcopal fuses are fast and more predictable in clearing time. The hermetically sealed case contains the spark while blowing. In fact, they have been developed for safe operation in explosive mixture of chemical vapors for the National Electrical Code Class 1 environment (hazardous).

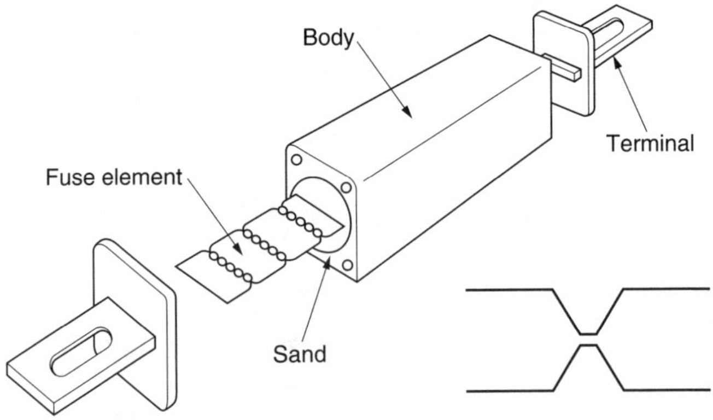

FIGURE 11.11 Typical semiconductor fuse construction.

The ground insulation requirements imposed on fuse vary with the voltage ratings. For example, a 100-V fuse would typically have:

- Dielectric withstand of 500 V_{rms}
- Ground leakage current less than 1 mA
- Resistance after the fuse clearing between 1 MΩ and 10 MΩ depending on the fuse rating and the fault current it interrupts.

11.10.1 Fuse rating

The ampere rating of the fuse is the current it would carry indefinitely on ground. The fuse clears (blows) only at higher currents as per its own *I–t* relationship. The fuse is required to carry 110% of rated current for at least 4 h without blowing, blow within 1 h at 135%, and within 2 min at 200% of the rated current. On the faster side, the fuse clearing time is in microseconds for 10 times the rated current. The general criteria for setting the current rating of the fuse are:

- The fuse carries 110% of the rated current for a long time in hours without exceeding 70°C above the room temperature.
- The fuse clears 135% of the rated current within a specified time in minutes, which depend of the type of fuse.
- The fuse withstands full recovery voltage, which is equal to the open circuit voltage that appears across the fuse terminals after clearing the worst fault.

The selected fuse rating is typically 1.2 to 1.3 times the current rating of the load circuit it protects, rounded upward to the next standard available rating. The fuse rating must be carefully selected because both the conservatively or liberally sized fuses are bad for the circuit protection. The ambient temperature has an influence on the current rating of the fuse as shown in Figure 11.12. It is desirable to up rate or down rate the fuse accordingly.

As for the voltage rating, the fuse can be operated at any circuit voltage as long as it can blow without suffering arc damage. When the fuse blows, the sharp break in the circuit current cause the full circuit voltage to appear across the blown fuse. If the load is inductive, the transient voltage across the fuse may be substantially high due to the inductive kick. A destructive arc may be formed across the fused element and continue to grow. The resulting heat and pressure may cause the fuse to explode in the worst case. The voltage rating of the fuse is conservatively given such that the fuse can interrupt a dead short without shattering or emitting flame or expelling molten metal.

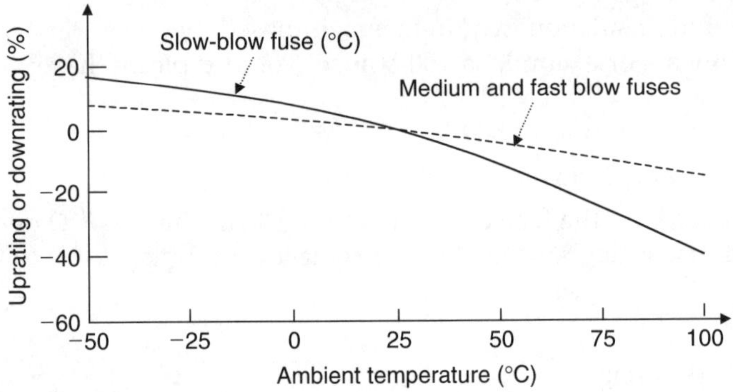

FIGURE 11.12 Effect of ambient temperature on current carrying capacity of fuse.

The equivalent circuit during fuse current build-up and interruption is shown in Figure 11.13. The total energy dissipated in the fuse body equals the energy stored in the circuit inductance plus the energy supplied by the main power source. It should not rupture the fuse body for safety. The current build up obeys the following equation.

$$\frac{di}{dt} = \frac{1}{L}\left[V_s - V_f - Ri\right] \tag{11.13}$$

where V_s = source voltage, V_f = fuse voltage, R = fuse resistance and i = fuse current. The fuse clearing has two distinct phases shown in Figure 11.14.

11.10.1.1 Pre-arc period

The pre-arc period lasts until the arc voltage appears across the fuse terminals. It is a function of the fault current, the elements neck section and

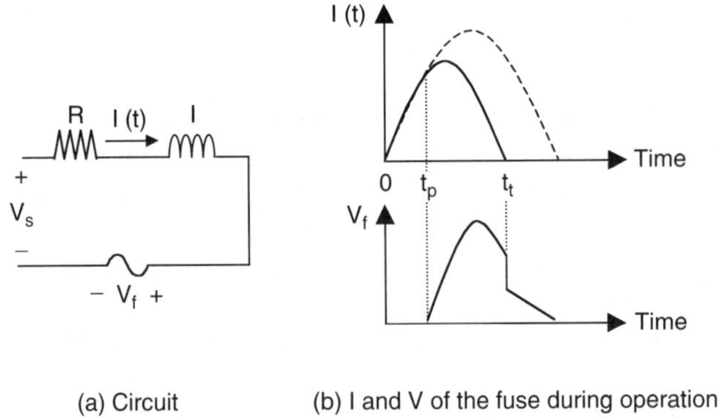

FIGURE 11.13 Equivalent circuit during fuse current build-up and interruption.

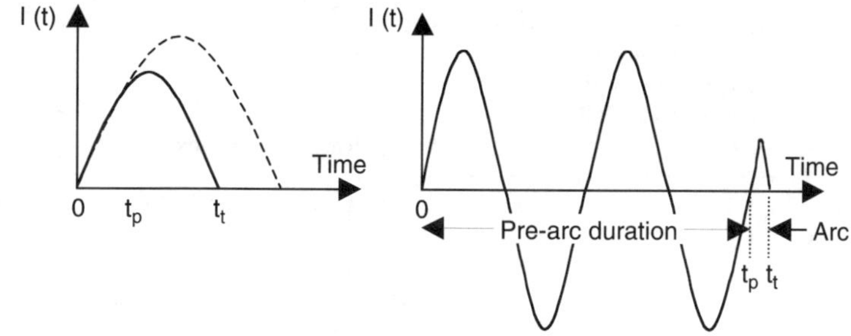

(a) High fault current (b) Low fault current

FIGURE 11.14 Fuse clearing process.

the material. In silver fuses, it lasts until the joule integral I^2ta^2 reaches 80,000 $A^2{\cdot}s{\cdot}mm^4$ with adiabatic heating, where I = rms current in ampere, t = pre-arc time in seconds, and a = fuse element cross-sectional area.

11.10.1.2 Arc period

The arc period lasts until extinction of the arc. The arc voltage during this period opposes the source voltage, limits the current, and eventually extinguishes the current. The higher the arc voltage, the shorter the arc duration. The arc energy is absorbed by the heat capacity and transformation of the surrounding material.

11.10.2 Types of Fuse

Although standard fuses are used in most load circuits, some loads may need relatively fast or slow acting fuses. Three types of fuse are available in the industry for different applications:

- Standard fuse for general use
- Time-delay (slow-blow) fuse designed to withstand inrush current drawn by most equipment for a short time after turn-on
- Current limiting (fast-blow) fuse, which limits the current by clearing fault before current builds up from normal to full fault value through inductive response of the circuit.

The typical blow-time and applications are listed in Table 11.6. Figure 11.15 shows current versus average blow characteristics of the three types. Even within a given type, significant variations exist around the average value, which must be considered in the system design to protect the circuit as required.

Table 11.6 Fuse types and their typical applications

Fuse type	Typical blow time (seconds) at 200% of rated current	Typical applications
Fast blow	< 1	Transistor circuits and meter protection
Normal blow	1–10	Most standard circuits
Slow blow	> 10	Circuits drawing high inrush current on switch-on, e.g., motors, transformers, heaters, capacitor charging, etc. It avoids nuisance blowing.

11.10.3 Fuse Characteristics

The fuse element generates heat at the rate of I^2R watts. Under a steady state condition, this heat is dissipated in the surrounding, keeping the element temperature and resistance stable. Under a fault, the heat generated at very high rate raises the element to melting temperature, typically in milliseconds. Assuming adiabatic heating (negligible heat escaping the fuse body in short time), the thermal energy balance requires I^2Rt to be equal to the heat capacity of fuse up to the melting temperature, which is constant for a given fuse. If the element resistance were constant, I^2t is also constant. The clearing time t is thus inversely proportional to I^2 in the first approximation. The fuse resistance rising with temperature alters this relationship to some extent. It is expressed as

$$R_f = R_o\,e^{-t/\tau} \tag{11.14}$$

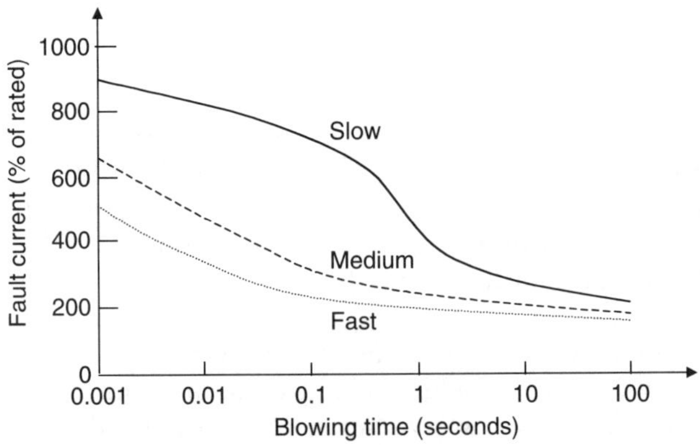

FIGURE 11.15 Current versus blow time for three types of fuses. (NASA/JPL Specifications ZPP-2061-PPL.)

where R_f = fuse resistance at time t after the fault, R_o = initial fuse resistance at normal operating temperature (generally at 25°C), and τ = thermal time constant of the fuse. Typically, the fuse clearing time $t_c = 1.5\tau$, giving $R_f = 4.5R_o$ when the fuse clears. Until then, $R_f = R_o\ e^{-1.5t/t_c}$. Denoting the fault current by I_f, we have

$$I_f^2 t_c = K$$

or

$$t_c = K/I_f^2 \tag{11.15}$$

which gives the fuse resistance at any time as

$$R_f = R_o \exp\left(\frac{1.5 I_f^2 t}{K}\right) \tag{11.16}$$

for $0 < t < t_c$. The fault current decays with the rising fuse resistance. The root mean square value of the decaying fault current should be used in the above expression for better accuracy.

The room temperature resistance of the fuse element has significant manufacturing variations, but it primarily depends on the current rating. However, the voltage drop across the fuse at nominally rated current is kept between 100 and 200 mV by design. The manufacturing variation in clearing time is also significant. For this reason, the *I–t* characteristic is generally depicted by two curves bounding the minimum and maximum times to clear the fuse at a given current (Figure 11.16). Such curves are derived by the vendor from numerous tests. Test data on recent manufacturing lots are often shown by dotted lines for guidance to the users on the latest design and manufacturing practices. Even than, the variations in clearing time are wide and must be accounted for in the power

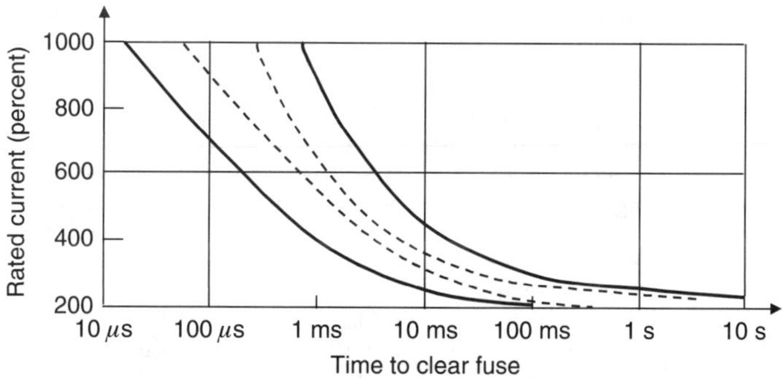

FIGURE 11.16 *I–t* characteristic band of normal-blow hermetically sealed fuses.

system design. Table 11.7 shows the variations in hermetically sealed fuses. Variations of 1:10 or 1:15 are common. They are even wider for traditional nonsealed fuses. The clearing time is less predictable in the normal and slow blow fuses, because the plasma lingers on for longer duration. The plasma behavior is difficult to predict with high certainty. The fast fuse and sealed fuse, on the other hand, have more predictable clearing time.

Many aerospace customers require transient survival analysis to demonstrate that each load's input current matches the blow curves of the fuse assigned to the load. With the actual parameters of the assigned fuse, detailed fault simulations are run on computer to verify that:

- The fuse **will** clear under the minimum fault current using the right hand side of the *I–t* envelope in the simulation run.
- The fuse **will not** clear under maximum inrush current during start-up of motor and heater type equipment using the left-hand side of the *I–t* envelope in the simulation run. In circuits where high start-up current lingers for a long time, posing a risk of unwanted fuse blow, one may use a slow-blow fuse of the same rating.

11.10.4 Fuse Derating

The arcing and heat transfer in space are much different compared to that on Earth where the fuse rating is established by tests. The space-qualified fuse is sufficiently derated from its normally rated current on Earth to obtain high reliability under the actual load current it would carry in service. The derating factor is defined as the ratio of the permissible load current to the nominal fuse rating. The purpose of derating is essentially to stay below the current when the resistance of the fuse element remains stable. Otherwise, it will experience an eventual thermal runaway leading to melting. Significant variability exists in the blow characteristic and the atmospheric pressure surrounding the fuse element in space. For this reason, nonsealed fuses for 1 A and smaller loads are typically derated to

Table 11.7 Typical variations in blow time at 500% rated current in hermetically sealed space qualified fuses

Current rating (amp)	Fuse type	Minimum blow time (second)	Maximum blow time (second)
3	Normal blow	0.0004	0.0065
3	Slow blow	0.002	0.02
10	Normal blow	0.0004	0.0065
10	Slow blow	0.003	0.045

35%, and 70% for 5 A and larger loads. Since the blow characteristic of hermetically sealed fuse is more predictable, the corresponding derating factor is 80% for fuses larger than 0.5 A rating. The largest hermetically sealed fuse available at present is 10 A. Further derating can come from higher than normal ambient temperature. It follows the following rules:

- 0.5% derating per °C ambient above 25°C for time-delay fuses
- 0.2% derating per °C ambient above 25°C for regular fuses

Yet additional derating may be imposed based on the specific mission environment. If the total fuse derating in space is say 40%, then a fuse with nominal ground rating of 10 A must be used in 4-A load circuit in space. This, however, creates a protection problem in ground testing, where a soft fault may carry the current for a long time and burn the wires. For this reason, an additional 4-A fuse (40% of the nominal 10-A rating) is inserted in series all the way up to the launch pad. It is then removed before the launch.

Most faults in space are hard faults that clear the fuse quickly. The fuse closer to the main bus clears even quicker under fault. However, secondary circuit fault at the end of a long line may clear slowly due to the voltage sag and the resulting low fault current.

11.10.5 Fuse selection criteria

The following considerations enter the fuse selection process:

- Fuse current rating after the required derating must be equal to the nominally rated circuit current after allowing for the overload margin and the harness gage.
- Fuse voltage rating must be greater than the source voltage for a.c. or d.c. circuits.
- Maximum current interruption capability of the fuse must be greater than the maximum possible fault current at the fuse location.
- Fast fuse is not necessarily better in all applications. However, it is required to protect semiconductor devices with $I^2t_{fuse} < I^2t_{semiconductor}$. With a fast fuse, the clearing time is less than the first peak of the fault current, and is more closely governed by constant I^2t because the rise to melting temperature is much closer to being adiabatic.

11.10.6 Redundant and Parallel Fuses

If a load circuit uses a redundant wire pairs for reliability, then each wire pair must have a separate fuse. Even with single wires, some customers

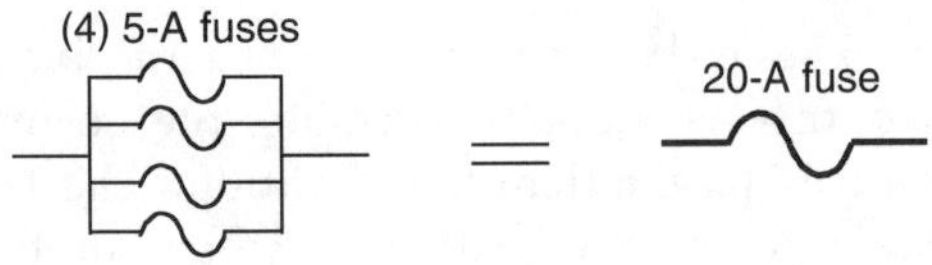

FIGURE 11.17 Parallel fuses increase the current carrying capacity of fuse assembly.

such as NASA require redundant fuses in most load circuits. If one fuse fails randomly under normal operation, the redundant fuse can carry a full load for the remainder of the mission. A redundant fuse in a 5-A circuit cannot be achieved just by placing two 5-A fuses in parallel, which essentially derates the fuse by a further factor of one half. Instead, an additional series resistance equal to several times the fuse resistance is inserted in series with the redundant fuse as was depicted in Figure 11.10. Since the current sharing between the redundant fuse branches is inversely proportional to the branch resistance, the primary fuse normally takes about 90% of the load current.

Parallel fuses (Figure 11.17) are used to obtain a fuse assembly with current rating in multiples of the individual fuse rating. Parallel fuses are not to be confused with the redundant fuses. Parallel fuses are required to share equal current, and therefore do not have additional series resistance in one branch. As a matter fact, two parallel fuses must be matched for equal resistance at operating temperature. Parallel fusing is not feasible with nonsealed fuses due to a great variability in their resistance at the operating temperature. One of the nonsealed fuses would always be stressed beyond its derated value and blow prematurely. Two heretically sealed fuses with matching resistance, however, can be connected in parallel to share almost equal current. Even with hermetically sealed fuses, it is difficult to balance the current with more than two fuses in parallel due to a possible zipper effect. For example, consider a 20-A fuse assembly consisting of four 5-A fuses in parallel (Figure 11.17). If one 5-A fuse blows prematurely, all will blow one after another by taking progressively higher current. If the 5-A fuses are highly derated to avoid such effect, then the protection level of the load circuit is compromised. The zipper effect, however, can be used to avoid a large fuse blow transient voltage by using a few small fuses in parallel with about 10% of the fuse resistance in series in each fuse branch. In case of an internal equipment fault, these fuses blow in zipper sequence, reducing the fault current as they blow progressively faster in turn. The last fuse blows fast and under smallest current, thus resulting in less severe voltage transient.

11.10.7 Fuse Testing

Fuses are tested as follows for proper resistance after the installation, but before powering the assembly. A four-arm resistance bridge and a suitable voltage source are used for these tests:

- A cold test with minimum current to assure minimum self-heating (as opposed to hot test at full rated current).
- The maximum test current is generally 1 mA for fuses rated 1/8 A or less, 10 mA for fuses rated 1/8 to 1 A, and 100 mA for fuses rated greater than 1 A. The measurements are compared to the expected value given in the fuse specification document to verify that the fuse is within tolerance, as opposed to merely checking the circuit continuity.
- When two fuses are installed in parallel, the d.c. resistance of each fuse must be within 50% of the value specified for the individual fuse. The test current in this case is chosen based on the fuse assembly resistance. The design engineer must be consulted if more than two fuses are installed in parallel for special measuring sequence or other instructions.

11.10.8 Unfused Loads

Critical loads are not fused for reliability, and some parts of the power system cannot be practically fused. For example, fuses cannot protect the main bus or the harness between the bus and the fuse board, or in current limiting designs. In such circuits, all wires are double insulated between the lead and return lines and between the lead and the structure. The double insulation is implemented by placing two physical layers of insulation between two isolated conductors. An isolated conductor layer is placed between both the insulation of the double insulated wire to verify the integrity of each insulation after the assembly. The insulation resistance is then measured between the intermediate conducting layer and each insulation. Often, connections via a daisy chain or other method is provided to test all double insulated points at the spacecraft level.

11.11 Remote Power Controllers

Numerous solid state remote power controllers (RPCs) are used on the ISS instead of fuses for protection against faults in various items of load equipment. The RPC has a programmable trip setting, controls the load power on and off and provides dead-facing. When the RPC is dead-faced, it grounds both terminals to zero voltage for personnel safety. On the other hand, when the load is turned off or has tripped, the output terminals may still be live with voltage. All three functions of the RPC are remotely programmable. The basic design features of the RPC used in the ISS are as follows:

- Solid state switches in series with a dead-facing relay, which is magnetically latched. The relay coil voltage can be 28 V or 120 V

- Uni-directional or bi-directional power flow
- Programmable trip setting in 256 levels (8-bit data word) with 1553 protocol for data interface
- Various maximum current ratings, such as 3, 10, 25, and 50 A uni-directional, and 25 and 130 A bi-directional. Several small RPCs are packaged together such that each RPC box has mass of about 2 kg and dimensions of about 15 × 15 × 7 cm. There is only one 130-A RPC available in this size box.
- Trip time of 50 to 100 μs with approximately constant I^2t characteristics
- Control circuit power consumption of about 5 W at ±15 V and +5 V with quiescent power less than one watt
- The on-resistance of RPC varies with rating such that the series voltage drop at the rated current is in the 0.25 to 0.50 V range in each rating.

11.12 Early Fault Detection

Many faults start as soft high impedance arcing faults and then develop into hard faults. It is desirable to detect a high impedance fault early, locate it, and remove it using the redundant paths. This is particularly important in high-power systems (>100 kW). Because of the high voltage, the fault current is high in such systems.

The conventional fuse and RPC protect the system from hard faults only. The high impedance soft faults, on the other hand, can be detected by monitoring the system voltage. The bus voltage would have high frequency hash if an arcing were present. A continuous frequency spectrum analysis of the bus voltage can be useful in an early detection of an arcing fault. Once detected, the fault can be located and bypassed before it develops into a hard fault.

Chapter 12

Ancillary Components

12.1 Introduction

The major components of a PV–battery power system are covered in Chapters 8 to 11. All other components, including the solar array drive, deployment devices, deployment controller, thermal controller, relays, battery monitors, current meters, and shielding equipment, are covered in this chapter.

12.2 Solar Array Drive

The solar array drive and array drive electronics (SAD/ADE) together provide the capability of rotating solar panels with respect to the spacecraft body. The operation of the SAD is controlled by decoded uplink commands from the on-board computer (OBC). It receives a 128-Hz timing clock and 0.25-Hz synchronizing signal from the OBC. In the 3-axis stabilized two-wing geosynchronous satellite, one axis is always aligned with the local normal to the sun and another axis along the orbit normal. Two sets of open loop (clock controlled) solar array drive motors maintain the sun orientation. A brushless d.c. stepper motor rotates each panel separately. The slip rings on the SAD shaft provide an interface between the rotating panel and the fixed Earth-pointing spacecraft body. One SAD controls the north panel and the other controls the south panel. The two are interchangeable in design, where the 'forward' direction of rotation is separately selected by external means. Each SAD has only one mechanical assembly, but has a redundant motor winding and redundant position telemetry potentiometer. Only one set of the winding is powered at a time. The redundant windings are fully isolated to prevent failure propagation. A reliability of 0.99 is typical for both sides combined over a 15-year mission. The mass of each SAD/ADE assembly is about 5 kg in GEO communications satellites.

The SAD/ADE provides telemetry defining its status and that of the solar panel. Each unit provides a potentiometer voltage signal which is directly proportional to the angular position of the panel shaft, ranging from 0 V at 0° to +5 V at 360°. A typical SAD uses a four-phase, 16-pole permanent magnet switched reluctance stepper motor to drive a zero backlash harmonic drive. Each phase coil resistance is in the 50 to 100 Ω range. A vanadium–cobalt steel stator core and neodymium–iron–boron permanent magnet rotor provide high torque per unit mass. The rotor is typically on 440C stainless steel ball bearings and titanium case. Each SAD/ADE draws power from the essential battery bus at −22 V via a latching relay drive. The peak input power is about 10 W and the average power about 1 W. Switch-mode power pulses applied to the stator coils at a suitable frequency drive the motor. The grounding scheme for ADE uses four separate grounds, one each for the power, signal, and communication circuits, and one for all equipment chassis.

The ADE controls all three operating modes of the solar array, namely the normal forward, slew, and the reverse. The SAD is capable of rotation in the forward and reverse directions at speeds from 0 (stop) to 8 Hz as commanded by the ADE. A typical stepping motor provides the stepping rate of 1.8°/step at 0.25 Hz in the normal mode and 1.8°/step at 8 Hz in the slew mode. The solar panel inertia, which can be in 50,000 to 500,000 $\text{lb}\cdot\text{in}^2$ range, determines the response of the SAD.

The motor is capable of stepping at a rate of 0 to 16 pulses per second, with the resulting array rotation of about 0.02° per step. It can have a direct harmonic gear drive that can drive backward and does not need a slip clutch. The other alternative is to use two orthogonal gears, which require a slip clutch since it cannot drive backwards. The slip clutch allows slippage which avoids breakage should there be backlash on the solar panel. The shaft speed is clutched down by two orders of magnitude through a helicon gear. This protects the shaft from backlash stresses.

The solar array is initially aligned by ground. The ADE maintains the alignment in an open loop control system. It takes the time signal from the spacecraft clock and turns panels to the sun continuously as per the clock time without further command. The SAD is fairly accurate, but accumulates small errors, typically a couple of degrees per month. The sun sensor detects the errors, which are corrected by periodic commands from the ground or from the spacecraft computer. In some satellites, the solar array is slewed every 6 AM until the error is zero. The spacecraft computer commands can be overridden by ground for maneuvering or other reasons.

The control signals for the required rotation rate comes from the TT&C system, which also selects the rotation direction and one of four speeds — slow (0.013 mradian/s), normal (1.03 mradian/s), fast (1.047 mradian/s), or slew (5.236 mradian/s). In some LEO satellites, such as EOS-AM, the ADE provides control signals and power to the SAD for the commandable drive rates of $\omega \pm 2.5\%$ at 0.1% increment and $3\omega \pm 12\%$ at 0.5% increment, where $\omega = 1.060$ mradian/s. The ADE accepts commands to select the SAD rotation rate and the direction. The ADE also accepts commands to select primary or redundant side. In some satellites, the solar array has two rates of rotation, one steady at one revolution per day and another slew at one revolution per 45 min.

The slip ring assembly, also known as the power transfer assembly, transfers power and telemetry signals to the spacecraft body. Each power and telemetry lead wire uses a separate slip ring, but some returns may share a common ring within its derated current capability. The current collecting slip rings and carbon brushes make the rotary joint (α joint) between the sun-oriented solar panels and the Earth-oriented spacecraft body as the spacecraft revolves around the Earth. The slip ring brushes may be traditional monofilament brushes, or high conductivity fiber brushes made with graphite or silver fibers for enhanced current collection with low power loss. The resistance of a 5-A slip ring–brush contact at the operating temperature at the BOL is about 5 mΩ, which degrades to about 10 mΩ at the end of life. With EOL current of say 5 A, the power loss is 0.25 W per slip ring contact, or 0.50 W per slip ring lead and return circuit.

In a typical SAD, there are 20 to 30 slip rings each rated 5 A for power circuit, and 10 to 20 rings rated less than 1 A for signal circuits. Some units come with heavier slip rings (Figure 12.1). Slip rings are used such that all positive rings are on one side and the negative rings on the other, with possibly one empty ring in the middle. The positive and negative rings are not placed side by side to avoid the possibility of a short. Advanced slip

Power rings:	Motor:
– 6 × 8A	– Stepper with redundant windings
– 2 × 40A rated current	– 0.0375 deg./step
	– 1.3W max. power
Signal rings:	Weight:
– 12 × 0.25A rated current	– 5 kg/assembly

FIGURE 12.1 Power transfer assembly of slip rings and drive motor.
(Source: Dornier Satellitensysteme Gmbh, Daimler-Benz Aerospace. With permission.)

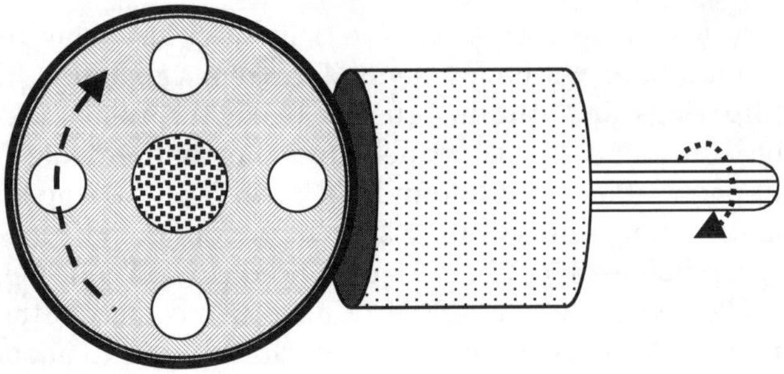

FIGURE 12.2 Biaxial solar array drive and power transfer assembly.

rings, such as those made by Litton Polyscientific Corporation, are lighter and need no lubrication, thus extending the mission life.

Most communication satellites have only one-axis gimbals, the α gimbals driven by the SAD. As a result, the solar array is normal to the sun on equinox days, but varies from normal by $\pm 23.45^\circ$ on solstice days. The solstice power generation is reduced, but the battery charging power requirement is near zero (no eclipses) and the heater load is also less. Some satellites, such as *Iridium*, use biaxial gimbals having two degrees of freedom in rotation (Figure 12.2), one for the daily α rotation and the other for the seasonal β rotation. Biaxials are used for both sun pointing and antenna pointing in satellites. However, their use is limited because of the complexities introduced in the joints, the mechanisms and the structure. Accepting the cos β reduction in power generation is often preferred over adding the dual-axis gimbals, at least in GEO.

12.3 Deployment Devices

Three main methods of deploying various spacecraft components are discussed below.

12.3.1 Electro-Explosive Device

This is the traditional method of deployment, which is also known as the pyro-technique ordnance. It uses an electro-explosive device (EED), in which electrical energy is supplied to ignite the squibb of explosive powder. The resulting force then deploys the component, typically under a spring-loaded force. The EED is ignited when a certain current deposits a certain amount of energy in a specified time. For safety reasons, it must withstand a certain minimum energy without igniting. A typical EED is rated at 1A–1W no fire, and 4A–4W sure fire.

The EED requires heavy shielding and great care with regard to EMI pickup. Moreover, the EED explosive is thermal and shock sensitive.

Therefore, installation of sensitive EED explosives is sometimes done at the launch pad, which costs much more than at the factory. A pyro harness is routed separately from the power and signal harness to minimize the EMI concerns for safety. EED deployment technology, although used for several decades, have some disadvantages, such as:

- All spacecraft components must be designed to withstand severe pyro shock excitation.
- High safety related costs in documenting, reporting alerts, and all other stringent regulations on EEDs must be complied with.
- Some manufacturers install EED squibs at the launch pad, while most install at the factory. In the latter case, the spacecraft is transported as a class 1 explosive (sensitive to thermal and shock environments), and this adds expense.

12.3.2 Laser-Initiated Deployment

Figure 12.3 shows the laser-initiated ordnance schematic, where the light emitted by diode D_1 is transmitted to the squib by an optical fiber. The photon energy deposited in the squib fires the explosive. After the firing, a sense light is sent by pulsing D_2 and sensing the transmission, reflection or refraction of the sense light. This is to confirm that the squib has been fired. This technology has been space qualified and flown. It has the following advantages over EEDs:

- It is insensitive to EMI, RFI, ESD, EMP, and cross talks. Hence, it eliminates accidental firing and related human safety issues. This reduces the cost not only in flight, but also in the factory, in transportation, and at the launch pad.
- insensitive to harness and explosive placements

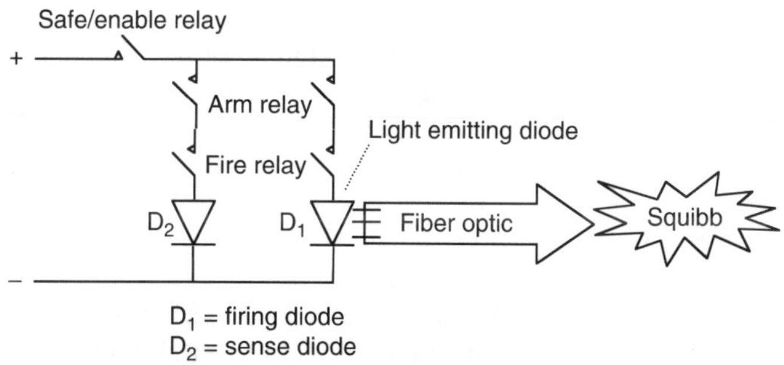

FIGURE 12.3 Laser-initiated ordnance schematic.

- Requires no shielding, hence the harness mass is significantly reduced.
- Lower integration and test cost, since no EMI silence clearance is required in the factory or on the launch pad.
- Insensitive explosive allows factory installation, which costs less.
- Higher reliability.

12.3.3 Shape Memory Metal

Recently the shape memory metal deployment technology has been developed for some satellites, such as EOS-AM and PM. In the working principle, a certain metal, when heated, regains tthe shape it possessed before forming into a new shape. In a way, the metal 'remembers' its original shape and returns to it when heated. The heat is applied by a traditional heater, which is insensitive to EMI. The most prominent benefit of using the shape memory metal deployment is the savings in safety related cost. More recently, a new plastic with shape memory has been developed at the Massachusetts Institute of Technology. It can be shaped as a string and inserted into a body through a hollow tube. When heated, it changes into a sheet or other shapes by manipulating the temperature and stress.

12.4 Deployment Controller

The deployment controller, also called the ordnance controller, delivers a constant current to the deployment devices. In the traditional pyro bridge wire, the wire resistance rises with temperature. The required voltage must therefore rise accordingly to supply the specified minimum current, typically a few amperes. If the bridge wire is supplied with a constant voltage source, it must carry the specified minimum current at the peak temperature around 800°C. The peak current is typically several times the minimum current specified for 10 to 20 ms. The relay must be capable of handling the peak current for sufficient number of times without overstressing. Figure 12.4 is a typical ordnance controller circuit powered directly by the battery through the pyro enable plug. The safe and arm plugs are placed between the ordnance controller and the EED squibs. Power is supplied via a redundant path to a redundant EED, one from each battery. When the enable and arm relays are commanded on, the EED is ready for the fire command. The fire command turns on the EED cavity relay, sending the pulse of power to the EED squib, which activates the deployment mechanism that cuts a rope, or a bolt.

The ordnance controller can be powered directly from the battery or from the bus via PRU (Figure 12.5). For reliability reasons, the deployment power is generally drawn directly from the battery by unfused circuits. In mid-

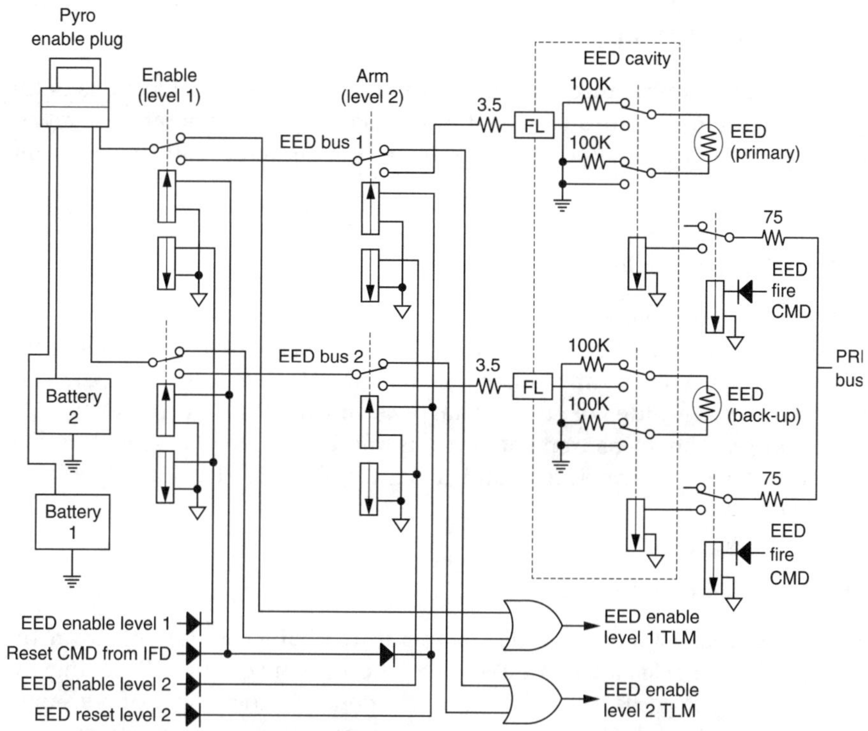

FIGURE 12.4 Ordnance controller circuit schematic.

sized communications satellite, each battery is typically capable of supplying a peak pyro load greater than 10 A for 10 s, 25 A for 5 s, or 50 A for 5 ms with 70 to 100% state of charge. The battery bus impedance is typically greater than 100 mΩ for d.c. to 10 kHz. Two independent pyro busses are tapped from the 22nd cell of each battery without cross-strapping. The voltage ranges from 22 to 35 V with one cell failed. Each pyro bus — also known as the essential bus — powers the primary and back up pyro loads, respectively.

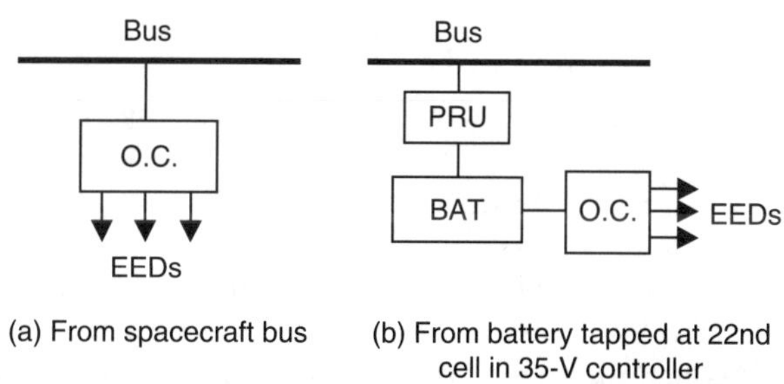

FIGURE 12.5 Ordnance controller powering options.

12.5 Thermal Controller

Various probes are required to sense temperature at various locations of the spacecraft for use with telemetry-thermometers and temperature controllers. A probe may contain any one of the following temperature measuring elements.

12.5.1 Thermocouple

A thermocouple utilizes the Seebeck effect, which occurs at the junction of two dissimilar metal wires. Thermally energized electrons in each metal generate a d.c. voltage at the hot junction, which is measured at the cold end. The magnitude of the voltage is linearly proportional to the temperature. The measurement error is minimized by keeping the cool end at a constant temperature and using a voltmeter that has high input resistance.

12.5.2 Metallic Resistance Wire

In a metallic resistance wire the temperature change is derived from the change of wire resistance compared to the reference value $R_{o,}$ using the relationship $\triangle R = \alpha R_o \triangle T$, where α = temperature coefficient of resistance per degree centigrade. Platinum is one element that is often used in space as it not only has high resistance, but also a high temperature coefficient of resistance, which is linear and stable over a wide temperature range. For platinum, $\alpha = 0.003$ per °C at 20°C.

12.5.3 Thermistor

A thermistor is a semiconductor in which a small change in temperature causes a large change in electrical resistance. Precise measurement of the resistance gives direct reading of the temperature at the thermistor location. This allows a simple readout instrument and the use of long leads without loss of accuracy. The temperature coefficient of resistance for the thermistor is negative and two orders of magnitude greater than metallic wires. The resistance of a typical thermistor at various temperatures is 75 kΩ, 7 kΩ, 3 kΩ and 1 kΩ at −40°C, 0°C, +20°C, and +40°C, respectively. The thermal time constant ranges from 5 to 15 s. The interchangeability tolerance is within ±0.1%.

The resistance wire and thermistor probes require applying a current and measuring the voltage across the element, from which the resistance is derived using Ohm's law. Since the thermal time constants are relatively long, the temperature is measured only periodically by passive telemetry to save energy. A constant current pulse is sent and the voltage is measured periodically across the element. The heaters are turned on or off as needed

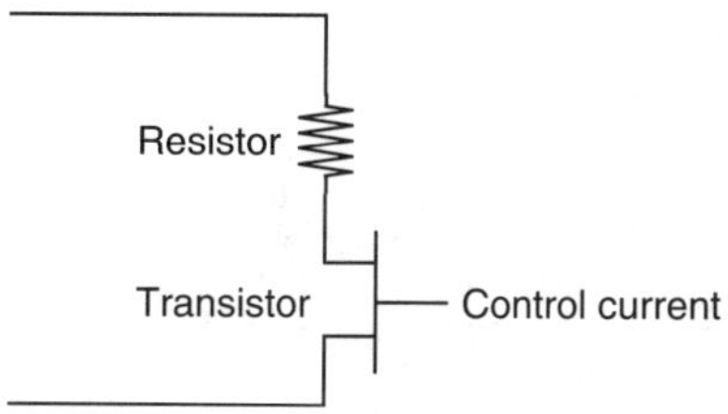

FIGURE 12.6 Proportional heat control circuit.

in response to the temperature telemetry. Instead of using conventional thermostats for turning the heat completely on or completely off, the proportional heat control is often used. It employs a transistor in series with a fixed resistor (Figure 12.6). The control current is changed in response to the average heat required. The heat generated equals the heat required less the payload-generated heat. This avoids frequent on–off cycling with the conventional thermostat where the heater is either on or off. The proportional heat control provides a steady current with no on–off cycles, thus eliminates the EMI. It is particularly useful near EMI sensitive equipments, such as science instruments.

12.6 Relays

An electromechanical relay has physical metal contacts that open or close on command as required. It offers certain advantage over transistor switching, such as negligible power loss, and it can be used singly without auxiliary circuit or with very simple circuit. On the negative side, it has a limited life due to a moving contact that wears. If arcing is present under normal or abnormal operation, it results in EMI and a much shorter life.

Relays come in a holding or latching variety. Most space-qualified relays are rated for 28 V and up to 30 A d.c. Larger current ratings come from Leach and smaller ratings from Genicom and Teledyne corporations. They are typically hermetically sealed in nitrogen at 1 atm and life tested for tens of thousands of cycles. Typical weights of 1, 5, and 10 A d.c. double pole–double throw (DPDT) 28-V relays are 3, 15, and 45 g, respectively. All relays for space applications are derated to 50% of the ground base current rating. Contacts are designed such that the voltage drop at rated current is 0.1 to 0.2 V.

Relay contacts for switching high currents around 50 A are difficult to procure, are heavy and vulnerable to vibration and shock environments during the launch phase. In a relay having spare poles, two poles can be connected in parallel to double the current capacity of the relay. Multiple pole space-qualified high-voltage relays are also difficult to procure. For this reason, many engineers find it necessary to use relays at voltages higher than rated values after derating to a lower current. Since the arcing voltage

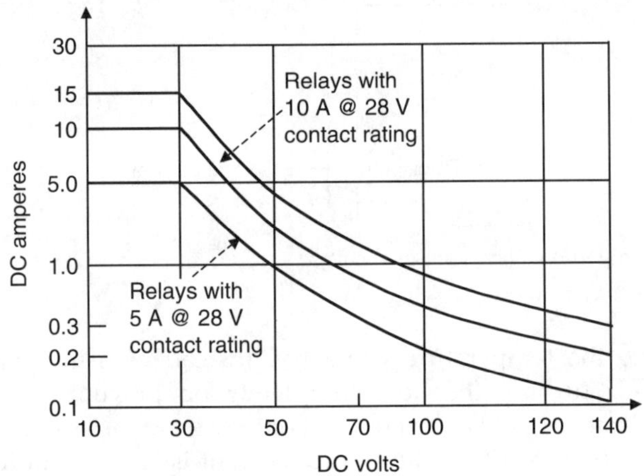

FIGURE 12.7 Typical 5-, 10-, and 15-A, 28-V relay derating for d.c. voltages above normal rating without arc suppression.

on breaking the current is equal to $L di/dt$, the contact ampere rating at higher than rated voltage decreases linearly in theory. However, manufacturers' test data suggest a square function, i.e.,

$$\frac{\text{actual current capacity}}{\text{rated current capacity}} = \left(\frac{\text{rated voltage}}{\text{actual application voltage}}\right)^2 \tag{12.1}$$

The derating data for 5, 10 and 15A latching relay is depicted in Figure 12.7. It is important to avoid arcing by absorbing and dissipating the inductive energy in the relay coil and the load circuit. Various schemes shown in Figure 12.8 for this purpose are: (a) steering and suppression (freewheeling and series) diodes, (b) bifilar coil with one side shorted, (c) diode and Zener, and (d) series R–C circuit. The relay is often required to "make before break," or "not to make before break." A relay failure in maintaining this sequence can cause damage.

Relay coils interfacing with the command decoder discrete command is isolated from all circuitry. Connections between the command relay coil in the user equipment and any primary or secondary returns are prohibited. Steering and suppression diodes are located in the user equipment (Figure 12.9).

12.7 BPM and BCVM

The battery pressure monitor (BPM) and the battery cell voltage monitor (BCVM) are part of the PRU. Each BPM strain gage along with its dedicated amplifier is adjusted for offset and gain to normalize the output. The output

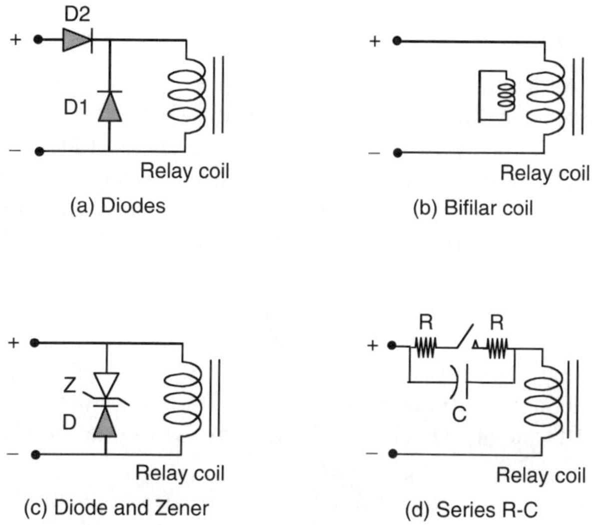

FIGURE 12.8 Relay contact protection schemes for absorbing inductive energy.

is adjusted at the battery vendor to meet the specified offset tolerance (in ± microvolts) at room temperature and zero pressure, as well as at room temperature and 1000 psig. The strain gage voltage is nulled at zero pressure and is typically 6 mV at high pressure. A reading is also taken at 500 psig for reference. The strain gage signal is brought to the BPM circuit, which contains a differential amplifier and a local regulator. Each BPM amplifier is adjusted via select-in-test resistor for offset and gain for normalizing the 6 mV strain gage signal to the 0 to 5 V level. In operation, the strain gage offset voltage and cell pressure follow the linear relation $P = mV_c + b$, where b is the zero pressure offset and m is the slope of the line, i.e., $\triangle P/\triangle V$. Having recorded the zero and full pressure parameters, they can be provided for the look-up and offset as described earlier. Typically, a couple of cells in each battery have the pressure and temperature telemetry.

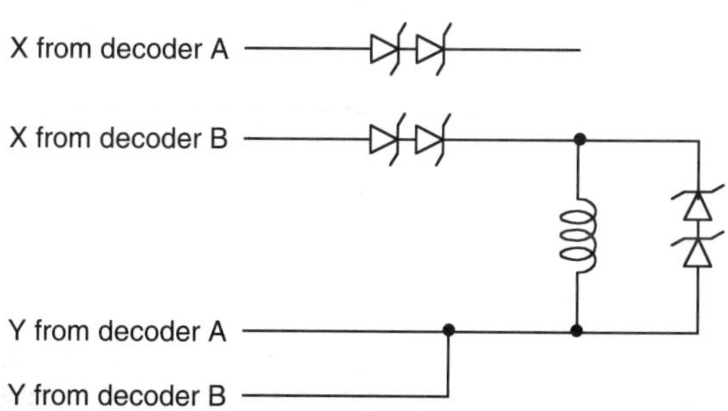

FIGURE 12.9 Typical relay load circuit.

12.8 Battery Let-Down Unit

The battery letdown unit (BLU) is a resistor assembly used to drain the battery completely during storage on ground, for transportation to the launch pad, or on orbit for reconditioning. For on-orbit let down, two drain rates are used, high in the beginning and low at the end. The drain resistance per cell is typically 1 Ω, or 0.5 Ω with a thin high-resistance harness. Three alternative design configurations of the BLU are shown in Figure 12.10. The approach (a) is often used. It works well as long as cell voltages are equal. However, cell voltages often diverge near zero voltage at the end of draining. The unequal loop currents under this condition cause unequal voltage drops in the harness, leading to errors in the cell voltage measurements across the resistors. A measured cell voltage that is positive could in fact be negative. This problems can be remediated by:

- Software compensation for uneven voltage drops in the harness for real-time corrections.
- Dedicated resistor for each cell with its own harness shown in (b). This adds some harness mass.
- Single-resistor design shown in Figure 12.10(c). It saves some harness mass, but also loses the ability of monitoring each cell voltage and driving it to zero individually.

12.9 Current Meters

Three types of current meter are used in space: the shunt, the Hall-effect transducer, and the magnetic amplifier.

12.9.1 Shunt

This sensor is essentially a known precise resistance inserted in series with the load. The voltage drop measured across it is linearly proportional to the

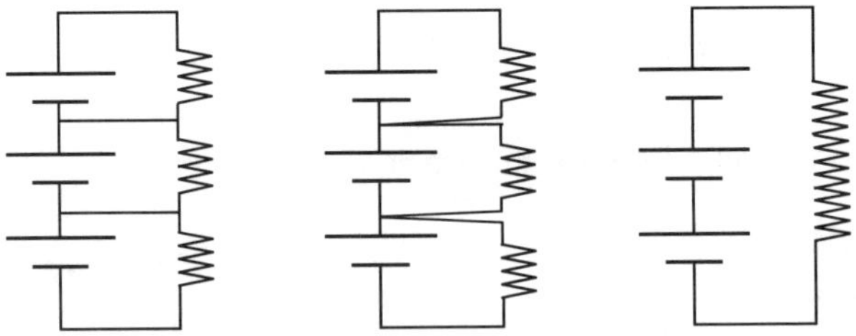

(a) Separate resistors with common wires (b) Separate resistors with separate wires (b) One resistor for all cells

FIGURE 12.10 Battery letdown resistor network options.

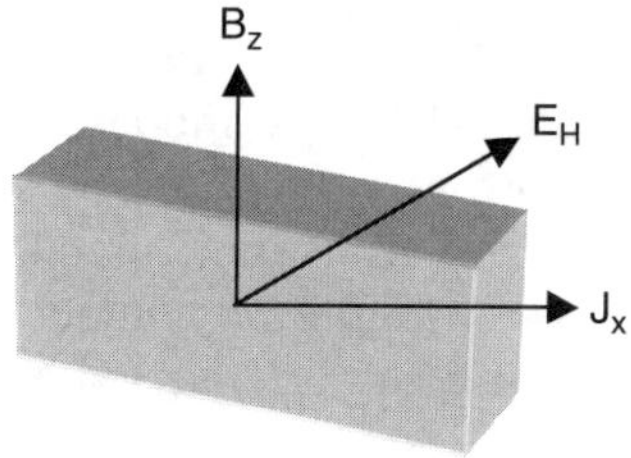

FIGURE 12.11 Hall-effect transducer as a current sensor.

current. It has low cost, but inserts a power loss and a reliability risk of failing open. Moreover, it offers no electrical isolation, which may be an issue in high-voltage busses.

12.9.2 Hall-Effect Transducer

It is an electromagnetic transducer that generates voltage when electrical current and magnetic field are simultaneously imposed on the conducting transducer material. With reference to the flat plate transducer of Figure 12.11, a current density J applied in the x-direction and a magnetic field B applied in the z-direction result in an electric field E_H in the y-direction with magnitude $E_H = R\ (B \times J)$, where R is known as the Hall coefficient. The voltage measured across y-thickness is calibrated in terms of the current in the x-direction.

12.9.3 Magnetic Amplifier

Figure 12.12 shows a schematic of the magnetic amplifier current sensor. The current I through a magnetic core causes the core to saturate, the effect of which is measured as the increased voltage across the coil. Magnetic ampliers are often used in measuring the output current of the north and

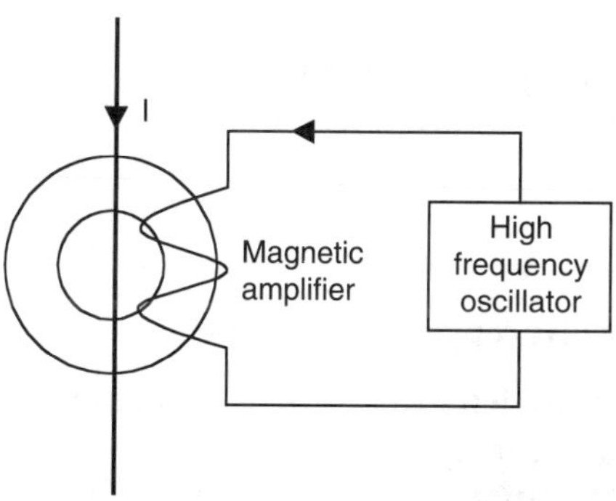

FIGURE 12.12 Magnetic amplifier current sensor.

south solar array wings separately to provide a health check. In this use, the amplifier has no mission critical failure mode. A loss of sensor is merely that — loss of a sensor — with no impact on the bus power. It costs more, but has no power loss in the transducer itself. However, it needs an oscillator that requires power. Having no contact with the current path, it provides electrical isolation and can be inserted as an afterthought.

12.10 Capacitors

Numerous capacitors are used in a typical space power system. In the bus voltage ripple filter, they store and deliver energy over one cycle. In snubber circuits, they absorb and dissipate transient voltage energy in solar circuit, PRU, LPC, etc. The energy storage per unit mass of the capacitor varies with construction and the voltage withstand capability. However, typically low voltage (28-V) capacitors have specific energy of 20 J/kg, medium voltage (1 kV) capacitors 100 J/kg, and high-voltage (20 kV) capacitors 200 J/kg. The mass density of high voltage capacitors is around 2 kg/l (0.07 lb/in^3). Various types of capacitor and their applications are as follows.

12.10.1 Tantalum Capacitor

The tantalum capacitor provides relatively high-density low-cost energy storage. The maximum voltage rating of standard production line tantalum capacitors is 100 V with 50% military derating. Thus, a 50-V bus is the limit for 100-V capacitor applications. The tantalum capacitor is generally heavy and comes in two constructions:

- **The foil (film) type** has a high ESR that gives high-power loss. Naturally, this type finds applications in circuits where damping resistance is a design requirement.
- **As a solid sintered powder** bonded in one solid porous body. It has low ESR and finds applications in ripple filters where pure energy storage is required.

12.10.2 Metalized Polypropylene Capacitor

The metalized polypropylene capacitor is available in higher voltages. A 300-V rated capacitor of this type used on a 120-V bus provides an adequate margin after 50% derating. However, it is bulky and offers no mass advantage over the tantalum capacitor.

12.10.3 Ceramic Capacitor

The ceramic capacitor is made of chunks of ceramics with metal foil bonded on two faces. It can be operated at very high temperature, with the ceramic–metal bond as the only weak link. It has a low ESR and is compact, but offers no mass advantage per microfarad. Ceramic capacitors are available in high voltages. The 160 to 120 V ISS bus has used ceramic capacitors.

12.11 Filters

A filter reduces the conducted EMI entering or exiting the equipment. A low-pass filter is most commonly used to suppress conducted high-frequency interference using inductance, capacitance or a combination of the two, as shown in Figure 12.13. The insertion loss I_L of a filter is defined as the noise voltage on load with the filter to that without the filter. It is usually expressed in decibels (dB). The capacitor is an efficient low-pass filter when connected in parallel to the high resistance load. It has $I_L = 20 \log(\pi RC)$. The inductor is an efficient low-frequency filter when used in conjunction with a low resistance load. It has $I_L = 20 \log (\pi L/R)$.

When L or C alone cannot provide the desired insertion loss, the L-C filter is constructed in L, π or T configuration shown in Figure 12.14. These filters are associated with suppression of signals in the frequency domain, such as ripple harmonics of the switch mode power converters. They would also suppress signals in the time-domain, such as transients spike voltages from ESD, EMP, and lightening type sources. The response of such filter is obtained by summing the pulse and frequency spectrum curves at each

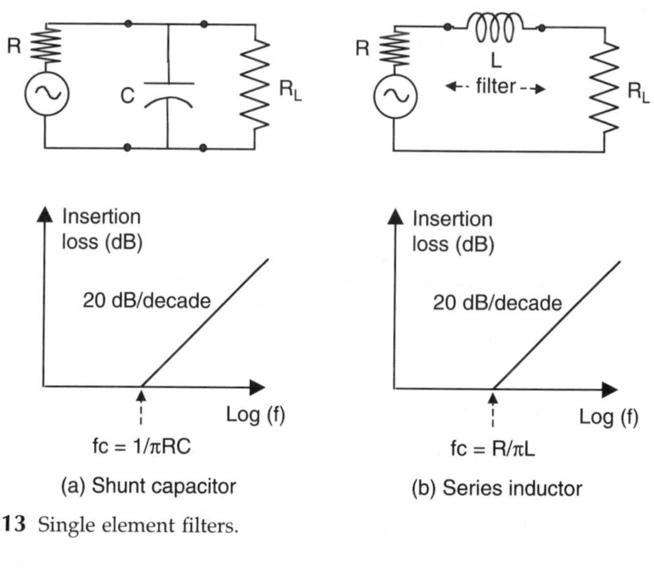

FIGURE 12.13 Single element filters.

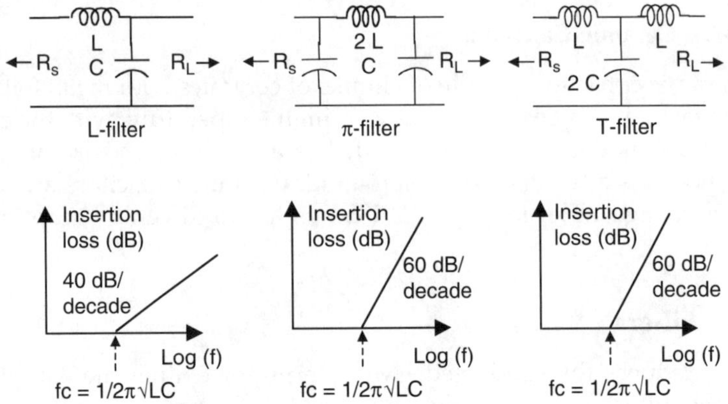

FIGURE 12.14 Basic L-C low pass filters for power circuits.

frequency. The π filter is commonly used at the input of a switch mode converter to reduce the reflected ripple current. The filter is usually built inside the converter. Very high-frequency noise (hash) is effectively filtered by using one or more filters in the following configurations:

- Feed-trough capacitor
- Ferrite bead around the conductor
- Ferrite bead with capacitor

12.12 Telemetry and Commands

Sufficient telemetry must be provided to operate the spacecraft in a safe manner in all phases of the mission. The following is generally provided:

- The on-off status of all EPS relays and switches with positive identification
- Load currents
- Bus currents and voltages
- Array currents and voltages
- Shunt voltages and currents with shunt switch status on or off
- The open circuit voltage and short circuit current of the solar array test coupon(s)
- Charge and discharge currents and terminal voltage of each battery
- Temperature and pressure co-located on two or more cells in each battery
- Individual battery cell voltage as often required by the customer.

A typical power system provides the TT&C system with about 60 analog and 40 discrete telemetry. The telemetry accuracy is generally ±3% of the full-scale value of all analog signals, and ±2°C of temperature. The types of telemetry used on the spacecraft are listed in Table 12.1.

The bus voltage is controlled automatically, so the telemetry is only for the information. The EPS telemetry is scanned every several seconds. The battery current, pressure, and temperature are scanned less frequently depending on the nature of the load and the battery SOC management requirement. There is no criticality in the telemetry time intervals. More frequent is better, but must be traded with the computing requirement. Most telemetry is for information purpose, except the battery current, pressure and temperature, in that order of criticality.

The OBC uses redundant telemetry to assure that the batteries are safe in presence of a single EPS failure and its own failure. Its is not required to continue automatic control after a failure without ground reconfiguration.

Commands are used to implement various functions and are typically of three types:

- Digital serial
- Discrete logic
- Discrete relay

Ground commands are provided to override and inhibit autonomous functions and to manage and control the EPS. Redundant units are individually commanded. Block-switching of redundant units in response to a detected failure is prohibited.

The electrical power system in spacecraft like GPS may have 70 to 80 discrete commands, 30 to 35 for EPS and 40 to 45 for pyro; and 35 to 40 serial telemetry points, 20 to 25 in EPS, 10 in SAD, and 2 in pyro. A typical 5000-W communications satellite may have 50 to 100 commands and a similar number of telemetry signals related with the power system

Table 12.1 Types of telemetry used in spacecraft

Telemetry type	Nature	Description
Digital serial	Serial	Software driven
Digital logic	Bilevel	+5-V logic pulse or zero
Discrete relay	Bilevel	+28-V power pulse or zero
Active analog	Analog	Voltage and current sensors which are active all the time
Passive analog	Analog	Temperature sensors, normally inactive, but periodically pulsed with 1 A. The measured voltage is then converted in temperature

monitoring functions. Most of them interface with PRU, which controls the bus voltage.

The EPS command and telemetry interface with OBC via the TT&C system. However, the EPS is capable of remaining in a safe mode in absence of TT&C support for several minutes. It is also capable of operating indefinitely in a mode in which all commands are issued from ground.

12.13 Electronic Packaging

The electronic packaging has multiple functions. It provides an internal structural support, the thermal conductivity of waste heat, and the enclosure that also provides protection against EMI and charged particle radiation. Since about 20 to 30% of the electrical assembly mass is in packaging, significant mass reduction potential exists by exploring alternative packaging materials. Aluminum (0.10 lb/in^3) has been the material of choice for its high strength-to-weight ratio, good electrical and thermal conductivity, low cost, and easy fabrication. Other materials which have been used are magnesium (0.06 lb/in^3) and magnesium–lithium (0.04 lb/in^3). Although lightweight, magnesium–lithium is expensive, flammable, and a poor thermal conductor. In recent years, fiber–epoxy composites have been developed with properties tailored to meet a wide variety of requirements. Aluminum filled with graphite fibers or carbide particles may prove to offer reduced mass with acceptable thermal conductivity and stiffness. A composite fiber enclosure with a thin layer of aluminum foil for EMI shielding is about 50% lighter than an aluminum-based enclosure.

If a hermetically sealed component is specified, then the following is used in the construction:

Hermetically sealed connectors with connector cases welded to the hermetically sealed enclosure.

All circuit boards are conformally coated with an opaque polyurethane-type compound by dipping, and then properly cured with no cracks in the finished card. Wide cracks due to mishandling damage or poor manufacturing is a risk. In a good circuit board, any spark due to a loose contact or arcing under a fault current would remain inside the conformal layer.

The circuit boards add significant mass to the electronic assembly. Some of the mass reduction potential may lie in using lower density Kapton (0.06 lb/in^3) versus glass (0.08 lb/in^3), and thinner and narrower copper traces (0.005 versus 0.010 inch). Circuit board fabrication is done on fiberglass board, copper plated on one side that works as the common return plane. All return leads are locally soldered to the copper back. A transformer weighing less than a pound is almost always toroidal, not potted. It is screwed to the board with a washer of the same outer diameter as the transformer and 1/8 inch diameter machine screws with no silicon or other adhesive. A transformer weighing less than ¼ pound, however, is glued directly on the board with silicon adhesive tab at the bottom and also at the top. Aluminum heat sinks are bonded with the base plate to dissipate heat.

Nonmagnetic components may be potted using urethane resin filled with alumina for thermal conduction. Large magnetic components are potted using magnesia, which penetrates well. Small magnetics are merely bonded to the board with alumina filled urethane resin.

Most electrical parts are required to be electrically isolated while maintaining good thermal contact for heat transfer. The insulators must be mechanically strong enough to withstand the short circuit forces. Most commonly used washers as spacers are beryllia (0.10 lb/in^3) and alumina (0.14 lb/in^3), 20 to 40 mil thick.

Advances are continuously being made in heat sinks. A detailed look at what is currently available in the market can prove beneficial.

12.14 Radiation Shield

Energetic particles passing through material lose energy by interaction with the atomic structure of the subject material. Electrons become highly scattered by the material, while the protons, being heavier, penetrate deeper into the material. A particle loses all its energy after penetrating through a certain thickness of the material. This thickness is known as the range, which approximately varies linearly with the energy level of the particle and inversely with the mass density of the material. For this reason, the range is commonly expressed as the product of the depth and density (g/cm^2) or in units of depth in aluminum as the basis of comparison as shown in Figure 12.15. A practical range, R_p, is defined as the most probable range of particle of a given incident energy.

Normal space-qualified devices are generally used in commercial space applications. These devices generally have sufficient shielding against the natural radiation of charged particles. This is because the device is in its own case, then often behind a card, and also in an overall metal enclosure. A more sensitive electronic device is protected by placing it in a heavy metal enclosure, such as cast aluminum of several millimeters thickness. Under a man-made radiation threat, such as a nuclear detonation around a defense spacecraft, the device may need additional radiation hardening. Such radiation hardened semiconductor devices are expensive.

The metal enclosure further reduces damaging effects of the charged particles to a level that is tolerable by the built-in radiation hardness of the electronic part inside. Selection of the enclosure metal and its thickness is a function of estimated fluence and the tolerance dose. For example, the radiation fluence entering the enclosure decreases with increasing product

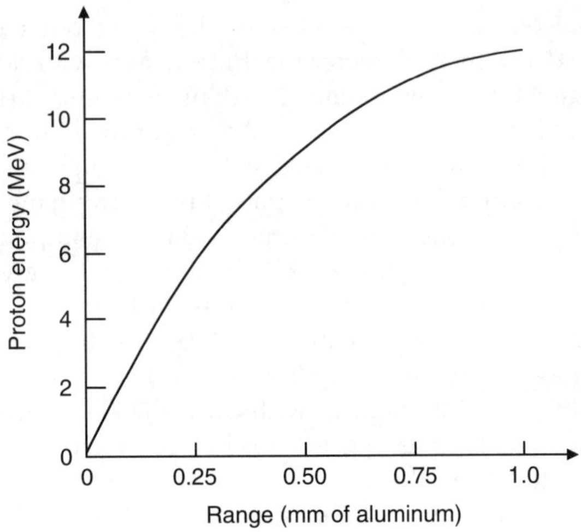

FIGURE 12.15 Range of radiation damage in aluminum.

of the wall thickness and mass density. Since aluminum is the most widely used enclosure material, the fluence penetration versus thickness curve is generally plotted taking aluminum as the reference (Figure 12.16). The dose absorbed by an electronic part inside the enclosure therefore depends on the wall thickness. For example, the total ionization dose comprising mainly the electrons and solar flare protons versus equivalent aluminum thickness in three different orbits is shown in Figure 12.17. Such curves are used to determine the required enclosure wall thickness as described below.

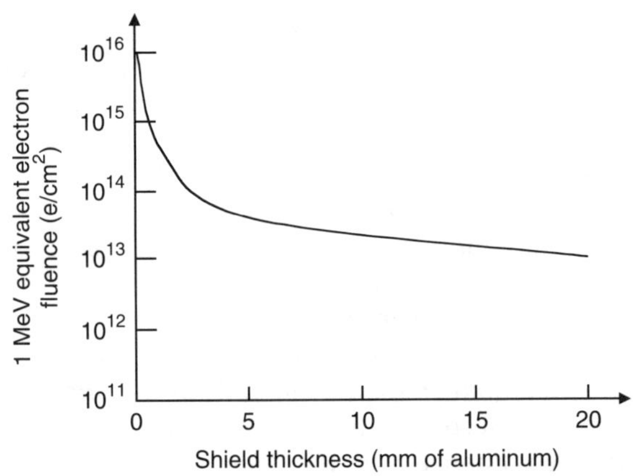

FIGURE 12.16 Radiation fluence penetration over 10-year mission versus shield thickness.

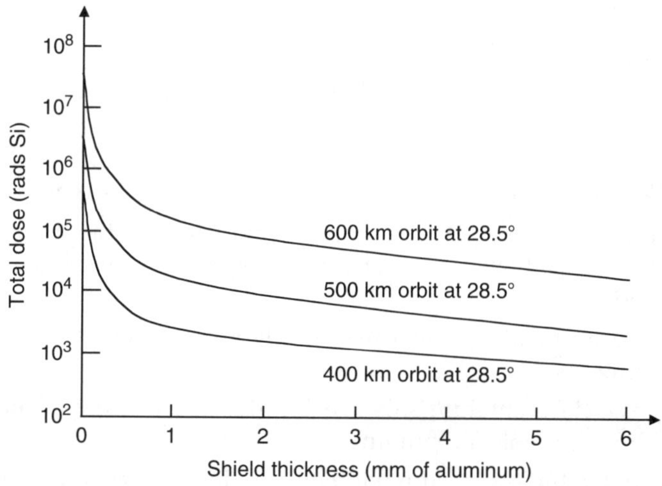

FIGURE 12.17 Total radiation dose over mission life versus shield thickness for three orbits.

12.14.1 Example of Radiation Shield Design

Referring to Figure 12.18 as the charged particle radiation dose versus enclosure wall thickness curve, say we wish to protect a piece part having a tolerance for a dose of 2×10^4 rads of ionizing radiation. The part has its own case that is equivalent to 1.5 mm of aluminum, which will result in a radiation dose 1×10^5 rads. In order to reduce the dose by a factor of 10, i.e., bring it down to 1×10^4, we must have a total shield thickness of 4 mm.

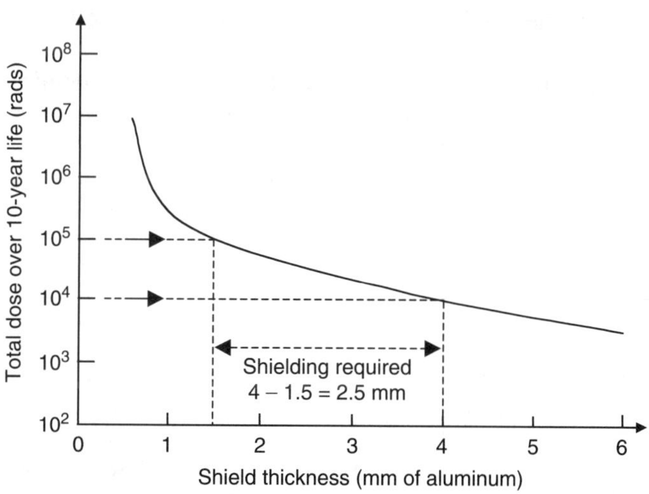

FIGURE 12.18 Example of shield thickness determination for protection against charged particle radiation.

This requires a 2.5-mm-thick aluminum enclosure on the top of its own 1.5 mm case.

The following are general guidelines in shielding the electronic devices against radiation.

- Shield the device individually if only a few devices require shielding.
- Place a shield only on one side if major dose of the radiation comes from one direction. Flat and deep packages are shielded with strips on top and bottom.
- Provide a shielding enclosure if a large number of devices require shielding.
- May use thin tantalum shield using the density equivalency of 1 mil tantalum = 5 mils aluminum.
- Investigate the common mode noise under solar flare radiation.

The enclosure of electrical component works as a dual-function shield against the charged particles radiation as well as the EMI radiation. Its design as the radiation shield is discussed above, whereas the EMI shield is discussed below.

12.15 EMI Shield

The conducting metal enclosure shields the equipment inside from radiated EMI. The metal walls provide eddy current shielding and control the near and far field coupling of the radiated electromagnetic fields. It attenuates the electrical and magnetic fields from leaving or entering the enclosure. Performance of the shield can be viewed in terms of the transmission line analogy as shown in Figure 12.19.

When an electromagnetic wave traveling along a transmission path in free space with the characteristic wave impedance Z_o impinges on a metal surface (load impedance Z_L), the surface absorbs some wave power, some reflect back and some continues beyond the shield wall thickness and reaches the victim equipment. The reflection coefficient, $\Re$, is given by

$$\Re = \frac{Z_L - Z_o}{Z_L + Z_o} \tag{12.1}$$

The higher the value of $\Re$, the more wave energy is reflected from the shield surface. For total reflection, i.e., for $\Re = -1$, Z_L must be small compared with Z_o. On the other hand, $Z_L = Z_o$ gives $\Re = 0$, there is no reflection and all the wave energy is absorbed by the shield wall thickness. This condition is known as the impedance matching. The far field transverse electromagnetic wave in free space has $Z_o = 120\pi = 377\,\Omega$. The electric dipoles have

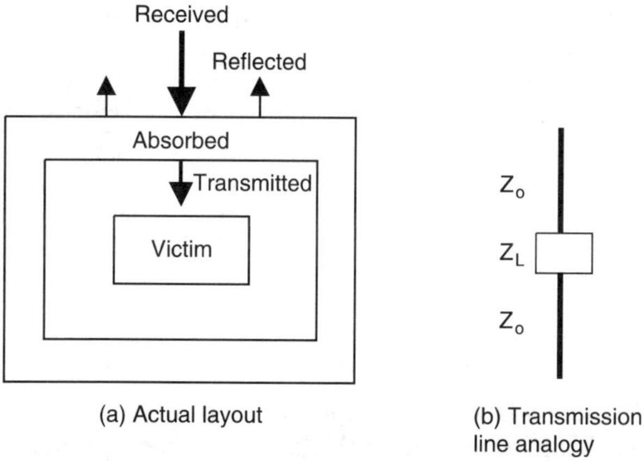

FIGURE 12.19 Reflected, absorbed, and transmitted EMI energy.

dominant *e*-field and higher Z_o. The current loops have dominant *B*-field and lower Z_o. The load impedance of a metallic surface is given by $Z_L = (2\pi f \mu \rho)^{1/2}$ where f, μ and ρ are frequency, relative permeability, and resistivity of the surface material, respectively. Their values for commonly used enclosure metals are listed in Table 12.2. With these values, all metals except stainless steel have values of Z_L much lower than 377 Ω. Therefore, they all provide good shielding from high impedance sources (dipole sources), but provide poor shielding from low impedance sources (loop currents).

The reflection loss is defined as the ratio in dB of the wave field intensity passing through the shield to the field impinging upon it. For aluminum and steel, the two most widely used metals for enclosures, the reflection loss at various frequencies is listed in Table 12.3.

Once the outside interference enters the shield box, a part is absorbed in the metal wall as a resistive loss, and the remainder propagates further. The I^2R loss absorbed in the wall thickness depends on frequency and the skin depth of the wall, and is given by

$$\delta = [\rho/(\pi f \mu)]^{0.5} \tag{12.2}$$

Table 12.2 The relative permeability and resistivity of commonly used shield metals

Material	Relative resistivity	Relative permeability
Copper	1	1
Aluminum	1.6	1
Mild steel	10	100–1000
Stainless steel (no-mag)	50	1

Table 12.3 Reflection loss in aluminum and steel at various frequencies

Frequency	E-field loss (dB)		B-field loss (dB)		Plane wave loss (dB)	
	Alum	Steel	Alum	Steel	Alum	Steel
60 Hz	280	240	20	0	150	115
1 kHz	240	200	30	10	140	100
150 kHz	180	130	50	20	115	80
15 MHz	115	80	75	40	95	70

If the wall thickness is five times the skin depth, all the energy entering the metal wall is absorbed. For other thickness, the absorption loss, A (in dB), is given by

$$A = 0.134t(f\mu_r/\rho_r)^{0.5} \tag{12.3}$$

where t = wall thickness in mm, ρ_r = resistivity relative to copper, and μ_r = permeability of the box interior relative to free space. The relation shows what is intuitively expected, namely the greater the wall thickness, the greater the power absorption. The absorption loss values for aluminum and steel enclosures at various frequencies are given in Table 12.4.

Table 12.3 and Table 12.4 show that the reflection and absorption of the energy depends inversely on the resistivity. The permeability, on the other hand, has a beneficial effect on absorption, but a detrimental effect on reflection. This property can be used in low-frequency magnetic shielding. The reflection and absorption losses in dB are added to obtain the total loss in dB. The table values are theoretical for a perfect shield. In reality, the total dB loss is lower by a few fold due to seams and ports for connector and displays. However, most practical boxes provide adequate shielding except for very low frequency magnetic fields, as long as they are metallic and thick enough to support the parts inside.

Table 12.4 Absorption loss in aluminum and steel at various frequencies

Frequency (Hz)	Absorption loss in aluminum (dB/mm)	Absorption loss in mild steel (dB/mm)
60	0.8	13
1000	3.2	56
150 k	40	680
15 M	400	4240

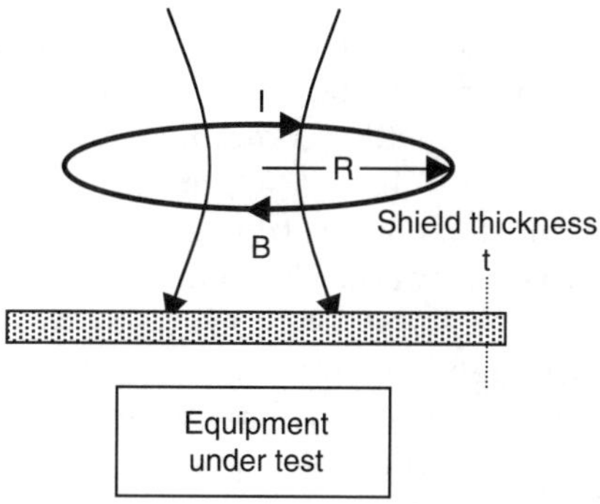

FIGURE 12.20 B-field of current loop entering metal shield.

The overall shielding effectiveness (SE) is often measured by testing the noise signal amplitude A with and without the shield, and using the expression

$$\mathrm{SE} = 20\log\frac{A_{with\ shield}}{A_{without\ shield}} \tag{12.4}$$

Enclosures made of solid aluminum or aluminum backed composite laminates provide 50 to 100 dB shielding effectiveness for high frequency noise, in megahertz. The metallic backing can be less than 0.1 mm in most applications. Other conductive materials may be used, such as metal mesh and conductive fabrics. The shield does not have to be grounded for the

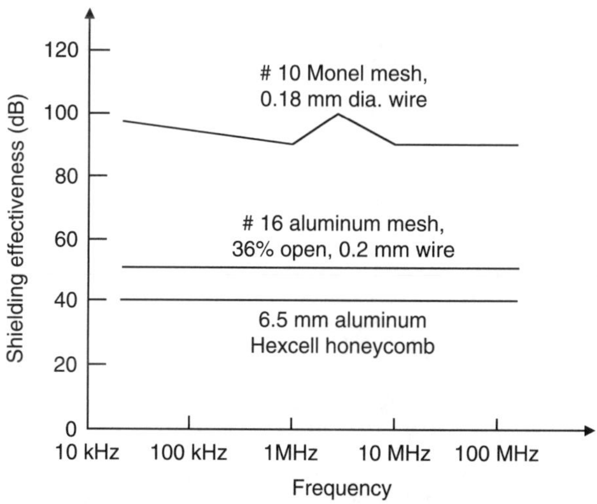

FIGURE 12.21 EMI shielding effectiveness of various materials.

shield effectiveness. It is grounded for safety reasons in case of a fault to the chassis and for protection against electrostatic buildup.

If a solid plate is used to shield a component by eddy currents (Figure 12.20), not only is it required to have $t > \delta$, but also $Rt >> \delta^2$ for good low-frequency shielding against radiation from small current loops. Here t = shield thickness, ρ = resistivity of the shield material, and δ = skin depth for the signal frequency to be filtered. Only magnetic metal can provide low frequency shielding, although it is rarely required. However, when needed, high permeability magnetic metal tape having the required saturation flux density is considered. The tape loses permeability after bending and straightening. Therefore, it often needs laser annealing after the wrapping process. Shielding effectiveness of various metal meshes and honeycomb are shown in Figure 12.21.

Part C
Power System Performance

Chapter 13

Energy Balance and Power Management

13.1 Introduction

When we speak of designing the spacecraft electrical power system, we really mean designing the electrical energy system. The satellite has a limited time in orbit to generate power, but the loads need to be powered all the time. The battery stores the required energy during sunlight and delivers to the loads during eclipse. The energy balance between the battery charge and discharge over one orbit period must be, on average, positive with some margin. In a simple case of constant load shown in Figure 13.1, the lightly dotted area must exceed the dark area with an acceptable margin. Otherwise, the battery would become totally depleted after some time. The power to and from various components must therefore be managed in order to maintain the energy balance in both the transfer orbit and the operational orbit.

In LEO satellites, the orbit period is short, about 1.5 h. The eclipse duration, although short in clock time, is much longer as a percentage of the orbit period. Consequently, the sunlight duration as a percentage of the orbit period is much shorter, imposing a stringent demand on the power system. It must not only collect the required solar energy in a relatively short time, but also manage it efficiently over the entire orbit period. The energy balance over the orbit period is maintained by a combination of hardware and software control as described in this chapter. It amounts to maintaining the following integrals equation:

$$\int_{sunlight} (\text{solar flux} \times \text{PV conversion efficiency}) \cdot \mathrm{d}t = \int_{sunlight} (\text{loads} + \text{charge power} + \text{shunt power} + \text{losses}) \cdot \mathrm{d}t + \int_{eclipse} (\text{loads} + \text{losses}) \cdot \mathrm{d}t \tag{13.1}$$

In most satellites, the energy balance is maintained autonomously on board. However, the ground control intervenes when it is necessary to (a)

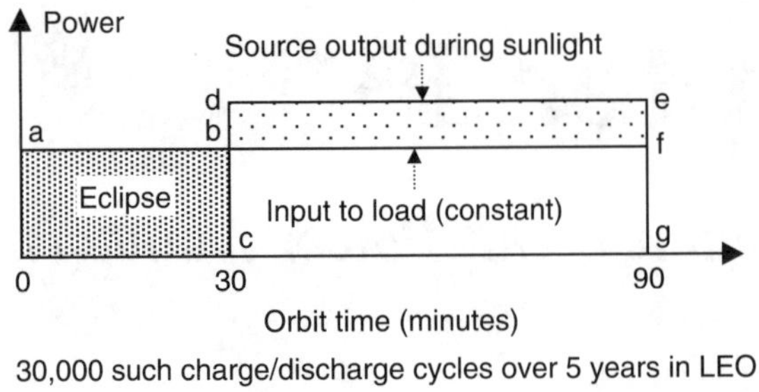

FIGURE 13.1 Energy balance over one orbit period with constant load power.

override the autonomous controls for any reason, (b) trim the system to account for predicted or unpredicted degradation in the performance, or (c) recondition the battery, if required.

The battery telemetry available — all with redundancy — for this purpose are:

- Two cell temperatures per battery, one at the hottest spot and one at the coldest spot
- Two cell pressure readings per battery, temperature compensated
- All cell voltages for comparative performance analysis
- Each battery voltage and charge/discharge current

The charge rate is selected between the full charge and trickle charge rates as controlled by the PRU mode controller to maintain the bus voltage. Typical charge rates in GEO satellites are:

- $C/120$ for trickle charge during solstice periods (no-eclipse seasons)
- $C/60$ for trickle charge following eclipses around equinox to maintain high state of charge, also for recharge during solstice after N–S station keeping
- $C/20$ normal charge rate
- $C/10$ maximum charge rate to charge the batteries sequentially for high efficiency if the power budget requires so

During the transfer orbit, continuos ground control is not possible and the on-board computer (OBC) automatically controls by the battery state of charge (SOC). The OBC can adjust all charge rates from default values and can inhibit automatic functions. At launch, the battery charger's minimum rate is set at $C/120$ for charge maintenance, and the maximum rate is set at $C/10$ for sequential charging for high efficiency up to 85% SOC.

In operational orbit, normally the batteries are charged concurrently at $C/20$ when they are under the autonomous control of the OBC. The ground control has the option of charging them sequentially at $C/10$ by programming the OBC. Charging at $C/10$ is more efficient up to a certain SOC, beyond which it is switched to $C/20$.

Using the telemetry data, the OBC estimates the battery SOC using the battery cell pressure telemetry compensated for the telemetered temperature value. Using the hysteric dead-band control, the OBC initiate charge if the cell pressure falls below the preset value selected for charge initiation less the hysteresis level of about 2 to 3%.

During charging, the OBC compares the cell pressures against the predetermined state, 85% being a typical default value. The OBC terminates the charge when the SOC rises above a preset value, and commands the chargers to one of the two trickle charge rates pre-selectable by ground command. The ground command also has an option to preset an additional amount of charge to reach *full charge* without excessive temperature and pressure rise, followed by automatic charge termination. All preset values in the battery charge and discharge management are alterable by the ground control.

13.2 Energy Balance Analysis

The energy balance analysis is performed at the design stage by simulating the power flow and energy account on the computer. The program is generally structured to allow analyses on the baseline design and its derivatives, and to answer many *what if* questions in normal and abnormal operation.

The program basically determines whether the EPS is in energy balance over a desired period, e.g., over every orbit or over several orbits, for a given load profile, solar array and the battery. Such analysis is performed during all phases of the mission for a given launch date (year, month, and day). Various fault conditions, including battery cell failures (short or open) and loss of a solar array circuit, are simulated to determine energy balance under the worst-case condition(s).

Other equally important uses of the energy balance computer program are:

- Determine and/or optimize the load capability of a given EPS
- Derive the system component ratings based on the maximum power flow through each component
- Determine the power dissipation for thermal design of each component, particularly the battery, since its performance is highly temperature sensitive

13.3 Computer Program Structure

The computer program is generally developed around variable parameters with no *hard-coded* numbers. Such a structure allows greater flexibility in using this tool for a wide variety of applications. Figure 13.2 depicts EPS components contributing to the energy balance program. The basic equation of currents in the energy balance analysis is:

during sunlight,

$$I_{sa} = I_{load} + I_{charge} + I_{shunt} - I_{discharge} \tag{13.2}$$

during eclipse,

$$I_{load} = I_{discharge} \tag{13.3}$$

The battery DOD at any time is computed typically every second or so as follows:

$$\text{DOD}(t) = \text{DOD}_{\text{initial}} + (\text{AH}_{\text{delivered}}/\text{AH}_{\text{actual}}) \tag{13.4}$$

where $\text{DOD}_{\text{initial}}$ is the initial DOD, $\text{AH}_{\text{delivered}}$ is the sum of AH delivered, and $\text{AH}_{\text{actual}}$ is the actual AH capacity. The entire program is divided into several software modules, each representing various components. Due to the nonlinear nature of the battery cell and the solar array performance parameters, the program uses *static* lookup tables to determine the cell performance characteristics as a function of the battery current, temperature, state of charge and the solar array operating voltage. Programming with a computer language, instead of modeling on a spreadsheet, significantly improves the capability of the program.

Major software modules of the programs can be structured as follows.

13.3.1 Battery Module

The *battery_module* determines a number of battery parameters, including the state of charge *(battery_soc)*, battery voltage *(battery_volt)* and battery

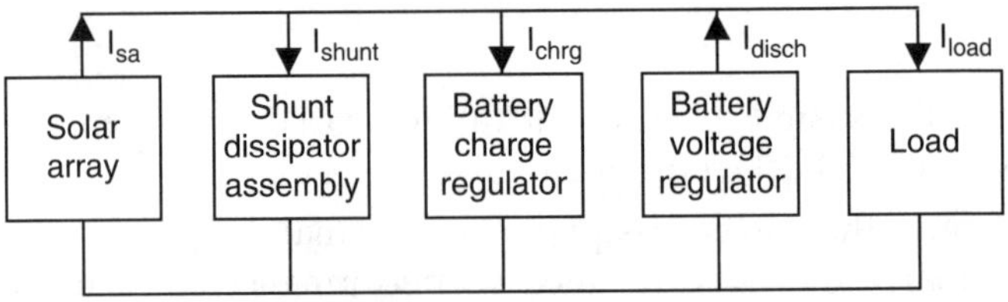

FIGURE 13.2 Power system components participating in the energy balance.

dissipation *(battery_disp)*. The ampere-hour (Ah) integration is performed to predict the battery SOC in terms of actual capacity. Ah returned to the battery and the Ah removed from the battery are calculated at each program *integration step* which is typically in seconds. The battery C/D ratio is determined by the ratio of the ampere-hours put into the battery to the ampere-hours taken out of the battery. The battery voltage is determined using a lookup table, which provides the battery cell voltage as a function of the charge or discharge rate (+ for charge and − for discharge), battery temperature, and the battery SOC.

The battery dissipation is determined as a function of the difference between the cell voltage and the thermal neutral potential *(therm_neut_pot)* of the battery cell, the battery current, and the battery charge/discharge efficiency. The thermal neutral potential for the NiH_2 battery cell is 1.51 V. The battery cell efficiency during discharge can be assumed 100%. The battery efficiency during the charge portion of the orbit varies during charge as a function of the battery charge rate, state of charge and the temperature. It is determined using a static lookup table and performing linear interpolation between points.

The following features are desirable in the battery module:

- Variable charge management, *V–T* and/or *V–P* limiting, and full, cut back, and trickle charge rates
- Variable maximum allowable charge/discharge currents
- Variable battery configuration (number of cells, number of open/short cells)
- Variable battery temperature
- Tracking the performance of each battery individually
- Dissipation in individual batteries and cell bypass diodes
- Dissipation in battery charge and discharge converters
- Variable bus voltage regulation profile
- Constant or time varying efficiency from test data

13.3.1.1 V–T Limit for overcharge control

Controlling the battery overcharge is extremely important for maintaining the battery temperature, which has a strong influence on the battery life. The battery when fully charged reaches the maximum voltage beyond which all input power is converted into heat. For every electrochemistry, there is a definite relationship between the voltage and temperature (*V–T*) at the end of charging. One can use that relation to determine that the battery has been fully charged, and then to cut back to the trickle charge rate equal to the self-discharge rate. The *V–T* relation varies with age of the battery. The fully charged voltage is higher when the battery is new, and then slowly degrades with age. For each battery, several V–T curves, each representing an age period, is stored in the computer. For example, Figure 13.3 shows six curves for a mission divided into six periods. The top six curves are for healthy battery with all cells working, and the bottom six are with one cell failed and bypassed by the diode.

13.3.1.2 Charge cut-back and trickle charge

When the full charge is approached, the battery charge current is often tapered down to the trickle rate linearly or in a few steps. The battery current and voltage profile over one orbit period shown in Figure 13.4 is typical for LEO satellites. A constant charge rate is maintained up to 95% SOC, typically in 50 to 60% of the sunlight duration. The charge rate is then tapered off to trickle charge in the $C/100$ to $C/50$ range.

Figure 13.4 is an example of the battery current analysis results in a LEO satellite. It starts at the beginning of eclipse. During eclipse, the discharge current rises (becomes more negative) as the battery voltage drops because of the constant power nature of the load. As the battery is placed in charge at the beginning of sunlight, the battery voltage rises under constant charge current. About half-way in the sunlight, the charge rate starts tapering and is reduced to the trickle rate at the end to maintain full voltage until inception of the next eclipse.

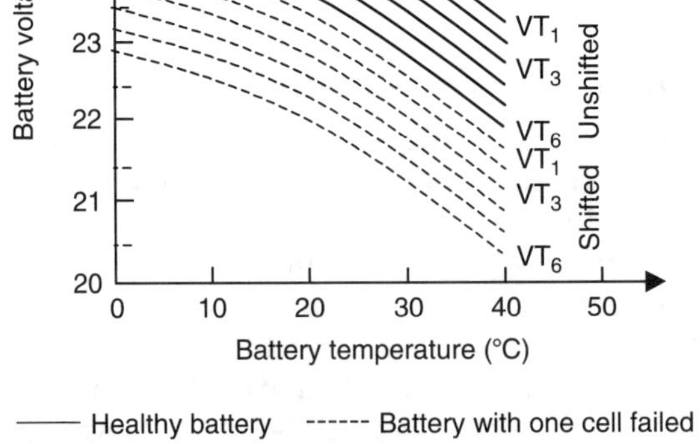

FIGURE 13.3 V–T limits for overcharge control in a 22-cell battery.

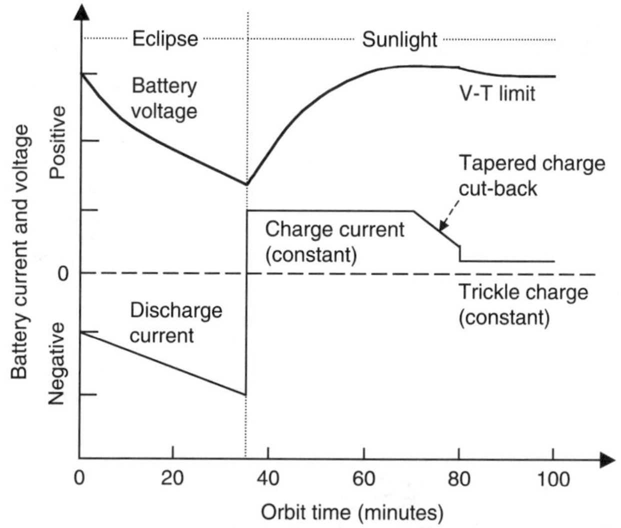

FIGURE 13.4 Battery current and voltage versus time for energy balance over one orbit period.

13.3.1.3 AH counter and sensor

The battery voltage is not a reliable indicator of the SOC. In the past, the battery SOC had been monitored and managed by analog controls using various sensors on-board. Many present-day satellites employ software instead, which is a mass-efficient approach. The software keeps the account of the ampere-hours in and out of the battery to determine the SOC and then proceed accordingly. The accumulated error can be detected by temperature and/or pressure sensors, and the ultimate safety of the battery is maintained by a suitable hardware circuit. One such self-explanatory computational flow chart often used in communication satellites is depicted in Figure 13.5 as an example.

In recent developments, monitoring the battery SOC using fuzzy logic analysis has been tested with coulomb counting by digital software and with the battery impedance measurement by analog hardware. The coulomb counting alone does not allow variations in the internal battery characteristics to be monitored; hence the advance warning of premature failure of the battery is not possible with this technique. The SOC monitoring with impedance measurement alone is not useful in practice either, because of the complexity of instrumentation and analytical techniques required to derive the equivalent circuit parameters of the cell. However, the combination of the fuzzy logic data analysis and the internal impedance technique for determining the battery SOC has a potential of providing a powerful approach for determining the battery SOC.

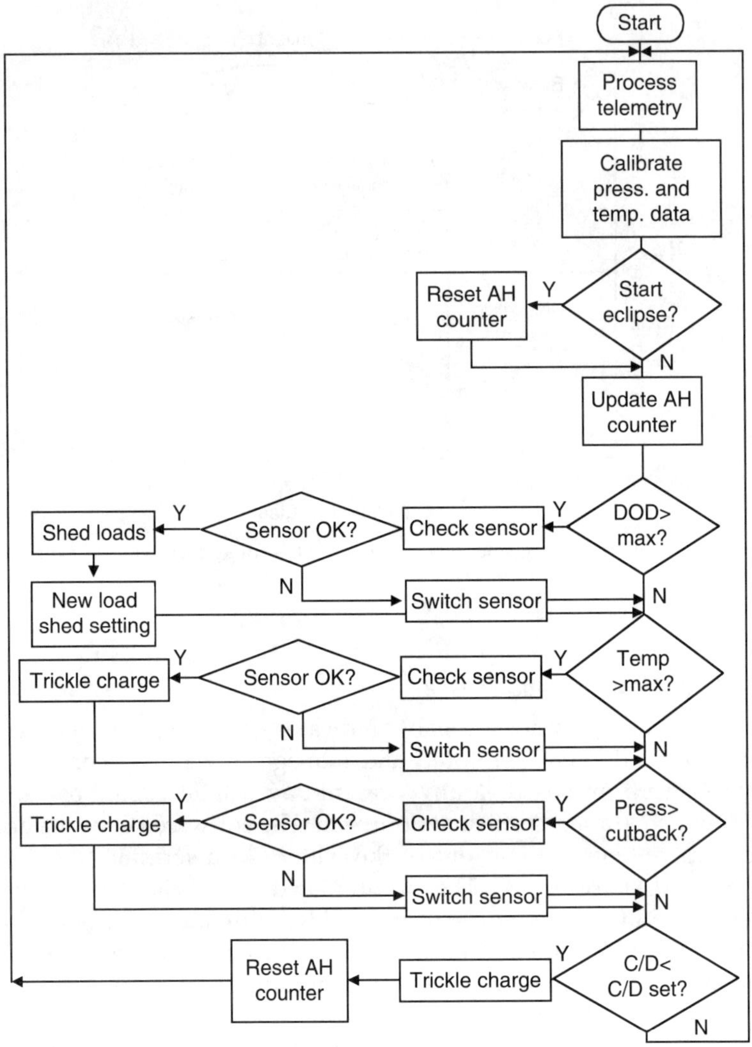

FIGURE 13.5 Logic flow chart for energy balance software.

13.3.2 Solar Array Module

The *sa_module* is programmed to incorporate the following:

- Solar cell *I–V* characteristics
- Array configuration (number of series cell × number of parallel circuits)
- Cell and array performance degradation factors
- Array temperature

13.3.3 Bus Module

The *bus_module* sums all currents available at the bus. They are the discharge current, charge current, solar array current, load current, and the shunt current, as shown in Figure 13.2. The bus voltage is determined by using the net current as a *derivative offset* to the previous *bus_volt*. The finite difference equation used to calculate the bus voltage at the kth time step in the numerical iteration process is:

$$V_{bus(k)} = V_{bus(k-1)} + [(I_{saa} + I_{discharge} - I_{charge} - I_{shunt} - I_{load})_k - t/C_{bus\ capacitors}] \tag{13.5}$$

where $\triangle t$ is the time interval.

The bus voltage regulator typically maintains the bus voltage as the source and the load currents vary to maintain the dc equilibrium of the two currents. A bus capacitor bank is generally used to absorb the leading and trailing edge transients of fast changing loads or sources.

13.3.4 Input Variables

It is desirable to have the energy balance program flexible in incorporating the following system inputs in various simulation runs:

- Time variable load, plus a constant load, over one or multiple orbits
- BOL or EOL conditions
- Off-pointing of the solar array
- Launch and descent phase of the mission

13.3.5 Output Format

Several output files are produced on running the program, such as the battery parameters (*battery_out*), EPS performance parameters (*eps_out*), BPC parameters (*bpc_out*) and solar array parameters (*sa_out*). The output files contain variable names and values and may be loaded into a spreadsheet program for subsequent plotting or a word-processing program for inclusion in reports.

The output files print out numerous performance parameters at selected print interval over the study period. For example, the *battery_out* file prints out all the following parameters:

Orbit	Sequential number of the orbit being printed
Minutes	Time from the beginning of the study period
V_{bat1} and V_{bat2}	Voltage across Battery 1 and Battery 2

I_{bat1} and I_{bat2}	Currents through Battery 1 and Battery 2
DOD_1 and DOD_2	Depth of Discharge in Battery 1 and Battery 2
Q_{bat1} and Q_{bat2}	Dissipation in Battery 1 and Battery 2
Q_{cell1} and Q_{cell2}	Dissipation in cells in Battery 1 and Battery 2
Q_{open1} and Q_{open2}	Dissipation in cell-open bypass diode in Bat1 and 2
V_{open1} and V_{open2}	Voltage across cell-open bypass diode in Bat1 and 2
η_{bat1} and η_{bat2}	Charge efficiency of Battery 1 and Battery 2, (from lookup tables representing measured data)
AH_{out1} and AH_{out2}	Ampere-hour going out of Battery 1 and Battery 2
AH_{in1} and AH_{in2}	Ampere-hour going in Battery 1 and Battery 2

Table 13.1 is an example of the *battery_out* file. It prints key battery performance parameters such as the battery voltage, current, depth of discharge, power dissipations in the whole battery, in individual cells and in cell bypass diodes of Battery 1 and Battery 2 in every orbit every minute or two.

13.3.6 Program Verification

Before using any such energy balance program, it must be calibrated and verified using a simple load profile for which the performance can also be calculated by hand. Figure 13.6 is an example of the computed battery DODs compared with the hand-calculated values. They must compare well within the limitation of hand calculations. As seen in the figure, the assumption of constant battery charge efficiency in hand calculation results in slightly different DOD as the battery approaches zero DOD.

13.4 Energy Balance Simulation Runs

On ***run*** command, the program initializes all input parameters, including the battery lookup tables and the solar cell lookup tables by calling the *init_module*. In this module, the simulation start and end times and the number of orbits (single or multiple) are specified. The BDR/BCR efficiencies and the harness resistances are entered in this module. The print interval is also specified, allowing data to be printed to an ASCII file in the designated increment (say, 1 min).

The *load_module* determines the load current requirement at the bus voltage according to the load power requirement specified in the input file. The load power can be either constant or can vary as a function of time with a time increment of 0.1 to 1 min.

Static lookup data files (battery cell data and solar cell *I–V* data for example) are located in separate files.

Table 13.1 Typical output format of the energy balance computer simulation (partial list)

Orbit	Minutes	Vbat1	Ibat1	DOD1	Qbat1	Qcell1	Qopen1	Vopen1	nbat1	AHout1	AHin1
1	2.0	23.89	−13.81	0.0082	24.64	1.449	0.000	0.000	1.00	−0.466	0.000
1	4.0	23.73	−13.90	0.0163	26.97	1.587	0.000	0.000	1.00	−9.28	0.000
1	6.0	23.58	−13.98	0.0244	29.27	1.722	0.000	0.000	1.00	−1.392	0.000
1	8.0	23.43	−14.07	0.0326	31.59	1.858	0.000	0.000	1.00	−1.860	0.000
1	10.0	23.27	−14.16	0.0409	33.94	1.996	0.000	0.000	1.00	−2.331	0.000
1	12.0	23.12	−14.25	0.0492	36.33	2.137	0.000	0.000	1.00	−2.804	0.000
1	14.0	22.98	−14.34	0.0576	38.55	2.267	0.000	0.000	1.00	−3.281	0.000
1	16.0	22.87	−14.41	0.0660	40.41	2.377	0.000	0.000	1.00	−3.760	0.000
1	18.0	22.76	−14.48	0.0744	42.18	2.481	0.000	0.000	1.00	−4.242	0.000
1	20.0	22.67	−14.53	0.0829	43.53	2.561	0.000	0.000	1.00	−4.725	0.000
1	22.0	22.59	−14.58	0.0914	44.89	2.641	0.000	0.000	1.00	−5.210	0.000
1	24.0	22.51	−14.63	0.0999	46.18	2.716	0.000	0.000	1.00	−5.697	0.000
1	26.0	22.44	−14.68	0.1085	47.47	2.792	0.000	0.000	1.00	−6.186	0.000
1	28.0	22.40	−14.70	0.1171	48.08	2.828	0.000	0.000	1.00	−6.675	0.000
1	30.0	22.37	−14.72	0.1257	48.64	2.861	0.000	0.000	1.00	−7.166	0.000
1	32.0	22.34	−14.74	0.1343	49.14	2.891	0.000	0.000	1.00	−7.657	0.000
1	34.0	24.73	12.00	0.1391	13.36	0.786	0.000	0.000	0.92	−8.012	0.088
1	36.0	24.78	12.00	0.1327	16.46	0.968	0.000	0.000	0.91	−8.012	0.488
1	38.0	24.83	12.00	0.1264	19.53	1.149	0.000	0.000	0.90	−8.012	0.888
1	40.0	24.85	12.00	0.1200	22.33	1.314	0.000	0.000	0.90	−8.012	1.288
1	42.0	24.88	12.00	0.1138	25.25	1.485	0.000	0.000	0.89	−8.012	1.688
1	44.0	24.90	12.00	0.1076	28.26	1.662	0.000	0.000	0.88	−8.012	2.088
1	46.0	24.93	12.00	0.1015	31.26	1.839	0.000	0.000	0.87	−8.012	2.488
1	48.0	24.95	12.00	0.0954	34.37	2.022	0.000	0.000	0.86	−8.012	2.888
1	50.0	24.98	12.00	0.0894	37.84	2.226	0.000	0.000	0.85	−8.012	3.288
1	52.0	25.01	12.00	0.0835	41.34	2.432	0.000	0.000	0.84	−8.012	3.688
1	54.0	25.05	12.00	0.0776	44.80	2.635	0.000	0.000	0.83	−8.012	4.088
1	56.0	25.08	12.00	0.0718	49.18	2.893	0.000	0.000	0.82	−8.012	4.488
1	58.0	25.11	12.00	0.0661	54.2	3.194	0.000	0.000	0.80	−8.012	4.888
1	60.0	25.15	12.00	0.0605	59.31	3.489	0.000	0.000	0.79	−8.012	5.288
1	62.0	25.18	12.00	0.0551	66.35	3.903	0.000	0.000	0.77	−8.012	5.688
1	64.0	25.20	12.00	0.0498	74.89	4.405	0.000	0.000	0.74	−8.012	6.088
1	66.0	25.21	12.00	0.0447	83.86	4.933	0.000	0.000	0.71	−8.012	6.488
1	68.0	25.23	12.00	0.0398	92.37	5.433	0.000	0.000	0.68	−8.012	6.888
1	70.0	25.25	12.00	0.0351	100.49	5.911	0.000	0.000	0.66	−8.012	7.288
1	72.0	25.29	12.00	0.0306	107.20	6.306	0.000	0.000	0.64	−8.012	7.688
1	74.0	25.30	10.98	0.0264	115.41	6.789	0.000	0.000	0.58	−8.012	8.075
1	76.0	25.30	0.85	0.0233	17.12	1.007	0.000	0.000	0.20	−8.012	8.413
1	78.0	25.30	0.85	0.0232	17.15	1.009	0.000	0.000	0.20	−8.012	8.442
1	80.0	25.30	0.85	0.0231	17.17	1.010	0.000	0.000	0.20	−8.012	8.470

After editing input file(s) to reflect the case to be simulated, the energy balance program is recompiled using the new input file(s). The new executable code is then **run**. The simulation is run for the following three distinct phases of the mission.

13.4.1 On-Orbit Simulation

On-orbit power flow analyses and energy balance is carried out for the worst case at the beginning of life and also at the end of life. Since the BOL

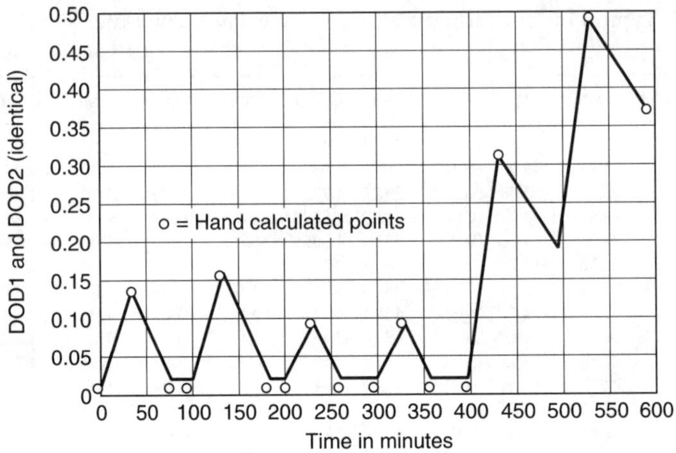

FIGURE 13.6 Calibration example for battery DOD 1 and DOD 2. (—— Computed,$^{\circ\circ\circ}$ Hand calculated.)

power generation is much greater than that at the EOL, the BOL power analysis establishes the current rating of the sliprings and the shunts, both of which always carry the greatest currents in the beginning of life. The end of life energy balance, on the other hand, must be verified using the worst-case orbit conditions, namely the yearly minimum solar flux, minimum operation altitude and the maximum eclipse time. If applicable, the solar β angle of 0° usually results in the longest eclipse time.

Besides being a design tool, the energy balance analysis is also carried out on board the spacecraft to detect and correct potential problems. This is done by performing the following forward computations.

For a given load power profile the spacecraft is about to enter, the battery DOD is projected at the end of the eclipse. If the projected DOD exceeds the maximum allowed, an alert is sent to the spacecraft computer for an approaching battery limited conditions and for a possible load shedding.

For the allowable DOD and the required charge power during the next sunlight duration, the maximum allowable load is computed. If it is less than the planned load, an alert is sent to the spacecraft computer for an approaching solar array limited condition and for a possible load shedding.

13.4.2 Transfer Orbit Simulation

This phase of the mission in communication satellites begins with separation of the spacecraft from the launch vehicle (approximately 30 min after lift-off) and ends with the spacecraft in full 24-h orbit around the Earth. The first orbit after separation is nominally a 10.5-h elliptical orbit having perigee at the spacecraft–launch vehicle separation altitude and apogee at the geosynchronous altitude. During a transfer orbit period spanning 1 to 2 weeks, a series of timed apogee engine burn raises perigee to the geosynchronous height. During this period, the solar panels are stowed against the north and south equipment panels, with only the outer panel of each wing exposed. The GEO spacecraft is slowly spun around roll or yaw axis typically at 0.1 rpm to maintain acceptable component temperature. Because the solar array panels are folded against two opposite sides of the spacecraft body which is spinning, the power output approximates a full-wave rectified sinewave. Consequently, the batteries are alternately charged and discharged at twice the spin rate (Figure 13.7). The selection of a high charge rate permits use of all available solar array power more efficiently.

The following are three basic spacecraft orientations in transfer orbit during which the solar panels of communication satellite are typically subjected. They experience just about every conceivable level of sun illumination as follows.

13.4.2.1 Roll spin

At times during the transfer orbit the spacecraft spins around the roll axis (east–west axis). The north and south outboard solar panels are alternately illuminated. The sun angle (the angle between the spin axis and the sun-pointing vector) is anywhere between 50 and 130°. Shadowing of solar panels during this phase does not occur. Each outboard panel sees a half-wave rectified sun input profile with the peak ranging from full sun to cos 40°.

13.4.2.2 Yaw spin

At times the spacecraft spins around the yaw axis (nadir—antinadir axis). As with the roll spin mode, the north and south outboard solar panels are alternately illuminated. For most of the orbit, the sun angle is held within 85

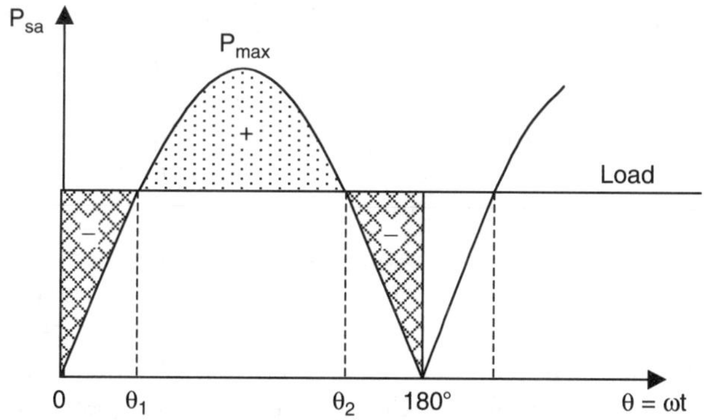

FIGURE 13.7 Transfer orbit energy balance with full-wave rectified sine wave variation in power generation.

to 95°. Each outboard solar panel sees full sunlight once per spin. During apogee engine firings, however, the sun angle range increases to 60 to 120°. Peak sunlight during rotation can be between full sun and cos 30°. In this orientation, the communications antenna reflectors can create shadows on the solar panels at spacecraft spin angles of 40° from the sun-normal.

13.4.2.3 Initial attitude fix

The spacecraft initiates an Earth search and lock maneuver approximately 1 h after separation from the launch vehicle. The spacecraft is spun slowly until the Earth sensors find the Earth. Once the Earth lock is achieved, the spacecraft remains in an inertially fixed attitude for up to 10 min. During this phase, each solar panel can see full sun, no sun, or anything in between. If one of the solar panels is pointed directly at the sun, the solar cell temperature may exceed 115°C.

13.4.2.4 Maneuvers to improve the energy balance

In case the transfer orbit energy balance with any one of the above three sun orientations is in deficit, the following changes in the operating maneuvers can improve the energy balance:

- Install more efficient soar cells on the outer panels. Such cells can be selected from the given lot by cell selection without added cost, or by procuring a class of more efficient cells at added cost.
- Partially deploy the solar wing in transfer orbit. The risk with this approach is a plume impingement by the control thrusters and some propellant penalty.
- Make outboard panel larger than the inboard panels. This may make the packaging difficult and add some cost.
- Improve the sun angle. For example, GPS II spacecraft spins at 55 rpm in the transfer orbit, and the worst-case sun angle is 16°. This can be improved by 10° by expending half a kilogram of fuel (one time cost). This cost may be recovered at the end of the transfer orbit when the spacecraft needs to move to apogee kick motor angle, which may be, say, 90°. Thus, not all fuel is wasted in orienting the spacecraft in transfer orbit for a better sun angle.
- Power-down the transfer orbit operations as much as possible. Ask whether we need all that heater power? Can we turn some loads off at some times during the transfer orbit? Is the power usage by all other equipment reasonable? Can they be reduced by better design or smart utilization? Is it possible to design them with a special low-dissipation pre-operational mode without adding much cost? Such questions may lead to a solution or a partial relief.

- Use smart energy management and operational work-around. For example, pausing the spin for a while when the sun angle is favorable to allow the outer panel collect more energy as shown in Figure 13.8 can help the transfer orbit energy balance. The dwell time in such a maneuver is determined by the allowable temperature limit on the panel or inside the body, which can become roasted under a long dwell time. Such stop-and-go operation, however, consumes additional fuel. The mass of this fuel is charged to the power system.

13.4.3 Transfer Orbit Load Budget

Since only one panel is facing the sun at a time as the spacecraft spins in the transfer orbit, the power output of the array is a full wave rectified sinewave as was shown in Figure 13.7. The power generation varies with sun angle as

$$P = P_{\max} \sin\theta \tag{13.6}$$

where

$$P_{\max} = \frac{P_{bol}}{N_p} f_{cs} \tag{13.7}$$

and θ — sun angle measured from the horizon (90° gives maximum power), N_p = number of panels on both wings, and f_{cs} is the improvement in outboard panel performance due to better cell selection (about 1.02). The sun angle is tightly constrained between 80 and 100°, except during maneuvers, when it can be between 30 and 140°.

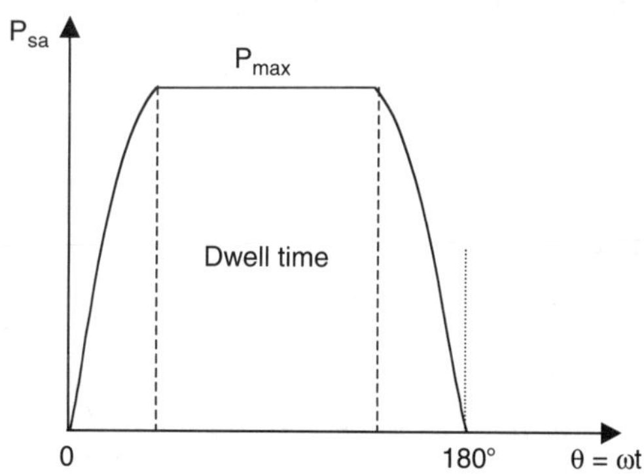

FIGURE 13.8 Dwelling at favorable sun angle to improve energy balance during transfer orbit.

The average power during transfer orbit, ignoring the thermal effects, is given by

$$P_{t.o.avg} = P_{\max} \frac{2}{\pi} \tag{13.8}$$

Also denoting P_L = load power, including the orbit average losses

E_{dis} = total energy discharged during half a turn (dark shaded area)
E_{chg} = total energy charged during the same half turn (light shaded area)
θ_1 = crossover point between the charge and discharge mode

we have

$$\theta_1 = \sin^{-1}\left(\frac{P_L}{P_{\max}}\right) \tag{13.9}$$

$$E_{dis} = 2\int_0^{\theta_1} \frac{P_L - P_{\max}\sin\theta}{\eta_d}\, d\theta \tag{13.10}$$

$$E_{chg} = \int_{\theta_1}^{\theta_2} (P_{\max}\sin\theta - P_L)\,\eta_c\eta_b\, d\theta \tag{13.11}$$

where θ_1 is in radians, and E_{dis} and E_{chg} are in watt-hours. For the satellite to be in energy balance during the transfer orbit, E_{chg} must equal or exceed E_{dis} with a suitable margin compatible with the uncertainty at a given design stage. Such analysis determines the required peak power $P_{\max}$, the spacecraft inclination or the maximum permissible load during the transfer orbit. It does not account for eclipse during the transfer orbit, which must be considered separately, if applicable.

13.4.4 Launch and Ascent Simulation

During the launch and ascent and until separation, the battery powers the computers, EEDs and other essential equipment until sunlight is acquired. In an LEO mission, the battery designed for a shallow (30 to 35%) DOD during normal operation orbit can be discharged to a higher (80 to 90%) DOD for one time during the launch/ascent phase. However, once the sun is acquired, the battery is first fully charged before undertaking other maneuvers. Figure 13.9 depicts a typical battery SOC for an LEO satellite.

If the battery is short of energy in this phase, two options are:

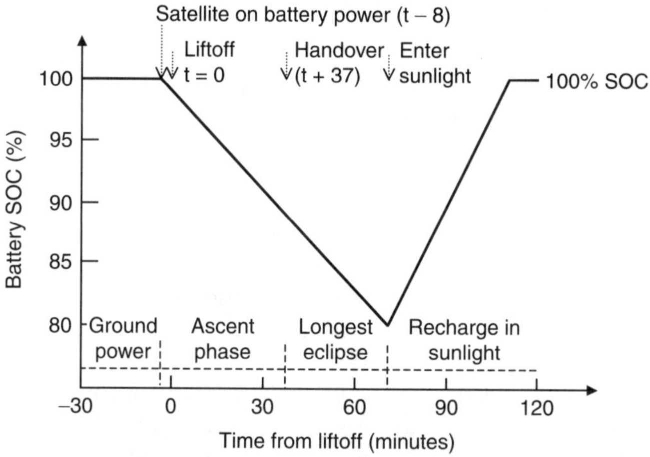

FIGURE 13.9 Energy flow and battery SOC during launch and ascent phases.

Cool the battery to a lower temperature by using a jet of cool air or nitrogen on the launch pad just before launch. This allows charging the battery to a higher Ah capacity. This is a widely used option.

If that is not enough, a small primary battery may be used to augment the main battery. The primary battery having an order of magnitude greater specific energy gives a mass efficient solution. Carrying a larger main battery just for this phase is almost never done, as it adds significant mass, and also costs much more.

13.5 Battery State of Health Monitoring

Since the battery is a critical component in the spacecraft power system, monitoring its state of health and the remaining life is important. The data collected for the on-board energy balance management can also be used for this purpose. The battery overall state of health on board is generally monitored by the following two key indicators.

The pressure strain gage readings monitor the battery's overall condition. Any unexpected reading is taken as an alert flag. Initial matching of the cells ensures that two cell pressures represent all other cells.

Cell divergences are detected using the individual cell voltage monitored by the BCVM. Trend analysis of the cell voltages is performed to detect individual cell voltage divergence. Any such divergence flags a battery weakness.

The battery state of health indirectly measures external behavior such as a capacity loss. It also measures internal behavior such as severe corrosion, the remaining useful life, and the remaining useful energy. The changes in

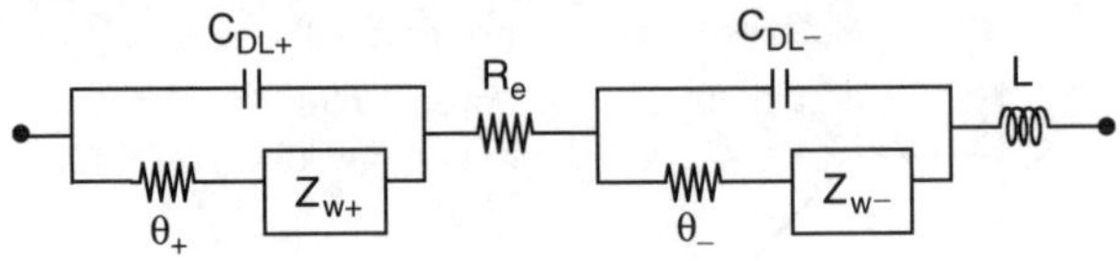

FIGURE 13.10 Circuit model for battery with wiring resistance and inductance.

the electrode surface, diffusion layer, and electrolyte are not measurable. However, the terminal voltage and body temperature is used to indirectly determine the performance. Accurate modeling of the transport mechanism inside the cell and the electrochemical and thermal models are needed for this purpose. The cell internal impedance frequency spectrum technique developed by Kozlowski et al[2] can retrieve useful information about the internal electrochemical conditions. The internal impedance of the cell is measured by applying a small signal excitation to the cell and measuring the response. A range of frequencies is used to account for the frequency dependence of the of many electrochemical impedance parameters. The impedance model of the cell between positive and netgative electrodes separated by an electrolyte is shown in Figure 13.10,[3] where R and L are the wiring resistance and inductance, respectively. It represents the physical electrochemical process via electrolyte resistance, R_e, charge transfer resistance, θ, double layer capacitance, C_{dl}, and the Warburg impedance, Z_w. These parameters represent the physical electrochemical process, such as the charge and mass transfer during cycling. For example, the model identifies the electrolyte resistance, R_e, versus the state of charge for a NiCd battery as shown in Figure 13.11. Such information is then used in a neural network and fuzzy logic state of charge prediction model. This technique is

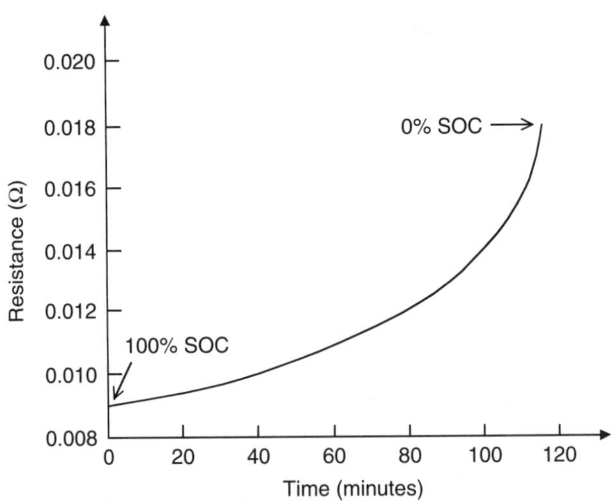

FIGURE 13.11 Electrolytic resistance of NiCd battery from the model.

under development at the U.S. Naval Research Center and holds promise for monitoring the state of charge, state of health, and remaining life of the spacecraft batteries.[3]

13.6 Battery Latch-up in Sun-Regulated Bus

In the sun-regulated power system, the difference between the stable operation of the solar array and that of the battery often leads to an operating condition known as the battery latch-up. It occurs when the satellite is coming out of an eclipse. At this time, although the solar array is capable of delivering full power, the battery continues to deliver the load power, instead of going into the charge mode. The bus remains latched to the battery voltage and is unable to recover to the shunt controlled voltage even during sunlight. Subsequently, it is not possible to maintain the energy balance, either by hardware or software. This is obviously an anomaly that should be immediately corrected to prevent the battery from further discharge. The condition is typical only in the sun-regulated system. The fully regulated system is inherently free of such latch-up.

The phenomenon is explained here with reference to Figure 13.12. The dotted parabola is the constant power load line for the spacecraft. The heavy solid line is the *V–I* characteristic of drained battery at the end of eclipse. The system operates at the stable point of their interaction, A_1. The other point of their intersection, U, is unstable. The faint dotted curve B is the *I–V* characteristic of the solar array soon after entering the sunlight. This curve keeps rising as the spacecraft moves from eclipse to penumbra to full

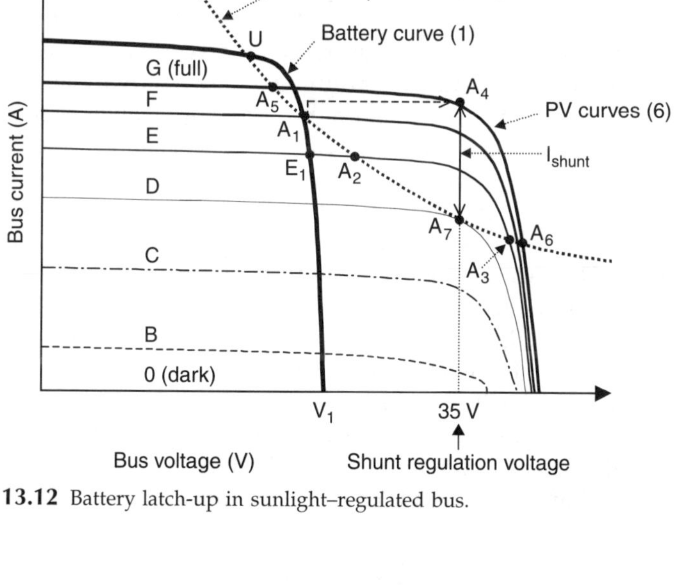

FIGURE 13.12 Battery latch-up in sunlight–regulated bus.

sunlight. Until it rises to curve D, the battery continues discharging to the load at point A_1, as curve D and load line have no point of intersection.

When the illumination rises to full sunlight curve E, there are intersection points A_2 and A_3 with the load line. But, since point A_1 is stable, it does not move to a higher voltage point A_2. The system thus remains latched at the battery voltage A_1. At this point, the current drawn from the solar array is V_1E_1 and the current drawn from the battery is E_1A_1. With curve F, the battery and the solar array match in their current capabilities. The system operates at point A_1, where it is stable because of being latched with the battery (the solar array is unstable at A_1).

When the illumination rises to curve G (full sun), the operating point A_1 moves to point A_5 on curve G, but since it is unstable, it starts moving towards the stable point A_6. On the extreme right hand side, however, the sunlight shunt regulation control set at 35 V limits it at point A_4, where $I_{shunt} = A_4A_7$. The operating point with load becomes A_4 with the shunt in control of the system now.

Thus, the battery latch-up can occur under the following two conditions at the end of eclipse:

- The battery voltage drops below certain low point, drawing heavy current to power constant power loads, and
- The solar array current output upon entry to the sunlight is less than the current drawn from the battery at the end of eclipse.

The remedy for avoiding the battery latch-up is to size the solar array more than that required for meeting the end of life load power with sufficient margin. In Figure 13.12, the *I–V* curve of fully illuminated solar array should be at least above curve F (point A_1) to unlatch the system from the low battery voltage during eclipse. This generally required 5 to 10% extra power margin compared to that in the fully regulated bus. The added mass and cost in the solar array constitutes a major disadvantage of the sun-regulated system. For this reason, large power systems, where this penalty can be high, find fully regulated system advantageous. If the battery voltage is too low during eclipse, even more solar array power is required to unlatch the system. On the other hand, if the battery voltage is too high during sunlight, the charge converter works inefficiently at low duty ratio (close to zero). The design engineer trades these two features.

Another remedy for the battery latch-up is to add one (rarely two) cell. With a 23-cell battery instead of the traditional 22-cell battery, the higher battery voltage prevents the latch-up. That extra voltage of one cell is often enough to avoid the latch-up. However, the 23-cell battery becomes fully charged at $23 \times 1.6 = 36.35$V, which is outside the 22-35 V limits on the bus. This may be allowed in a system where the battery is charged directly from a dedicated solar array without a charge regulator.

If the battery charge power is derived from a dedicated section of the solar array with a charge regulator, yet another remedy could be connecting the charging array on the bus until the latch is cleared up.

After exhausting all possible remedies, if the bus is still latched on the battery, there is nothing one can do except to shed some non-essential or noncritical loads temporarily. Some heaters can always be momentarily shut off. The reduced load current then would allow the sunlight voltage regulating shunts to regain control of the bus voltage. The loads can then be brought back into service.

References

1. Singh, P., Gaddam, V., Arey, S., and Yang, Z., Battery state-of-charge meters for high performance batteries based on Fuzzy logic methodology, in *Proceedings of the 34th Intersociety Energy Conversion Engineering Conference*, SAE, 1999, Paper No. 01-2467.
2. Kozlowski, J.D., Crawley, T., and Byington, C.S., *Model Based Predictive Diagnostics for Primary and Secondary Batteries*, Pennsylvania State University, Applied Research Laboratory, June 1999, Technical Report No. 99-076.
3. Kozlowski, J.D., Watson, M.J., Byington, C.S., Garga, A.K., and Hay, T.A., Electrochemical cell diagnostics using on-line impedance measurements, state estimation and data fusion techniques, *Proceedings of the 36th Intersociety Energy Conversion Engineering Conference*, ASME, 2001, pp. 981–986.

Chapter 14

Dynamic Performance and Stability

14.1 Introduction

The dynamic performance of power system under an internal or external transient perturbation is influenced by the dynamic bus impedance and the control loop gain. Key performance attributes emerging from the dynamic study are the bus voltage ripples, transient deviations, fault and fuse blow transients, and the control loop stability under harmonic ripple excitation. On the other hand, the static performance under a slow change or after the dynamic response has settled in time is largely influenced by the static impedance of the bus. The voltage regulation long after a load change is an example of static performance. Since the dynamic and static bus impedances are similarly defined, they are jointly covered in the following section.

14.2 Bus Impedance and System Stiffness

A complex electrical network having a number of sources and loads between any two terminals can always be reduced to a simple equivalent source, known as the Thévenin equivalent source, consisting of one source voltage V_s with an internal series impedance Z_s (Figure 14.1). The two source parameters are determined as follows.

With the system operating at no load between points 1 and 2 in Figure 14.1, but with all other parameters at rated values, the voltage between

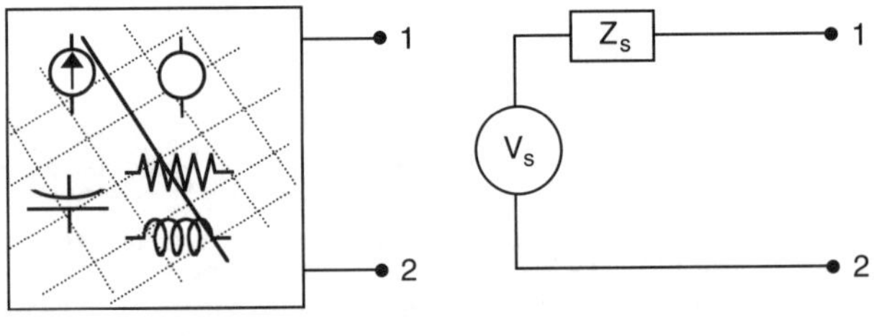

FIGURE 14.1 Complex electrical network reduced to Thévenin equivalent source.

terminals 1 and 2 equals the source voltage V_s (since the internal voltage drop is zero). Therefore,

$$V_s = \text{open circuit voltage of the system at the load terminals} \quad (14.1)$$

With terminals 1 and 2 shorted, the internal voltage is now totally consumed in driving the current through the source impedance only. Therefore,

$$Z_s = \text{open circuit voltage/short circuit current at the load terminals} \quad (14.2)$$

The short circuit current is determined by calculations in the actual circuit. If it is desired to determine by tests, a reduced voltage is applied to limit the current to rated value. The full short circuit current is then calculated by scaling to the full rated voltage. Any nonlinearity, if present, must be accounted for.

The source impedance of most spacecraft bus architectures is highly nonlinear due to the use of multiple bus regulators along with dead-band regions. The dynamic performance of such systems is largely driven by the transient nature of switching from one mode to another, such as solar array shunt control to battery discharge control. Linear modeling may be acceptable for steady-state regulation and small signal load changes within the control range of each controller, but not for modeling mode change transition, which must use transient models for the individual controls.

Two alternate ways of determining Z_s are as follows:

- Set all independent voltage and current sources in the system to zero and determine the impedance between open terminals 1 and 2.
- With the system energized and the load connected between points 1 and 2, measure the voltage drop $\triangle V$ for a slow load change $\triangle I$. Then, Z_s is the ratio $\triangle V / \triangle I$.

The Thévenin equivalent source model derived under the steady state static condition gives the static bus impedance Z_s. The source impedance derived under the dynamic condition — that is, for an alternating or incremental load — is the dynamic bus impedance Z_d. The Z_d varies with frequency and can be either calculated or measured with the test set up shown in Figure 14.2. With the bus in operational mode delivering the rated load, a small high frequency ac current I_h is injected into the bus using an independent current source. The value V_h, the high frequency voltage perturbation in the bus voltage, is measured. The dynamic bus impedance at that frequency is then

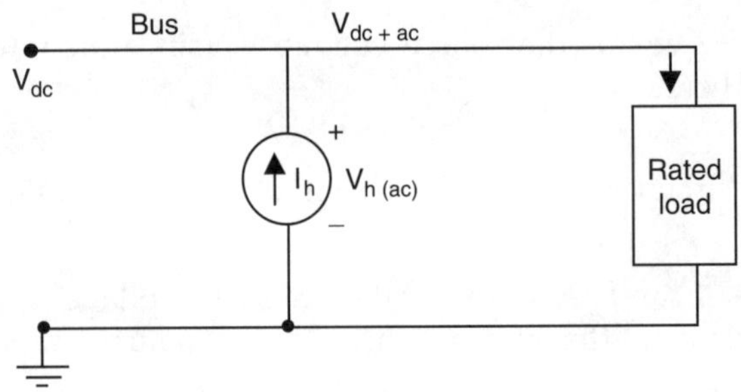

FIGURE 14.2 Dynamic bus impedance measurement by injecting ripple current I_h.

$$Z_d = \frac{V_h}{I_h} \tag{14.3}$$

Since Z_d has a strong influence on the system's dynamic performance, it is kept below specified limits. Figure 14.3 and Figure 14.4 are examples of the main bus and the battery pyro bus source impedances for mid-size communication satellites.

One way of evaluating the relative fuse blow speed in a power system is to evaluate the system stiffness, which is inversely related with the static bus impedance. After the initial inductive rise of dc fault current takes place in five time constants (L/R), the steady fault current magnitude is inversely proportional to the static impedance of the source looking back from the fault location. We now define the following terms leading to the system stiffness:

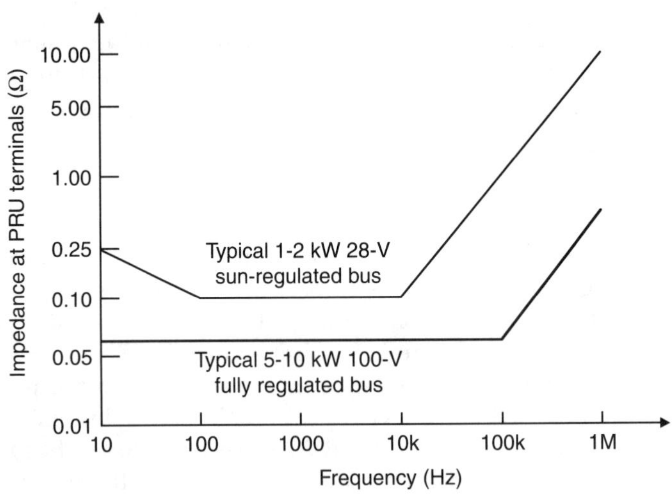

FIGURE 14.3 Main bus source impedance requirement for typical GEO communications satellite.

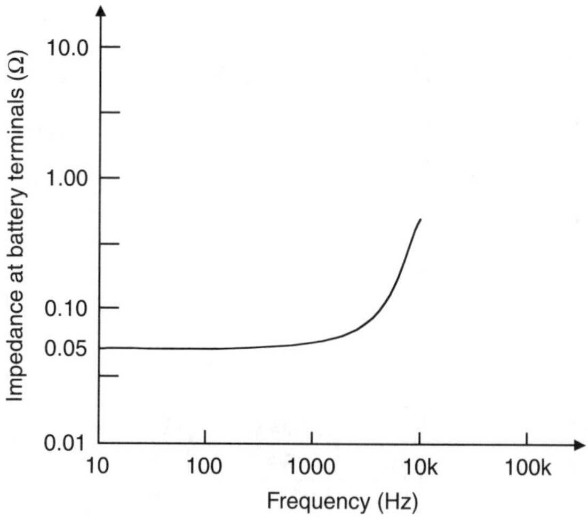

FIGURE 14.4 Pyro (battery) bus source impedance requirement for typical GEO communications satellite.

Rated system impedance $Z_{rated} = V_{rated}/I_{rated}$
Normalized static impedance $Z_n = Z_s/Z_{rated}$
System stiffness = fault current in multiples of the rated current

$$= 1/Z_n \tag{14.4}$$

The stiffer the system, the quicker the fuse blow. This concept is useful in top-level comparison of two power systems of different designs.

14.2 Voltage Regulation and Transients

The steady-state voltage rise on removal of the full rated load current is then $\triangle V = I_{rated}Z_s$, and the static voltage regulation (as a percentage) is defined as

$$\text{voltage regulation} = \frac{\Delta V}{\Delta V_{rated}} \times 100 \tag{14.5}$$

The dynamic voltage regulation is defined as the change in the steady state voltage after the feedback control dynamics have taken effect. It primarily depends on the control loop design, rather than on the dynamic bus impedance.

Under a load step change — partial or full — the voltage oscillates until the transient settles to a new steady-state value as depicted in Figure 14.5. A good design limits and damps the oscillation in a short time to avoid the

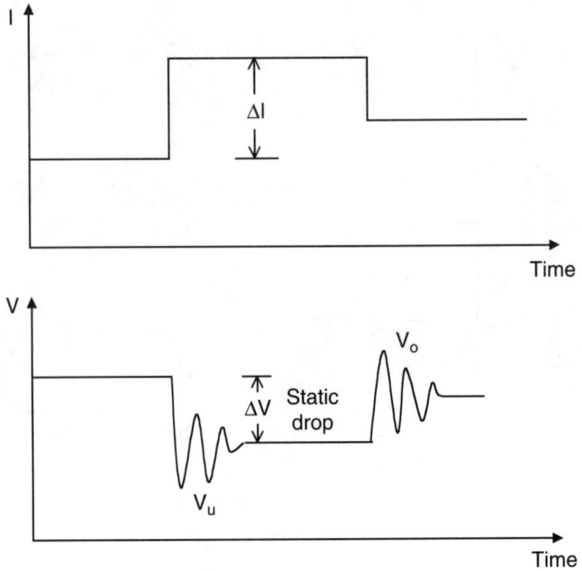

FIGURE 14.5 Transient voltage undershoot and overshoot following load step change.

system becoming unstable. If the load current rises in step, the voltage oscillates before settling down to a lower steady state value. The steady-state change in the bus voltage is then given by $\Delta V = \Delta I \cdot Z_s$ if there were no feedback control. The feedback voltage control loop, however, responds to bring the deviated bus voltage back to the rated value as shown in Figure 14.6. The bus voltage is maintained within specified limits at a specified bus

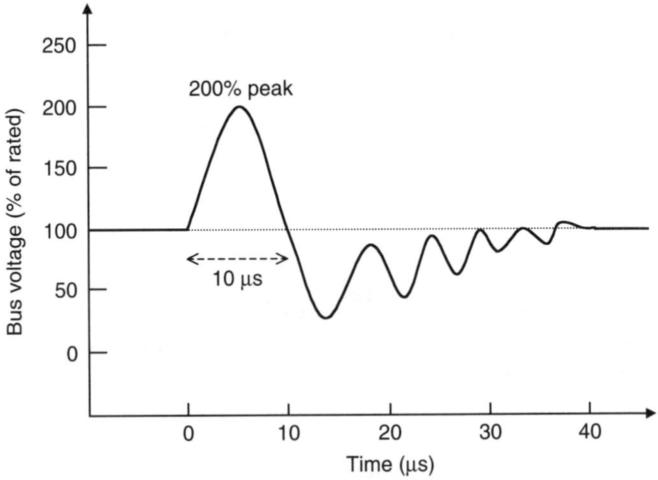

FIGURE 14.6 Main bus transient voltage limit for typical 100-V bus.

sense point, generally the power distribution point inside the power regulator unit. In order not to flutter the system more than necessary, the control system is designed with suitable deadbands. For example, Figure 4.9 in Chapter 4 illustrated a 120-V PV–battery system with two deadbands in the control loop.

The bus voltage can deviate from its nominally rated value due to many reasons. Computers and microelectronic circuits are more susceptible to voltage transients than rugged power equipment such as motors, transformers and heaters. The deviation that can be tolerated depends on its magnitude and the time duration. Small deviations can be tolerated for longer time than large deviations. The tolerance band is generally defined by voltage versus time (V–t) limits. The system voltage must be maintained within the specified V–t envelope shown in Figure 14.7 as an example. The right-hand side of the band comes primarily from the steady-state performance limitations of the load equipments. The middle portion comes from the voltage flicker considerations. And the left-hand side of the band comes from the electronic load susceptibility considerations, which is in microseconds range based on the volt-second capability of the power supply magnetics and the dielectric withstand capability of the insulation.

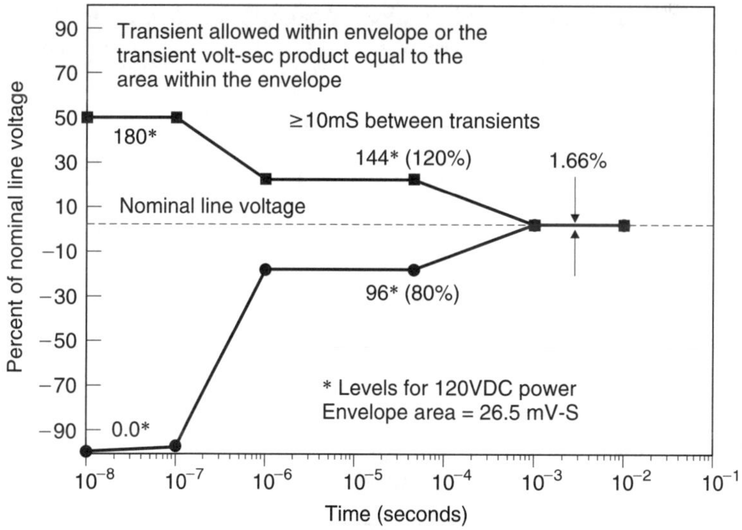

FIGURE 14.7 Bus voltage deviation limits versus duration for typical 3-kW, 120-V LEO satellite. (CE07 limits for EOS-AM.)

14.4 High-Frequency Ripples

The ripple is the term used to describe peak-to-peak voltage deviation from the normal steady state operating voltage under repetitive current pulses. The ripple is sometimes expressed in terms of the rms value also, defined as

$$V_{r(rms)} = \frac{V_{r(p-p)}}{2\sqrt{2}} \tag{14.6}$$

The ripple is caused by switching a current on or off, such as in the PWM operation of solar array shunts, battery power converters, and communication load modulation. Figure 14.8 illustrates the ripple voltage generation under repetitive on-off pulses of current. The PWM switching of shunt current is a major source of the bus voltage ripples. The ripple frequencies are integer multiples of the switching frequency. The ripples are periodic but not sine wave, and are superimposed on the fundamental voltage, a.c. or d.c. The ripple voltage induced on the bus due to ripple current of certain frequency is given by

$$V_{ripple} = I_{ripple} Z_d \tag{14.7}$$

where Z_d is the dynamic bus impedance at the frequency of interest. The impedance is generally in hundreds of milliohms and the ripple voltage limit varies over a wide range in hundreds of millivolts. Figure 14.9 is a typical voltage ripple limit imposed on a main bus of a mid-size communications satellite.

The ripple is minimized by bus filter capacitors connected to the bus, or preferably at the component terminals causing the ripples. The ripple

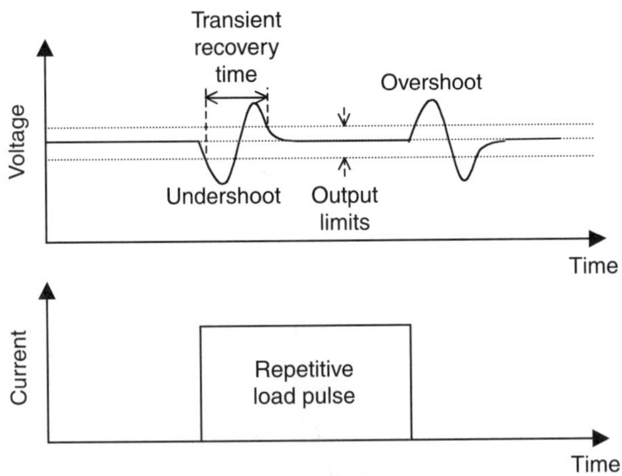

FIGURE 14.8 Bus voltage ripple under repetitive load pulses.

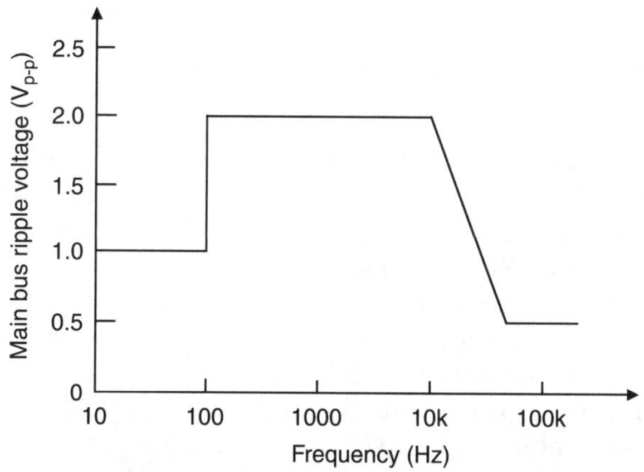

FIGURE 14.9 Bus voltage ripple limits for typical 5- to 10-kW, 100-V GEO communications satellite.

current is then supplied or absorbed by the capacitor, rather than by the bus, thus improving the quality of power. For a 28-V sun-regulated bus for communications satellite with 4000-W payload, a typical bus filter may be about a 4000-μF capacitor bank with an ESR less than 60 mΩ at 25°C. The filter design is made two-fault tolerant by using a bank of many capacitors connected in parallel with internal fuse protection for each.

Equal output source impedance of the PRU in the sun and the eclipse operations can benefit the bus design for meeting the ripple limit. Otherwise, the higher of the two must meet the limit, while not receiving any benefit from the lower ripple in the other mode of operation. Controlling the input ripple to the PRU is also important to avoid excessive common mode noise at the battery monitoring circuit.

Multiple ripple frequencies are often present in the system. The system performance may suffer if ripples of numerous frequencies having about the same p–p value are present, although each individually meets the p–p ripple specification. For this reason, it is a general practice to specify not only the p–p limit at any one ripple frequency, but also the total ripple distortion factor (TRDF) due to all ripples combined. The TRDF is defined in terms of a Fourier series, where a periodic voltage of any shape is decomposed in a series of sinusoidal components, i.e.,

$$V(t) = V_1 + \sum_{n=2}^{\infty} V_n \sin(n\omega t + \alpha_n) \tag{14.8}$$

The first component on the right-hand side of the above equation is called the fundamental component, whereas all other higher frequency terms ($n = 2,3 \ldots \infty$) are called the harmonics. The TRDF is then defined as

$$\mathrm{TRDF} = \frac{\sqrt{V_2^2 + V_3^2 + \ldots\ V_n^2}}{V_1} \tag{14.9}$$

The TRDF is useful in comparing the quality of power with respect to the ripples between two or more power systems, or at various locations of the same power system. In a pure ripple-free voltage source, TRDF = 0. The greater the value of TRDF, the more distorted the bus voltage. The major operational concern arising from the ripples is the EMI. Additionally, the interaction of ripple voltage and current do not contribute to the average power. The ripple currents merely result in more I^2R loss for the same useful power delivered. Poor quality of power therefore adversely affects the power system efficiency also.

The harmonic distortion on bus voltage caused by the nth harmonic current I_n drawn by any nonlinear load is given by $V_n = I_n Z_n$. This distortion in the bus voltage that causes the harmonic current to flow in other loads is called the victim load. A harmonic-rich bus voltage supplies distorted current to all loads, linear or nonlinear. Therefore, harmonics are generally filtered out before feeding power to the bus. A bus voltage having the TRDF less than 3 to 5% is generally considered acceptable.

14.5 Ripple Measurement

Two simple test methods to measure ripples are shown in Figure 14.10. They consist of connecting the oscilloscope probe to the output terminal of the equipment under test either (a) directly, or (b) with a capacitor. The set-

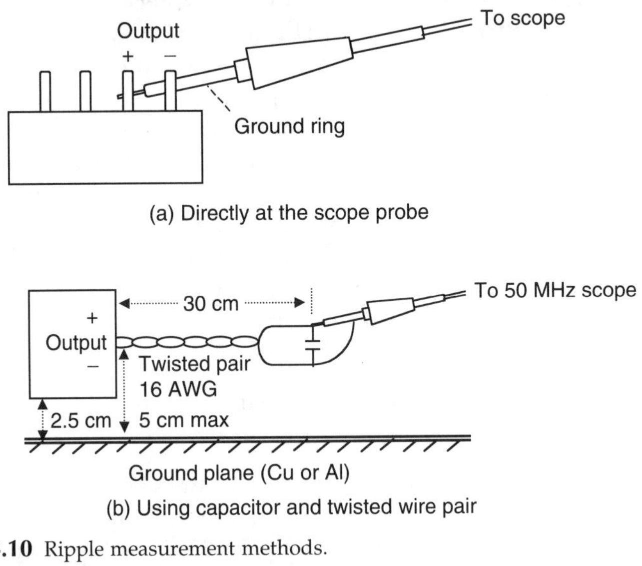

FIGURE 14.10 Ripple measurement methods.

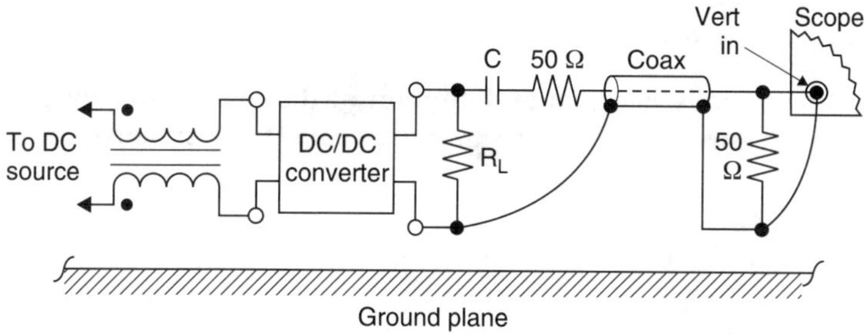

FIGURE 14.11 Ripple measurements using terminated coaxial cable.

up (b) is widely specified method for power converters. It typically uses AWG 16 twisted pair of output wires terminated into a 50 μF capacitor and an oscilloscope of bandwidth greater than 50 MHz.

Figure 14.11 shows another way to measure output ripple voltage using a terminated 50 Ω coaxial cable going directly to the oscilloscope. A ground plane is used between the power converter and the scope. Extraneous noise pick-up is minimized in this low impedance shielded measurements. In addition, an input balun is used with the d.c./d.c. converter to minimize any common mode noise pickup. Any noise pickup in this set up may be checked by connecting both input and ground leads to the same converter terminal and looking for noise on the scope. With the termination resistance matching with the coaxial cable characteristic impedance, the measured ripple voltage is equal to one half the actual value of the p–p ripple.

It is sometimes required to measure the reflected ripple current from a d.c./d.c. converter back into the source. Such measurements are done with a circuit shown in Figure 14.12. It uses a wide-band current probe and oscilloscope, both of bandwidth at least 20 MHz to give accurate wide band measurements. A non-inductive current sampling resistor to make a voltage measurement is another method, but it takes considerably more care to get accurate results.

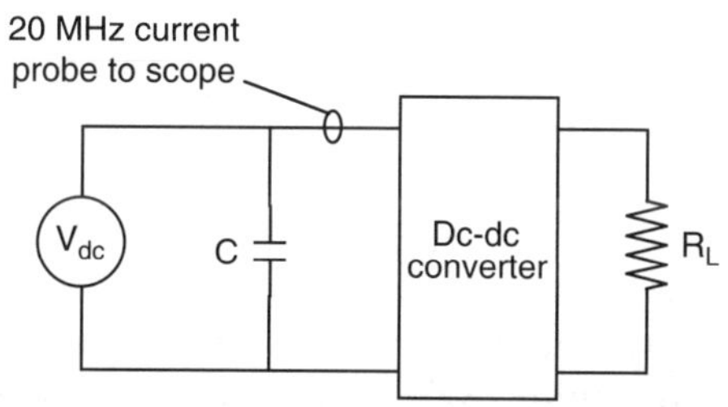

FIGURE 14.12 Reflected ripple current measurements.

14.6 Quality of Power

There is no generally acceptable definition of the quality of power as yet. However, the power quality is regarded to have three major performance criteria:

- The total ripple distortion factor generated by the power electronics
- The transient voltage sags caused by system disturbances and faults
- Periodic voltage flickers

The power quality specifications are generally written in two parts. Part I address the source and the distribution system. It specifies the following attributes at the output of PRU that various loads can expect:

- Steady state voltage limits
- Transient voltage limits
- Source impedance
- Ripple voltage at the bus

Part II, on the other hand, specifies the requirements the loads must comply with in order to be connected to the bus. Specifically, it limits the:

- Steady state voltage deviation withstand capability
- Transient voltage deviation withstand capability
- Load impedance
- Isolation voltage and resistance
- Turn-on in-rush current
- Ripple emission

The systems engineer ascertains that the source power quality specified in Part I is compatible with the load power quality specified in Part II. For example, the voltage transient created by the source should be less than the voltage transient the load is required to withstand with some margin.

14.7 Minor Fuse Blow Cross-Talk

A voltage spike occurs in the system when a fuse blows. If many load equipment are connected to a common fuse board, fuse blowing in one equipment may generate a voltage spike in another. This is often known as the cross talk between equipments. Various equipments should be individually fused to minimize cross talk between them. Full-scale circuit simulation can predict the cross talk voltage spike. When the loads are not well defined in an early stage of the program, the energy approach presented below may give approximate, but quick and directly usable, upper limit on the cross-talk spike. It is based on a simple circuit model of Figure 14.13 using only the cable parameters shown in (b).

Various wires and other resistances in series determine the steady state fault current just before the fuse blow. With that current, the wire inductance would store energy equal to $\frac{1}{2}LI^2$. When the current is interrupted on the instant of the fuse blow, this energy is dumped on C, the capacitance of the wire or the EMI filter, or both. Thus, the cable inductive energy gets transferred to the capacitance, and the capacitor voltage rises. Ignoring the energy loss in the transfer and the initial energy in the capacitor at rated voltage, the inductive and capacitive energy must be equal, i.e.,

$$\tfrac{1}{2}LI^2 = \tfrac{1}{2}CV^2 \tag{14.10}$$

Thus, the bus voltage on the instant of current interruption would rise to the value given by, in the first approximation,

$$V = I\sqrt{\frac{L}{C}} \tag{14.11}$$

If the cable capacitance is insufficient to absorb the inductive energy, an external capacitance can be added so that the total capacitance keeps the voltage spike under the specified limit.

14.7.1 Example

Consider a 28-V bus cable 5 m long, made of four parallel wires, each with inductance of the 0.375 μH/m and capacitance of 0.077 nF/m. If a fault current in each wire before the fuse blow is 70 A, the energy equation gives $\frac{1}{2} \times (0.375 \times 5\text{ m})\, 10^{-6} \times 70^2\ 4 = 0.018\text{ J} = \frac{1}{2} \times C \times V^2$. To keep the voltage

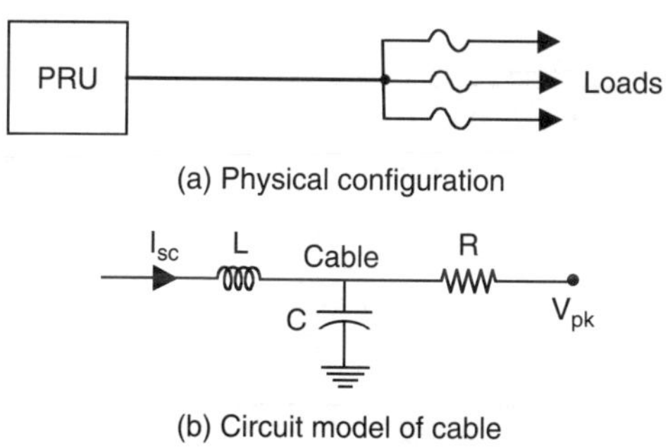

FIGURE 14.13 Cross-talk between equipment following a small fuse blow.

rising above say 80 V, we need a capacitance $C = 2 \times 0.018/(80^2) = 5.6\,\mu F$. Since the estimated wire capacitance ($0.77 \times 5 \times 4\,nF$) is much less than what we need (5.6 μF), we must ascertain that the load converters' input EMI filters have at least 5.6 μF at the front end. Alternatively, one can place 5.6 μF in the fuse board to limit the fuse blow voltage spike from rising above 80 V.

The hermetically sealed fuse blows suddenly because no plasma is generated in the sealed vacuum. Therefore, it can cause severe voltage spike following a fuse blow, and may need the above estimated capacitance at the fuse location. The regular nonsealed fuse may require less capacitance because it blows slowly due to the lingering plasma. An alternate method of spike suppression often used is a freewheeling diode in parallel with the fuse. This technique, however, cannot be used to suppress a transient generated by a source inductance common to the affected power distribution. The most effective means of such transient control is to minimize the source inductance and use capacitive damping.

Some load faults can result in large but short duration transient currents in hundreds of amperes. Such fault currents would result in bus voltage transient that are much greater than the example above. They may momentarily collapse the bus as discussed in the next section.

14.8 Major Fuse Blow Transient

The bus voltage rapidly drops under fault in large equipment connected to the bus. For example, a fault in one of the arcjets with a 50 A fuse is simulated as $Load_1$ in Figure 14.14. The solar array, battery and the flitter capacitors, all feed current to the fault as shown by arrows in the figure. After the 50 A fuse blows, the available solar array power starts recharging the battery and the capacitors. The voltage takes a sharp v-turn and starts rising until it reaches the nominal bus voltage. The power system control loop then stabilizes the bus voltage within the tolerance band. The fall and

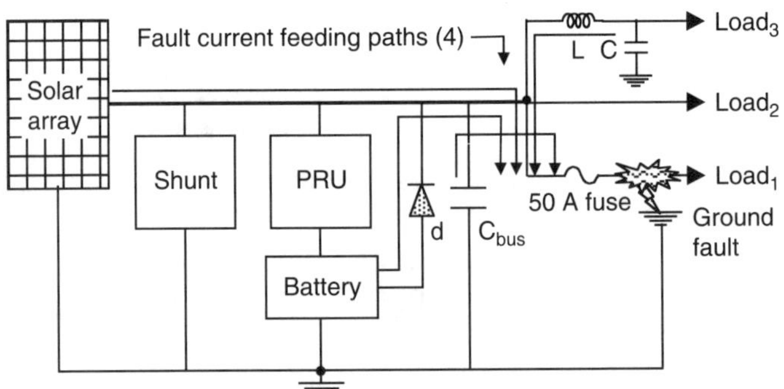

FIGURE 14.14 A circuit model for a major fuse blow transient analysis.

rise of the bus voltage under such major fuse blow is depicted in Figure 14.15. Thus, the entire power system is involved in clearing a large fuse. The larger the rating of the affected fuse, the deeper the voltage sag. The power system specifications generally limit such voltage sag to some minimum value set by a critical equipment requirement. For example, the central computer on a 70-V system cannot perhaps tolerate the voltage sag below 35 V for more than 5 ms. The power system design engineer must therefore run the fuse blow simulation analysis and show that such a requirement is met.

Most faults are between the line and chassis involving the ground, rather than between the line and return. The fuse-blow transient is sensitive to the ground return resistance, which may vary from fault to fault. For this reason, the fuse blow circuit simulation is carried out with various possible fault resistances. The results are parametrically plotted as shown in Figure 14.16. This assures that under the worst-case fault resistance the bus voltage meets the requirement. In addition to the random fault resistance, there is always some minimum ground impedance depending on how the system is grounded to the structure. For example, the structure with graphite composite face-sheet has higher ground contact resistance. The resulting spreads of the current give higher inductance also. For this reason, the ground contact impedance of the graphite face-sheet used in some modern communication satellites is about three orders of magnitude higher than that with aluminum face-sheet. This significantly changes the transient characteristic for faults involving ground. High ground impedance reduces the final fault current only slightly, but greatly increases the $L\,di/dt$ when the fuse blows, thus raising the EMI/EMC concerns.

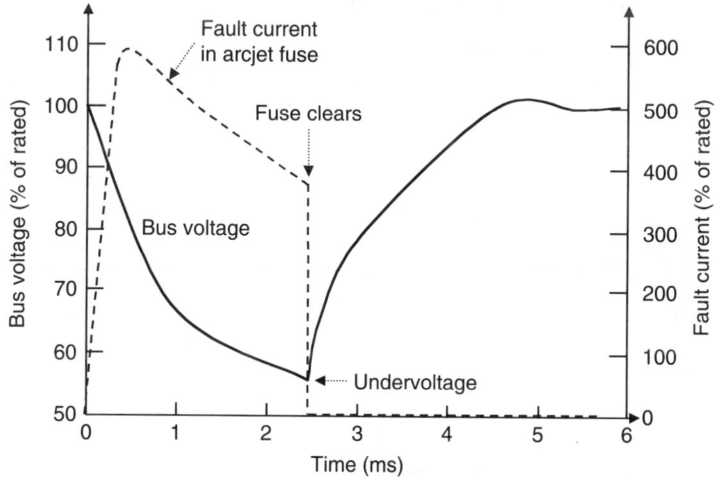

FIGURE 14.15 Bus voltage decay and recovery transient during a major fuse blow.

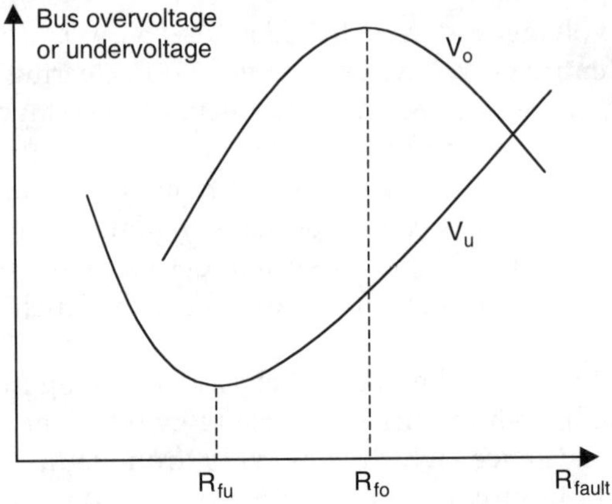

FIGURE 14.16 Bus under-voltage and over-voltage versus fault resistance during major fuse blow transient.

A typical main power bus fault response requirement is shown in Figure 14.17. In the past, fuse transient has not been a major problem, except some telemetry may give an erroneous reading on that instance, but recovers soon. Such a glitch is generally tolerable in commercial satellites, but not in military and NASA satellites, because, in an extreme case, it may upset the CMOS computer gate.

The EPS is generally designed to accommodate payload shedding in case the bus voltage drops to 90–95%, and is capable of powering essential spacecraft functions up to 75–80% of the nominal bus voltage.

14.9 Stability and Bus Impedance

The steady state operating point is stable if it gets restored after an arbitrarily small momentary disturbance in the source or the load. For example, at operating point A_1 under constant power P_1 in Figure 14.18, suppose a disturbance causes a drop of $\triangle V$ in voltage. At the new voltage, the source current is less than the load current. Consequently, the voltage decreases again and the operating point moves further away from A_1, and again further away, and again in a runaway situation. Similarly, a small increase in voltage due to a disturbance will make the system move away from point A_1 in the other direction. Thus point A_1 is an unstable operating point, where the system works like a positive feedback system.

On the other hand, at point A_2, a small change in voltage in any direction, due to any reason, produces counterchanges to bring the system back to its original operating point A_2. Thus, point A_2 is a stable operating point, where the system works like a negative feedback system. Analytically, we can derive the following condition for static stability. An

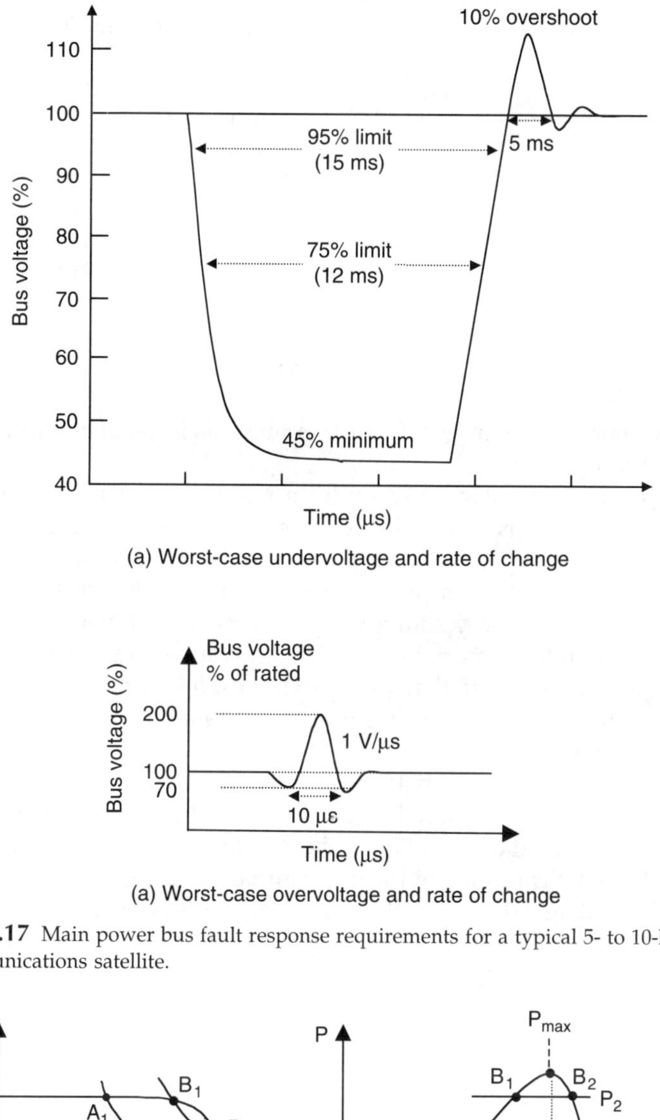

FIGURE 14.17 Main power bus fault response requirements for a typical 5- to 10-kW, 100-V GEO communications satellite.

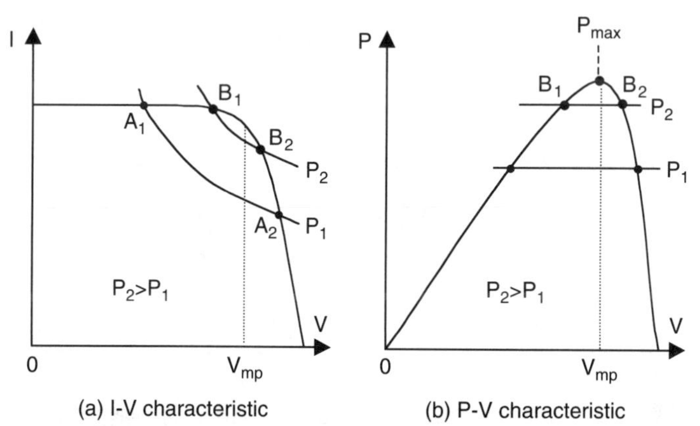

FIGURE 14.18 The stability of a solar array operating point under a constant power load.

operating point is stable if and only if the following conditions are met at that point

$$\left[\frac{dV}{di}\right]_{source} > \left[\frac{dV}{di}\right]_{load} \tag{14.12}$$

If we think in terms of the absolute values of the dynamic impedances and define Z_s as source output impedance and Z_L as the load input impedance, then the system is stable only if

$$Z_s < Z_L \tag{14.13}$$

at all frequencies. When $Z_s > Z_L$, further analysis is needed to determine the system stability.

The operation of solar array under a resistive load is shown in Figure 14.19. The load line is a straight line from zero with the slope equal to the resistance value. For such a load, the above condition is always met, hence the solar array powering a purely resistive load is always stable.

A nonlinear solar array source coupled with a nonlinear load may result in multiple equilibrium points under a given operating condition. The actual equilibrium point that the system resides at is determined by the stability nature of the equilibrium points and the past history of the operating point.

During eclipse when the battery directly supplies a constant power load, the intersection of the source line and the load line always gives the stable operating point as shown in Figure 14.20. The unstable point in this case is outside the practical range of the constant power load, which its tripped off before the voltage drops to the point where the battery operation may become unstable.

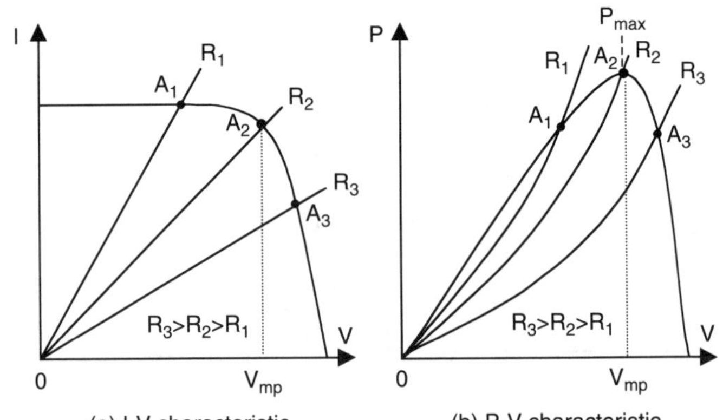

FIGURE 14.19 The stability of a solar array operating point under a resistive load.

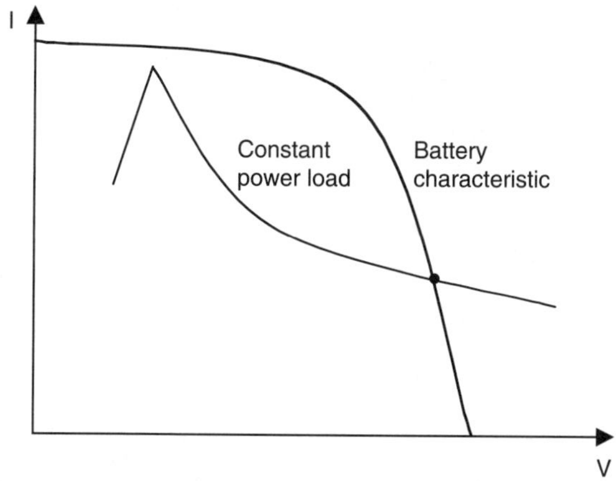

FIGURE 14.20 The stability of a battery operating point under a constant power load.

Figure 14.21 shows an example of the load and source impedances overlapping and the unstable operating region. The source and load impedance specifications are primarily set to ascertain the system stability under all conditions. This task can be complex even for a small spacecraft. It can be extremely complex for a large space platform with a variety of loads that can be connected, not all of them known in advance, such as on the ISS.[1] In those cases, the load impedance specification for every load can be established from the fundamental systematic considerations presented below.

Any complex system with a number of sources and loads can be reduced to the equivalent network with Thévenin parameters V_s and Z_s. Any additional load with impedance Z_L must be greater than this value of Z_s for stability. A critical factor in the design of a large platform such as the ISS is the stability of the system under all expected conditions of loading and transient perturbations. The ISS power systems engineers have developed a set of impedance criteria for both the source converters as well as the load converters to assure stability under significant variations in the load connectivity over the life of the station. When a new load is added on the large platform, the old source and load impedances combine in parallel to form the new source impedance (Figure 14.21). Thus, the new Z_s will always be less than the old Z_s. The new system would therefore be stable as long as the new load impedance is greater than the greatest load impedance in the old system. This logic may be used to set the dynamic input impedance of any load that can be connected by any user on the ISS type platform.

If the system is stable, then any additional load in parallel will keep the system stable.[2]

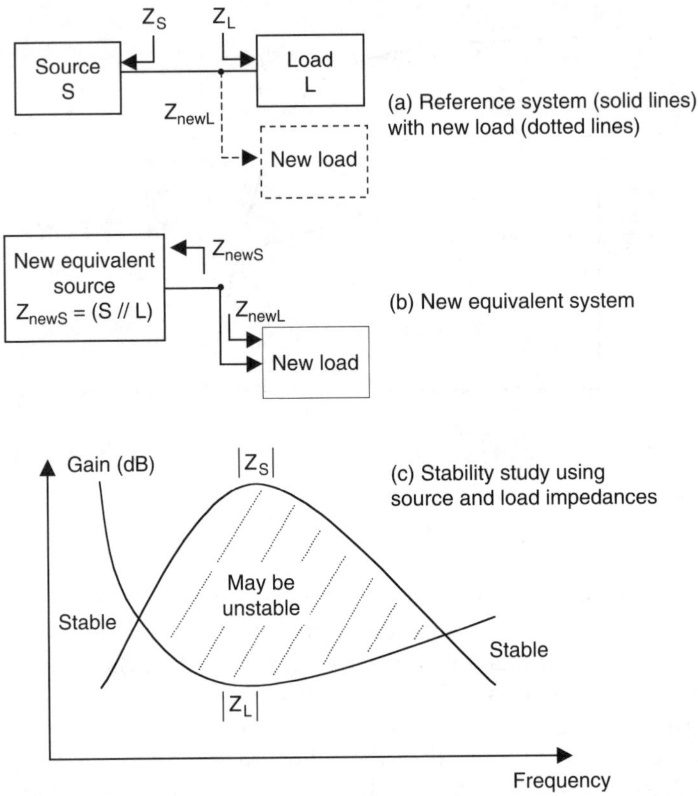

FIGURE 14.21 Thévenin equivalent source and dynamic load impedance for a stability study in a large power system.

14.10 The dynamic Stability of a Control System

A dynamic stability study determines whether the system will be stable under sustained small harmonic perturbation coming internally from the power system or from an external source via conducted EMI. Verifying the stability involves modeling the total system with all control loops. A detailed theory of the control system is outside the scope of this book, but a short review follows.

The negative feedback control system is widely used for controlling many system, subsystems and components, often with multiple nested loops. Figure 14.22 represents a feedback control system in terms of the transfer functions (TFs), where G is the main equipment TF and H is the TF of the feedback control loop. It can be shown that the TF of the entire system is

$$\text{TF} = \frac{\text{Laplace transform of the output } y(t)}{\text{Laplace transform of the input } x(t)} = \frac{G(s)}{1 + G(s)H(s)} \tag{14.14}$$

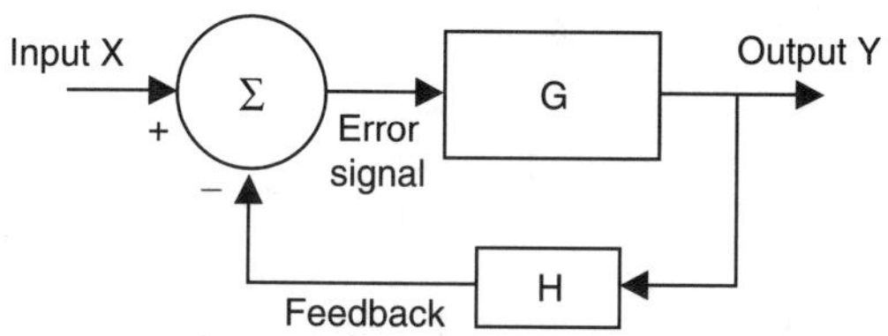

FIGURE 14.22 A classical feedback control system model.

This equation indicates that $1 + G(s)H(s) = 0$ results in an infinitely large output for an arbitrarily small input. The solutions (roots) of Equation 14.14 are called the system poles, which are in general complex, say $s = \alpha + j\beta$. We say that the output blows up, i.e., the system is unstable, if

$$G(s)H(s) = -1 \tag{14.15}$$

In other words, the open loop gain of 0 dB and the phase angle of 180° give an unstable system. The stability is often judged two ways. In one method, the gain in dB and phase angle in degrees are plotted with respect to frequency in what are known as Bode plots shown in Figure 14.23. The system is generally considered stable if the gain is below zero with about a 30-dB margin, and the phase angle is away from 180° with about a 30° margin. These margins vary with the spacecraft requirements and heritage data on similar bus. In one 3000-W, 120-V LEO satellite, the PRU and mode controller amplifier were required to provide stability phase margin of at least 45° and gain margin of at least 10 dB.

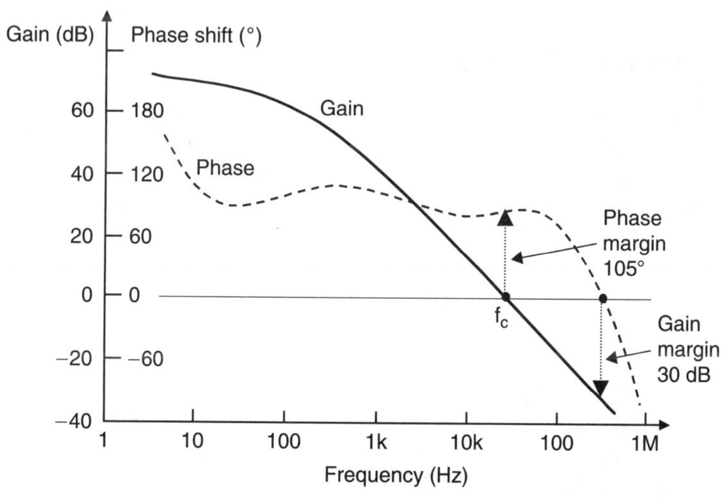

FIGURE 14.23 Dynamic stability criteria from open loop Bode plots.

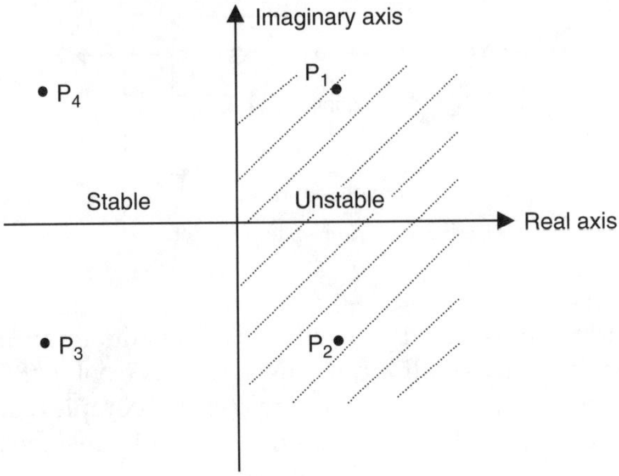

FIGURE 14.24 The dynamic stability criteria from poles plot in complex plane.

In another method for determining stability, the poles are plotted in the complex plane. The system is stable only if all poles have negative real part which results in damped oscillations. Or, in other words, all poles must be in the left-hand side of the complex plane (Figure 14.24). In case the system is found unstable, a compensating lead–lag network is introduced in series with the main equipment as shown in Figure 14.25. The combined forward TF is then C · G, which is used in evaluating the stability of the compensated feedback system. Since the circuit parameters of the compensating network may vary with temperature and age, it is customary to check the stability under the worst compensation and load conditions as illustrated in Table 14.1. The table in this example shows that the system is stable with comfortable margins under the worst conditions.

14.11 Dynamic Simulation Model

Considering the size and complexity of modern spacecraft power systems, the digital computer is the only viable tool for system modeling and simulation. Two general classes of simulation programs appear in the literature. They are: (1) generalized circuit and system simulation models,

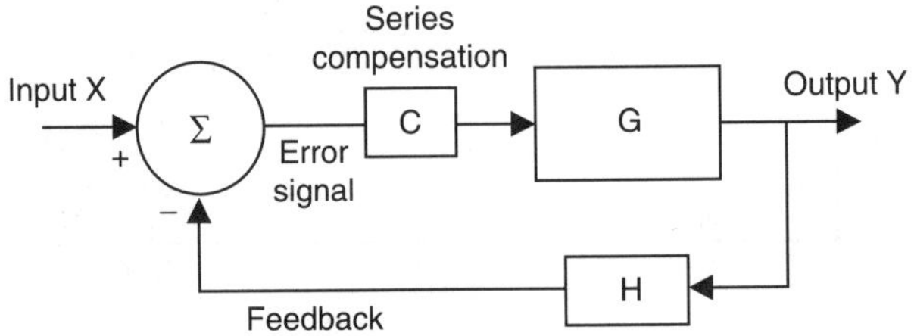

FIGURE 14.25 Series compensation (lead–lag network) added for stability.

Table 14.1 An example of a stability study under the worst compensation and load combination for a 100-V bus

Payload configuration	With maximum compensation gain			With minimum compensation gain		
	System band-width (Hz)	Phase margin°	Gain margin (dB)	System band-width (Hz)	Phase margin°	Gain margin (dB)
All loads off	1700	105	40	1100	70	46
1 TWTA on	1400	104	40	1000	70	44
2 TWTAs on	1400	103	40	900	69	48
3 TWTAs on	1200	102	40	900	70	46

or (2) dedicated models of specific system. The generalized computational models as a class are built using commercial programs with tradenames such as PSPICE, SABER, SYSCAP, ICAP, and a number of others. All have similarities in that they offer a large library of component templates capable of analog, analog/digital mixed signal and mixed technology simulation (electrical, mechanical, thermal, optical, and even chemical). The templates are assigned parametric values to simulate the actual component used in the system. The simulation can be at any level: component, behavioral, functional, or control level. The scope of total simulation model generally consists of developing the following models:

Component level models of

- Solar array
- Battery

Behavioral or component level models of

- Solar array shunt control
- Battery charge and discharge converters
- Load types: constant power, constant current or resistive
- Power distribution and protection: cables, relays, fuses, RPCs
- Control loops: usually nested in multiple layers

14.11.1 Solar Array Model

Two approaches can be taken for modeling the solar array: micro and macro. In the micro, each cell is modeled by its *I–V* characteristic equation presented in Chapter 8. The advantage of this approach is that cell-to-cell variations internal to the array can be modeled. For example, a local cell failure or shadowing can be inserted into the model. However, the number of cells needed for a full simulation in a large satellite can be prohibitive and inefficient. The macro model, on the other hand, can represent the entire array as one component with the total *I–V* characteristic of all cells combined in series–parallel connections.

The shunt control loop may show some sensitivity to the capacitance of the solar string circuits. However, in practical shunt control loop designs, the gain and phase margins in the stability analysis are found to be insensitive to the cell capacitance varying over a wide range around the nominal value. The solar cell capacitance can be calculated from the solar array impedance measurements under sweeping frequency with d.c. voltage and current bias conditions. The capacitance of 2 × 4 cm silicon solar cell is typically 0.25 μF when in dark. Under illumination, it is about 0.50 μF near short circuit current and 0.75 μF near the peak power point. A typical dynamic a.c. model for one 5-A, 70-V solar circuit is shown in Figure 14.26.

Lindmyer[3] gives the a.c. impedance (in ohms) of the solar circuit as

$$Z_{ac} = \frac{R_d}{\sqrt{(1 + \omega\tau}} \tag{14.16}$$

where $R_d = \mathrm{d}V/\mathrm{d}I$, the slope of the solar cell *I–V* curve, i.e., the dynamic resistance, and τ = diffusion time constant of electrons in the p-region of the cell. The Z_{ac} includes the a.c. resistance of the equivalent diode as well as the R_s and R_{sh} components shown in Figure 8.5. It varies with the d.c. operating point, and also the cell temperature via the τ term. The capacitance, C_s (in farads), can be analytically obtained from NASA/JPL data[4] as follows:

$$C_s = A\sqrt{\frac{q\,\varepsilon\,N_a}{2(V_b - V_a)}} \tag{14.17}$$

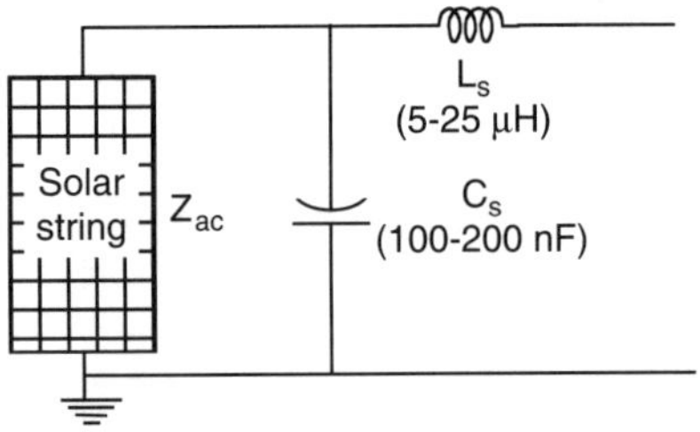

L_S = maximum harness inductance from outboard section to boom connection
C_S = effective shunt capacitance with array structure connected to –terminal

FIGURE 14.26 The solar array circuit dynamic model for system simulation.

where

A = cell area,
q = electron charge,
ε = permittivity of silicon,
N_a = acceptor density in the p-region,
V_b = barrier voltage,
V_a = applied voltage (positive in forward bias).

A typical 2 Ω·m silicon cell with $\varepsilon = 1$ pF/cm, $N_a = 1{\cdot}10^{16}\,\text{cm}^{-3}$, and $V_b = 0.8\,\text{V}$ at room temperature gives a value for C_s, in μF/cm², of

$$C_s = \frac{0.0288}{\sqrt{0.8 - V_a}} \tag{14.18}$$

The maximum and minimum values of C_s occur at open circuit and short circuit, respectively. Not much has been published on the variation in C_s with temperature, probably because the changes in d.c. operating voltage far outweigh the temperature effects. However, C_s increases with increasing temperature at about 0.1% per °C.

14.11.2 Other Component Models

The battery model can be simply a constant voltage source with an internal resistance shown in Figure 9.36, or a complex nonlinear voltage source shown in Figure 9.37. The choice depends on the degree of precision desired from the model. For a 100 Ah NiH_2 cell, the battery internal inductance is about 0.35 μH/cell.

The solar shunt regulators, and the battery charge and discharge regulators can be modeled at their own component levels in the micro sense, or at their behavioral levels in the macro sense. Both approaches are valid, although the latter is perhaps simple and yet sufficiently accurate for a wide variety of analyses.

The communications loads (TWTA and SSPA) are constant power load. Many bus loads such as arcjets and reaction wheels are also constant power loads. They are cut off when the voltage drops below certain values. The time for which they stay on before tripping off is a function of the undervoltage as shown in Table 14.2. During on-time, the constant power load is modeled as a variable resistance of value $R = V^2/P$, where V is the actual bus voltage at any given time. If the voltage remains below the left-hand column value for time greater than that in the right-hand column, the load is cut off. On the other hand, constant resistance loads like heaters remain connected all the time. Before trip-off, the constant power loads

Table 14.2 The trip-off voltages and the duration of under-voltage for typical constant power load

Bus voltage transient minimum level (% of nominal bus voltage)	Time duration before trip-off (μs)
85	10.0
71	2.6
58	1.6
45	1.1
31	0.9
18	0.7
5	0.6
0	0.5

reflect as negative resistance on the converter, while all other loads reflect as positive resistance.

14.12 Simulation Runs

The labor involved in building the dynamic model is rather extensive. Once the model has been built, independent calibration with known solutions is the next major task. It is used for the analysis only after verification has been carried out satisfactorily and reviewed by a team of independent engineers.

The operating point of the system is determined by the source (solar array or battery), the power conditioning equipment and the load characteristics. Since the solar array output power and the load demand vary over a wide range, several modes of operations exist, and all must be analyzed. Some uses of the dynamic simulation are:

- Stability studies
- Impedance plots versus frequency
- EMI–EMC compliance verifications
- Trade studies and parametric studies
- Failure effects and worst-case analyses
- Fuse blow under-voltage and over-voltage
- Troubleshooting

However, the dynamic simulations are performed primarily to determine that all four primary control loops are stable and have adequate gain and phase margins. In shunt mode, the regulated bus voltage control loop (shunt loop) and the battery control loop (BCR loop in charge mode) are two loops that are independently active. Also in shunt mode, the battery charge current is a constant current load, which is a part of the overall solar array load. In discharge mode, the bus voltage regulation (BVR) is performed by the BVR loop. For these loops, linear circuits can be modeled using PSPICE or SABER.

There are two primary areas of interest. The small signal analysis is to obtain phase/gain characteristics of each control loop under various conditions and to obtain the output impedance characteristic. The large signal analysis is to assure that each loop meets its specified performance characteristic for the specified disturbances, i.e., the loop intra-mode large signal response. Often the inter-mode computer simulations may be necessary to combine two or more control models to simulate the inter-mode response, i.e., the transfer of control from one loop to another.

Spacecraft level tests can detect inter-modal transient problems, which may require re-compensation of the BCR and BVR control loops. Examples of changes that can be implemented to remedy such problems are:

- Addition of output filter for each BVR
- Mode controller error amplifier to reduce inter-modal transients
- BVR switching circuit that affects the average switch model
- BVR error amplifier

References

1. Wilde, R.K., Aintablian, H.O., and Gholdston, E.W., International Space Station U.S. major elements laboratory power quality test, in *Proceedings of the 34th Intersociety Energy Conversion Engineering Conference*, SAE, 1999, Paper No. 01-2434.
2. Fassburgh, H.D., Gholdston, E., and Mong, A, Development and implementation of stability requirement for the International Space Station electrical power system, in *Proceedings of the 36th Intersociety Energy Conversion Engineering Conference*, ASME, 2001, Paper No. AT-48, pp. 281–288.
3. Lindmayer and Wrigley, *Fundamentals of Semiconductor Devices*, D. Van Nostrand, 1965, pp. 47–48.
4. NASA/JPL, *Solar Cell Radiation Handbook*, 3rd edition, 1982, Report 82-69, pp. 1–19.

Chapter 15

Electromagnetic Interference and Compatibility

15.1 Introduction

All spacecraft systems are required to be compatible with the electromagnetic interference expected from internal and external sources. For decades, the electromagnetic interference (EMI) and electromagnetic compability (EMC) requirements have come from MIL-STD-461. It specified the maximum emission limit of the potential culprit, and the minimum susceptibility level of the potential victim equipment. The companion MIL-STD-462 defined the test methods for verifying that MIL-STD-461 requirements are met, and MIL-STD-463 defined the applicable terms and units. The first two standards are now merged into one, MIL-STD-461. For complying with the EMI/EMC requirements, the contractor of commercial and defense spacecraft is required to develop the following three documents and submit to the customer as deliverables:

1. EMC control plan containing the analysis of expected EMI in the spacecraft versus the imposed requirement, and the steps to be taken for making the two compatible.
2. EMI test plan with detailed test setups, cables to be used, the grounding details, frequency range, point of monitoring, and the pass–fail criteria.
3. EMI test report, summarizing the findings of the tests, and steps to be taken to rectify the failures, if any.

The EMI requirements broadly fall in two general groups, the conducted EMI and the radiated EMI. In addition, some spacecraft-specific requirements may also be imposed. The EMI can enter the equipment either by conduction via wires, or by radiation in space. For verifying that the spacecraft would meet such requirements, the first task is to determine the conducted EMI and the radiated EMI from potential sources, and the degree of coupling to the victim equipments.

15.2 Sources of EMI

In the commercial world, radiated EMI comes from radio and TV towers, mobile radio vans, lightening, and nearby power lines. The conducted EMI comes mainly from power cables. In space systems, the main sources of EMI are:

- Switching large currents or voltages at high frequency, causing large dI/dt and dV/dt
- Electrostatic discharge
- Nuclear detonation

15.2.1 Conducted EMI

Active components, such as switches and diodes, are the main sources of conducted EMI in the switch mode power converter. As was seen in Section 14.4, the switching frequency voltage ripple is primarily influenced by the dynamic impedance of the equipment. The high-frequency EMI, however, arises due to the parasitic ringing in voltage following a current switching. Therefore, refined component models are needed to predict accurate ringing frequency, amplitude overshoot, and rising or falling transients. Such models must be verified by tests. The commercially available circuit modeling tools, such as PSPICE and SABER, may not be accurate enough for the EMI prediction. Specifically, accurate circuit models for the following high frequency components are needed in the applicable frequency range.

- Active components (switches, diodes, etc.)
- Passive components (transformers, capacitor, inductor, resistor, etc.)
- Printed circuits board patterns used in the spacecraft

Accurate models of these components cannot only predict the high-frequency EMI level, but also can identify the cause. An optimized EMI solution can then be achieved before fabricating the hardware.

15.2.2 Radiated EMI

The power system harness is a major source of low frequency electromagnetic radiation and coupling. Around a short wire (< 10% of the signal wavelength), generally one type of field dominates, either the magnetic field (B-field) around high-current lines or the electric field (E-field) around high-voltage lines. The B-field dominates around a typical power cable. It is derived from Ampere's law. In the commonly encountered geometries shown in Figure 15.1, it is given by the following expressions:

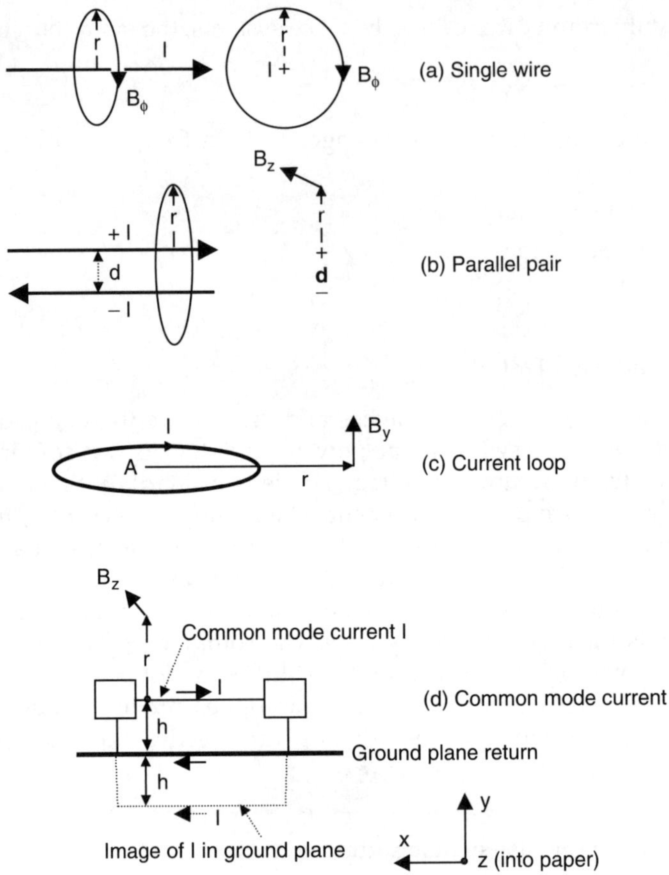

FIGURE 15.1 Magnetic field due to current carrying conductors in various configurations.

For a single wire,

$$B_\phi = \frac{\mu_o I}{2\pi r} \tag{15.1}$$

For a parallel pair,

$$B_{z \text{ max } at\, z=o} = \frac{\mu_o I\, d}{2\pi\, r(r+d)} \cong \frac{\mu_o I\, d}{2\pi\, r^2} \tag{15.2}$$

at far distance.

For a current loop,

$$B_{y\ \max\ at\ y=o} \cong \frac{\mu_o I \times \text{loop area}}{r^3} \tag{15.3}$$

at far distance.

For a common mode current,

$$B_{z\ \max\ at\ z=o} = \frac{\mu_o I\, 2h}{2\pi\, r(r+2h)} \tag{15.4}$$

The B-field is negligible around a coaxial cable with small eccentricity, and also at some distance from a twisted wire pair carrying a differential mode current. For the wires depicted in Figure 15.1, the B-field falls off as $1/r$ for a single wire, $1/r^2$ for two parallel wires carrying equal and opposite currents, and $1/r^3$ for a current loop.

The high-frequency electromagnetic field travels at the speed of light in the medium of propagation. The field is characterized into two types: the far field and near field in terms of the wavelength λ, which is the ratio of speed of light over frequency.

15.2.2.1 Far field at $r > \lambda/2\pi$

This is also known as the radiation field. It travels long distances in wave motion with a characteristic wave impedance Z_o. In vacuum or air, Z_o is 377 Ω. In coaxial cables it is about 50 Ω. The field strengths, both electric and magnetic, fall off as $1/r$.

15.2.2.2 Near field at $r < \lambda/2\pi$

This is also called the induction field. Its intensity at a point is determined by the characteristic of the source. The E-field falls off as $1/r^2$ if the source impedance $Z_s << Z_o$, and as $1/r^3$ if $Z_s >> Z_o$. On the other hand, the B-field falls off as $1/r^2$ if $Z_s >> Z_o$, and as $1/r^3$ if $Z_s << Z_o$.

15.2.2.3 Transition field at $r = \lambda/2\pi$

At this distance, the field characteristic changes from the near field to the far field.

15.3 Modes of Coupling

The electromagnetic field coupling between the two components can be via one or more of the following parameters:

- Stray capacitance
- Mutual inductance
- Ohmic conduction

The degree of coupling can be quantified in terms of the following parameter:

$$\text{transfer impedance} = \frac{\text{voltage across victim (succeptor)}}{\text{current through culprit (emitter)}} \tag{15.5}$$

Figure 15.2 depicts capacitive coupling between various conductor configurations on a relative basis. The E-field coupling for an ESD type of transient is governed by the mutual capacitance between the source and the victim cable. For two straight conductors shown in Figure 15.3, the circuit analysis leads to the capacitive coupling between conductors 1 to 2 to be

$$\frac{V_2}{V_1} = \frac{j\omega\left[\dfrac{C_{12}}{(C_{12}+C_{2G})}\right]}{j\omega + \dfrac{1}{R(C_{12}+C_{2G})}} \tag{15.6}$$

The B-field coupling between a culprit and a victim cable can be strong around the high-current bus, transformer or motor. The voltage induced is given by $V = -A\ \mathrm{d}B/\mathrm{d}t$, where A is area of the loop formed by the victim cable. If the radiation is sinusoidal at frequency f, then $V = 2\pi fAB$ volts, or V

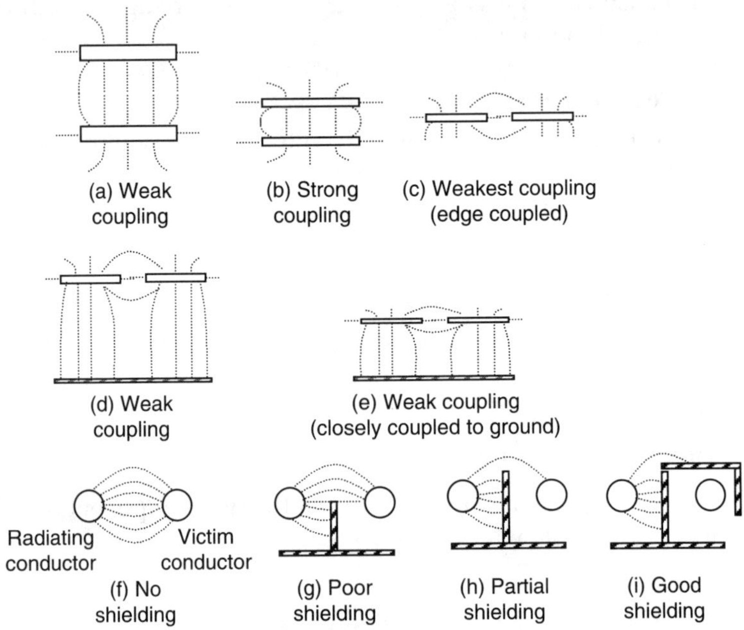

FIGURE 15.2 Capacitive coupling between various conductor configurations.

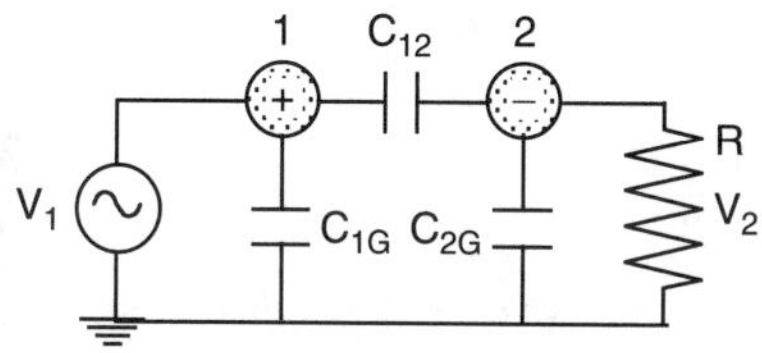

FIGURE 15.3 Capacitive coupling between two straight conductors.

(in dB·μV) $= K + B$ (in dB·μT), where $K =$ constant. In terms of the mutual inductance M, the term V (in volts) is then

$$V = M\mathrm{d}I/\mathrm{d}t = 2\pi fMI \tag{15.7}$$

The total inductive and capacitive couplings between two parallel unshielded wires over a ground plane are determined by the mutual inductance and capacitance, which, in turn depend on the geometry of the cables. The two mutual couplings are shown in Figure 15.4, where C and M are determined from the classical transmission line theory. The capacitive coupling circuit between an unshielded culprit and shielded victim wires is shown in Figure 15.5. At very low frequency, the capacitive coupling is

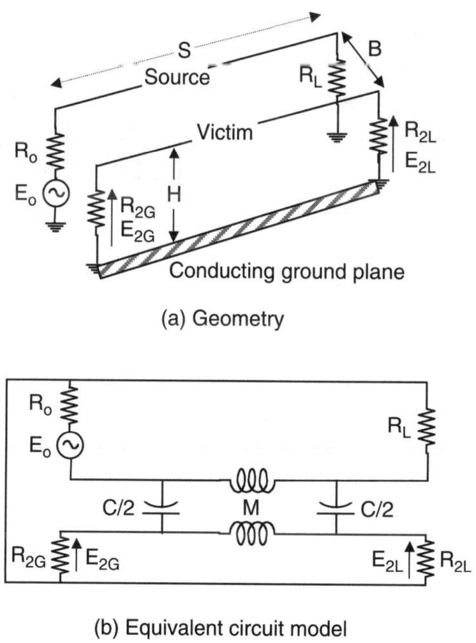

FIGURE 15.4 Capacitive and inductive coupling between two parallel wires over ground plane.

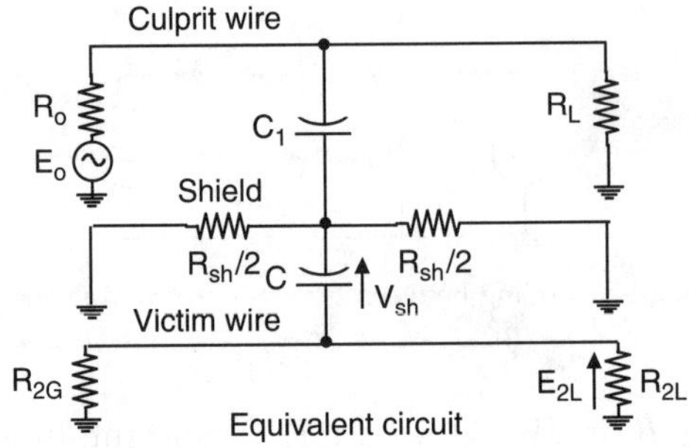

FIGURE 15.5 Capacitive coupling between unshielded culprit and shielded victim wires.

dominant when the termination resistances are high, so that the current flow is low and the magnetic coupling (inductance) is ignored.

At power frequencies up to 100 kHz, the coupling is primarily magnetic. In this frequency range, the following are some practical design guidelines for controlling the cable coupling:

- The coupling increases with the common run length and frequency and decreases with an increase in spacing.
- The noise voltage of the victim's termination load is a function of the load impedance. A capacitive filter connected from the victim line to ground can be used to decrease the load impedance at noise frequency, provided the victim's interface can accommodate such capacitance.
- The cable shield characteristic is defined by the transfer impedance. Lower impedance results in more effective cable shielding. Good impedance values are in the milliohm per meter range.
- The twisted shielded wire pair is effective up to 2 MHz, above which the coaxial cable is required.
- High current power lines and low impedance loads are prime sources of low frequency magnetic coupling. Therefore, the power line leads are always twisted with their returns.
- The twisting effectiveness increases with the number of twist per meter and decreases with the frequency. It is effective only on balanced or single point ground system and up to 100 kHz. Above this frequency, capacitance from the wires to the ground causes ground current to circulate, which generates common mode coupling. Practical twist pitch is 30–50 per meter.
- The capacitive coupling of the electric field is best avoided by shielding the culprit, victim or both cables, and grounding the shield (Figure 15.6).
- Single point ground is used at low frequency where the wavelength of the frequency to be suppressed is longer than 10 times the shield length.

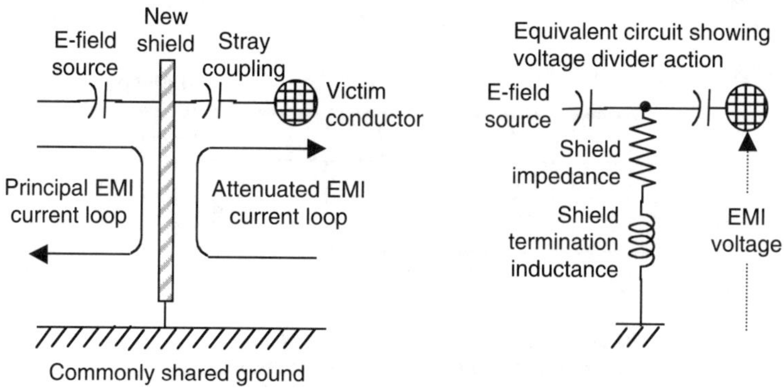

FIGURE 15.6 Shield protection from capacitive coupling.

If the culprit wire is shielded, the ground point is at the exit from the culprit source. If the shield is on the victim wire, the ground point is at the entry to the victim's electronics (Figure 15.7).

- The shield grounding at high frequency should be at both ends of the shield. This directs the return current away from the ground and into the shield, thus decreasing the ground loop currents and common mode radiation pickup (Figure 15.8 and Figure 15.9).
- Single point grounded shield does not reduce the common mode EMI (Figure 15.10).
- A 360° peripheral low-resistance bond should be used via a connector backshell assembly to assure low-impedance shield grounding in high-quality shield cable. This bond resistance limits the shield performance.

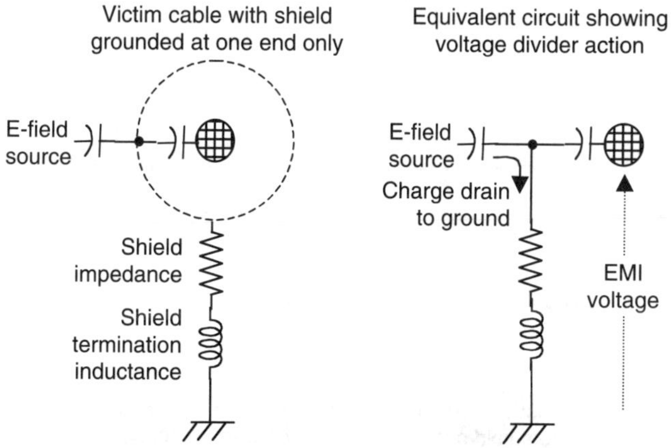

FIGURE 15.7 Victim cable with a shield grounded at one end.

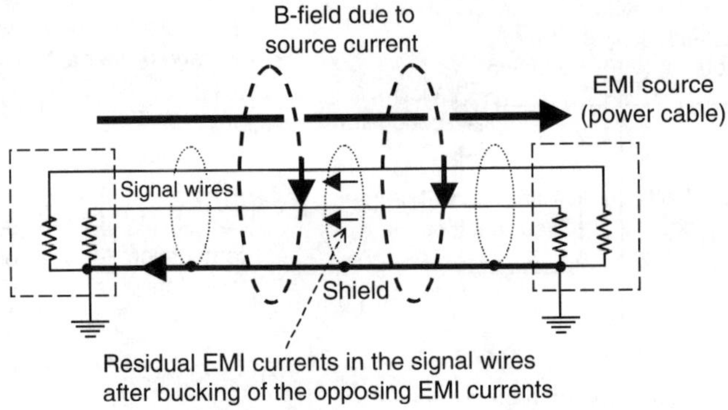

FIGURE 15.8 A shielded cable with both ends grounded.

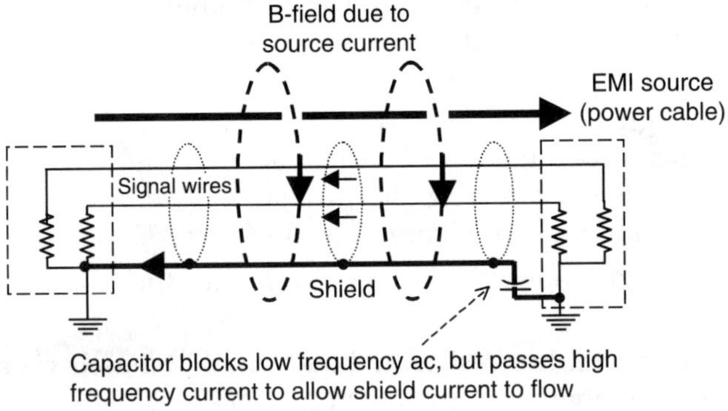

FIGURE 15.9 A cable with shield grounded at one end for low frequency and at both ends for high-frequency signals.

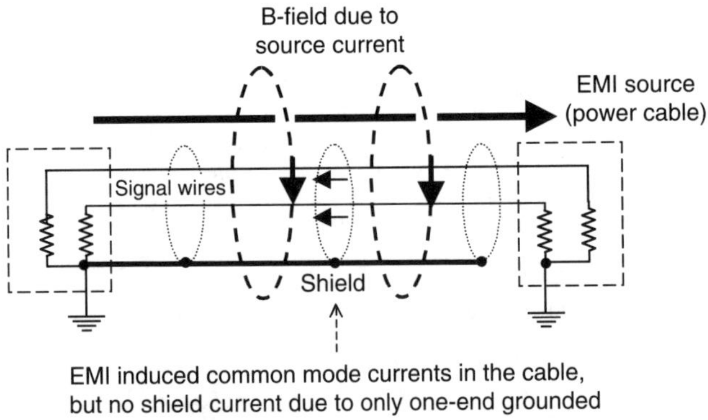

FIGURE 15.10 A shielded cable with only one end grounded.

15.4 EMI/EMC Specifications

The EMI standard establishes the maximum interference an equipment or a system, such as the power system as a whole, can conduct or radiate into the load equipment. The EMC standard, on the other hand, establishes the minimum conducted and radiated interference that an item of equipment connected to the power system must withstand without being susceptible to a functional impairment. Obviously, the susceptible level must be higher than the emission level with comfortable margin, usually tens of dB depending on the criticality of the equipment in the system.

The EMI/EMC levels are specified in dB, which equals 10 $\log_{10}$ (signal power/reference power), or 20 $\log_{10}$ (signal voltage or current/reference voltage orcurrent). Thus, 3 dB doubles the power, or 6 dB doubles the voltage or current.

The EMI/EMC requirement is specified in two parts, the conducted and radiated emission levels of the culprit equipment, and the conducted and radiated susceptivity level of the victim equipment. The general requirements for all spacecraft and launch vehicles come from MIL-STD-461. Those applicable to power equipments are listed in Table 15.1. The spacecraft specific requirements are also specified. Some such representative requirements are shown in Figure 15.11 through Figure 15.17.

The safety margin specified between the emissions levels and the susceptibility levels are generally:

- 20 dB for critical equipment
- 6 dB for all other equipment

In addition, certain self-derived EMI/EMC requirements that have been established in the past on some spacecraft are listed below as examples of what may be required of the power equipment:

- The EMC specification applies at the load points, not at the PRU location.

Table 15.1 MIL-STD-461 requirements applicable to spacecraft and launch vehicle power equipment

Requirement	Conducted (C)	Radiated (R)
Emission (E)	CE01	RE01
	CE03	RE02
	CE07	–
Susceptibility (S)	CS01	RS02
	CS02	–
	CS06	–

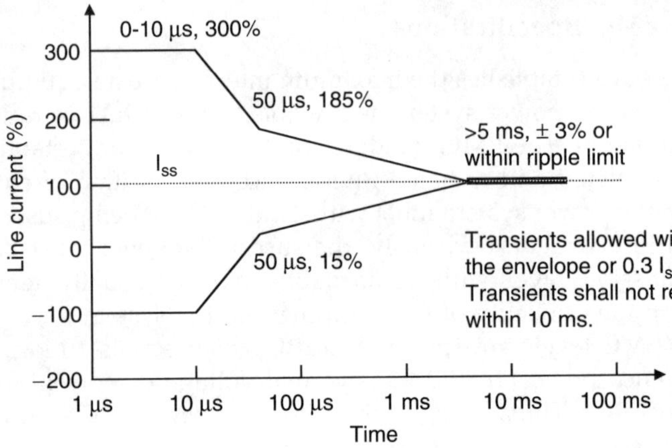

FIGURE 15.11 Repetitive transient current limits for equipment on a typical 10-kW communications satellite.

- The bus output impedance is estimated excluding the bus loads. However, for estimating the bus ripples the effect of major loads on the bus impedance is taken into account.
- The main bus power return, the signal return, and the structure are connected to a single reference ground point. All electrically conductive

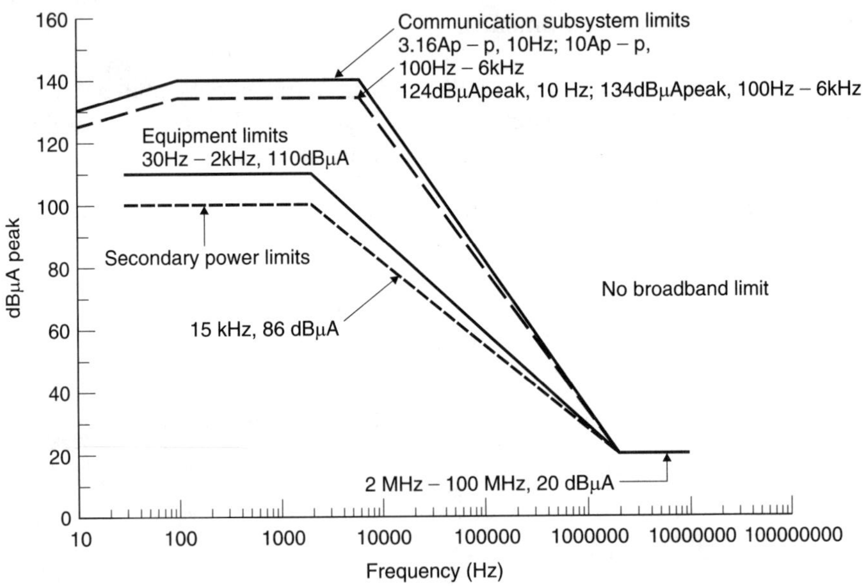

FIGURE 15.12 Conducted emission limits for equipment on a typical 10-kW communications satellite.

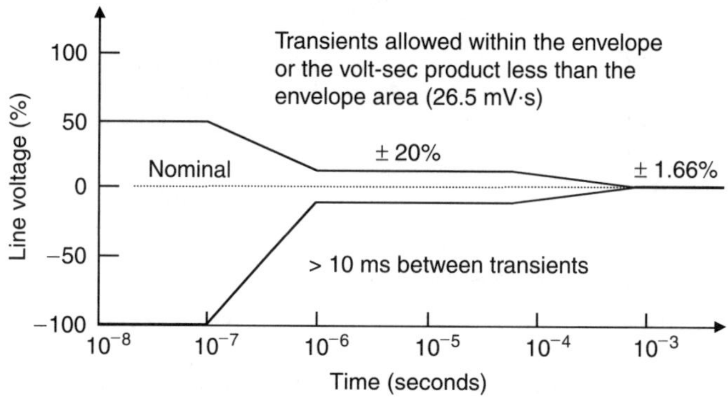

FIGURE 15.13 The CE07 limit for main bus voltage for a 3-kW, 120-V, LEO science satellite.

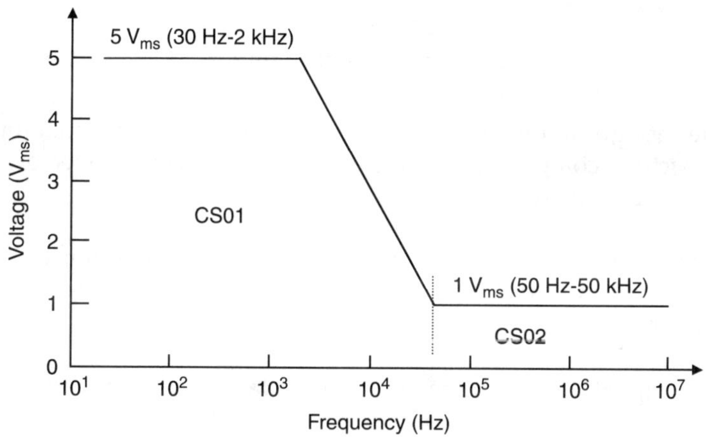

FIGURE 15.14 CS01/CS02 susceptibility requirements.

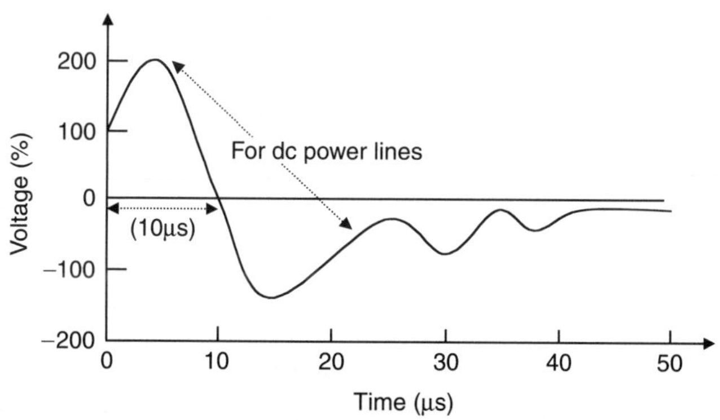

FIGURE 15.15 CS06 transient susceptibility requirements.

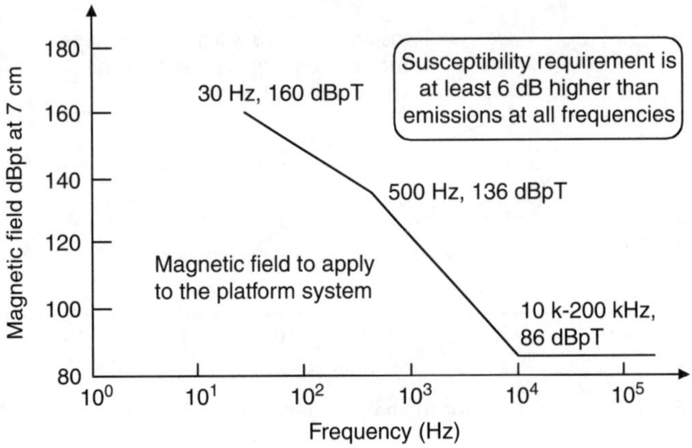

FIGURE 15.16 RS01 B-field susceptibility requirements for a 3-kW, 120-V, LEO science satellite.

surfaces are connected to the bus structure to form a common satellite ground cage.

- The case ground lead for each component is connected to a pin on the respective component multi-pin connector. The connector shell is always conductive and bonded to the case.

Typically, large commercial communications satellites designed for a 5- to 10-kW load and 100-V bus generally have the following self-derived requirements:

- The magnetic dipole moment of each solar array wing is limited to 3 to 5 A·m^2.

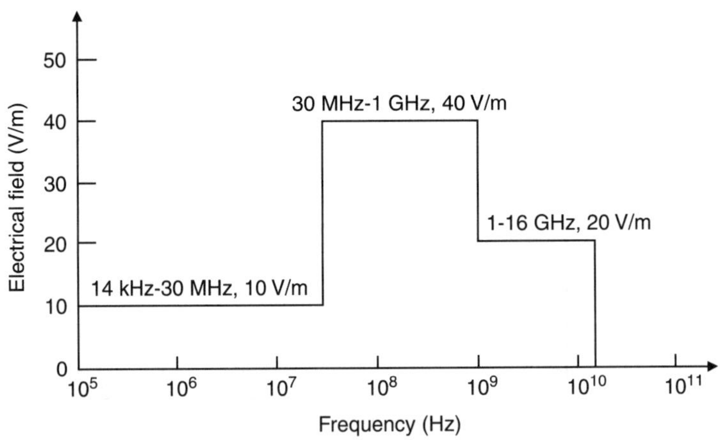

FIGURE 15.17 RS03 E-field susceptibility requirements (14 kHz to 16 GHz) for a 3-kW, 120-V, LEO science satellite.

- The magnetic moment of each battery is limited to 20 A·m^2, which is aligned with the pitch axis of the spacecraft. In two-battery systems, the magnetic dipoles of the two batteries oppose each other to cancel out at a distance.
- The magnetic moment of the whole spacecraft is limited to 5 to 10 A·m^2.
- Any load step is limited to less than 5% of the total load current capability of the bus, and the dI/dt is limited to a few tens of milliamps per microsecond. The undershoot and overshoot voltages are kept below 1% of the bus voltage.
- The load is required to withstand the peak-to-peak ripple voltage of 2 to 4% of the bus voltage, depending on the load equipment sensitivity.
- The dynamic impedance is limited to 250 mΩ in the 50 Hz to 20 kHz range and 500 mΩ for frequencies below 50 Hz.
- The bus voltage ripple at any load point resulting from the load current ripple is kept below 2 V_{p-p} from d.c. up to 50 MHz.

In the class of small satellites, as an example, one 28-V fully regulated defense satellite bus with load less than 1 kW, the ripples were limited to 500 mV_{p-p} at 36 kHz due to the battery discharge regulator. The user turn-on or turn-off dI/dt were kept below 10 A/μs for load current changes exceeding 2 A. Any transients that did not meet this rate had the I–t product limited to 100 μA·s. The bus broadband ripples were required to be 6 dB below that specified in the EMI/EMC specifications for the spacecraft. User equipment dynamic impedance was kept below 8 Ω from 3 to 10 kHz.

For EOS-AM type platform delivering a 3000-W load at 120-V bus, the bus source impedance at the capacitor bank was kept is 0.1 Ω d.c. to 100 kHz, and the ripples were limited to 0.5 V_{p-p}.

15.5 EMI Suppression Methods

Various methods of controlling and/or suppressing the EMI, in the order of their importance, are listed below and some discussed further in the next section:

- Minimize the EMI generation in the first place by
  - Minimizing the current loop area in switching circuits
  - Minimizing the switching transients' dI/dt rate in large current loops
  - Using snubber capacitors to minimize the voltage transients' dV/dt rate
- Minimize the E-field and B-field couplings between the culprit and the victim equipments by

- Minimizing the inductive coupling by twisting wires or using coaxial cables
- Minimizing the capacitive coupling by using shields and by reducing area of exposed metal and keeping it far from the ground (since $C = A\varepsilon/d$)

- Divert the energy impinging on the victim equipment to ground by using
    - Proper grounding scheme
    - Faraday shield, single or double
- Protect the equipment from the coupled energy by using
    - L-C filters for conducted EMI
    - Enclosure shield for radiated EMI

15.5.1 Twisting Wires

The twisting not only reduces the spacing between conductors, it also produces self-contained B-fields of opposite polarities around the adjacent spans. Thus, it results in a much lower B-field at far distance due to the cancellation effect. It does not, however, reduce the local B-field close to the cable.

Note that twisting two wires carrying a common mode current cannot null the B-field. The only effective way to eliminate the common mode EMI is to block the common mode current by breaking the ground loop. An alternative is to use a common mode choke (balun) discussed in Section 15.5.

15.5.2 Grounding

Two configurations of the single point ground (SPG) used for d.c. or a.c. power systems up to 20 kHz are shown in Figure 15.18. Configuration (a) is preferred. Configuration (b) is simple but less effective for high frequency EMI, since the ground wire inductance increases the Z_{ground} between two pieces of equipment. The grounding at a single point in the system effectively minimizes the radiated EMI by avoiding ground current loops.

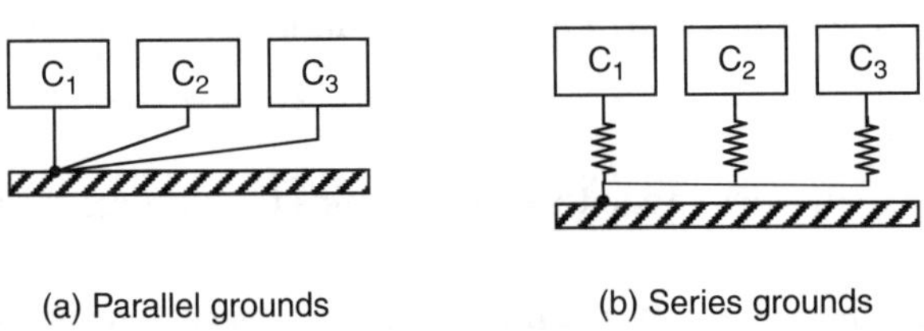

FIGURE 15.18 Single point grounding schemes.

Power return is grounded only at one point. All ground leads are directly connected to one point, which is then solidly connected to the structure. Single-point ground reduces the ground inductance. It is used when the circuit dimensions are less then 3% of the wavelength to be suppressed.

As shown in Figure 15.19(a), a single point ground is used to suppress low frequency EMI. On the other hand, multiple point ground (MPG) shown in Figure 15.19(b) is used for radio and higher frequencies EMI suppression. MPG is used when the circuit dimensions are greater than 15% of the wavelength at the operating frequency. Each ground lead is connected to the closest ground conductor. The ground plane minimizes Z_{ground} between two pieces of equipment. In a box having circuit boards sensitive to different frequencies, a hybrid of the SPG and MPG approaches may be needed.

15.5.3 Cable Shielding

Military and science satellites having stringent EMI requirements cannot tolerate any unshielded segment of the harness, or even an open fuse board. All harness groups — power, signal, and pyro — are kept in separate bundles with separate shielding. Commercial satellites, on the other hand, may have fuse boards or some segments of the harness unshielded. All shields (power and signal) are grounded about every meter to a SPG.

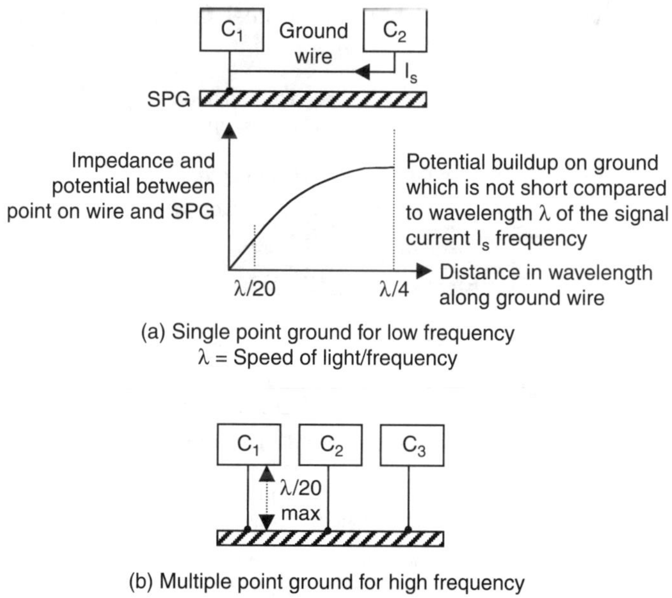

FIGURE 15.19 Single versus multiple point grounds.

The cable shielding alternatives are the braids versus tape and copper versus aluminum. The braids are used when extreme flexibility is required. The shield mass as compared to the wire conductor is about 40% for flat cables and 20% for round cables. In tapes, 2-mil-thick copper vapor deposited on Mylar or Kapton tapes is widely used. The tape is applied on the cable with insulation touching the cable, followed by another half-lap layer on the top (Figure 15.20). The shield mass with such tapes is roughly 80% of that with braids. The bend diameter must be less than about 10 cm to avoid unwrapping of the tape. Aluminum backed tape can be used, but being brittle, it needs greater bend radius to keep it from cracking at cold space temperatures. It does not create oxidation-related problems, as long it does not come in contact with copper or other dissimilar metals.

15.5.4 Bonding

In fabricating a shielding enclosure or in making ground connections, the conductors are bonded together by various processes. The bond is schematically shown on the circuit diagrams by a heavy dot before the conducting part meets the ground symbol. The solid bond generally requires having impedance less than a few milliohms. Movable parts may be bonded using a strap having resistance less than a few milliohms and inductance less than a few tens of nanohenries. The mechanical and electrochemical stability is required in the bond. Welding is recommended for magnetic enclosures, while brazing and soldering are recommended for conductive enclosures. The corrosion in bonding two dissimilar metals, such as a copper strap to steel box using steel fasteners, is controlled by

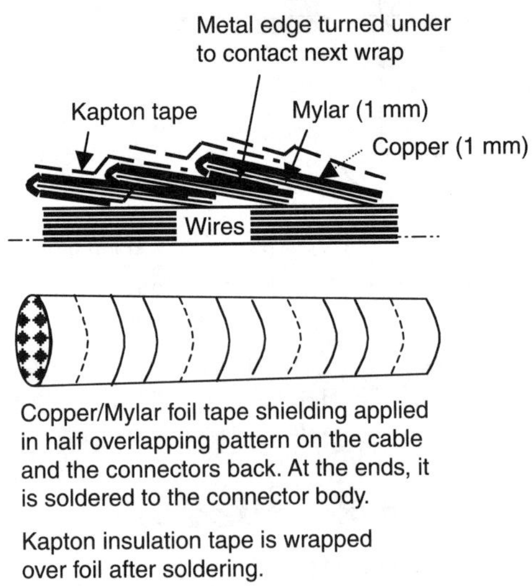

FIGURE 15.20 The cable shielding method using metal backed insulating tape.

selecting two materials in contact with about equal galvanic potentials. This requires the electric potential differences between them to be near zero. Otherwise the condensed humidity between the two acts as an electrolyte, creating a voltaic cell. Copper, silver, platinum, and gold are cathodic, having positive galvanic potential, while many other commonly used metals are anodic, having negative galvanic potential.

15.6 Common Mode EMI

Even a small common mode current can cause trouble. We recall that the induced voltage is given by $e = -L\ dI/dt$. A typical value of the wire inductance L is about 80 nH/m. If the rise time of say 100 mA is 1 ns (ESD type discharge), it induces voltage equal to $80 \times 100/(1 \times 10^{-9}) = 80$ V/m length of the cable. This can upset many sensitive electronic and control circuits.

Common mode noise can be detected by a circuit shown in Figure 15.21. The noise can be minimized by using a balun filter as shown in Figure 15.22. A transformer or a coupled inductor inserted between components 1 and 2 reduces the common mode propagation. Its magnetic core provides two large inductances to the common mode current, but provides zero net inductance to the differential mode current. Therefore, the inserted impedance in this mode equals two times the self (mutual) impedance of the transformer plus the self-impedance of the wire with respect to the ground. Bifilar winding or twisting of wires does not reduce the common mode noise. This is in contrast to that in the differential mode noise, where the impedance to the noise signal is equal to the leakage impedance of the transformer plus the wire, which is very small and can be further reduced by using bifilar windings and twisted wires.

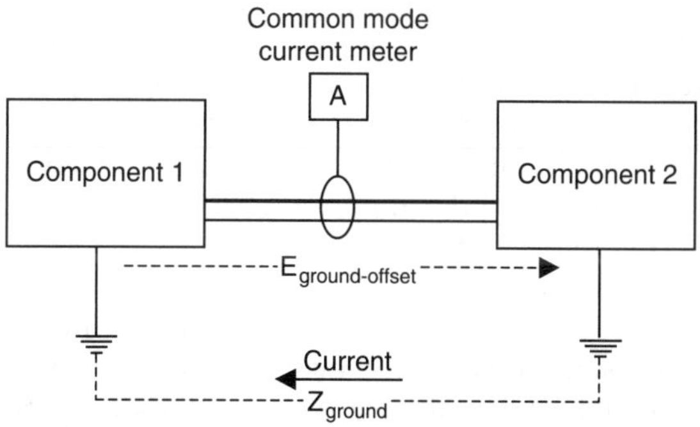

FIGURE 15.21 A common mode noise detection method.

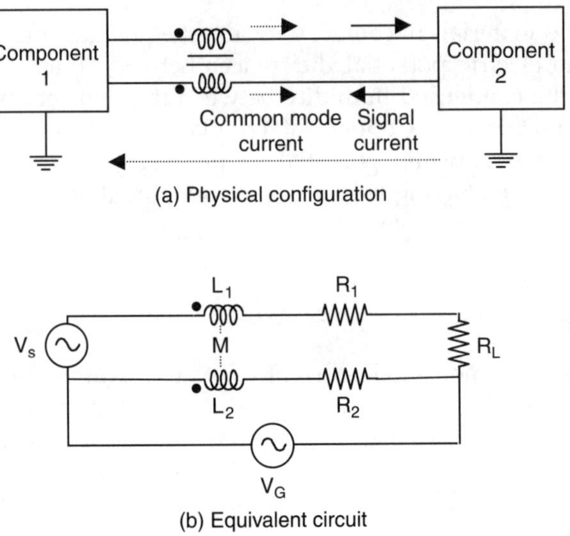

FIGURE 15.22 A common mode noise filter (balun.)

15.7 Broadband EMI

EMI is classified as narrow-band or broadband, depending on the signal's spectral distribution relative to the resolution bandwidth of the receiver used for the measurement. It is essential to know which type of signal is being measured to avoid erroneous test results and the wrong interpretation of data. In this context, the knowledge of the instrument's impulse bandwidth is necessary to correctly determine the absolute amplitude of an emission. Determination of the receiver's impulse bandwidth can be done with a series of measurements, but not simply by multiplication of its 3 dB or 6 dB bandwidth with a constant factor. Only the accurate knowledge of this parameter allows a meaningful comparison of the measured emission levels to a limit line and verification of a product's compliance with an EMI standard.

When measuring impulsive signals, it is necessary to know the impulse bandwidth of the interface filter of the EMI receiver or the spectrum analyzer. Due to the broadband nature of impulsive signals, some of the spectral components are not contained in the interface filter pass-band at any one time. As the filter bandwidth becomes wider, more spectral components are included and the amplitude of the signal at the filter output increases until the filter is wide enough to enclose the entire signal spectrum. Filters implemented in EMI receivers or spectrum analyzers are usually defined by their 3 dB or 6 dB bandwidths, which is suitable for measuring narrow-band signals. However, this specification is insufficient for measuring broadband signal because the filter's amplitude and phase response affect the results.

A comparison of test data taken with different EMI receivers is very difficult (if not impossible) to make. A study conducted in Japan in 1996 showed that the difference in measurement results of up to 6.4 dB is possible when measuring broadband impulsive signals, based on a 3-dB or 6-dB bandwidth specification. This could lead to situations where equipment under test passes or fails a compliance test, depending on the receiver being used in the tests.

15.8 Commercial Off-the-Shelf Equipment

In recent years, NASA and defense agencies have considered using commercial-off-the-shelf (COTS) equipment in space for reducing procurement cost. In Europe, interest in using commercial equipment in space has resulted in the EMC Directive issued within the European Union (EU) and enforced since 1996. The EU undertook a revision of the Directive in 1998, and recently has published its final revised draft for review and discussion among its 25 member countries. The fully and finally revised new Directive was published in mid-2004. It will become effective 30 months after the date of publication, possibly with a 2-year transition period. During transition, products can comply with the old or new Directive as advantageous to the manufacturers.

The EMC Directive requires commercial manufacturers to use EMI test methods similar to those required for aircraft, military, and space equipment. However, this requires acceptable methods to interpret and compare space system EMI test data with commercial test data, and that is difficult.

For example, NASA Specification SSP30237A establishes the electromagnetic emission and susceptibility requirements as well as design requirements for electronic, electrical, and electromagnetic equipment and subsystems designed or procured for use on the International Space Station. The applicability of the emission and susceptibility requirements is wholly dependent on the intended location or installation of the equipment or subsystem within the space station. SSP30237A designates the equipment or subsystem as an internal or external equipment depending on the intended location site. Internal equipment applies to equipment located inside a module or node. On the other hand, external equipment applies to all equipment located outside the modules and nodes. The SSP30237A relies on MIL-STD-461 as its framework document, which has been tailored based upon previous work on known electromagnetic environment for the space station. The MIL-STD-461 limits were adjusted based on power requirements and receiver sensitivity as well as margin for safety and to limit system-level compatibility concerns. The detailed test procedures for SSP30237A are contained in SSP30238A — Space Station Electromagnetic Techniques.

NASA has performed tests as per SSP30237A and has compiled EMI test data on various components. The system designers use this database for locating and co-locating various components within the space vehicle. If SSP30237A data cannot be correlated with data obtained from the corresponding commercial test requirements, then the equipment may either be placed in an environment where it can be susceptible or can cause susceptibility without violating the system level performance requirements. In addition, if correlation is not possible, the system designer may decide to remove the component from a questionable environment to a benign environment. This may cause an extensive redesign and impact on cost and schedule of the entire system.

15.8.1 EMI Requirements in Commercial Equipment

The following are three major EMI test requirements used extensively in the commercial market.

15.8.1.1 Code of Federal Regulations — Part 15 (FCC Part 15)

This Code addresses conducted and radiated emissions in the frequency range of 450 kHz to 1 GHz. The FCC divides the EMC environment into Class A and Class B requirements for office and household environment, respectively.

15.8.1.2 EMC Directive 89/336/EEC

This Directive is for use in the European Community (EC), which specifies both emission and susceptibility (immunity) requirements. The EC documents divide the EMC environment into household and light industrial environment in one class and industrial environment into another class. The European EMC requirements rely on a series of test methods developed by the International Electrotechnical Commission (IEC) and designated as IEC-1000-X-X, the European Union's emission and immunity requirements designated as EN-550XX, and the immunity testing documents designated as EN-610XX.

15.8.1.3 RTCA/DO-160C

This Standard is used in avionics industry for use in commercial airlines. The original version of the document was tailored from MIL-STD-461, but has been updated over the years and contains requirements beyond the MIL-STD-461 test methods, such as lightening and environment tests. Since the space station requires equipment similar to that used in commercial airlines, this standard is important when purchasing navigational and communications equipment.

15.8.1.4 Other requirements

There are areas where the commercial requirements do not fully cover the International Space Station requirements specified in SSP30237A. For example, when comparing CE01 to IEC-1000-3-2, the commercial standard does not fully cover the frequency range. The CE01 requires the measurement of conducted emissions from 30 Hz through 15 kHz, whereas IEC-1000-3-2 requires the measurement of the harmonic current emission only up to the 40th harmonic. Therefore, comparison between IEC-1000-3-2 certified equipment and SSP30237A qualified equipment can only be applicable if the inter-harmonic emissions are not of concern and if the equipment does not have local oscillators less than 10 kHz such as in d.c.-to-a.c. switching mode power supplies. The SSP30237A requires CE01 testing for d.c. power leads only, whereas IEC-1000-3-2 is not applicable to d.c. systems. Since NASA may procure COTS equipment designed for 115 or 220 V, 50 or 60 Hz operation, analyzing a.c. equipment against the CE requirements of SSP30237A to provide a reference point for NASA engineers is necessary. The scope of that work must include identifying differences between various measuring instruments as well as components used in test setups, such as the line impedance stabilizing network (LISN) and feed through capacitors. Measuring instruments are generally similar in related specifications. However, there are some differences that require careful examination. For example, the LISNs called out in the NASA specifications are different from the LISNs in the DO-160-C specifications and the LISNs specified in the majority of IEC-1000-4-X requirements are still of different design.

To narrow the difference, one must include the cable parameters, such as the length of the cable, height above the ground plane, whether the cable is run behind, in front or alongside the equipment under test, and other configuration issues. These all can be addressed by including the cable capacitance, inductance, resistance and ground conductance (C, L, R, and G, respectively) for the type and length of the cable used in the actual laboratory setup. These parameters come from standard high-frequency transmission line theory that covers the entire frequency range from low to high frequencies. The parameters can be lumped for short cables, or can be distributed if necessary for better accuracy in long cables.

15.8.2 NASA Versus Commercial Test Methods

The basic approach must be to analyze corresponding test methods and limits from SSP30237A and the applicable requirements in the FCC regulations, the EC EMC requirements, and/or RTCA/DO-160C. Then it is necessary to develop fitting transfer functions based on the comparison of the two sets of data for translating one of the commercial data into the NASA SSP30237A electromagnetic environment envelope.

We first consider the NASA CE01 and CE03 requirements, and the corresponding commercial requirements, which are listed in Table 15.2. The table shows how each of the SSP30237A requirements are analyzed against the various commercial standards. The analysis must be performed by evaluating various instrumentation, measurement techniques, and applicable limits for each of the test methods. This includes evaluations of different line impedance stabilization networks (LISNs), feed-through capacitors, spectrum analyzers, oscilloscopes, and other test instrumentations that are used in different test methods, which could lead to differences between measured quantities. For example, CE01 and CE03 require the use of a 10-μF capacitor on all power lines and measuring the RF current directly on the power line using an RF current probe and EMI receiver. FCC Part 15 and EN55022 each require an LISN on the power line under test and the measurement of voltage across a resistor in the LISN. Therefore, the two limits are in different units and cannot be simply compared. This requires an analysis to determine the relationship between the measured quantities and thereafter for developing the transfer function between the two.

15.8.3 Testing COTS for Space Compliance

One of the major concerns in comparing different EMI test requirements is the test setup. For instance, the majority of the C-1000-4-X requirements specify that the equipment under test be placed on an insulated surface approximately 80 cm from the ground plane, whereas SSP30237A requires the unit to be placed approximately 5 cm above the ground plane. This would cause a major difference in test results that must be carefully investigated. For the purpose of conducted emission testing, the cable can be modeled as a transmission line having *C*, *L*, *R*, and *G* parameters in an equivalent T network.

Table 15.2 NASA versus commercial EMI requirement

SSP 30237A requirement	FCC requirement	ECLIPSE requirement	DO-160C requirement
CE01	–	IEC-1000-3-2	–
CE03	Part 15	EN55022	Section 21
CS01	–	IEC-1000-4-13	Section 18
CS02	–	IEC-1000-4-6	Sec. 18 & 20
CS06	–	IEC-1000-4-5	Section 17
RE01	–	–	Section 15
RE02	Part 15	EN55022	Section 21
RS02	–	IEC-1000-4-8	Section 19
RS03	–	IEC-1000-4-3	Section 20

The standards being considered use very similar measurement techniques. They use measurement receivers to measure voltages, and current probes to measure currents. They also use current probes to measure current across a known resistor to determine voltage with respect to a given frequency range. Conversion factors are then applied based on calibration of the equipment used to perform the measurement and then a comparison is made to a given emission limit. However, each method has various differences in the approach for maintaining consistency within testing. The IEC-1000-4-X series of documents recommend the use of coupling/decoupling networks to match the typical installation impedance of various cables to the test cable configuration. The idea here is to minimize the distortion between the laboratory measurement and a typical field measurement. The SSP30238A relies on proper calibration of the equipment being used to perform the measurements. The values obtained from the various test specifications require adjustments in order to compare limits and test results. These adjustments need to be handled on a test method basis.

As for the frequency range, the test requirements must be reviewed to determine if they overlap the entire specifications. If not, the differences are taken into account as needed.

The simulations and the laboratory results on the four test setups have shown that the type of cable and its configuration and layout used in the actual setup make significant difference in the results. A parametric study made with PSPICE simulation showed sensitivity of the results to one more parameter, the cable resistance, in determining the peak of the response. This is expected, as low resistance results in a high-Q circuit having high and sharp peak at the resonance frequency.

A seemingly minor variation in the laboratory test setup, in particular the cable length, routing and types such as coaxial, twisted or randomly laid out, and the cable resistance makes significant difference in the correlation between any two test setups. This indicates that for properly comparing similar specifications, a careful and robust analysis of the similarities and differences in the setups is required.

Since the cable length and layout are equipment specific, it is extremely difficulty, and may perhaps be misleading, to correlate one test method to another in space systems, unless the cable length, layout and types are specified in the test standards.

15.9 Electromagnetic Pulse/Nuclear Threat

The electromagnetic pulse (EMP) is a man-made threat that arises from a nuclear detonation around the spacecraft. It can destroy satellites in low and mid Earth orbits. (Figure 15.23). A nuclear burst can upset electronic systems hundreds of miles away. Therefore, it is usually included in the

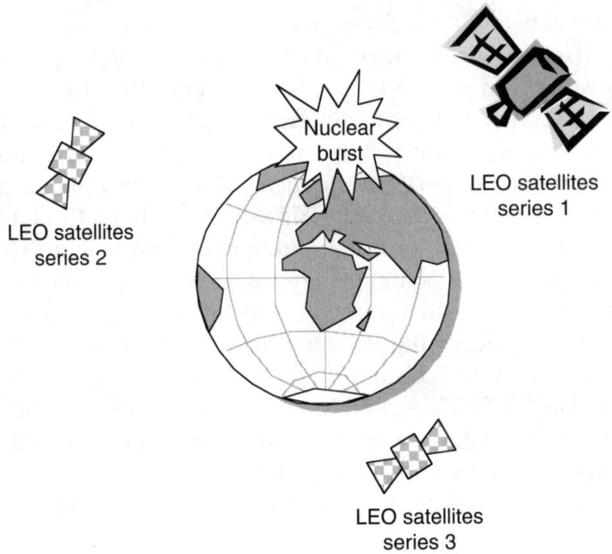

FIGURE 15.23 A nuclear burst in LEO altitude can destroy many satellites.

defense spacecraft power system requirements. The nuclear threat requirement is generally included on defense satellites in any orbit. For example, it was included in the original GPS satellites, which were procured exclusively for defense use of the U.S. Air Force until mid-1990s. It was removed from GPS-IIF and later versions, which were made available for commercial use also.

The EMP is basically the burst of electron flow resulting from the gamma radiation. It is an intense time-varying electromagnetic field of very short time pulse in the nanosecond to microsecond range. The radiation and propagation of the electromagnetic waves in space obey Maxwell's equations. The wave carries little energy to the equipment at far distance, but is deposited in a fraction of a microsecond, making the rate of energy deposition (power) extremely high. As a result, it can cause the following effects in the power system:

- Radiation dose degradation, particularly on solar array
- Common mode noise in power system equipments
- Temporary upset of sensitive electronic parts
- Semi-permanent upset (latch-up), requiring power shut-off and restart
- Permanent damage to sensitive semiconducting devices

The most dominant damage comes from the charged particle radiation dose on solar array and sensitive electronic parts. A high-altitude burst test named STARFISH conducted in 1962 produced a large number of beta electrons, which became trapped in the Earth's magnetic field. Its yield was 1.4 MT at 400 km altitude above Johnson Island in the South Pacific. The intense man-made radiation belt created by this burst ended the functional life of seven satellites within 7 months, primarily by solar cell degradation and command decoder failures. This radiation lingered for 10 years until 1972. Today, since LEO space is crowded with satellites, a severe nuclear detonation can cause the loss of hundreds of satellites resulting in enormous economic loss and national security concerns. It is estimated that a 10-kT burst over Japan at 150 km altitude can shorten the life of many LEO satellites such as *Iridium* and *NOAA* from 5 years to 5 months. The size, magnitude and duration of the nuclear radiation is calculated by the Air Force computer code that models the nuclear burst in space and calculates its effects on satellites. Using this unique code, the Air Force and the Institute to Defense Analysis predict that a 20-kT burst over Iraq would accumulate a radiation dose of 100 krad on *Iridium*-type LEO communications satellites in about 10 months, and the enhanced radiation would persists for several years. In some cases, the electron level would remain over 100 times higher than the natural level even 2 years after the burst. This makes the rapid replacement of destroyed satellites infeasible because they will also fail permanently due to the lingering radiation. For this reason, the replacement must wait and still use a much thicker radiation shield.

The EMP effects can be studied by quantifying the energy deposited on the victim equipment, which is the time integral of power. The power in an electromagnetic wave is given by the Poynting vector, which is the cross-product of E and H fields. The energy is therefore,

$$energy = \int power \cdot \mathrm{d}t = \int (ExH) \cdot \mathrm{d}t \tag{15.8}$$

We then think in terms of energy-coupling to various systems component. Wires exposed to EMP can couple damaging amount of energy into the system. The degree of coupling is a complex function, and is outside the scope of this book. However, it depends on the cable configuration, geometry of the cable run, and its orientation with respect to the electromagnetic field vectors. Once the energy is coupled to the wires, the high frequency transmission line theory applies for propagation along the wires. A part of the coupled energy is absorbed by the load (victim equipment). The remaining energy is reflected, the amount of which depends on how the load resistance matches with the characteristic impedance of the incoming wire. A good conductive paint on the surface can reduce the coupling when the EMP energy impinges directly on the equipment's enclosure surface, rather than entering via cables.

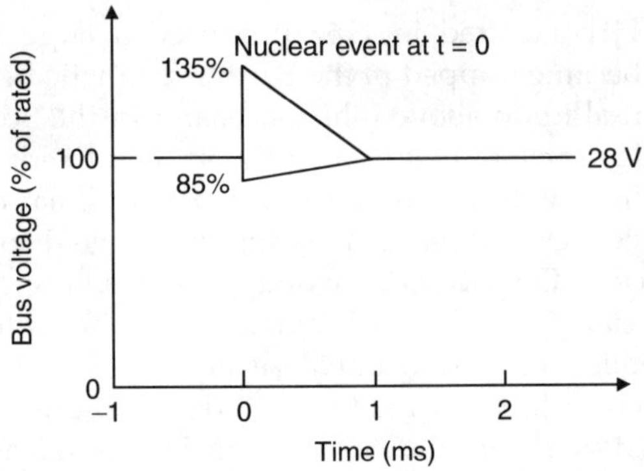

FIGURE 15.24 Main bus voltage transient requirement following nuclear burst on a 28-V fully regulated MEO communications satellite.

As an example of the power system requirement arising from a nuclear threat, the main bus voltage transient following a nuclear burst on a fully regulated 28-V MEO communications satellite was required to be kept within the envelope of Figure 15.24. In another example, the bus voltage of a 28-V fully regulated GEO communications satellite was required to be within the envelope shown in Figure 15.25. All loads must survive and return to the operating state after such abnormal transient is over.

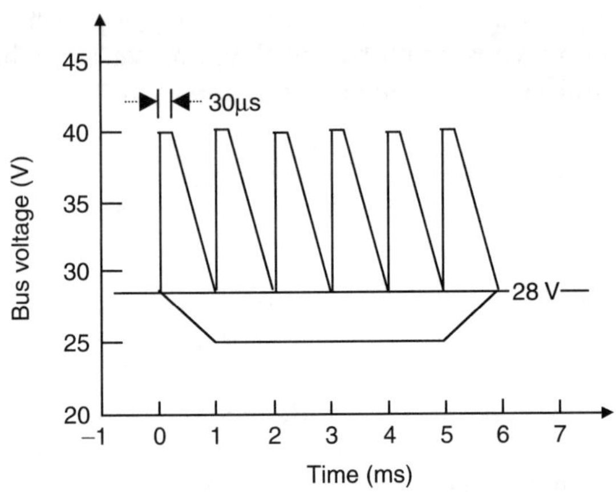

FIGURE 15.25 A main bus voltage transient requirement following a nuclear burst on a 28-V fully regulated GEO defense communications satellite.

Chapter 16

Electrostatic Discharge

16.1 Introduction

Electrostatic charge accumulates on any probe in space regardless of being inside or ouside the Van Allen belts. The accumulated charge raises the electrical potential of the probe, causing a current flow from the probe to the surrounding plasma (Figure 16.1). If the current cannot maintain a balance of charge, the probe potential will keep rising until arcing takes place. Such a problem can occur particularly when the spacecraft leaves or enters an eclipse when the interaction with space suddenly changes. Arcing can also arise due to differential charging of insulated surfaces electrically not connected together. Each isolated surface acts as an independent probe in space, which floats to a potential that results in no net current to or from the space plasma. That potential is of the order of the plasma kinetic energy. Insulating surfaces do not distribute surface charge, hence can charge up to a much higher differential potential until discharge takes place by way of arcing and/or flashover. The charging, the subsequent electrostatic discharge (ESD), and their remediation in GEO and in LEO are significantly different as discussed below.

16.2 ESD in GEO

In GEO's high-energy plasma environment, the major concern is the differential charging of different parts of the spacecraft, leading to high electric fields and subsequent arc discharges between components. The

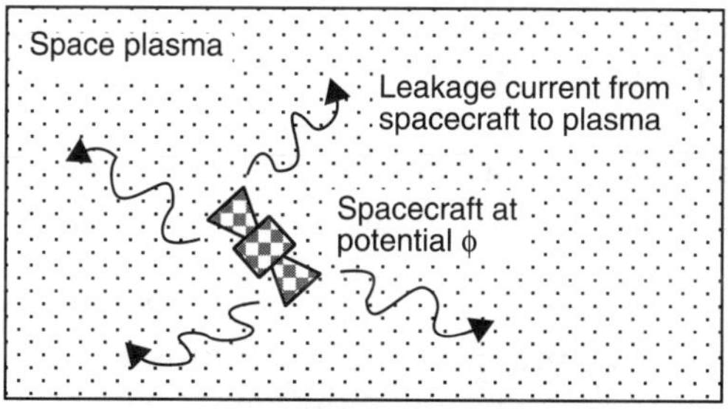

FIGURE 16.1 A spacecraft floating in space plasma.

ambient thermal plasma current densities are insufficient to discharge spacecraft surfaces rapidly. However, the electron charge accumulated on insulating surfaces increases the electric field to adjacent conductors above the breakdown level, leading to arcing. The resultant arcing currents traveling through conductors can upset electronic components and induce spurious signals. A common design solution for GEO is to coat all outside surfaces of the spacecraft with conducting materials. This prevents differential charging by distributing the charge over all surfaces and equalizing their potentials. Coating materials having surface resistivity less than 5 kΩ/square is considered adequate to eliminate differential charging. On a solar array, an indium oxide type conductive and transparent coating is applied on the coverglass when required.

16.3 ESD in LEO

In LEO, because of high thermal plasma current density, surfaces do not ordinarily collect much differential charge. The major concern in LEO is the absolute charging of spacecraft surfaces with respect to the surrounding plasma. Normally, the collected plasma current bleeds off the absolute potential rapidly. However, some spacecraft impose differential charging on themselves by using distributed high voltages. If conductors at different voltages are not exposed to space plasma, all spacecraft surfaces float within a few volts of the surrounding plasma potential. However, if conductors of similar area at different voltages are exposed to space plasma, a rule of thumb is that the most negative surfaces float negative of the plasma about 90% of the total voltage difference between the surfaces. This is true, for instance of solar arrays, where interconnects between solar cells or cell edges are exposed to the plasma, yet operation of the array depends on the voltage distribution. Spacecraft designs not using solar array also place different surfaces at different potentials. High voltage spacecraft in LEO environment have experienced significant interactions with their environment. The Upper Atmospheric Research Satellite (UARS), with the array voltage reaching 100 V when coming out of eclipse, was seen to charge to 90 V negative of its surrounding plasma because of plasma interactions with the solar array. This made data interpretation from some of its instruments difficult. The International Space Station floats at 140 V negative of its surrounding plasma. Because this make it liable for arcing and sputtering damage, a plasma contactor to control the floating potential is incorporated in the ISS design. The payload of SP-100 space nuclear power system as designed would float at about 100 V negative of LEO plasma because its structure ground is discontinuous between the power system and the payload. Locally, however, insulating surfaces would charge to only a few volts negative of the surrounding plasma. Therefore, arcs in LEO can occur from conductor to insulator junctions (including

Table 16.1 Electrostatic charging in LEO and GEO compared

Parameter	LEO	GEO
Plasma environment	Low energy $E \sim 0.1$ eV	High energy $E \sim 10{,}000$ eV
Potential difference due to spacecraft motion in the Earth's magnetic field	$V \times B$ high (~ 0.3 V/m)	$V \times B$ negligible (~ 0)
Potential difference on insulating surface	~ 1 V	> 1000 V
Potential difference on conductors	5-5000 V	~ 1 V
Remedy	Cover exposed conductors with insulation coating	Drain surface charge off insulators using conductive coating

holes in the cable insulation) when the conductor is highly negative of the surrounding plasma or from anodized or other dielectric surfaces when the underlying conductor is at a negative potential higher than the dielectric breakdown strength of the coatings. Micrometeoroid and/or debris impact, dielectric breakdown, etc may expose conductors. Arc currents may flow out into the surrounding plasma, with the return currents distributed over wide areas of other surfaces of the spacecraft. Arcs also can occur through the plasma between closely spaced conductors at differing voltages. Conducting surfaces highly negative of the plasma will attract high-energy ions, and are liable to sputtering. Nearby surfaces may acquire a sputtered conducting coating, changing their electrical, optical, and thermal properties. At high positive potentials, electrons are collected, leading to localize heating and significant power drain.

The LEO has low energy plasma environment (E $\sim$ 0.1 eV), but high electric field generated in the conductor moving in the Earth's magnetic field (V $\times$ B $\sim$ 0.3 V/m). This can add up over a long conductor and may cause arcing. The remedy is not to expose bare parts of large conductor, and apply insulating coating on all large conductors in LEO. In GEO, on the other hand, conductive coating and grounding the structural parts together at one point equalizes the potential of the parts and eliminates the ESD. Thus the causes and remedies of EDS in LEO and GEO are quite different as compared in Table 16.1.

16.4 Dielectric Breakdown

The electrostatic discharge from one surface to another involves dielectric breakdown. It is not like lightening, as no discharge can ionize the vacuum. However, the rise time is much faster than lightening, in the order of a nanosecond as shown in Figure 16.2.

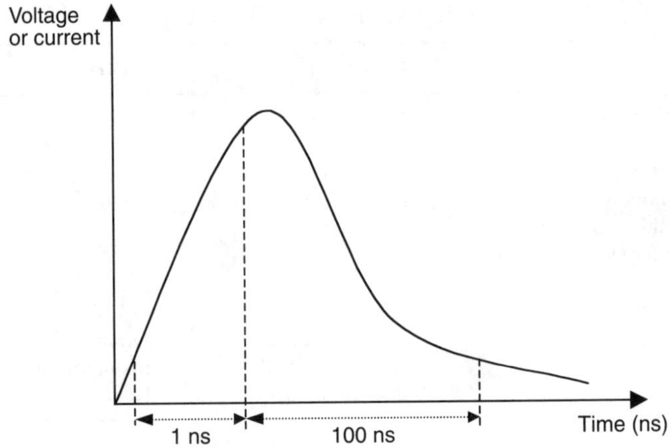

FIGURE 16.2 Typical ESD rise and fall times.

Ground tests for the International Space Station found that dielectric coatings often break down at 100 V in LEO plasma. Rated dielectric strengths and the strengths measured were not consistently reached before breakdowns occurred in simulated space plasma. It is suspected that this is due to porosity of the coatings, allowing plasma ions to reach much closer to the underlying conductor than the nominal dielectric coating thickness. Even coatings with rated dielectric strengths of much more than 100 V seemed porous enough to the plasma ions to have real dielectric strengths of less than 100 V. Thus, for surfaces that must be exposed at high negative potentials, it is important that dielectric coatings be used. Or, its outer surfaces are strong enough to stand off the full system voltage in plasma. Care must be used in selecting surface coatings of high dielectric strength. In particular, the chromic acid anodization commonly used on aluminum exposed to space often have insufficient strength to stand off 100 V in plasma. Thickening the coating usually influences its thermal properties, so a thicker coating may not be a good answer. Sulfuric acid anodization coatings have greater dielectric strengths, but very different thermal properties. For LEO missions, the sulfuric acid anodization coating is a preferred remedy where thermal control is not an overriding concern. On all other exposed high voltage surfaces, AO-protected aluminized Kapton of 1000-V dielectric strength is preferred.

16.5 ESD Entry Points

The following are the ways the ESD energy can enter the sensitive electronic equipment and the ways of avoiding the potential damage.

16.5.1 Strike Through Gaps

The ESD striking to the enclosure's outside wall causes voltage to rise locally until the charge distribution equalizes. Small gaps from the enclosure to the inside circuit boards may breakdown and arc under the resulting high voltage. A remedy is placing a grounded metal barrier between the striking point and the vulnerable board. The air breaks down at about 3000 V/mm, and ESD specifications require testing up to 15,000 V. Therefore, all air gaps less than 5 mm are critical spots during ESD testing on ground.

16.5.2 Puncture of Solid Insulation

Most power electronic equipment use solid insulation of puncture strength 7 to 8 kV/mm. For 15-kV withstand, all insulation less than 2 mm thick must be carefully evaluated for a possible strike during ESD tests.

16.5.3 Indirect Interaction

ESD current flowing through the enclosure couples energy into wires, which is then conducted to the circuit boards inside. Large dE/dt and dB/dt associated with the ESD-induced transients cause electromagnetic radiation, which in turn propagates to the inside circuits. This must be analyzed along with the withstand level for every component with ample margin in order to avoid circuit board damage in case of an ESD failure during tests or in operation. A safety factor of 10 is considered adequate for ESD-protection.

16.6 Effect of ESD

The ESD failure can affect the spacecraft equipment in the following ways:

- Human safety
- Operational disturbance
- Component damage

The operational disturbances and component failures can be as follows:

- Noise in radio communication and flicker in video display
- Switching of logic gates
- Component failure by over-voltage on semiconductor junction insulation

- Ignite spacecraft EEDs (critical failure)
- Ignite fire in hazardous areas (Class 1, explosive fuel vapors)

16.7 ESD Mitigation

The primary options for arc discharge mitigation in the spacecraft are:

- Controlling the spacecraft floating potential in order to eliminate the initiating arcs
- Limiting the bus voltage and/or string current that could feed an arc to level below the threshold for sustained arcs
- Use of proper grounding scheme

The above options can be implemented as follows:

- Increase conducting area on the spacecraft, thus lowering the floating potential.
- Use plasma contactor, thus lowering the spacecraft potential (as done in ISS).
- Lay out the solar array strings to reduce the potential difference between the adjacent cells below 60 V.
- Use grout between cells to insulate the gaps between cells.
- Limit the string current below 1 A, and use diode isolation.

In circuits using transformers, one of the windings and the chassis are always grounded to the spacecraft structure ground. The core, on the other hand, is kept floating with respect to the chassis by maintaining some gap between the chassis and the core. It is possible that the charge from the spacecraft can enter the core by flashover from the chassis, but cannot enter the winding unless the major winding insulation punctures. The floating core design is required by military standards, but the issue is debatable as to its superiority over the grounded core.

The electrostatic discharge effects are analyzed and controlled in the following steps:

- Examine every surface of the spacecraft. If the electrical resistivity of any surface is greater than 10 kΩ/square, apply conductive coating (ITO) if practical on that surface.
- Apply grounded shield on all cables outside a Faraday cage.
- Determine the expected ESD-induced transient voltage and the withstand level for every component. If the analysis shows a safety factor of less than 10, ESD testing is performed to establish the survivability. If

the component does not survive with acceptable margin, additional shielding and/or filters are added in the design.

An extensive study by Ferguson[1] on environmental effects on spacecraft gives the following recommendations for the power system design:

- Wherever there are discontinuous structure potentials of more than 100 V on the spacecraft, they must have an insulating material of sufficient thickness between them.
- Attention must be paid to geometry and materials where there are exposed conductor–insulator junctions in LEO. In particular, conductors known to have plasma arcing threshold higher than the conductor potential relative to the surrounding plasma must be used. Plasma testing to identify such materials is desirable.
- There is some evidence that arc rates are higher for materials that adsorb water and/or other volatile materials. Such materials should be avoided near conductor–insulator junctions, such as the solar cell edges.
- Geometry can influence plasma arcing as well. The geometry that helps prevent ions from entering the conductor–insulator region is preferred.
- For LEO missions, the structure and payload surfaces at high negative potentials should avoid exposed conductors, or at least exposed conductor–insulator junctions in order to avoid the same sort of arcing considered above.
- For GEO missions, fully conductive surfaces should be used everywhere to avoid differential charging.

Implementing a proper grounding scheme is important in controlling the surface potentials in the spacecraft structure relative to the surrounding plasma. A negatively grounded power system collects sluggish ions, and even the large surface area may only shift the floating potential away from the most negative system voltage by a few volts. On the other hand, positively grounded high voltage solar array in LEO may control its floating potential. However, space-qualified high voltage electronics for this polarity is scarce, slow, and may be prohibitively expensive.

16.8 ESD Control in Solar Array

The entire array (the cells, coverglass and substrate) is designed to perform at full specification under the effects of electrostatic charging. In GEO, this often requires indium–tin–oxide (ITO) coating, which reduces the solar array power output by a couple of percent or more. For example, 700

Angstroms (0.07 μm) coating reduces the cell power output by approximately 4%.

Regarding the grounding and conductivity of the array, the following are noteworthy:

- The back of the array is grounded to the aluminum honeycomb core of the panels in at least three places on each panel. The honeycomb from each panel is then connected to a ground lead from the boom.
- The panel end and the SAD end of the boom are both electrically tied together and connected to the spacecraft structure via a set of two 20-kΩ bleed resistors, as shown in Figure 16.3.
    - If a conducting tape is used on the boom, it is grounded by overwrapping the boom with the bus wire, which is grounded to the array structure in the same way as the boom.
    - In an early design stage, the best option is the array layout to eliminate closely spaced cells with large voltage gradients.

As a protection against ESD, the following design and/or construction features are generally implemented on the solar array:

- The back surfaces of the panels are coated to have electrical resistivity of less then 100 MΩ/square.
- The booms have conducting outer surfaces. The use of non-conducting tapes or paints over the conducting boom surfaces is minimized. The conducting tapes used on the boom are grounded.
- Blankets and loose surfaces are grounded to the array structure by means of the appropriate number of bonding straps.
- Snubber capacitors are used in the SAD before the slip rings and/or in the PRU.

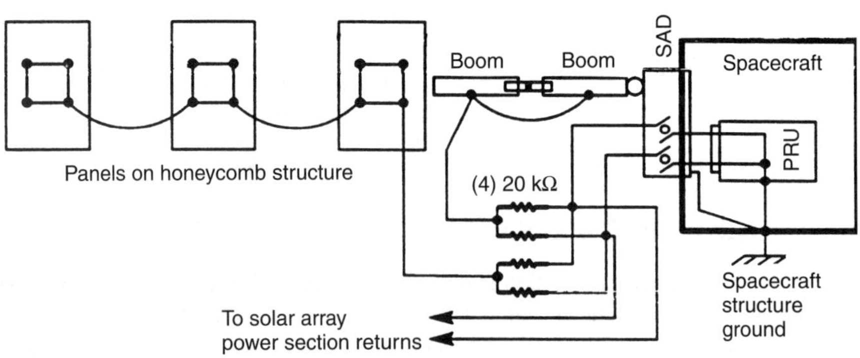

FIGURE 16.3 A solar array grounding scheme for ESD protection.

16.9 Part Sensitivity

The ESD results in high potential (up to 20 kV) of short duration (100 to 1000 ns). Energy levels depend on the surface areas involved. The current injected into the spacecraft depends on local structure configuration and impedance of the propagation paths. The surge impedance of the biaxial and twin-axial cables is 50 to 60 Ω. For twisted shielded wire pairs, it is about 60 Ω for AWG 20 wires and about 120 Ω for AWG 30 wires.

Although the energy level in the ESD is small, typically in several millijoules, the energy required to damage sensitive components such as CMOS, FET and pin diodes is even smaller (microjoules to millijoules). Robust component such as magnetics and vacuum tubes can withstand much higher energy levels. On-line components are more vulnerable than off-line components.

The ESD sensitivity is generally measured in terms of the voltage the part rises to under the charge. MIL-HDBK-263 lists ESD sensitive parts in terms of the ESD voltage ranges. The list is in three classes, class I parts sensitive to less than 1000 V, Class II sensitive to 1000 to 4000 V, and class III sensitive to 4000 to 15,000 V.

Table 16.2 lists the ESD damage threshold of some electronic piece parts commonly used in the spacecraft power system. The upset usually takes about 10% of the damage threshold voltage.

ESD specifications require testing up to 15,000 V. An ESD simulator, such as shown in Figure 16.4, is used for this purpose. It basically consists of charging a capacitor from a d.c. source via high resistance (slow charge) and then discharging to the equipment under test via low resistance (fast discharge).

Table 16.2 ESD damage threshold levels of power electronic devices

Device	ESD damage threshold (Volts)
MOSFETs	10–200
GaAsFET	50–2000
JFETs	150–7000
Bipolar transistors	300–7000
Laser diodes	100–2000
PIN diodes	200–1000
Schottky diodes	300–2500
Schottky TTL	500–2500
CMOS	150–3000
Operational amps	200–2500
Film resistors	300–3000

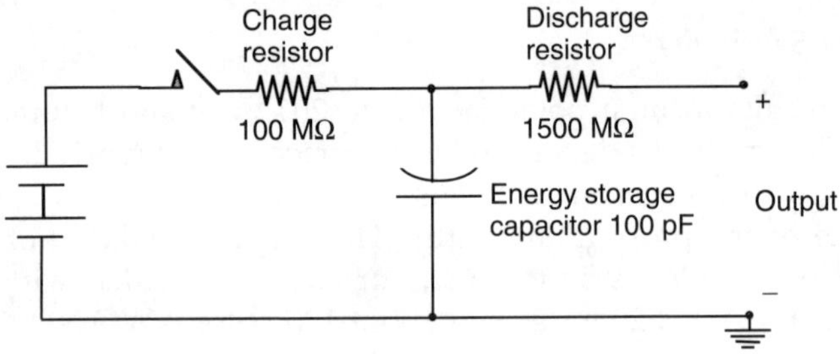

FIGURE 16.4 ESD simulator for typical circuit board testing.

16.10 Safety During Tests

While testing sensitive electronic parts, particularly the circuit boards on a test bench, care must be exercised for protecting the parts from ESD caused by human body charging from the static charge in air. The body potential can rise significantly as listed in Table 16.3. Sensitive parts can get damaged if picked up without draining the charge to the ground via conducting bracelet worn by the test personnel. ESD training and certification is, therefore, a part of the test floor standard practice.

Table 16.3 Typical electrostatic voltages on human body

Means of static generation	Electrostatic voltage	
	10–20% Humidity	**65–-90% Humidity**
Walking across carpet	35,000	1500
Walking over vinyl floor	12,000	250
Working at bench	6000	100
Opening vinyl envelop for work instruction	7000	600
Picking up common polyvinyl bag from bench	20,000	1200
Working on chair padded with polyurethane foam	18,000	1500

Source: MIL-HDBK-263.

Reference

1. Ferguson, DC., Interactions between spacecraft and their environment, in *Proceedings of the Intersociety Energy Conversion Engineering Conference*, 1992, Paper No. 0092-X.

Chapter 17

Reliability and Derating

17.1 Introduction

The economic success of any system depends on its availability for the intended use. The overall availability has two components: the reliability and the maintainability. For satellites once launched, the availability depends only on the reliability, as the maintenance is generally beyond reach.

The system reliability is based on probability of numerous components working together successfully over the entire mission duration. The reliability of a given competent is derived from tests on a large number of identical components under actual operating conditions. In the spacecraft industry, an actual flight history is available on only a small number of satellites. This makes the reliability estimate difficult, sometimes debatable, or even questionable. However, the estimate is extremely useful, at least on a relative basis, in identifying and correcting weak links in the chain of reliability. The perspective of the system reliability developed during the process is perhaps more important than the final reliability number.

The probability of failure of a component comes from four sources:

- **Random failure,** which is just a matter of chance.
- **Wear-out failure,** which is well understood and accounted for in the design.
- **Design failure,** which occurs if operating stress exceeds the design limit under an anticipated or unanticipated disturbance. It is avoided by allowing adequate design margin.
- **Manufacturing failure,** which depends on the consistency and sensitivity of manufacturing processes and the quality assurance procedure in place. Two manufacturing problems that can impact the power system reliability are:
    - Nicked wire that may occur during removing the enamel insulation from wire strands. This degrades the wire both electrically and mechanically, and may lead to a short or open circuit.
    - Protruding screws in a box, puncturing the surrounding wire insulation and shorting to another piece part. It may also cause dielectric stress concentration, leading to arcing under ESD.

The component reliability estimate is based on its failure rate in time (FIT), also known as the hazard rate, which is quoted in number of failures in million hours. The failure rate of any component varies with age and follows the well-known bathtub shape shown in Figure 17.1(a). The initial high failure rate — the infant mortality rate — is caused by manufacturing defects. The rate decreases rapidly with time in use, particularly in electronic components. Therefore, newly manufactured components are placed in operation for some time in a process called 'burn-in' to weed out infant failures. For power electronic components, most early failures occur in the first 12 h of burn-in operation.

A low constant rate of failures for a long time in service makes the flat stable part of the bathtub curve. In this region, the failures are random in nature, which are treated probabilistically in engineering designs. The rising failure rate near the end of life is due to wear out mechanisms that are generally well understood and accounted for by the design engineer.

The flat part of the curve is considered to be the useful life of the component. Only the random failure rate is used in developing the reliability estimate over the useful life. Electronic components with no moving parts have a relatively longer flat part of the tub. On the other hand, mechanical components with moving and physically wearing parts have a relatively shorter flat part as shown in Figure 17.1(b).

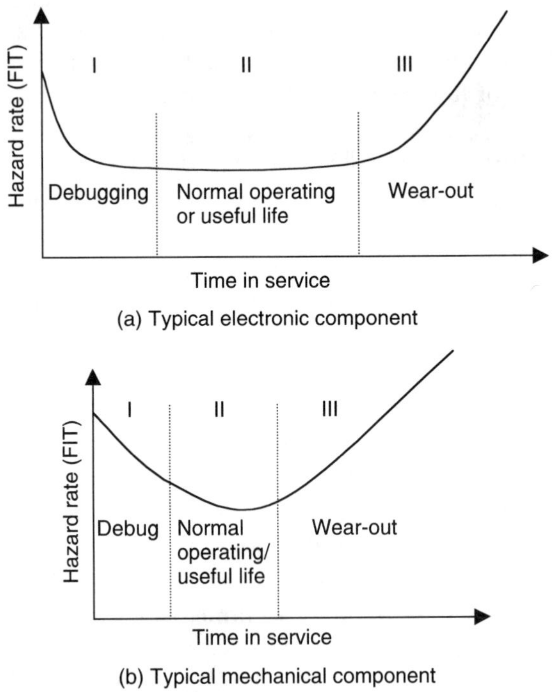

FIGURE 17.1 Component hazard rate versus time in service.

17.2 Random Failures

If a large population of the component is put to test simultaneously, the number of random failures in a given time duration is constant regardless of the beginning or end of life. In other words, under random (chance) failure, if the component has not failed at a given time, it is as good as new. However, under the accumulative nature of chance failures, the probability of component still working decreases exponentially with laps of time. The reliability is, therefore, an exponential function of time, known as the Poisson distribution function, expressed as

$$R(t) = \mathrm{e}^{-\lambda t} \tag{17.1}$$

where λ = failure rate, i.e., the number of failures in unit time, also known as the hazard rate. The probability of failure in time t, i.e., unreliability, U, is given by

$$U(t) = 1 - R(t) = 1 - \mathrm{e}^{-\lambda t} \quad \text{and} \quad \frac{\mathrm{d}U}{\mathrm{d}t} = \lambda\,\mathrm{e}^{-\lambda t} = f(t) \tag{17.2}$$

The probability of component failing during a time interval Δt at any time t is $f(t)\Delta t$. For this reason, $f(t)$ is known as the failure probability density function. For a constant hazard rate λ shown in Figure 17.2(a), $f(t)$ is an exponentially decaying function as shown in Figure 17.2(b). The component unreliability is the area under $f(t)$ curve from time 0 to t, and the reliability is the shaded area from t to ∞.

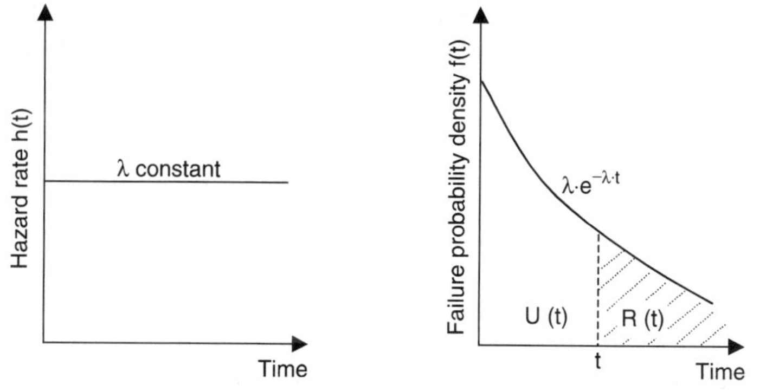

(a) Constant hazard rate (FIT) (b) Failure probability density function

FIGURE 17.2 Constant hazard rate and failure probability density function versus time in service.

The expected mean time between failures (MTBF) in a large population of a given component is

$$\text{MTBF} = 1/\lambda \tag{17.3}$$

That is, if numerous identical parts were put to test under identical conditions, then the mean time between failures would be $1/\lambda$, always the same because λ is constant. The probability of component working at any time t in terms of the MTBF is therefore,

$$R(t) = \exp(-\lambda t) \;=\; \exp(-t/\text{MTBF}) \tag{17.4}$$

and the probability of component not working at any time t is

$$U(t) = 1 - R(t) = 1 - \exp(-\lambda t) = 1 - \exp(-t/\text{MTBF}) \tag{17.5}$$

17.3 Wear-Out Failure

Components such as bearings, solenoid valves, frequently operating relays, and battery electrodes wear out. The life under wear-out failure mode is expressed in terms of the mean life μ and the standard deviation σ. If $f(t)$ represents the probability of failure in time t, it follows the normal bell-shape probability density expressed by the Gaussian distribution function

$$f(t) = \frac{1}{\sigma\sqrt{2\pi}} \exp\left\{-\frac{1}{2}\left(\frac{t-\mu}{\sigma}\right)^2\right\} \tag{17.6}$$

The probability of a wear-out failure at time t is the integral of $f(t)$ from 0 to t. The reliability at time t, or the probability of not failing in time t, is 1 minus that integral, i.e.,

$$R(t) = 1 - \frac{1}{\sigma\sqrt{2\pi}} \int_0^t \exp\left\{-\frac{1}{2}\left(\frac{t-\mu}{\sigma}\right)^2\right\} \mathrm{d}t \tag{17.7}$$

17.4 Fundamental Theorems of Reliability

The reliability of any complex system having components in series, parallel or any combination thereof, is determined by the following two fundamental theorems of reliability:

1. If A and B are two independent events, with their individual probability of occurrence $P(A)$ and $P(B)$, respectively, then the probability of both A and B occurring is

$$P(A) \times P(B) \tag{17.8}$$

2. When there are n identical components each with reliability R, then the probability that there will be exactly k components working at any time t is given by the binomial theorem,

$$P_k = \frac{n!}{(n-k)!\,k!} R^{n-k}(1-R)^k \tag{17.9}$$

Using the first theorem, the total reliability of a component under all failure modes is the product of the reliability under each mode separately. For example, the overall reliability under the four failure modes identified above is

$$R_o = R_r R_w R_d R_m \tag{17.10}$$

where the four reliabilities on the right-hand side account for the random, wear-out, design, and manufacturing failures, respectively. The component is designed with an adequate design margin. The reliability under wear-out mode is made almost equal to 1.0 by making the mean wear-out failure time much longer than the design lifetime. Eliminating potential failures using consistent manufacturing and quality assurance procedures leaves only the random failures to account for in the total reliability estimate, and so is done in practice.

17.5 Series–Parallel Reliability

Two or more components are said to be working in series in the reliability sense if all components have to work for the whole assembly to work successfully. On the other hand, two or more components are said to be working in parallel in the reliability sense if only one component has to work for the assembly to work successfully. The series–parallel working of various components for reliability of an assembly is not to be confused with the series–parallel connection in the electrical circuits. Electrically parallel components can be in series in the reliability sense, and vice versa.

If two components having individual reliability R_1 and R_2 are working in series as shown in Figure 17.3, then using the theorem (17.8), reliability of the assembly is given by

$$R = R_1 R_2 = \exp(-\lambda_1 t)\exp(-\lambda_2 t) = \exp[-(\lambda_1 + \lambda_2)/t] \tag{17.11}$$

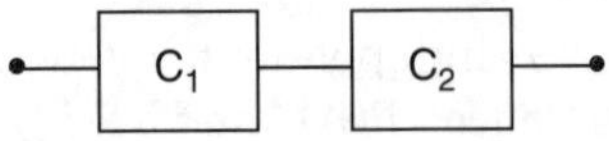

FIGURE 17.3 The assembly of two series components.

Or, the unit failure rate $\lambda = \lambda_1 + \lambda_2$, and

$$\text{MBTF} = \frac{1}{\lambda_1 + \lambda_2} \tag{17.12}$$

If two components having individual reliability R_1 and R_2 were working in parallel as shown in Figure 17.4, then the assembly would not work only if both were not working. That is, $U = U_1 \; U_2$, and $R = 1 - U$, or

$$R = 1 - U_1 U_2 = 1 - [(1 - e^{-\lambda_1 t})(1 - e^{-\lambda_2 t})] \tag{17.13}$$

By using some algebra, it can be shown that for such unit,

$$\text{MTBF} = \frac{1}{\lambda_1} + \frac{1}{\lambda_2} - \frac{1}{\lambda_1 + \lambda_2} \tag{17.14}$$

17.5.1 Examples

Applying the series–parallel reliability principles presented above we examine a few examples below:

- The reliability of an assembly made of two series component, each having a reliability of 0.99, is $0.99^2 = 0.9801$.
- Another assembly made of the same two component in parallel has the reliability of $1 - 0.01^2 = 0.9999$.
- A system of 15 series component, each having 0.95 reliability, has the overall reliability of $0.95^{15} = 0.4633$.
- If a system made of 150 components in series, each having equal reliability R_1, has the overall system reliability requirement of 0.99, then the minimum requirement on the component reliability is derived from equation $0.99 = R_1{}^{150}$, which gives $R_1 = 0.999933$.

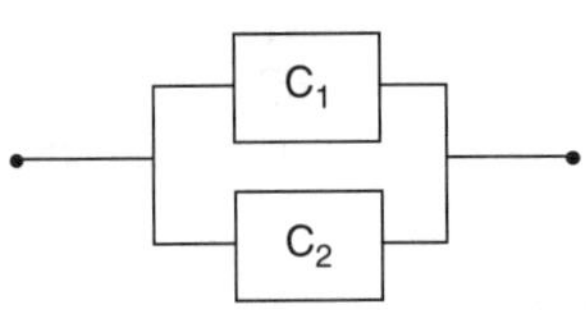

FIGURE 17.4 The assembly of two parallel components.

Practical systems are made with some components in series and some in parallel. For example, we calculate the reliability of the overall system shown in Figure 17.5 as follows. Dividing the system in sub-assemblies A, B, C and D, we determine their unreliabilities and reliabilities as

$U_A = (1 - 0.90) \times (1 - 0.90) \times (1 - 0.90) = 0.001$, and $R_A = 1 - 0.001 = 0.999$

$U_B = (1 - 0.80) \times (1 - 0.80) \times (1 - 0.80) = 0.008$, and $R_B = 1 - 0.008 = 0.992$

$U_C = (1 - 0.992 \times 0.99) \times (1 - 0.97 \times 0.98) = 0.000885$, and $R_C = 1 - 0.000885 = 0.999115$

and the sub-assembly D, having only one part, has $R_D = 0.998$.

The overall system reliability, R_s is therefore,

$$R_s = R_A \times R_c \times R_D = 0.999 \times 0.999115 \times 0.0998 = 0.99612$$

For an assembly such as a battery and solar array made of numerous identical series cells having equal reliability, Figure 17.6 depicts the assembly reliability versus number of components. The rule of 72 is noteworthy here:

The reliability of the assembly reduces to one-half when the product of the number of series component and the percent unreliability of each component equals 72.

The assembly reliability increases with the number of parallel components, but with rapidly diminishing rate of return. This is seen in the last column of Table 17.1, which gives the percent improvement in reliability by adding parallel components starting with a single unit. It is given by $(R_n - R_1)/R_1$,

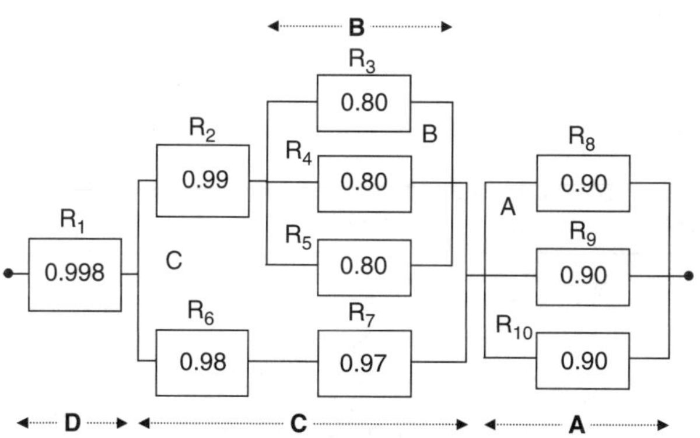

FIGURE 17.5 The assembly and sub-assemblies of series and parallel components.

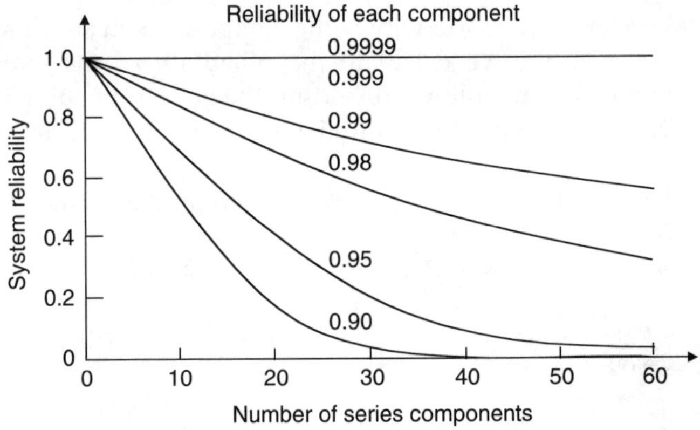

FIGURE 17.6 System reliability versus the number of series component and the reliability of each component (rule of 72.)

where R_n is the reliability with n component in parallel, each having the reliability of R_1. If each component has 80% reliability, then two components in parallel would improve the assembly reliability by 20%, three by 24%, and it would take infinite many components in parallel to improve by 25%. This clearly shows that using two components significantly improves the reliability, and more than three component in parallel is almost fruitless.

The cross-strapping of solar array sections shown in Figure 17.7 is often used to improve the array reliability. It is also used for other assemblies if required. The improvement is quantified here for a system shown in Figure 17.8. It is made of two identical parallel paths, each having four series components with reliability indicated in the component boxes. In configuration (a) without cross-strapping, the reliability of the assembly is

$$R_a = [1 - (1 - 0.99 \times 0.98 \times 0.97 \times 0.96)^2] = 0.991$$

Table 17.1 Diminishing rate of return in system reliability using parallel units

Number of components	Overall system reliability	Incremental system reliability	Improvement in reliability over single component (%)
1	0.800000	Reference	Reference
2	0.960000	0.160000	20.00
3	0.992000	0.032000	24.00
4	0.998400	0.006400	24.80
5	0.999680	0.001280	24.96
6	0.999936	0.000256	24.99

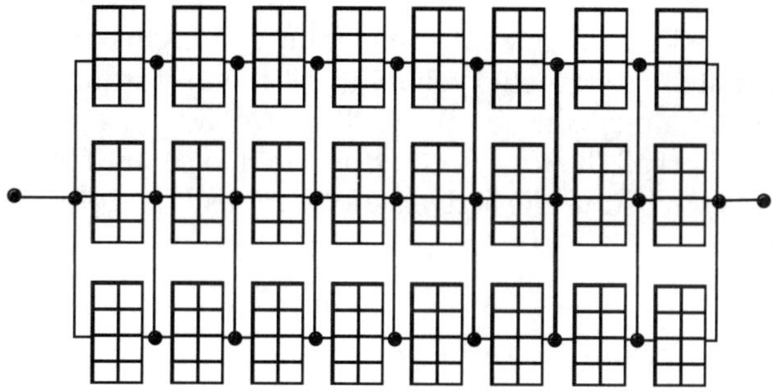

FIGURE 17.7 Cross-strapping of solar array sections.

With cross-strapping, the assembly reliability increases to

$$R_b = (1 - 0.01^2) \times (1 - 0.02^2) \times (1 - 0.03^2) \times (1 - 0.04^2) = 0.997$$

The cross-strapping thus reduces the assembly failure rate to one third, from 9 to 3 in 1000.

17.6 Redundancies

Redundancy is obtained by placing more units in parallel than necessary for the specified operation. Redundancy has a value only if the failure is instantaneously detected and acted upon to switch over to the backup unit. It is therefore important to continuously monitor the unit performance. Deviation in one or more performance parameters can be interpreted as

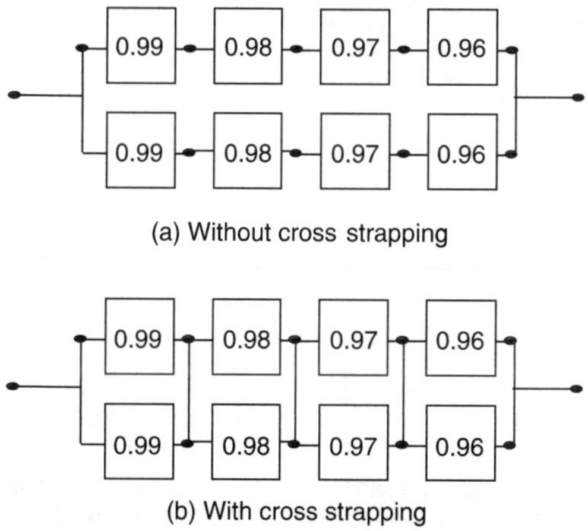

FIGURE 17.8 System without and with cross-strapping.

failure. It is needless to say that the failure detector must be of high reliability to match the targeted reliability of the unit it monitors.

The redundant units can be operated in two ways as shown in Figure 17.9: (a) active (hot) all the time, or (b) dormant (cold) until the primary unit fails and then switched to the active state. Various types of redundancy are described below.

17.6.1 Active n for $(n - 1)$ Units

For an assembly made of n identical units active all the time in parallel, where $(n - 1)$ units must work — or any one may fail — for the assembly to perform the specified function, the probability of assembly failing, i.e., all units failing, is

$$U = U_1 \times U_1 \ldots n \text{ times} = U_1^n$$

where U_1 is the probability of failure for each unit, assumed to be the same for all units. The reliability of the assembly is then

$$R = 1 - U = 1 - U_1^n \tag{17.15}$$

For a given series reliability, using two large units ($n = 2$) in parallel, one active and one spare, would result in a heavier assembly than three small units, two needed and one spare. On the other hand, using 10 small units would also be mass heavy, as there is always some fixed mass for each small unit. Obviously, there is an optimum number of units in between for a

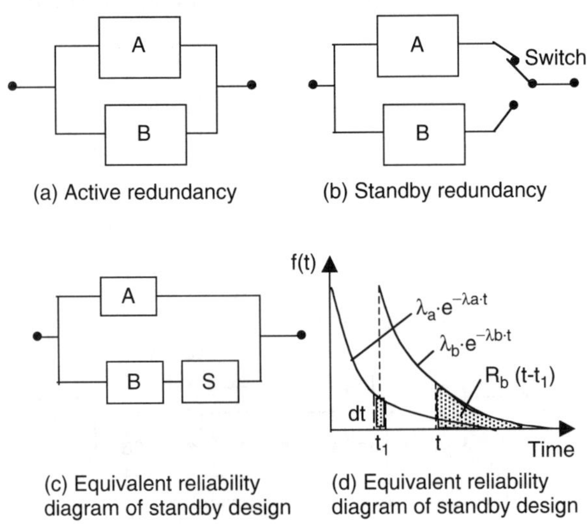

FIGURE 17.9 Active and standby redundancies.

given reliability with minimum assembly mass. We examine this for a battery charge or discharge converter made of multiple small units working in parallel.

The mass of power converter in the 100 to 200 W range is approximately given by $500 + 3P_o$ g, where P_o is in watts. It has little sensitivity to the voltage and bus voltage regulation. The unit mass in general is expressed as mass $= \alpha + \beta P_o$, where α is the fixed overhead and β is the power proportional mass constant. If n units are to be used for $(n - 1)$ needed to meet the converter power requirement with one unit failed, the optimum value of n for minimum mass can be shown to be

$$n_{optimum} = 1 + \sqrt{\frac{\beta P_{out}}{\alpha}} \tag{17.16}$$

Thus, if the fixed mass α is small, many small units will result in the minimum assembly mass, although it would cost more. Practical experience suggests that $n = 3$ is the optimum for all values of $\beta\ P_{out}/\alpha$ between 2 and 7.

17.6.2 Active *n* for *m* Units

If a total of n units are active in parallel, and m units must work to meet the full rated operation, the probability of exactly k units working is given by the binomial distribution

$$\frac{n!}{(n-k)!k!} R_1^{n-k}(1 - R_1)^k \tag{17.17}$$

The probability of at least m units working is the sum of all terms from $k = m$ to n, i.e.,

$$R = \sum_{k=m}^{k=n} \frac{n!}{(n-k)!k!} R_1^{n-k}(1 - R_1)^k \tag{17.18}$$

where R_1 is the reliability of each unit.

17.6.3 Dormant *n* and Active *m* Units

For an assembly with $(m + d)$ units in parallel, m units active which must work and d units dormant in spare, any one of which can replace a failed active unit, the mean time between failures is given by

$$\text{MTBF} = \left(\frac{d+1}{m}\right)\text{MTBF}_1 \tag{17.19}$$

where $MTBF_1 = 1/\lambda_1$ for each unit.

For example, if there were three for two battery chargers (two active and one spare), the MTBF of the assembly would be $[(1 + 1)/2]$ $MTBF_1$, which is the same as that for each unit. If there were only two for two units with no spare, the MTBF of the assembly would be ½ $MTBF_1$. Here all units are assumed to have identical failure rates, and the reliability of the switching element is assumed to be one. In reality, the dormant unit under no operating stress would have much lower failure rate until switched on.

With two parallel units, active A and backup B, with a switch at one end as seen in Figure 17.9(b), the reliability is estimated with the switch in series with the backup unit. Because, for the assembly to continue working after one unit fails, the unit B and the switch both have to work at the time of switching.

17.6.4 Shared Versus Separate Battery Chargers

The battery charge/discharge converters can be configured two ways as shown in Figure 17.10(a) and (b). In (a), each battery separately has two chargers, each rated 100% of each battery load, or 50% of the total system load, allowing one converter failure. In (b), both batteries jointly share three chargers; each rated 50% of the total system load so that any two converters will meet the full system load. Shared chargers (three for two for both batteries) require fewer chargers, and hence result in lower mass. However, if one cell fails (open or short) in one battery, say B_1, one cell must be removed from battery B_2 also and placed on trickle charge to keep the removed cell alive. Otherwise, B_1 will share a much-reduced load due to its loss of 1.25 V since the internal battery resistance is small. This is true for any two voltage sources working in parallel. Their voltages must be equal for equal load sharing. For this reason, the actual implementation of shared chargers becomes complicated; hence it is seldom used in practice. Separate

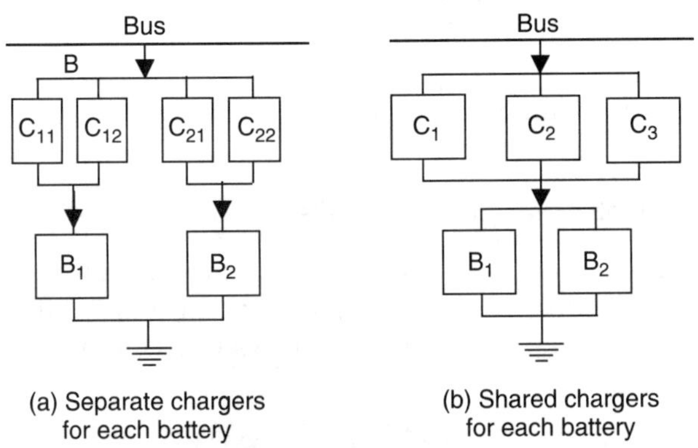

FIGURE 17.10 Battery charge/discharge converters configurations.

chargers for each battery require more parts resulting in greater mass. However, being simple in implementation, it is the most widely used redundancy configuration for battery charge and discharge converters.

17.6.5 Diode O-Ring for Active Redundancy

The primary and backup units often use a diode o-ring circuit as shown in Figure 17.11. Such a circuit makes the primary unit take almost all the load and the backup unit remains active in parallel taking only 5 to 10% of the load. The efficiency of the assembly is optimized by making the primary unit work more efficiently near full load. The uneven load sharing is achieved by R_1, R_2 and R_3 settings in the control circuit of the voltage dividers of the primary and backup units. The value of $R_2 > R_3$ makes the control voltage $V_{c1} > V_{c2}$, so that the primary unit takes almost all the load.

Figure 17.12 displays the method of determining the load sharing between two units. The terminal voltages drop lines of both units, V_p and V_b, are plotted versus load in individual units between two vertical lines separated by the total load to be shared. Since the two units are electrically connected in parallel, their common terminal voltage must have the same value V_t at the point of their intersection. The load is then L_p on the primary unit and L_b on the backup unit, and the total load is $L_p + L_b$.

The diodes are necessary in this circuit to isolate the two units against the reverse current flow under two possible conditions, namely:

- An internal fault in any one unit with the other unit feeding the fault, or
- The drift in V_{c1} and V_{c2} can cause the primary unit to take, say, 110% load and the backup unit to take −10%, i.e., the backup unit works as a load.

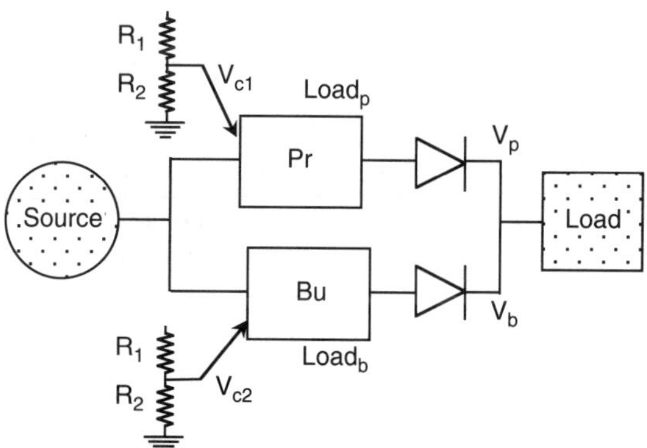

FIGURE 17.11 Diode o-ring circuit.

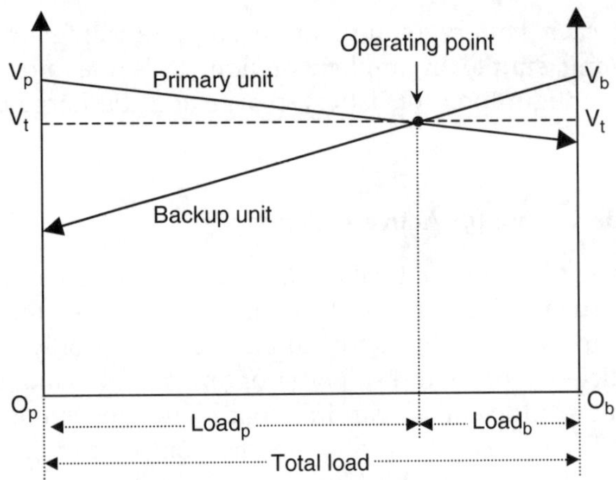

FIGURE 17.12 Load sharing between two parallel components.

17.7 Failure Rate Statistics

The failure rate is derived from tests on a very large population of identical components. It is not practical to carry out such tests to end until all components in the population fail, for it would take very long time. Instead, testing is done for a preset duration, failed components are replaced as they fail, and the testing is terminated after long enough time duration. This gives the results as intended because the failure rate is constant and the unfailed component at any time is as new as a brand new component. For a given population, the failure rate in unit time under the test environment is then simply given by

$$\lambda = \frac{\text{total number of failures}}{\text{total test duration}} \tag{17.20}$$

Statistics on failure rates under particular operating conditions are generally available from the manufacturer. Various organizations are also active in compiling such information and in publishing various documents. For example, MIL-HDBK-217 compiles voluminous data on various components under various operating environments. A comprehensive review of these data shows that passive components, such as capacitors, resistors, diodes, and connectors, are the most reliable. The least reliable parts are the traveling wave tube amplifiers (TWTAs) and components with moving parts such as sliprings, bearings, potentiometers, and relays. This is in line with what is expected.

17.8 MIL-HDBK-217

This military handbook establishes the uniform method of predicting reliability of military electronic parts, equipment, and systems. It lists the base failure rates of numerous parts under base thermal, electrical, and mechanical stresses. Any deviations from the specified operating conditions would alter the failure rate as given by the following expression:

$$\lambda_a = \lambda_b \pi_p \tag{17.21}$$

where

λ_a = failure rate under actual operating conditions in the design

λ_b = failure rate under base operating conditions specified in the handbook

π_k = factor to modify λ_b for environment, operating stress, and construction differences, if any, for $k = 1,2, \ldots n$

$\pi_p = \pi_1 \cdot \pi_2 \cdots \pi_n$ = product of all modifying factors.

Major factors that modify the failure rates of electrical and electronic parts are:

- π_E = Environmental factor to account for factors others than temperature.
- π_Q = Quality factor to account for difference in quality levels (class A, B, S, etc.) For example, this factor may be 1.0 for a class B part (commercial) and 0.01 for a class S part (space qualified.)
- $\pi_T = K^{\triangle T}$ = Operating temperature deviation factor, where $\triangle T$ is temperature deviation from the nominal rated value in °C, and K is a constant varying from 1.05 to 1.15 depending on the part construction. With an average value of K around 1.10, it is noteworthy that the failure rate factor doubles with every 10°C rise in the operating temperature, and cuts in half for 10°C drop.
- $\pi_D = (V_{op} / V_{rated})^{\alpha}$ = dielectric stress factor, where $\alpha = 2$ to 4 near rated voltage, and 5 to 8 above corona inception voltage. Obviously, the corona takes away life at a much faster rate. Therefore, corona in high-voltage equipment should be avoided.
- π_V = vibration factor, which depends on the fatigues life, which in turn depend on many factors such as the vibration amplitude, frequency, stress concentration, fracture toughness of the material used in the part fabrication, etc.

Table 17.2 Base failure rate for some selected electronic parts

Component	Failure rate, λ, per 10^6 h		Component	Failure rate, λ, per 10^6 h	
CAPACITORS			**TRANSISTORS**		
Ceramic	66	22	Signal, NPN	71	14
Film	78	26	Signal, PNP	100	20
Tantalum	200	63	Power, NPN	650	130
Electrolytic	1600	480	Thyristor, power	1800	360
			ICs, digital ≤ 20 gates	1050	18
RESISTORS					
Fixed, composition	45	45	ICs, linear ≤ 32 Qs	900	33
Fixed, wire wound	100	100			
Fixed, film	3	3	**TRANSFORMERS**		
Variable, cermet	2760	1380	Signal	10	3
Variable, wire wound	300	100	Power	82	32
			Switches, thermal	225	3
DIODES					
Signal	28	5.6	Connectors, contact pairs	32	4
Power	2160	432			
			Fuses	300	100

Source: MIL-HDBK-217.

There are other modifying factors listed in the handbook, which must also be accounted for in the reliability estimates. The base failure rate λ_b for some selected commercial and military power converter parts before applying the modifying factors are reproduced from MIL-HDBK-217 in Table 17.2.

17.9 Parts-Count Method of Reliability Estimate

For a bid proposal and early design stage, the unit reliability can be estimated by using a simple part count method outlined in MIL-HDBK-217 as follows. The information needed to apply this method are: (1) generic part types and quantities, (2) part quality levels, and (3) operating environmental of the unit. The general expression for equipment failure rate with this method is given by

$$\lambda_{eqp} = \sum_{i=l}^{i=m} N_i \lambda_{bi} \pi_{pi} \tag{17.22}$$

where

λ_{eqp} = total equipment failure rate (failures/10^6 h)

λ_{bi} = base failure rate for the ith generic part (failures/10^6 h)

π_{pi} = product of all modifying factors for the ith generic part to account for the environment, operating stresses and quality class

N_i = Number of ith generic parts used in the equipment, $i = 1, 2, 3, \ldots m$

m = number of different generic part categories used in the equipment.

The above method applies only if the equipment is used in one homogenous environment. Otherwise, it is applied to portions of equipment in each environment. Then the failure rates of those portions are added to determine the total equipment failure rate.

17.10 Derating for Reliability

The failure rate of a component depends on operating stresses, such as the voltage, temperature, etc. The electrical insulation at high temperature oxidizes, becomes brittle and may crack, leading to failure (short circuit). The oxidation is a chemical degradation, which follows the Arrhenius exponential growth. Data on numerous items of equipment show that failure rate doubles, or life shortened to one half, for every 7 to 10°C rise in the operating temperature. In the reverse, the life doubles for every 7 to 10°C reduction in the operating temperature. Similar degradation (wear) takes place above certain voltage, although not as well understood as that for the temperature. The rise in failure rate with the operating stress level is not to be confused with the wear-out failure rate described earlier in Section 17.3. It is to be seen as raising the flat part of the bathtub. The failure rate is still constant per unit time, although another constant at another operating stress level.

Derating is the reduction of electrical, thermal, and mechanical stress levels applied to a part in order to decrease the degradation rate and prolong the expected life. Lowering an operating stress on the component reduces the failure rate. Operating the component at lower than the nominally rated stress level is called 'derating'. It is routinely practiced to decrease the wear-out failure rates in military and space worthy designs. The derating in current is often done to lower the temperature. On the other hand, the current derating in some active devices, such as transistors, may be to control the di/dt rates, which can upset the semiconductor operation. Derating increases the margin of safety between the operating stress level and the actual failure level of the part. It provides added protection from system anomalies unforeseen by the design engineer.

The derating can come from many sources, and all should be considered. The primary derating comes from the environment and operating conditions of the subject part. The derating factor when multiplied by the normal rating gives a maximum recommended design stress level and does not preclude further derating from other factors. Parts having no derating factors listed on an approved part list may mean that inadequate data is available to make any recommendation. In that case, a judgment based on data from similar parts may be considered to derate the part in question with generous margin.

Most programs maintain their own preferred part list based on the failure rates. Such a preferred part list also gives derating factors, which must be applied to all designs on those programs. Both the selection of parts and the derating factors in the preferred part list are based on experience on the heritage designs successfully flown. Space agencies, such as NASA and the European Space Agency, maintain their own preferred part list. If a desired component is not in the preferred part list, it must be qualified by rigorous testing under the same environment as those in the list.

17.11 Quick Estimate of Failure Rate

The full scale reliability testing is time consuming and expensive. A quick estimate of the reliability for screening a new proposed part for a potential application can be obtained by testing several units in parallel until the first one fails in normal or accelerated testing. If T is the operating time to the first failure, obviously T is not the mean time between failures, since it is not derived from a large population. But it can be used to roughly estimate the MTBF as follows:

$$\text{MTBF} = 1/[1 - (1 - C)^{\frac{1}{T}}] \tag{17.23}$$

where C is desired confidence level, which is generally 60 to 90% in commercial world, and higher than 99.9% in defense and nuclear power equipment.

17.11.1 Example

To understand the above method of quick estimate, suppose we put to test 20 units of a new part, and the first unit failed at 3000 h. The total operating time of all units is then 20 × 3000 = 60,000 h before the first failure occurred. The MTBF of the population is then estimated using Equation 17.23 for different confidence levels as follows:

Confidence level C	50%	63.2%	90%	99%
MTBF for the population	72,130	60,000	26,058	13,030

Note that for exponential distribution, 1 – (1/e) = 0.632, and that is why the MTBF to the first failure has a confidence level of only 63.2%.

17.12 FMECA

The reliability analysis provides an estimate of the inherent reliability of the system. The failure mode and effect analysis (FMECA) on the other hand, goes beyond that. Its primary purposes is to identify not only the failure modes, but also their effects and criticality on the mission success. For example, criticality 1 means that if the component fails, it can cause injury to human, or cause the mission to fail completely. Obviously, such a component needs sufficiently high reliability. In other components, the probability of failure must be at the minimum possible level that is acceptable for the mission under design.

17.13 Non-Redundant Components

The top-level reliability specifications generally read as follows:

The power system reliability of meeting all the specified performance requirements shall exceed 0.95 over 15 years. No single point failure should cause loss of redundant paths for any critical function, nor should propagate to other systems. It should not connect the battery directly and permanently to the main bus.

To obtain such high EPS reliability of 0.95 over 15 years, many high-reliability components are used with redundancy. However, redundancy cannot be provided in the following power system functions, because it is either not practical even for a critical function, or is not really needed for a noncritical function:

- SAD shaft, bearing, stepper motor rotor, and gears cannot be practically made redundant from the spacecraft fabrication point of view
- Telemetry not critical to the spacecraft operation is nonredundant, such as
    - Battery cell voltage monitoring, and
    - Battery reconditioning circuit, if used.

It may be useful to cite some general design guidelines for achieving high reliability in the power system.

- For each unit, the wear-out failure rate is kept well below 10% of the random failure rate, and the MTBF is designed to be 40 to 60% longer than the mission life.
- The slip rings can fail in seizure because of the cold welding, which must be avoided.

- Laser welded connections in the bus and other wiring improves reliability.
- Wire break is a failure mode, often caused by a nicked wire that can occur during removal of the enamel insulation prior to soldering. Wire nicks should be avoided by a well-controlled manufacturing process.
- Single-point failure (short) in nonredundant wires is avoided by using double insulation.
- The American practice in the power system design has been to use hard low-resistance single-point ground. However, such systems would trip or clear the fuse under any fault involving the ground. The system reliability may be improved by soft grounding through a high resistance so that one fault may not shut down the power. This makes the system single ground fault tolerant. The concept of grounding to the structure through a several thousand ohm resistor was thoroughly investigated by NASA/Jet Propulsion Laboratory several years ago, but has not been implemented as yet in any spacecraft due to a some concerns, the major one being electrostatic buildup.

17.14 Example of Reliability Calculation

Consider a solar array that is required to have 99.2% reliability over a 15-year mission. Say it needs 30 shunt circuits to generate the require power. Each shunt circuit consists of three strings, 126 4 × 6-cm cells in each string, and has an estimated 15-year reliability of 0.95, including the string wiring, slip rings, shunts and the random cell failure at 1.0 FIT (failures per million hours). An array of 30 such shunt circuits would have the reliability of $(0.95)^{30} = 0.21$, way below the required value of 0.992. As seen in Table 17.3, for achieving 0.992 solar array reliability, we must use 32 circuits, two more than the needed 30 for the required power.

Table 17.3 Example of determining number of shunt circuits for achieving required reliability

No. of shunt circuits	Probability of failure	Solar array reliability
1 for 1	0.050	0.950
30 for 30	0.786	0.214
31 for 30	0.114	0.886
32 for 30	0.008	0.992
33 for 30	0.001	0.999

Chapter 18

Integration and Testing

18.1 Introduction

If an assembled spacecraft malfunctions, either on the test floor or in orbit, or anywhere between, a great deal of time and money are spent in detecting the problem, investigating why it happened, and in fixing it. A thorough testing during various fabrication and integration phases is required to weed out potential problems at an early stage, and to ascertain a timely delivery within the contracted budget. The spacecraft customer requires the prime contractor to prepare the Spacecraft Level Test and Verification Plan, which becomes the guiding document for all systems and components testing. The plan generally draws heavily from MIL-STD-1540 for environment testing and MIL-STD-1541 for EMC testing.

Once a component or an assembly is manufactured to the final design, acceptance testing is done on the floor to verify that no manufacturing errors or defects exist in the final product. Components are tested before they flow into their respective system assemblies, and the systems are tested separately before integrating into the spacecraft. The operating environment in space is simulated during the tests as much as possible and practical.

The tests typically performed on key power system component are listed in Table 18.1. If the thermal shock torture test on a component is required, it is often carried out by quenching the part from boiling water (100°C) to liquid nitrogen (−193°C). The conformal coating and ceramic parts may crack under such shocks.

Analyses may be used in lieu of testing to verify compliance to some requirements which may pose undue test difficulties. The selected analytical techniques typically include engineering analysis, comparative analysis, statistics, analog and digital modeling and computer simulation. The analyses must be preformed based on the data available at the critical design review stage.

18.2 Test Limits

All system components are subjected to extensive testing to prove their ability to perform reliably in the launch and orbit environment. Newly developed components undergo most stringent *qualification tests* at the

Table 18.1 Tests performed on key power system components

Solar cells	Battery packs[a]	Battery power converters	Pre- assembly
Cell performance	d.c. isolation[b]	Efficiency	Tri-channel control[d]
Deformation	Thermistors	Input voltage range	SAS1, SAS2 voltage drops
Solar panels	Strain gauges	Output voltage regulation	Close loop phase shift and gain
Circuit check	Ampere-hour capacity	Input/output ripple	Bus ripple
High-voltage isolation	Charge capability	Transfer function	**Command and telemetry**
Illumination and continuity	Retention of charge	Min/max loads	BPM accuracy
Thermographic map	Reconditioning	Local over-voltage protection	BCVM accuracy
Battery blankets	Electrolyte leakage	Command and telemetry checks	SAD functions
Grounding	Vacuum[c]	Him on/IUm off (?) i/face u?	SAD currents
Effective conductance	Heater operation	Battery current regulation	Auxiliary power redundancy and regulation
	By-pass protection		Failure detection

[a]Battery cells are tested in acceptance tests at the vendor site.
[b]Resistance of 100 MΩ minimum at in both pre-vibration and post-vibration Megger® tests.
[c]Exposure to $5 \times \cdot 10^{-5}$ torr for 8 h.
[d]Self-diagnostics test.

highest stress levels. The components previously flight qualified undergo routine acceptance tests before being placed in the spacecraft assembly. The test levels under these tests are higher than the expected stress levels under normal operating range. Typical temperature bands for various classes of tests are shown in Figure 18.1. Each spacecraft follows its own bands as specified in the requirements. But in all cases, the qualification temperature band is the widest, then the acceptance band, the design band, and then the normal operating band, in that order. Meeting these temperature requirements at various phases of testing assures adequate margins for the mission success.

The test temperature limits for components are derived from the following considerations:

- The flight heritage and temperature capability of the component
- Knowledge of expected temperature range within the spacecraft enclosures

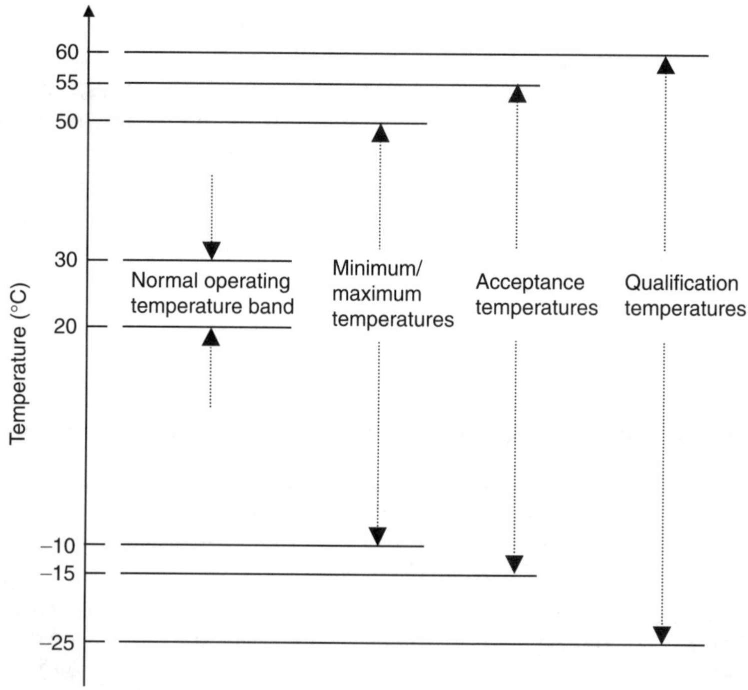

FIGURE 18.1 Temperature bands during normal operation and during acceptance and qualifications tests.

- Evaluation of thermal environment external to the spacecraft as a function of the thermal design variables
- Appropriate margins

The Test Plan is the contractual document that establishes the basis for derivation of component temperature limits. The plan generally requires that components be qualified at temperatures of 10 to 15°C more severe than the minimum and maximum levels expected during the entire flight mission. It requires a 5 to 10°C margin with respect to the same expected mission extremes for acceptance testing. The minimum and maximum temperature levels likely to be encountered during the flight mission are based on thermal analysis of the spacecraft design, environmental exposures, and internal dissipation profiles associated with all modes of the spacecraft operation. Because such analysis is subject to physical and mathematical modeling uncertainties, it is a standard practice to add 5°C uncertainty to the temperatures predicted by the analysis to derive qualification and acceptance test temperatures for the component. The customer specification defines expected environmental extremes and associated qualification and acceptance test levels for various components.

It also defines several other parameters for component thermal testing including the number of temperature cycles required, cumulative soak times, and the temperature transition rates. In certain cases, adequately sized thermostatically controlled heaters built in the design of a specific component permit reduction of the standard temperature margins.

Table 18.2 lists the component temperature limits for a typical GEO communications satellite. The basic considerations for defining the survival, qualification, acceptance, and design analysis temperature limits in the table are as follows.

Table 18.2 Component temperature limits for typical GEO communications satellite

Component	Survival (non-operational)	Qualification (operational)	Acceptance (operational)	Design analyses (operational)
Solar array	−180/75	−180/75	−175/70	−165/60[a,b]
Solar array drive	−35/70	−35/70	−30/65	−20/55[c]
NiH_2 battery	−15/45	−15/45	−10/30	−5/25[d–g]
Battery let-down unit	−35/70	−35/70	−30/65	−20/55
Battery cell voltage monitor	−35/70	−35/70	−30/65	−20/55
Bat pressure monitor	−35/70	−35/70	−30/65	−20/55
Power regulator unit	−35/70	−35/70	−30/65	−20/55
Ordnance controllers	−35/70	−35/70	−30/65	−20/55
Other power electronic boxes	−35/70	−35/70	−30/65	−20/−55

[a]60°C is for a 15-year mission. The PV conversion efficiency degrades more in a long mission, which increases the array temperature. For a 5-year mission, it could be 50°C.
[b]In the pre-operational mission phase, the array temperature can reach 110°C, so the array needs to qualify for 130°C.
[c]The maximum allowable instantaneous temperature difference between inner and outer races of associated bearings is 10°C in either direction.
[d]During trickle charge.
[e]Temperature margin for this item in cold conditions is provided by thermal control heaters with a minimum of 30% excess heating capacity.
[f]Maximum allowable battery cell temperatures during transient events are: 30°C during charge in transfer orbit, 30°C during reconditioning letdown, and 15°C during reconditioning recharge. The maximum allowable mission mean temperature of battery cells is 15°C.
[g]Maximum allowable battery cell temperature gradients are: 5°C within a cell surface, 7°C from the pack average to the dome of any cell, 7°C between pack averages within a battery assembly during charge, and 8°C between battery assemblies during discharge.

18.2.1 Survival (Non-operating Mode Only)

The component design should be capable of surviving exposure to the extreme survival temperatures such that:

- The component does not necessarily operate at the design performance level, but
- The structural integrity is maintained to prevent damage, contamination, or degradation of other spacecraft hardware, and
- Full functional performance is achieved when the temperature is subsequently brought within the qualification temperature range.

18.2.2 Qualification (Operating Mode Only)

The component design should be capable of providing functional performance as defined by the component specification at the qualification temperature limits.

18.2.3 Acceptance (Operating Mode Only)

Each component delivered to the spacecraft assembly should demonstrate the functional performance as defined by the component specification at the acceptance temperature limits.

18.2.4 Flight Design Analysis (All Modes)

The spacecraft thermal design should maintain the applicable interface or component element within this temperature range as demonstrated by the worst-case thermal analysis. Where no temperature range is shown in the non-operating column,

- Either the component is operational during the entire flight mission, or
- A specific temperature control is not required to preserve the integrity of the component and the spacecraft when the component is not operating.

18.2.5 Temperature Location

Location of the temperature must also be specified. It can be the area-average temperature at the interface of the mounting structure that is in direct conductive contact with the component, or the area-average temperature of a major structural element (usually a baseplate or mounting chassis) within the component. The location of this element is selected mutually by the electrical and thermal engineers.

18.3 Test Sequence

Each protoflight or flight unit, built in accordance with the documentation, procedure, and controls as reviewed and accepted during the unit's final design review is subjected to protoflight tests for qualifications or prior to acceptance for integration into the spacecraft. The input levels and range for the required acceptance and qualification tests are as specified in the component environmental test specification. The tests are carried out in certain sequences shown in Figure 18.2 and Figure 18.3, respectively. The description of each test follows.

18.3.1 Ambient Functional Performance Test

These tests include measurement of all performances in all modes of operation at ambient pressure and temperature. They are repeated a few times, starting with the initial performance tests, and ending with the final performance tests.

18.3.2 Leak

All pressurized or hermetically sealed units are subjected to leak tests. The leak test is conducted before and after the thermal vacuum, vibration, and pressure tests. The leak test is performed with the component pressurized at the maximum operating pressure and then at the minimum operating pressure if the seals are dependent upon pressure for proper sealing. The duration of the leak test is 6 h except for units which are part of the propulsion system. The duration of the propulsion component leak test should not exceed the time that the pressurized condition is normally experienced during propellant loading with a minimum of 6 h.

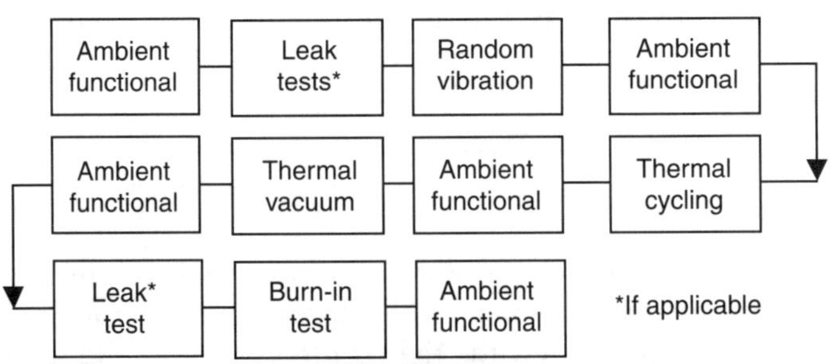

FIGURE 18.2 Component acceptance test sequence.

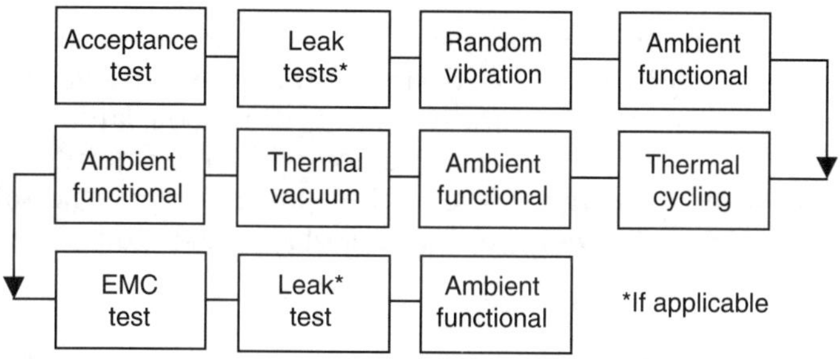

FIGURE 18.3 Component qualification test sequence.

18.3.3 Pressure

All pressurized units are subject to pressure tests. The pressure test is performed on the unit prior to and after successfully completed thermal vacuum and vibration tests. The pressure level for qualification is at least 1.5 times the maximum operating pressure. The hold time is at least 5 min. The test is then repeated for a number of cycles equal to at least 4 times the mission life at the maximum operating pressure.

18.3.4 Random Vibration

Each unit is subject to a random vibration test except for those tested for acoustic. The unit is mounted to a fixture through the normal mounting points and operated in the mode in which it is launched. The unit is vibrated once per axis in each of the three mutually perpendicular axes.

18.3.5 Sine Vibration

Each unit is subjected to a sinusoidal vibration test. The unit is mounted to a fixture through the normal mounting points and operates in the mode in which it is launched. The unit is vibrated once per axis in each of the three mutually perpendicular axes.

18.3.6 Post-Vibration Performance

These tests are to verify that there is no performance degradation caused by the above dynamic testing.

18.3.7 Acoustic

For units with large surface areas and relatively low density, it may be more practical to expose the qualification unit to an acoustic test instead of a random vibration test. If random vibration test is considered too severe, the unit is subjected to an acoustic test. The unit is installed in an acoustic chamber capable of generating the desired sound pressure level. During the test the unit is operated in the same manner as during the launch phase. It is monitored for malfunctions in accordance with the requirement of the applicable unit test procedure. In electrical power system, the solar array panels and the batteries are generally subjected to the acoustic test.

18.3.8 Thermal Vacuum Cycles

Each qualification unit is subjected to a series of ten thermal cycles in a vacuum less than 10^{-5} torr. During the thermal vacuum cycle, the unit's turn-on and turn-off capability is demonstrated, as well as the thermal design is verified. The performance test is conducted during the first and last cycles.

18.3.9 EMI/EMC

These tests are carried out using the test setup and procedures as per the EMI/EMC Test Plan.

18.3.10 ESD

Each unit is subject to an ESD susceptibility test. The unit is mounted to a graphite epoxy panel in a representative flight configuration. A current is injected into the panel in order to demonstrate that the unit is not susceptible to ESD induced transients. The test levels and duration is in accordance with the ESD Control Plan.

18.3.11 Life Test

All units which have not been previously life tested and are susceptible to noticeable wear, degradation, fatigue, thermally induced creep, or preload-induced creep during its projected lifetime is subject to life testing. During the life test, the unit is subjected to the conditions representative of those experienced on-orbit and during ground testing. When accelerated life tests are performed, each unit must demonstrate 50% life margin above its projected mission lifetime. In the power system, the battery cells, solar cells and SAD are subjected to life tests.

18.4 Solar Cell Tests

The incoming quality of solar cells is subjected to *I–V* performance, radiation, thermal shock, humidity, and peel tests as explained below.

The cell's *I–V* performance is tested under equivalent of one-sun illumination of a xenon arc lamp with infrared filters to match the sunlight spectrum. The voltage and current measurements are made over the complete range by varying the load resistor from zero to infinity. Since the cell performance is temperature dependent, the tests are made under controlled temperature. The short-circuit current, the open-circuit voltage, and the maximum power point are read from the *I–V* graph and compared with the specifications to which the cells were procured.

The performance degradation under space radiation is conducted on a test coupon using lab-generated radiation from a Van de Graaff electrostatic accelerator. The accelerator produces charged particle radiation at rates many orders of magnitude greater than those found in space, so that several minutes of test simulate the total equivalent 1 MeV fluence over the mission life. Such tests are conducted with and without the coverglass.

Thermal shocks are normal on solar array when going into eclipse (hot to cold) and coming out of eclipse (cold to hot). Test coupons are heated to high temperature (120°C) and then dipped in liquid nitrogen (−193°C) to simulate the first condition, and the reverse to simulate the second condition.

Humidity of around 95% is applied for many hours to verify that the moisture penetrating under the cell electrodes does not weaken the bond. Sticky tape is applied to the cell and then peeled off to test the bond strength.

18.5 Tests on Wires

Isolation testing between return wires of two circuits is conducted to verify that the following typical requirements:

- d.c. isolation test to verify that dc resistance of the insulation between wires is greater than 1 MΩ
- a.c. isolation test to verify that capacitance between the wires is less than 0.1 μF.

18.6 Power System Tests

The power system as a whole is subjected to the following major tests during fabrication and acceptance:

- Solar array illumination test under flash to verify that the bus power generation meets the BOL requirement

- Performance test to verify that all connections are made correctly, and the currents and voltages are measured as expected within the specified tolerance limits
- Electromagnetic interference and compatibility tests, with a particular emphasis on the conducted emission, which originates primarily form the power system
- Magnetic moment measurement on the assembled spacecraft. The solar array and electrical harness are primary contributors to the total moment
- Corona test under out gassing to demonstrate the minimum time to turn on and off high voltage and high frequency components when exposed to thermal vacuum
- Mechanical tests such as shock, vibration, and acoustic
- Thermal vacuum bake-out test to clean contamination, if any

The power system is again tested as follows when the spacecraft is fully integrated:

- Flash light go–no go tests on solar wings to check continuity of all connections in the power system
- Acoustic tests at wing level to check all joints, hinges, gimbals, etc.
- Continuity checks and insulation tests on batteries
- Continuity and insulation tests on all power electronics boxes

Major power system tests are described below.

18.6.1 Solar Array

The electrical performance of the solar array is measured at room temperature using a large-area pulsed sun-simulator. Demonstration of workmanship is accomplished by subjecting the solar array panels to 10 thermal cycles, followed by an electrical performance test. High and low temperatures correspond to the predicted highest and lowest temperatures in orbit. The solar array is subjected to the following tests after mounting the solar array drive:

- Electrical performance test
- Structural integrity by sine and acoustic tests
- Release deployment at zero gravity conditions

The electrical performance tests on solar array include the following series of tests.

18.6.1.1 Wiring loss

All wiring losses within the solar array assembly are predicted by analysis and verified by measurements.

18.6.1.2 Current–voltage

Using a solar simulator, an *I–V* curve is developed for each circuit on a panel, both before and after the vibration test. These data are used in predicting the solar array performance.

18.6.1.3 Power generation

The solar array panels on each wing are simultaneously illuminated by lamps at approximately one third the maximum solar flux. The array output and array shunts are monitored for qualitative operation. The telemetry circuits of the array are checked, and the spacecraft power system interfaces are verified.

18.6.1.4 Array life

The degradation of the solar array is computed by analysis to verify that the mission life requirements are met. Power distribution and other losses are included in the analysis.

18.6.1.5 Stability test

Worst-case conditions applicable to the solar array current, capacitance, constant power load, etc. are simulated to demonstrate stable regulation operation. This test is an alternate way to demonstrate stability margin since no direct measurement can be made with only the solar array assembly.

18.6.1.6 Shunt control

Measurements are made for each shunt control cut-in operation, voltage, and regulation for each unit and all units together. Bus stability simulation upon eclipse entrance and exit is demonstrated. The shunt drive capabilities at the unit level, system level, and the spacecraft level are demonstrated at various stages except at the launch site.

18.6.1.7 Shunt regulator and shunt

The solar array impedance measurements are made with the shunt regulator in the linear mode for each shunt unit. The test for the shunt demonstrates the ability to detect the detector failure and/or the controller failure and to switch over to the redundant detector and/or controller.

18.6.1.8 Thermal cycling test

The solar panel is installed and instrumented in a large temperature chamber or in some other facility with comparable performance. Electrical tests are performed to verify the thermal test instrumentation prior to closing the chamber door. After the chamber has been dehumidified, the panel temperature is increased until the highest reading thermocouple reaches +60°C. The panel temperature is held at this level for 16 h. It is now decreased until the highest reading thermocouple reaches the specified lower temperature level, and then increased until the highest reading thermocouple reaches the specified upper temperature level. The solar panel temperature is subsequently decreased to ambient conditions. This constitutes one complete thermal cycle.

18.6.1.9 Functional tests

The solar panels are subjected to pre-thermal functional test to determine whether manufacturing defects have been introduced into the fabrication, and whether the panels meet the specified performance requirements. Functional tests are performed before and after thermal cycling. As a minimum, functional tests include the following tests, unless specified otherwise.

- **Illumination tests:** All sections of each circuit and full panel are illuminated at AM0 equivalent intensity at +35°C with a standard solar cell selected in accordance with the power system specifications. The current–voltage (*I–V*) curves of the solar cell circuits are generated and electrical parameters such as I_{sc}, V_{oc}, and illumination intensity are recorded.
- **Diode tests:** The isolation diodes connected to the output of each circuit is tested for forward bias voltage drop at the operating current and for reverse leakage current at the operating voltage.
- **Shunt dissipator tests:** Each shunt dissipator section is tested to demonstrate its capability to meet all the applicable electrical performance requirements.
- **Temperature telemetry test:** Calibration data is acquired by recording temperature sensor data at ambient, hot, and cold temperatures.
- **Substrate isolation test:** The resistance between each solar cell circuit and the substrate to which it is bonded is measured and recorded.

18.6.2 Solar Array Drive

Testing each solar array drive individually includes the following:

- The rotational speed of each drive is tested.
- The drive torque for each redundant drive motor is demonstrated to satisfy the worst-case static load and temperature. This test demon-

strates both the step size and the drive rate to satisfy the panel orientation and the dynamic criteria.

- Maximum predicted current at maximum predicted operating temperature is passed through the sliprings for one complete revolution. The slip ring voltages are monitored to verify the slip ring operation and the contact resistance.
- Noise, vibration, acoustic, deployment shock, and steady state acceleration (spacecraft spin) are the mechanical tests performed on the solar array drive.

18.6.3 Battery Assembly

Each battery assembly undergoes acceptance level test sequence shown in Figure 18.4. In spacecraft with two battery assemblies, one is typically mounted on the north bus equipment panel and the other on the south bus equipment panel. Each equipment panel then undergoes the following acceptance tests with a nonflight battery that is used for in-house testing:

- Temperature and pressure sensors check-out
- Thermal cycle tests with charge and discharge cycles of the battery subassemblies as units
- Launch site capacity tests at spacecraft operating temperature
- Launch environment testing with the spacecraft
- A functional checkout of the complete battery system with the spacecraft during the system level launch environmental exposure, and the launch site–operations–integrated system testing. These tests include:
  - Thermal tests
  - Low- and high-temperature capacity tests

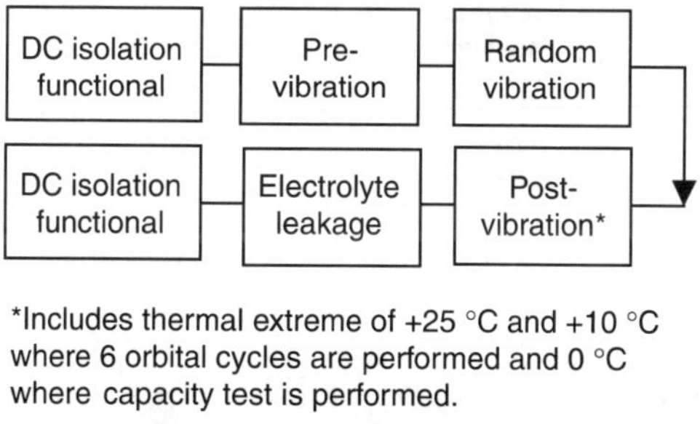

FIGURE 18.4 A battery acceptance test sequence.

- High discharge rate capacity test
- Overcharge and charge retention test
- Available capacity at minimum charge rate test
- Launch environment tests

18.6.4 Power Electronics

Power electronics, such as the control electronics and power converter assemblies, undergo acceptance testing in typical sequences shown in Figure 18.5.

18.6.5 Bus Impedance and Stability

The bus dynamic impedance is tested using a set-up shown in Figure 18.6. A small a.c. current of desired frequency is injected in the bus and the a.c. bus voltage at that frequency is measured. The dynamic impedance of the bus is then V_{ac}/I_{ac}. Such a test is repeated over the entire frequency range of interest using the frequency sweep of the instrument.

Figure 18.7 is a set-up for testing the small signal dynamic stability of the entire power system and the control loop. The power system and the load circuit are broken and a small a.c. voltage, V_{stab}, is inserted in the closed loop. The V_{ac1} and V_{ac2} are measured and plotted. The measurements are repeated over the frequency range of interest by the frequency sweep, and the gain and phase shift are plotted on Bode plots from which the gain and phase margins are determined and compared with the required values. This assures the power system stability under small signal perturbations in the test frequency range.

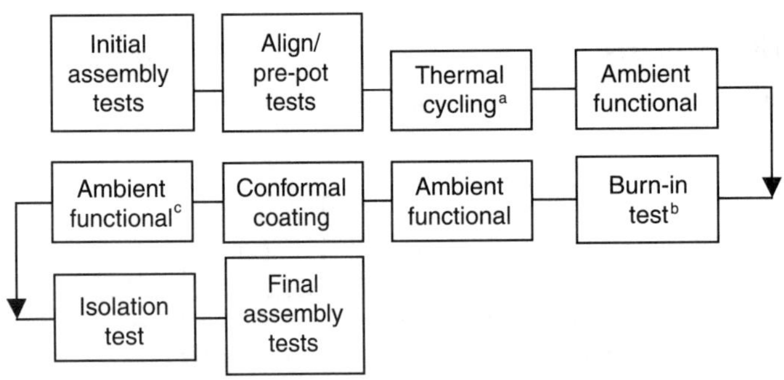

[a]8 cycles of {+55 °C, −20 °C}
[b]48 hours at +55 °C
[c]Optional

FIGURE 18.5 Power electronics assembly test sequence.

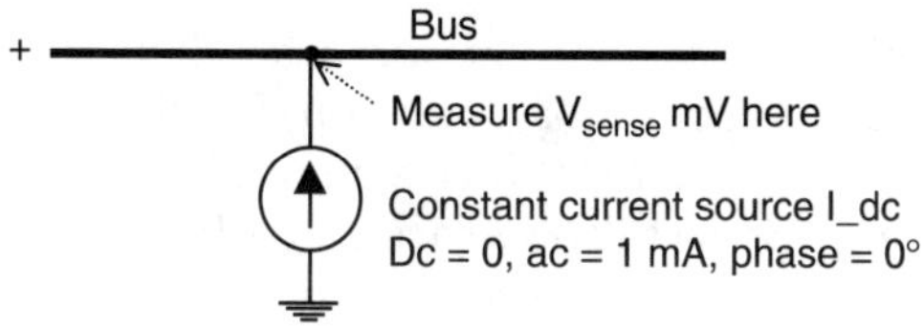

FIGURE 18.6 Dynamic bus impedance test set-up.

18.7 Spacecraft Level Tests

The spacecraft undergoes the following acceptance level tests after the integration of all major systems, including the power system.

18.7.1 Initial Power Turn-On Test

The initial power turn-on test is used to provide an orderly sequence for verifying that the application of power to the spacecraft does not damage previously integrated components or systems, and that the items being tested are not adversely affected. This test is used to initialize each component and system that makes up the spacecraft. A sequential turn on procedure is developed that allows initialization of each of the spacecraft's systems. Data sheets are developed on which the voltage and current values for each system are recorded. This allows a comparison between the current values and those obtained during the system testing and the component/system specifications. Where possible, breakout boxes and test tees are used for measurements of the electrical values.

18.7.2 Electrical Power System Test

This is the first full electrical test performed on the electrical power system after completion of the satellite assembly. This test is comprised of a performance baseline test conducted on each system while the remaining systems are active in their maximum EMI states. Both the primary and

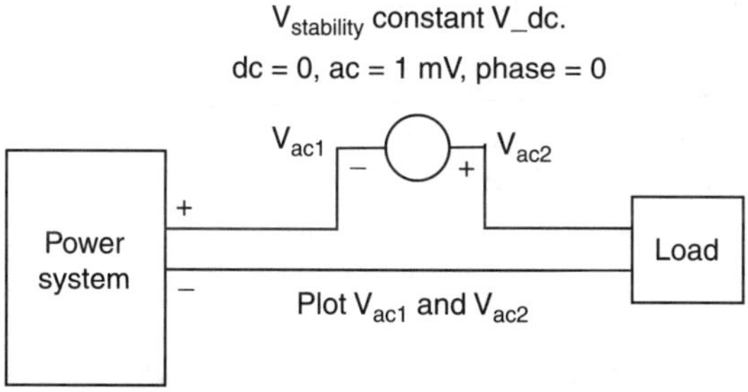

FIGURE 18.7 Small signal dynamic stability test set-up.

secondary path redundancies are measured for their compliance with the specification. Alternate modes of operation for the TT&C and payload communications systems are functionally verified and cross-strap switches are exercised during the test. The spacecraft RF signature is obtained in parallel with the electrical systems tests.

The objectives of the EPS tests at this stage are to:

- Verify that the spacecraft assembly operations are completed properly
- Demonstrate that the interfaces are operating and are interconnected properly between the bus and all other systems, and
- Establish the baseline payload performance data for the transponder/ antenna interfaces

Subsequent electrical system tests demonstrate that the spacecraft systems continue to meet all electrical specifications and provide a basis for trend analysis.

Each spacecraft system is initialized and checked to verify that it can be commanded into all its operational modes, and the electrical and/or mechanical performance are measured. These data, when compared with the data obtained during component or system testing, become the baseline for all following spacecraft tests and trend analysis.

Following the initial turn-on of the spacecraft, a part of the electrical system tests is performed during which the TT&C system is ON to ensure that a reliable communications link can be established between the spacecraft and the ground electrical equipment. This communications link is maintained with the spacecraft for the remainder of all electrical system testing. It provides a baseline test for the TT&C system.

18.8 Test Points

Sufficient test points are provided to permit testing of all on-line redundant and all failure detection circuits at the spacecraft level. Test points are also provided to permit open-loop stability testing (Bode plot test), large signal load transient testing on low-voltage and high-voltage busses, and bus impedance testing at the spacecraft level. All test points are buffered to protect from the external shorts, and are properly isolated so that a direct short to ground of a test point does not degrade spacecraft performance, nor overstress spacecraft hardware.

The following test points are generally provided as a minimum. The voltage test points are provided with sense lead carrying no current:

- Bus voltage test points typically at two places
- Battery voltage test points

- Battery charger out-of-limit current simulation test points
- Battery discharge converter failure simulation points
- Load current access port from the bus capable of carrying up to 10 A in satellites within a few kilowatts load
- Spacecraft star ground test points
- Spacecraft star ground access points for external load current return capable of carrying up to 10 A
- Load current access port from the low-voltage-regulated bus capable of carrying up to 10 A
- Shunt control system failure simulation test points, one for each control system
- Bus stability test points

18.9 Ground Test Equipment

External power sources, such as a solar array simulator and a battery simulator (or a workhorse battery), are needed for ground testing of the integrated spacecraft. Adequate protection must be provided to protect from erroneous input voltage to the spacecraft during these tests. Blocking diodes are placed in series with component input power leads for protection against application of the reverse voltage. The ground test set-up should also allow the spacecraft start-up and shunt testing for voltage regulation with the solar panel installed. Figure 18.8 is a typical ground test set-up. It shows only one simulator and one group of solar array circuits representing one wing of the spacecraft. Another group will be in parallel to represent the other wing.

The solar array is simulated with one or more current sources, each with several ampere current limit, and the voltage limit a few percent above the nominal bus voltage. Each source output is connected via umbilical connector to one of many solar array shunt circuit. A T-connection in the PRU connects the simulator to the shunt and the array circuits. Batteries are charged using the solar array simulator and the spacecraft battery charge converters. The initial power turn-on is accomplished with the solar array simulators providing power via the battery–PRU harness connectors without the battery connected. Once proper operation of the power system is confirmed, the workhorse battery is connected in its place. The battery relay turns the battery on and off.

18.10 Launch Site Testing

At the launch site, external power via an umbilical cable from the ground power source supports the spacecraft testing and the housekeeping loads

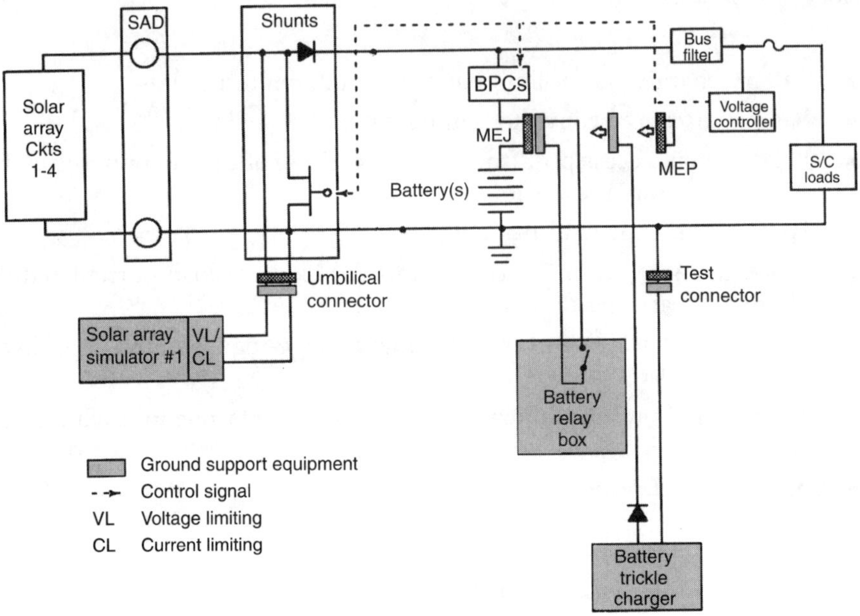

FIGURE 18.8 Ground test equipment set up.

FIGURE 18.9 *Hubble Space Telescope* satellite in integration and testing before going to launch site. (Source: NASA.)

prior to launch. The isolation diodes in the ground power supply prevent the current flowing backward when the ground source is de-powered or shorted. Typically, several amperes at nominal bus voltage are needed from the ground supply interface after the isolation diode. A battery trickle charge power supply may be connected to keep the battery fully charged when the spacecraft is turned off, such as for transportation between facilities. Such a feature minimizes the number of battery charge operation at the launch site.

Figure 18.9 shows the new *Hubble Telescope* spacecraft in the final integration and testing before going to the launch site for a planned launch in late 2004.

Part D
Special Power Systems

Chapter 19

Interplanetary and Deep Space Missions

19.1 Introduction

The power systems for interplanetary and deep space missions differ from those used in the Earth orbits because of the significantly different environments they encounter. The mission environment depends heavily on the spacecraft distance from the sun (Figure 19.1). Nine planets of the sun are divided into two groups: the inner planets (Mercury, Venus, Earth, and Mars) and the outer planets (Jupiter, Saturn, Uranus, Neptune, and Pluto). The inner planets are small and are composed primarily of rock and iron. The outer planets are much larger and consist of hydrogen, helium, and ice. The inner planetary missions are characterized by higher solar flux

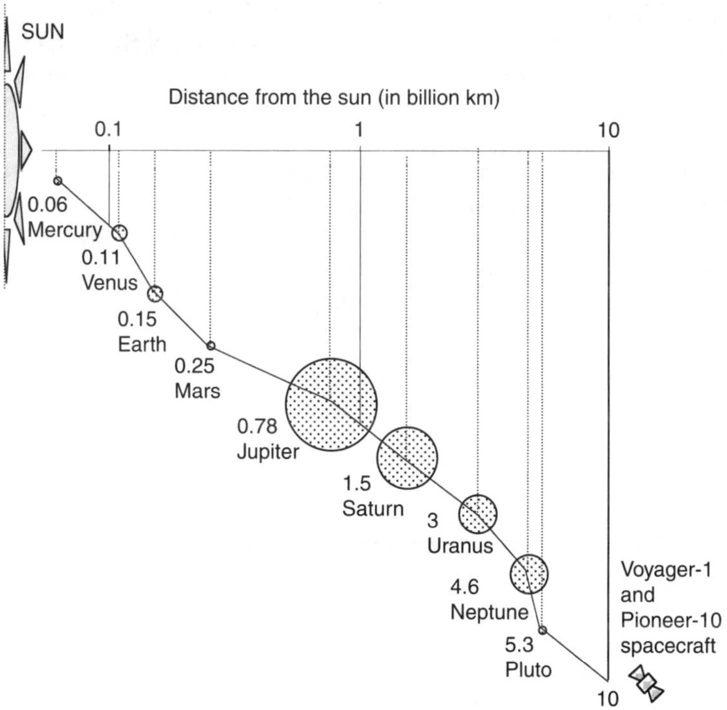

FIGURE 19.1 The planets in our solar system with the distance from the sun.

and elevated spacecraft temperature. Mercury orbit temperature can reach 300°C. Venus has a CO_2 atmosphere 90 times denser than on Earth, causing the greenhouse effect and the resulting temperature up to 477°C on the surface. Protecting the spacecraft from heat is a major design concern. On the other hand, missions to the outer planets, comets, asteroids, and deep space are characterized by lower solar flux and cooler spacecraft temperature. The temperature in Saturn orbit can be as low as −250°C. Pluto is even farther from the sun and so cold that methane freezes on its surface. Generating electrical power and protecting the spacecraft from extreme cold are major concerns in such missions.

If one could look at our solar system from far above the Earth's North Pole, the planets revolves around the sun in the counterclockwise direction. All planets, except Venus and Uranus rotate on their axes also in this same direction. The solar system is remarkably flat — only Mercury and Pluto have inclined orbits. The moons of various planets mimic the behaviors of their planets and move in the counterclockwise direction with a few exceptions. Some remarkable patterns also exist. For example, Mercury rotates on its axis three times for every two revolutions around the sun; and Uranus and Neptune are deficient in hydrogen.

The asteroids are small rocky bodies that move in orbits primarily between Mars and Jupiter. Numbering in thousands, they range in size from 1000 km diameter to microscopic grains. Comets circle the sun in two main groups. The Kuiper belt is a ring of debris that orbits the sun beyond Neptune. Many comets with periods less than 500 years come from the Kuiper belt.

For missions away from the Earth, either closer to the sun or farther away from the sun, the solar array, battery, and the power electronics designs differ significantly because the extreme temperature on either side, high or low, has a large impact on their performance.

The solar flux at any distance in deep space is given by

$$I = \frac{I_{Earth}}{R^2} \tag{19.1}$$

where I_{Earth} = solar flux in the Earth orbit (1358 W/m^2), and R = distance from the sun in astronomical units (AU). This equation assumes the sun to be a point source, and may give some error at distances less than a few sun radii. The solar array power output varies linearly with the incident solar flux. Therefore, as the spacecraft moves away from the sun, the power decreases inversely with the distance squared. In the PV power system, the array temperature also decreases in the same ratio, which results in a higher PV conversion efficiency. The combined effect of the flux and the temperature changes is such that the photovoltaic power generation varies not inversely with the distance squared, but to a lesser exponent in R^{α}, where

Table 19.1 The solar flux and PV power generation in the orbits of the various planets in our solar system (relative to those in the Earth's orbits)

Planet	Distance from the sun (AU)	Solar flux relative to Earth orbits	PV power generation accounting for temperature difference
Mercury	0.31–0.47	10.40–4.52	Severe loss of voltage
Venus	0.72	1.93	1.63
Earth	1.0	1.0	1.0
Moon	1.0	1.0	1.0
Mars	1.66	0.36	0.59
Jupiter	5.20	0.037	0.084
Saturn	10.08	0.0098	0.031

α is approximately 1.5. Thus, the power generation is less than two times with two times the solar flux, and more than one half with one half solar flux.

Table 19.1 lists planets in our solar system with their distances from the sun, in AU, and approximate power generation rates in their orbits considering both the solar flux and the temperature variations. The data transfer rate to Earth from outer space depends on both the distance and the power available as depicted in Figure 19.2.

19.2 Temperature in Deep Space

The Stefan–Boltzmann law determines the temperature of an object which receives and radiates heat. With the spacecraft surface radiating to absolute zero in deep space, the surface temperature is determined by

$$T = \left[\left(\frac{I}{\sigma}\right)\left(\frac{\alpha}{\varepsilon}\right)\right]^{\frac{1}{4}} \tag{19.2}$$

where T = absolute temperature in K, I = solar flux in W/m^2, σ = Stefan–Boltzmann constant (5.67×10^{-8} W/m^{2}K^4), α = absorptivity, and ε = emissivity of the surface. For a solar panel with one side absorbing and two sides radiating, the above equation becomes

$$T = \left[\left(\frac{I}{\sigma}\right)\left(\frac{\alpha}{2\varepsilon}\right)\right]^{\frac{1}{4}} \tag{19.3}$$

The α/ε ratio of various surfaces varies from 0 to 1 (Table 19.2). It is near zero for shiny mirror type surfaces, near 1 for flat black surfaces, and 2 to 8 for highly absorbing surfaces. Typical solar cells have this ratio near 0.8.

As an example of using the above equation, we determine below the solar array temperature in the Earth's orbit during sunlight after ignoring the

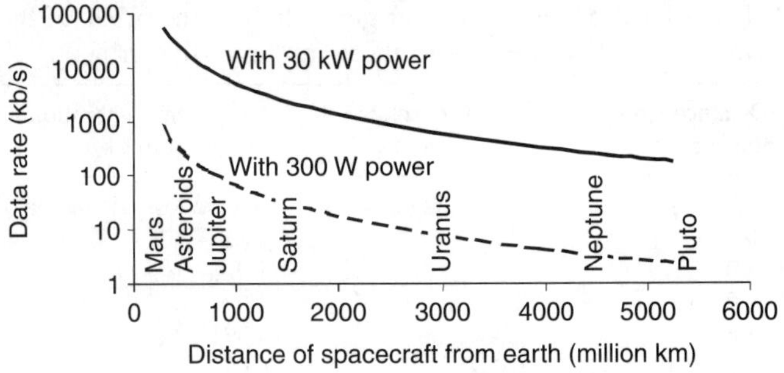

FIGURE 19.2 The data transfer rate from outer space at two power levels.

Earth albedo and infrared radiation on the spacecraft. The array with α/ε ratio of 0.8 would give rise to

$$T = \left[\left(\frac{1357}{5.67 \times 10^{-8}}\right)\left(\frac{0.8}{2}\right)\right]^{\frac{1}{4}} = 313\text{K or } 40^{\circ}\text{C} \tag{19.4}$$

And, the array with an α/ε ratio of 0.6 would give rise to

$$T = \left[\left(\frac{1357}{5.67 \times 10^{-8}}\right)\left(\frac{0.6}{2}\right)\right]^{\frac{1}{4}} = 291\text{K or } 18^{\circ}\text{C} \tag{19.5}$$

On the other hand, a spacecraft in deep space at 2 AU distance from the sun receives solar flux $1/4^{\text{th}}$ compared to that on Earth. Its solar array with an α/ε ratio of 0.8 would give rise to a temperature of

$$T = \left[\left(\frac{1357/4}{5.67 \times 10^{-8}}\right)\left(\frac{0.8}{2}\right)\right]^{\frac{1}{4}} = 221\text{K or } -52^{\circ}\text{C} \tag{19.6}$$

As for the power generation, missions beyond Mars cannot effectively generate PV power because of decreasing solar intensity. Moreover, the

Table 19.2 Absorptivity and emissivity of various surfaces

Type of Surface	Absorptivity, α	Emissivity, ε	Ratio α/ε[a]
Solar reflector (mirror)	Low	High	< 0.1
Solar reflector	Low	High	0.1-0.2
Flat black reflector	Low	Low	1
Flat black absorber	High	High	1
Solar absorber	High	Low	2-8

[a]A typical PV array has this ratio around 0.8.

greatly varying sun angle, illumination intensity, and temperature in such mission cause wide variations in *I–V* response. The *NEAR Shoemaker* spacecraft's PV power system shown in Figure 19.3[1] is designed for use at AU distances varying from 1 to 2.2, farthest from the sun to date and much beyond Mars. Although the peak power tracking architecture is well suited under such conditions, the direct energy transfer PV–SuperNiCd battery system were used in the *NEAR Shoemaker* spacecraft to overcome certain unique design challenges.[1]

Missions to the inner planets and near the sun would also be difficult to design with a photovoltaic power system because of the strong solar intensity and the resulting high temperatures. However, some examples where PV cells have been successfully used in the past follow. *Mariner-2* and *Mariner-5* went to Venus, and *Mariner-10* went to Venus and Mercury in 1973. A long duration *Magellan* mission went to Venus in 1989. For missions either too far or too close to the sun, the radioisotope thermoelectric generator (RTG) becomes a strong candidate, which is covered in the next chapter.

The power system related environmental aspects in some interplanetary and deep space missions undertaken by NASA in the past are discussed below. The interplanetary space is relatively free of charged particles. However, the charged particles trapped in the planet's orbit due to the planet's magnetic field do impact the power system design. If the planet has no magnetic field, its orbit is free of Van Allen type trapped particle belts. However, untrapped particles traveling to the surface of the planet unimpeded can cause some damage there. Since Mars has negligible magnetic field, it has no radiation belts. On the other hand, Jupiter has a high magnetic field, hence more trapped particles in its orbit.

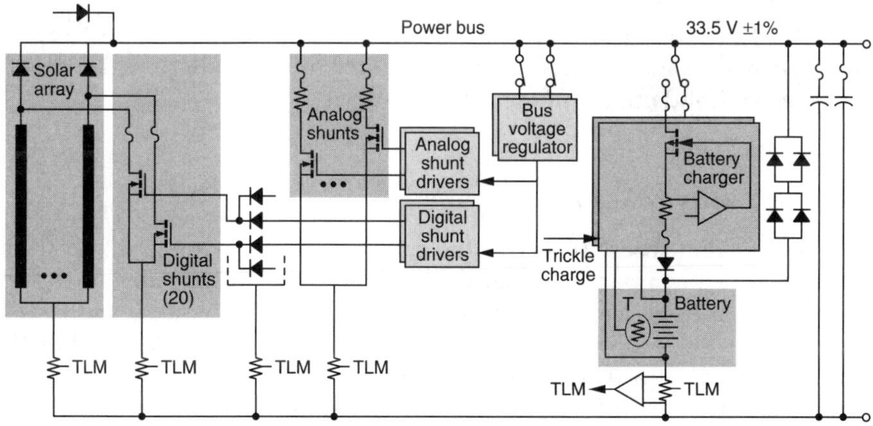

FIGURE 19.3 Power system for NEAR-Shoemaker spacecraft to asteroid.
(From: J.E. Jenkins and G. Dakermanji, Proc. 36th IECEC, ASME, p. AP-15, 2001. With permission.)

19.3 Lunar Mission

The moon, although not a planet, is the closest neighbor of significant size that has drawn our attention for thousands of years. More recently, it has been the subject of human exploration by actual landing on its surface. *Apollo-11* was the first manned mission that landed on the moon in 1969. In 2004, China announced its own unmanned lunar program with a goal of landing in 2010. Also in 2004, in the US, President Bush announced a long-term space program for human landing on the moon by 2015, and then on Mars. The following are the environmental features that impact the power system design for lunar missions.

- The moon, being the Earth's natural satellite, is approximately at the same distance from the sun as the Earth. The solar flux in the moon's orbit is therefore essentially the same as that in the Earth's orbit.
- There is no atmosphere around the moon. The pressure ranges from 10^{-8} to 10^{-2} torr. This is essentially a vacuum, which can support a high-voltage power system. However, out-gassing from the surrounding equipment can alter the local pressure. This should be accounted for in the high-voltage breakdown considerations.
- There is no space plasma. Plasma means ionized gas, and if there is no gas in the atmosphere, there cannot be any plasma.
- The temperature on the lunar surface ranges between -175°C during night and $+75^{\circ}$C during day, about the same as that in the Earth's orbit.
- The gravity is about $1/6^{\text{th}}$ of that on the Earth's surface.
- The moon has negligible magnetic field. This means that the sun's charged particles can reach the moon's surface unimpeded, without being deflected and trapped in belts. The solar array degradation rate due to charged particles would therefore be higher than on Earth.
- The meteoroid impact rate is higher on the moon, giving a higher rate of mechanical damage on the solar array.

19.4 Mercury Mission

Mercury is the closest planet to the sun, and has a diameter of 4880 km, gravity $1/18^{\text{th}}$ compared to that on Earth, and no air or water. It is very dense with a large iron core. Its surface has countless craters, the largest being about 1300 km in diameter. It has a weak magnetic field. It orbits the sun in 88 days at a distance varying between 0.31 AU and 0.47 AU. The solar intensity varies from 10.6 suns to 4.6 suns, with the resulting solar array temperature rising to 430°C at the equator. Since the planet has no atmosphere and rotates slowly on its axis once in 60 Earth days, most of the heat is lost at night with the temperature plunging to -160°C.

Due to its proximity to the sun, the electrical power system for a Mercury mission must meet the harsh thermal and radiation environment. Haines[2], of the European Space Agency, has recently reported the power system design for a mercury sample return mission. It consists of three independent power systems for each phase of the mission. For example, a 20-kW, 100-V high-power system shown in Figure 19.4 is used for electric propulsion, which is jettisoned just before the orbit insertion. After that, a 500-W, 28-V system is used for the orbiter, and a smaller power system for the surface landing, sample collection, and return phase of the mission. The electric propulsion power system array has the following design features:

- Peak power tracking power system in which the array is rotated as it approaches the sun in order to limit the temperature
- The solar array has a thermal shield along the length of the panel to assure that the panel edges are not overheated when tilted
- The solar cells are GaAs with a 100-μm-thick ceria microsheet cover-glass coated with infrared and Earth albedo reflectors
- Adhesives used in the solar array fabrication are compatible with the high temperature operation
- The cells are interleaved with about 100-μm ceria microsheet surface mirrors to reflect a part of the solar flux for controlling the temperature
- Solar panels are rotated ±180° around one axis to keep the operating solar cell temperature below 120°C from 1 AU to 0.375 AU and below 150°C at AU < 0.375. The sun shielded solar array drive is used with no

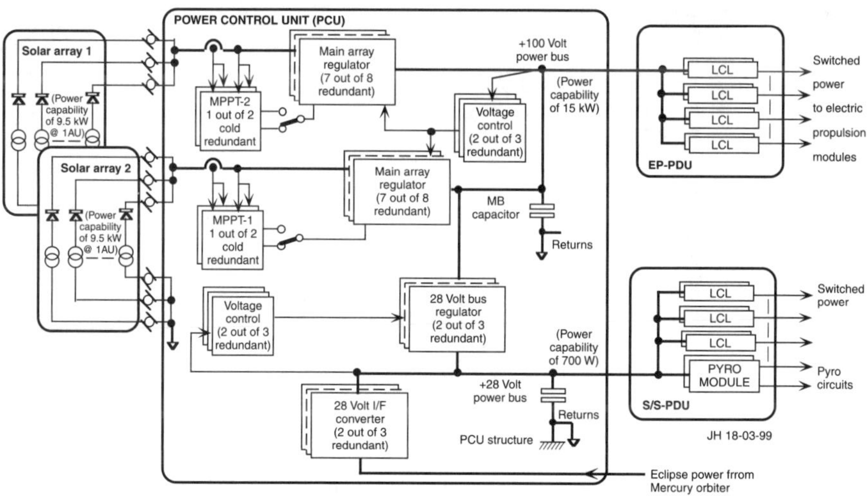

FIGURE 19.4 Power system for electric propulsion to Mercury. (Source: European Space Agency/J. Haines.)

slip rings, but with flex harness (coiled cable) sufficiently flexible for ±180° rotation

- Lithium ion battery is used to meet the orbit parameters

19.5 Near-Sun Mission

For missions near the sun, the solar intensity increases to 100 suns at 0.1 AU (~21 sun radii, 1 sun radius equals 0.00476 AU), and to 2500 suns at 0.02 AU (about 4 sun radii). The PV power system design for near-sun missions between Mercury and the sun needs special considerations due to high temperature. The PV cell loses power generation capability at such temperature due to loss in the open circuit voltage. Various options to limit the temperature below 1000°C include array tilting, adding mirrors on the surface to decrease absorptivity and increase emissivity, partially silvered cover glass, and various louvers and shades to control the solar flux. Moreover, the PV cell having high band gap is needed. Figure 19.5 from a paper by Brandhorst and Chen[3] shows effective power output as a function of AU distance and band gap of various PV cells operating below 1000°C. At distances greater than 0.5 AU, the bandage has no significant effect on power generation. At a distance of less than 0.5 AU, the higher band gap PV cell generates more power up to 0.1 AU. Closer than 0.1 AU, the PV cell becomes useless. The curves assume that the cell temperature is limited to 1000°C in all cases.

It is apparent from Figure 19.5 that the PV power system has limitations in approaching the sun at a close distance. An alternative approach may be

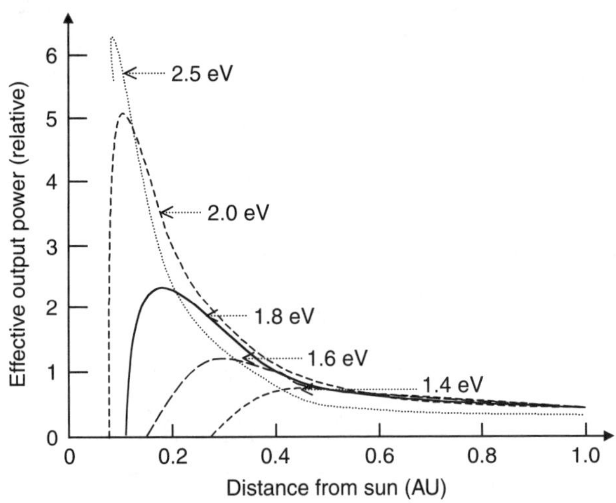

FIGURE 19.5 The distance from the sun versus power output of PV cells of various band gaps operating under 1000°C.

to use the thermo-photovoltaic (TPV) direct energy conversion. There are several advantages, including easy coupling to a thermal source operating above 2000 K. The feasibility of such an approach has been demonstrated under DoE funding, but has not been fully developed.

Another alternative is to use the thermoelectric converter (TEC) with the sun as the heat source. Such system is feasible for solar probes requiring instrument power under a few hundred watts. For example, NASA/JPL designed for launch in late 2003 the sun-TEC power system for a fly-by probe to Jupiter and then towards the sun to study the coronal heating and origin and acceleration of solar winds. It will become the first spacecraft to fly in the sun's *atmosphere*. In the power system design reported by Choi[4], the probe's distance from the sun varies greatly, from 5.2 AU near Jupiter (gravity assist orbit) to less then 0.1 AU (4 solar radii) near the sun. The corresponding solar flux varies over 5 orders of magnitude from $50\,W/m^2$ to $4 \times 10^6\ W/m^2$. The spacecraft bus is shaded by the primary sunshield blocking the sun (Figure 19.6). The shield outside temperature is estimated to be 2100°C at 4 solar radii. A high-temperature multi-layer thermal blanket keeps the spacecraft components cool. The shield and the blanket are made of carbon–carbon composite.

The solar probe has three different power systems: (a) low temperature solar array for AU > 0.5, (b) high-temperature solar array operating at or below 400°C between 0.5 and 0.1 AU, and (c) TEC for AU < 0.1 when the solar array is retracted inside the umbra of the sunshield.

Near the Earth, the low temperature solar array generates 2700 W of power. Since this is far more than needed, most of it is shunted in the power system. Between 0.5 and 0.1 AU, the power is provided by the high temperature solar array, which is retracted inside the umbra of the sunshield at 0.1 AU when the array temperature exceeds 400°C. Inside 0.1 AU up to 4 sun radii, 200 W of power is generated by using infrared heat from the primary sunshield on the TE converters. In a concept described by Choi (Figure 19.7), silicon germanium/gallium phosphorus (SiGe/GaP) TE converters use infrared radiation from the spacecraft primary heat shield as the primary energy source. A liquid sodium high-temperature cooling system to maintain the TE hot side to 1200 K. The (SiGe/GaP) converter efficiency is about 7% at 1200 K. The remaining 93% heat is radiated back into space by using carbon–carbon composite and liquid sodium heat pipe radiators.

19.6 Mars Mission

A spacecraft takes more than 7 months to reach Mars — also known as 'the red planet'. By contrast, going to the moon takes just 3 days. Although Mars, as a planet, is close to the Earth, it has proved to be very inhospitable to us. Of about 35 missions since 1960, more than 22 have crashed en route

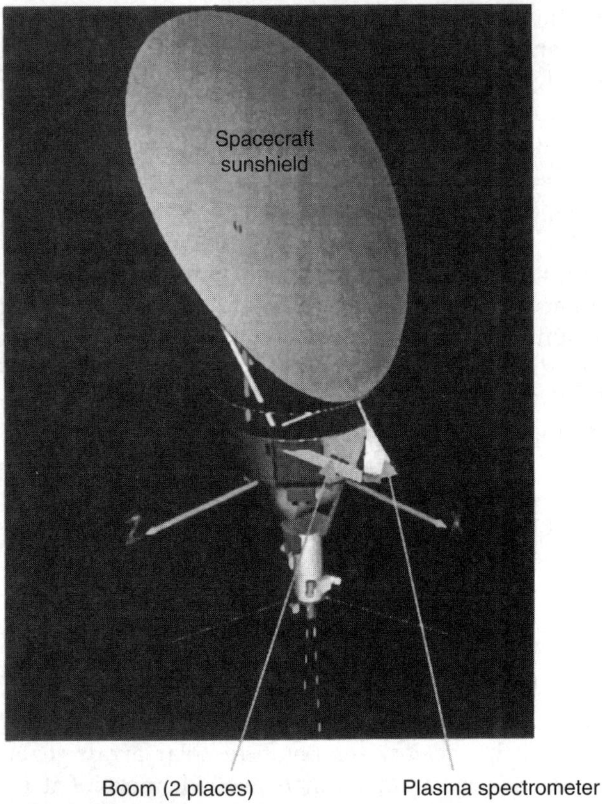

FIGURE 19.6 A solar probe in the intense heat of the sun at an AU distance of 0.1. (From M.K. Choi, Proc. 36th IECEC, ASME, p. TM-17, 2001. With permission.)

or otherwise failed before they could complete the mission. And yet, in June 2003, we launched four spacecraft, two by NASA and two by ESA, to study the history, geology, and biology of Mars. NASA's rovers, *Spirit* and *Opportunity*, reached Mars in January 2004.

Mars probably once had liquid water on its surface, but now its CO_2 atmosphere is so thin that the planet is dry and cold. The Mars environment has the following features:

- The distance from the sun is 1.66 AU, with the average solar flux 36% of that in the Earth's orbit.
- The typical operating temperature for large area solar array range between −100°C and 0°C. The low-temperature affects material properties and PV cell performance.
- The surface gravity is about 40% of that on the Earth.

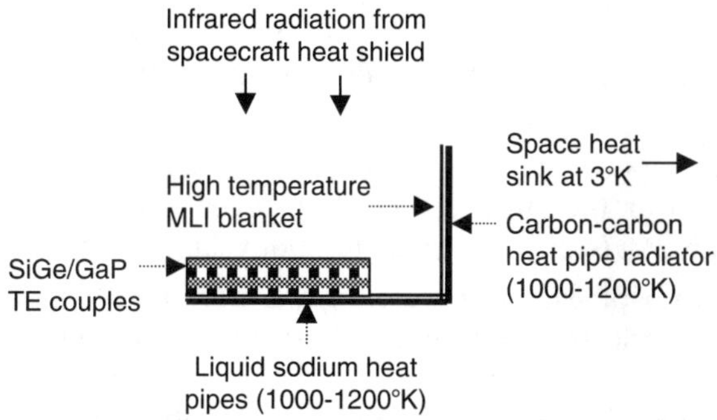

FIGURE 19.7 The TEC layout for NASA's solar probe.

- The atmosphere consists of 95% CO_2, 3% nitrogen, and 2% all other gases.
- The atmospheric pressure is low — in the order of 10 torr. This happens to be the worst possible value on the Paschen voltage breakdown curve for CO_2. At 10 torr, the breakdown voltage between bare conductors in CO_2 is 50 V. Therefore, the power systems design with exposed conductors must be limited to 35 V to leave some design margin. Higher voltages must be encapsulated or shielded in an enclosure with a suitable pressurized gas, which can conveniently be CO_2.
- The dry low-pressure atmosphere is conducive to electrostatic charging. The spacecraft surfaces are likely to collect a large amount of triboelectric charge via moving surfaces and wind-born dust particles.
- The temperature ranges between −100°C and −30°C, somewhat cooler than in the Earth's orbit.
- The great dust storms on the Mars surface cover an area greater than 1 million km^2 with durations over 100 days and a peak opacity exceeding 6. The dust is about 60% SiO_2. Such storms occur with a yearly probability of 30 to 80%. The atmospheric dust scatters and absorbs solar wavelength radiation to affect the magnitude, angular distribution, and the solar radiation spectrum.
- Local dust storms covering less than 1 million km^2 area occur with a probability of 5% in the equatorial regions. They have only minor impact due to their limited size, short duration of a few days, and opacity around 1. Frequent dust storms occur in gusts of 30 to 75 miles per hour.
- The dust haze, ground fog, CO_2, and water ice clouds obscure the sunlight. The dust particles precipitated and collected on surfaces following the dust storm activity reduce the transmission of light and decrease the surface emissivity. This raises the array temperature. The PV cell current loss of 0.28% per Martian day due to dust precipitation

was measured over a short period of time in 1997 during the *Pathfinder* mission. The dust collection problem makes the solar array effective only for 3 to 6 months on Mars.

- The wind speed in the upper atmosphere can exceed 100 m/s. However, near the surface boundary layer, the dust wind speed can range from 3 m/s at the surface to about 50 m/s at 5-m elevation.
- With an atmospheric pressure of only 6 torr, the PV array aerodynamic loading is modest in high wind, about twice the body force.

Mars has a negligible global magnetic field — about 3×10^{-5} times that of Earth. However, NASA's Mars global survey data suggest that the planet has pockets of magnetic areas that can be 10 to 20 times stronger than the Earth's magnetic field.

The electrical power system design issues in low Mars orbit are somewhat similar to that in LEO, including the AO degradation and arcing to plasma. On the surface of Mars and the moon, Paschen breakdown of atmospheric gases in a low-pressure neutral environment, and chemical and electrical interactions with dust may also be of concern.[5,6] The Mars environment is thick. It provides effective shielding from meteors and proton/electron radiation. Therefore, the PV array design can ignore the radiation degradation and meteoroid damage. The design must, however, consider the radiation protection of human crews and sensitive electronics.

NASA is studying a novel wind power system using the sand storms on Mars. Wind turbines designed to general electricity at Earth's South Pole and in remote regions of Alaska, where there are about 6 months of darkness each year, may some day lead to similar wind machines for Mars bases. One reason to propose wind power on Mars is that wind turbines could generate electricity during the month-long Martian global dust storms that can make days on the red planet as dark as night. Wind power and solar power may complement each other on Mars. When a large dust storm blocks the sunlight, a wind turbine can generate electricity.

19.7 Missions to Jupiter and Saturn

Jupiter is the largest planet, with an immense magnetosphere. Its hydrogen and helium atmosphere contains pastel-colored clouds. Saturn is similar to Jupiter in many ways. The sun is not bright enough near Jupiter and Saturn, making the PV power system ineffective. The electrical power is therefore generated using the RTG, which is covered in the next chapter. A few missions to Jupiter and Saturn are described below.

NASA's *Galileo* spacecraft to Jupiter at 5.2 AU distance from the sun was a mission in two parts, the orbiter and the probe. The probe was designed to enter the Jupiter environment. After delivering the probe, the orbiter was routed to the Jupiter's moon Ganymede. The probe recorded a temperature of −171°C above the clouds, 305°C near the surface, and the wind speed of 400 miles per hour. The probe was destroyed due to high atmospheric pressure. In 2003, the orbiter concluded its 14-year exploration of Jupiter and its moons. This 3000-pound spacecraft traveling at 108,000 miles per hour ran out of the on-board fuel used to steer it during its 35 orbits of Jupiter. *Galileo* was then purposely plunged into its own atmosphere to avoid crashing and contaminating Europa, one of Jupiter's watery moons.

The radiation hardening of the power system components is more important in Jupiter orbits than in orbits of other planets. This is because Jupiter has a strong magnetic field and strong Van Allen type radiation belts. Spacecraft encountering the powerful radiation belts of Jupiter have undergone single event upsets. This becomes a significant design issue.

Pioneer-10 spacecraft was built by TRW Corporation (Figure 19.8) and launched in 1972 to probe the environment of Jupiter. It was designed for a 3-year life, but kept ticking for more than 30 years. It passed Pluto's orbit in 1983. In 1997, it was some six billion miles away from Earth, much beyond Pluto, cruising at 27,700 miles per hour to become the farthest man-made object in deep space. It used RTGs with plutonium-238, and produced more than two thirds of its 1972 power in 1997. Citing diminishing return, NASA

FIGURE 19.8 The *Pioneer-10* spacecraft with two RTG power sources at the end of two booms. (Source: NASA Ames Research Center.)

halted all regular communications with *Pioneer-10* in 1997. The last regular signal received from it was 1×10^{-21} W, and took 9 h to reach Earth. In 2002, scientists at NASA's Jet Propulsion Laboratory sent a signal more than 12 billion kilometers away in an attempt to contact the spacecraft. After 22 h, they detected Pioneer's response on a 70-m dish antenna in Madrid, Spain. Even without communication to and from Earth, it will continue its journey in deep space. It carries pictures of terrestrial man and woman, and our address, i.e., the location of the Earth in our solar system. It cannot reach Proxima Centauri, the nearest star next to our sun, as it would take 100,000 years at the current speed.

Voyager-1 is another spacecraft, launch in 1977, which passed by Jupiter and Saturn. In 2004, it left our solar system and entered the interstellar space, when it was more than three times as far as Pluto from the Earth. However, it is still working after more than a quarter century since its launch. If it continues at its present speed of over one million miles a day, it could take 40,000 years before reaching the nearest star.

The *Cassini* spacecraft was launched in 1997 on a 7-year mission to Saturn. It is designed to send back data after entering the orbit of Saturn in late 2004 and making 60 orbits in various orientations around the planet over a 4-year period. The mission is a joint undertaking of NASA, the European Space Agency, and the Italian Space Agency. It carries RTG with 30 kg of plutonium on board. Because of such a large amount of the potentially deadly radioactive element, antinuclear groups opposed its launch. Public concerns have negatively affected the future of nuclear power in space, although Russians have built and flown many small nuclear reactor–TEC power systems similar to the U.S. built SNAP-10.

19.8 Deep Space Missions

The power system for a deep space mission typically includes RTGs, power electronics, and a small battery located inside the spacecraft body. The RTG heat may be sufficient to protect the power system from low temperature. If not, additional isotope heat is needed to keep the electronics at the required temperature. For example, an unheated interplanetary spacecraft launched to explore the rings of Saturn would experience an average temperature of about −190°C, the liquid nitrogen (LN) temperature. For this reason, low temperature power electronic circuits have the potential of finding applications in deep space missions, and also in cryogenic instrumentation in superconducting energy storage, which is covered in Chapter 26. Such circuits designed and operated at low temperature may result in a more efficient system layout than the room temperature circuits. The advantages include reducing or eliminating the thermal shutters and the need for an isotope heat, which can cause overheating during launch. Understanding the performance of power electronics at extreme low temperature is needed for this purpose. The following is known about the operation of power electronic components near liquid nitrogen temperature.

The performance of certain semiconductor devices improves with decreasing temperature down to liquid nitrogen temperature. At low temperature, the majority of carrier devices demonstrate reduced leakage current and reduced latch-up susceptibility. In addition, these devices show higher operating speed resulting from increased carrier mobility and saturation velocity. An example is the power MOSFET, which has lower conduction loss at a low temperature due to the reduction in drain-to-source resistance $R_{ds(on)}$ resulting from increased carrier mobility. NASA has tested other components such as resistors, capacitors, and magnetics that are needed for various power converters at liquid nitrogen operating temperature. Many of them have been found suitable for operating at unheated interplanetary spacecraft.[7]

The battery can be a roadblock at very low temperature. The Li-ion battery offers a somewhat favorable combination of the energy and power density. However, its temperature performance below −40°C is poor. Tests[8] have shown the following about the Li-ion cells:

- Between room temperature and −20°C, variations in electrolyte resistance and the anode to electrolyte resistance are negligible, but the cathode electrolyte interface resistance increases substantially.
- The cell voltage and Ah capacity fall to approximately one half at −40°C (Figure 19.9). As a result, practically no energy was delivered at 40°C. This is due to substantial increase in the total internal resistance.

Poor cell performance at low temperature can be attributed to the electrolyte becoming viscous or solid as the temperature is lowered.[9] It is

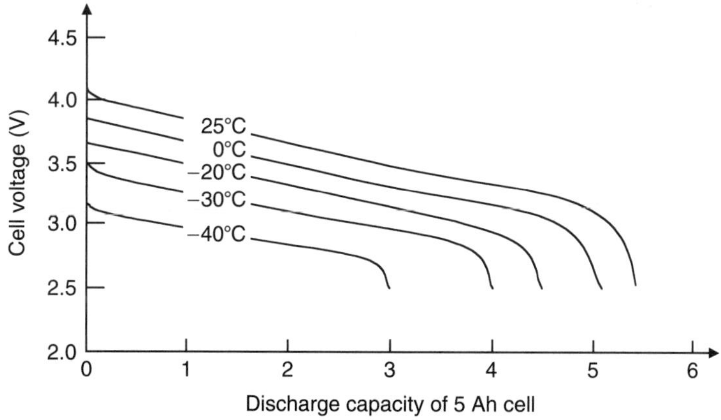

FIGURE 19.9 The capacity of a 5-Ah Li-ion cell at low temperature in interplanetary missions.

also attributed to the poor lithium diffusivity in the electrolyte. Work is under way to improve the low temperature performance of the Li-ion cell.

References

1. Jenkins, J.E. and Dakermanji, G., Near Earth asteroid rendezvous—*Shoemaker* spacecraft power system flight performance, *Proceedings of the 36th Intersociety Energy Conversion Engineering Conference*, ASME, 2001, Volume I, pp. 251–256.
2. Haines, J.E., Inner planets sample return missions, the challenge for power systems, *Proceedings of the 34th Intersociety Energy Conversion Engineering Conference*, SAE, 1999, Paper No. 2483.
3. Brandhorst, Jr, H.W. and Chen, Z., PV approaches for near-sun missions, *Proceedings of the 34th Intersociety Energy Conversion Engineering Conference*, SAE, 1999, Paper No. 2631.
4. Choi, M.K., Power and thermal systems with thermoelectric generators at 930°C for solar probe inside 0.1 AU, *Proceedings of the 36th Intersociety Energy Conversion Engineering Conference*, ASME, 2001, Vol. II, pp. 1161–1163.
5. Kerslake, T.W. and Kohout, L.L., Solar electric power system analysis for Mars surface missions, *Proceedings of the 34th Intersociety Energy Conversion Engineering Conference*, SAE, 1999, Paper No. 01-2482.
6. Kilecki, J.C. and Hilalrd, G.B., *Proceedings of the Electrical and Chemical Interactions at Mars Workshop*, NASA Report CP-10093, 1992.
7. Elbuluk, M.E. et al. Low temperature performance evaluation of battery management technologies, *Proceedings of the 34th Intersociety Energy Conversion Engineering Conference*, SAE, 1999, Paper No. 01-2543.
8. Nagasubramanian, G, Low temperature electrical performance characteristic of Li-Ion cells, *Proceedings of the 34th Intersociety Energy Conversion Engineering Conference*, SAE, 1999, Paper No. 01-2462.
9. Smart, M.C, Huang, C.K., Ratnakumar, B.V., Surampudi, S., and Sakamoto, J.S., Factors affecting Li-Ion cell performance, *Proceedings of the 37th Power Sources Conference*, Paper No. 239, 1996.

Chapter 20

Radioisotope Thermoelectric Generator

20.1 Introduction

Deep space and outer planetary missions cannot effectively use photovoltaic power generation due to insufficient solar flux. For those missions, an on-board nuclear energy source or a radioactive isotope is often used to generate electrical power. The radioisotope heat is directed at a thermoelectric junction, which generates electrical potential just as in a thermocouple. This concept was briefly presented in Section 3.6, and is covered in detail in this chapter.

The radioisotope thermoelectric generator (RTG) has been fully developed and used for decades for power levels in several hundred watts. Such a power source has an advantage of supplying power all the time, thus eliminating the need for a battery in a base load system having no peak power requirement. An obvious disadvantage is the heavy radiation shielding required around electronic components. Also, the nuclear fuels that are safe and easy to handle with little shielding, such as curium-244 and plutonium, are expensive. Inexpensive and easily available fuel, such as strontium-90, is unsafe.

High-energy particles emitted from the radioactive isotope material are the primary sources of energy, which heats the absorbing material. The mass of the isotope decay exponentially at a rate characterized by half-life, $T_{1/2}$. The thermal radiation decreases proportionally with the remaining mass. Therefore, the thermal power $P(t)$ radiated at any time decays exponentially from its initial value P_o, as given by

$$P(t) = P_o \exp(-0.7t/T_{\frac{1}{2}}) \tag{20.1}$$

Table 20.1 compares the isotope fuels presently used in RTGs with their half-life and specific power achievable.

The advantages of an RTG are:

- It provides power for a long period of time, independent of the spacecraft orientation and distance from the sun.
- It is suitable for missions far away from the sun, too close to the sun, or lunar missions with long eclipse periods.

Table 20.1 Characteristics of three alternative isotopes for RTGs

Isotope	Half-life (years)	Thermal power radiated per gram of isotope (W/g)	Radiation level (Curies/W)
Curium-242	0.45	~100	27
Strontium-90	28	~0.25	153
Plutonium-238	86	~0.55	30

- The power output is not affected by radiation damage in the Van Allen belts or due to man-made nuclear threats.

The RTG consists of numerous thermoelectric converter (TEC) cells connected in series–parallel combination to obtain the required voltage and current. Each TEC converts the isotope thermal energy into electrical energy. The power conversion efficiency, η, of the TEC is defined as

$$\eta = \frac{\text{electrical power output}}{\text{thermal power depletion}} \tag{20.2}$$

The efficiency depends on the material properties and the hot and cold junction temperatures T_h and T_c, respectively. The theoretical limit on this efficiency is the Carnot efficiency, which is

$$\eta_{\max} = \frac{T_h - T_c}{T_h} \tag{20.3}$$

Practical designs yield about one half the theoretical maximum efficiency. The most widely used material, plutonium-238 with Si–Ge TE cells, gives about 7% conversion efficiency. Removing the remaining 93% of the system energy as waste heat poses a significant design challenge. The specific electrical power output of RTGs is typically low. Based on the total power system assembly mass, it was 2 W/kg in *SNAP-19* spacecraft, and 5 W/kg in *Galileo* spacecraft. Plutonium-238 costs more than a couple of million dollars per kilogram. The DoE inventory of this isotope in 2002 was merely 10 kg. The U.S. currently does not produce it, but purchases it from Russia.

20.2 Thermoelectric Fundamentals

The working principle of the TEC is based on the Seebeck effect, which generates electrical potential when any two dissimilar materials are maintained at different temperatures. It involves electron or hole transfer between the two dissimilar materials under thermal energy. The two materials can be conductors or semiconductors. The TEC for space power applications uses semiconducting materials, one p-type and the other n-type, as shown in Figure 20.1. If two such dissimilar materials are held at temperature difference $\triangle T = T_h - T_c$, an electric potential difference V_{12} is produced at their junction. It is given by

$$V_{12} = \alpha_{12} \Delta T \tag{20.4}$$

where α_{12} is known as the Seebeck coefficient of the couple, generally expressed in μV/°C. The α_{12} is often called the TE power, although it is not really a power. It is a characteristic constant, which depends on the material properties. The coefficient α_{12} is considered positive if the Seebeck voltage polarity produces current in the p-type material from high temperature to low temperature. The current under this voltage then delivers electrical power to the external load connected in the loop.

Each arm of the two materials has electrical resistivity ρ and thermal conductivity k. The parameters ρ, k, and α all vary with the operating temperature. Since ρ and k cause loss of power and reduces the conversion efficiency, the figure of merit, Z, for each material in the couple is defined as

$$Z = \frac{\alpha_{12}^2}{\rho k} \tag{20.5}$$

Another figure of merit for each material is the power output per unit mass. It is defined as

$$Z' = \frac{\alpha_{12}^2}{\rho} \tag{20.6}$$

The figure of merit of both materials together in the couple is defined as

$$Z_{12} = \frac{\alpha_{12}^2}{\left(\sqrt{\rho_1 k_1} + \sqrt{\rho_2 k_2}\right)^2} \tag{20.7}$$

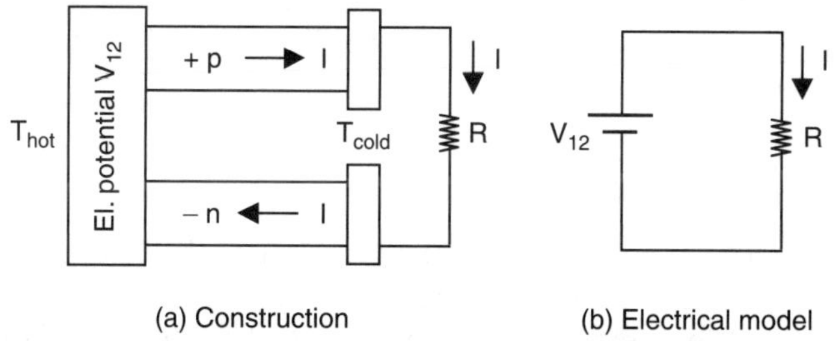

FIGURE 20.1 Single-stage unicouple thermoelectric converter.

where ρ and k are the electrical resistivity and thermal conductivity of the two materials 1 and 2, respectively.

Equation 20.7 suggests that a good TE junction has the following properties:

- High Seebeck coefficient α, which generates a high voltage.
- Low electrical resistivity ρ, which results in low ohmic loss.
- Low thermal conductivity k, which gives high $\triangle T$.

For a junction of two metallic arms, such as in thermocouples used to measure temperature, the Seebeck coefficient is low around 50 μV/°C. Both ρ and k are large, resulting in poor efficiency around 0.1%, which is unsuitable for power generation. Some semiconductors have high Seebeck coefficient in the 300 to 500 μV/°C range. Their conductivity is relatively low, resulting in relatively high conversion efficiency in the range of 5 to 7%.

The fraction of heat removed from the junction and delivered to the load in the form of electrical power is

$$P_e = \left(1 - \frac{T_c}{T_h}\right)\left\{\frac{\sqrt{1+Z\bar{T}} - 1}{\sqrt{1+Z\bar{T}} + \dfrac{T_c}{T_h}}\right\} \tag{20.8}$$

where

$$\bar{T} = \sqrt{T_h + T_c} \tag{20.9}$$

Low-grade commercial TECs have potential terrestrial applications such as recovering waste heat from various industrial processes and vehicle exhaust to decrease fuel consumption. Typical space-qualified RTGs uses high-grade SiGe TECs with plutonium-238 as the heat source. The SiGe material is doped with phosphorus for the n-leg and with boron for the p-leg. For widely used SiGe couples, Z_{12} is around 0.001 with ±30% variation depending on the material grade and quality of the manufacture. Various TEC configurations used in the RTG designs for space applications are described below.

20.2.1 Single-Stage Unicouple

This is the most basic elementary construction of the TEC converter that is shown in Figure 20.1. One end of the p and the n arms are kept in a common reservoir at hot temperature T_h. The other end of the arms is kept at cold temperature T_c. The d.c. voltage is generated between the p and n terminals at the cold end, which is proportional to the temperature difference $\triangle T = T_h - T_c$. A load resistance R would draw electrical current I and power I^2R. Any form of heat can be converted into electricity. The conversion efficiency obtained with the state of the art materials is about 7%.

If the cold temperature reservoir is a liquid metal bath, as is typical, the p and n terminals must be kept at two different reservoirs electrically isolated in order to withstand the generated voltage. Otherwise, the resulting electrical short would prevent the power transfer to the load.

This configuration derives its name because there is only one TE couple, which sees $\triangle T$ in one stage only. The heat energy at T_c is discharged from the TEC system as waste heat. The efficiency is therefore relatively lower than the other configurations described hereafter.

20.2.2 Single-Stage Multicouple

Two or more p–n couples are stacked together in the single-stage multicouple TEC assembly. Figure 20.2 show one such assembly with two couples, which are thermally in parallel and see the same temperature difference $\triangle T = (T_h - T_c)$ in one stage only. Electrically, however, they are in series. The total voltage output of the assembly is therefore twice that of one element. Material B in the figure is an electrical insulation with high thermal conductivity, so that the copper straps C are at the same temperature as the adjacent heat reservoir.

This configuration gives twice the voltage using twice the TE material. Therefore, the specific power per kilogram of the active material remains the same. However, the specific power of the system is higher due to the overhead mass less then double.

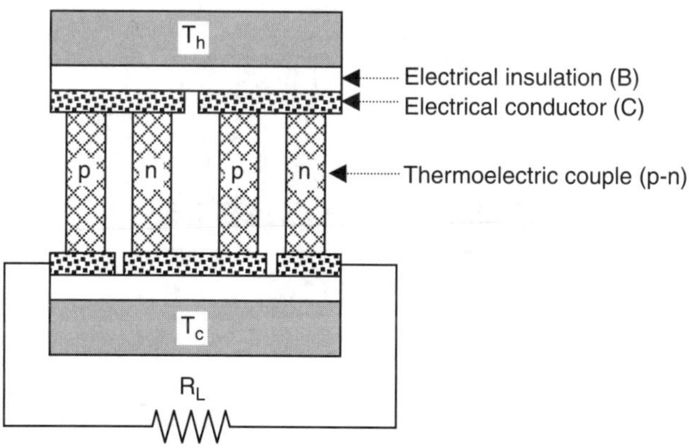

FIGURE 20.2 A single-stage multicouple thermoelectric converter.

20.2.3 Multistage Multicouple

In this configuration, the heat is extracted in two stages as depicted in Figure 20.3. The temperature drops from T_h to T_m in the first stage having two couples, and from T_m to T_c in the second stage also having two couples. The material B (unshaded) is an electrical insulation of high thermal conductivity to minimize the temperature drop across it. The copper strap C (checkered) is a material with high thermal and electrical conductivities. It conducts the electrical current and distributes the heat uniformly across the device so that all couples in the same stage see the same temperature difference.

20.2.4 RTG Assembly

The Seebeck voltage per junction is around 300 to 500 μV/°C. Numerous junctions are connected in series to obtain the desired voltage, and numerous strings of series junctions are connected in parallel to obtain the desired current. The series–parallel combination is a matter of design that results in the maximum power transfer to the load. The RTG is

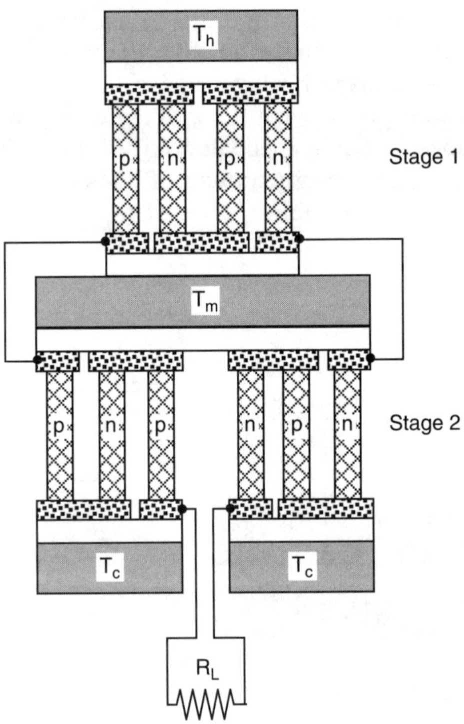

FIGURE 20.3 A multistage multicouple thermoelectric converter.

basically an assembly of all TECs, which is typically a cylindrical canister as shown in Figure 20.4. Once ignited, the isotope heat generation cannot be turned off. The refractory metal insulation shield and the graphite isotope capsule could be damaged if they were subjected to atmospheric oxygen while operating at high temperature, and could experience sublimation damage even before the mission starts. For this reason, the RTG is filled with pressurized xenon gas until after launch. The very low thermal conductivity of xenon makes it possible to store and launch the RTG hot and generate the needed pre-launch and launch power. Once in space, the xenon is vented.

20.3 Electrical Model of RTG

The electrical equivalent circuit of an RTG is shown in Figure 20.5(a). The total voltage generated due to the Seebeck effect works as an internal voltage source. With an open circuit (zero load current), the external terminal voltage V is the same as that generated internally. We designate this voltage as the open circuit voltage V_{oc}. When electrical current is drawn by load resistances R_L, there is an internal voltage drop. This is represented by an internal resistance R_i, which is approximately constant at a give temperature. The external terminal voltage V therefore decreases linearly with increasing load current, i.e.,

$$V = V_{oc} - IR_i \tag{20.10}$$

With the external terminals shorted, the maximum current flows to the load. We designate this current as I_{sc}. The above equation gives

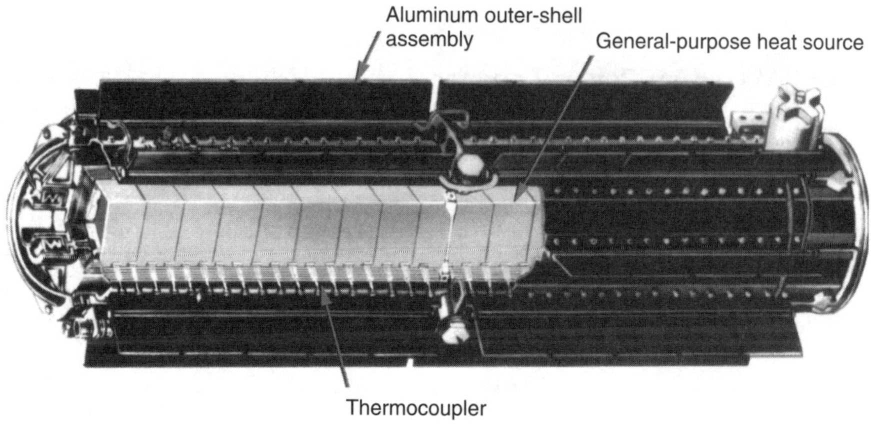

FIGURE 20.4 A radioisotope thermoelectric generator assembly. (From *IEEE Spectrum*, p. 63, March 1998. With permission.)

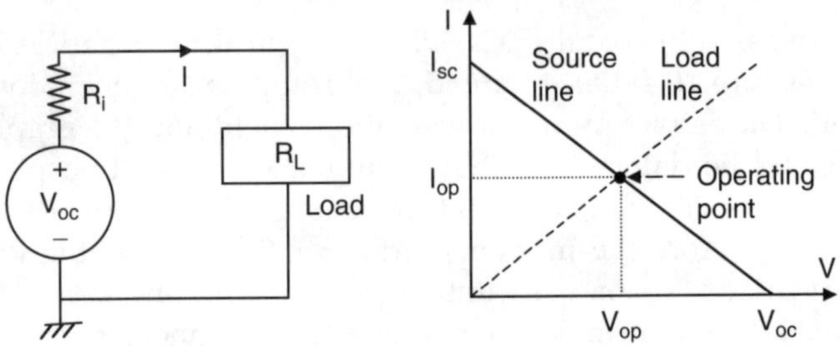

FIGURE 20.5 Electrical equivalent circuit model and operating characteristics of TEC.

$$I_{sc} = V_{oc}/R_i \tag{20.11}$$

Equation 20.10 can be rearranged using Equation 20.11 as

$$I = I_{sc} - \gamma V \tag{20.12}$$

where $\gamma = I_{sc}/V_{oc}$, a characteristic admittance of the RTG power source.

Equation 20.12 sets the I–V characteristic of the RTG. It is a falling straight line from I_{sc} at zero voltage to zero current at V_{oc}, as shown in Figure 20.5(b). On the other hand, the voltage, V, required to circulate current I through the load increases with the current in a straight line. This line, known as the load line, has a slope of R_L. An RTG having the source line shown by solid line and the load line shown by dotted line operates at the point of their intersection. At that operating point, the electrical power transferred to the load is

$$P = V_{op}I_{op} \tag{20.13}$$

20.4 Maximum Power Transfer

The power, P, transferred from TEC to the load at any operating point P in Figure 20.5(b) is

$$P = VI = V(I_{sc} - \gamma V) = VI_{sc} - \gamma V^2 \tag{20.14}$$

The power system design for extracting the maximum power from RTG to the load must operate at a voltage such that, at the operating voltage,

$$\mathrm{d}P/\mathrm{d}V = I_{sc} - 2\gamma V = 0 \tag{20.15}$$

Equation 20.15 gives the corresponding operating voltage $V_m = I_{sc}/(2\gamma)$, which is also $\frac{1}{2}V_{oc}$, and the current at that voltage is $\frac{1}{2}I_{sc}$.

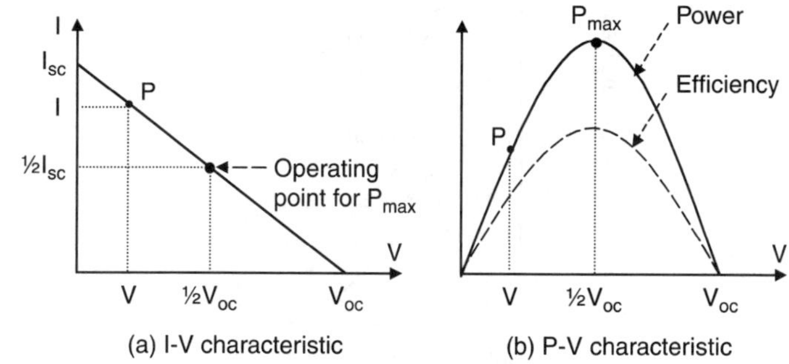

FIGURE 20.6 A TEC operation for maximum power transfer to load.

The maximum possible power transfer therefore is

$$P_{max} = \tfrac{1}{2}V_{oc} \times \tfrac{1}{2}I_{oc} = \tfrac{1}{4}V_{oc}I_{sc} \tag{20.16}$$

It occurs when the RTG is operated at the voltage that is one half of the open circuit voltage. This point is shown as P_{max} in the power versus voltage plot shown in Figure 20.6(b). The power at zero voltage is obviously zero. It is also zero at V_{oc} since the current is zero there. In between, the power rises with the operating voltage, reaches the maximum value P_{max}, and then falls to 0 at V_{oc}. The RTG conversion efficiency is maximum at the maximum power transfer point. The I–V, P–V and η–V curves with the operating voltage as an independent variable are shown in Figure 20.7.

20.5 The Effect of Temperature and Aging

The characteristics displayed in Figure 20.7 are for a given $\triangle T$ across the TE junctions. The I–V line of the RTG shifts upward for a higher $\triangle T$, and

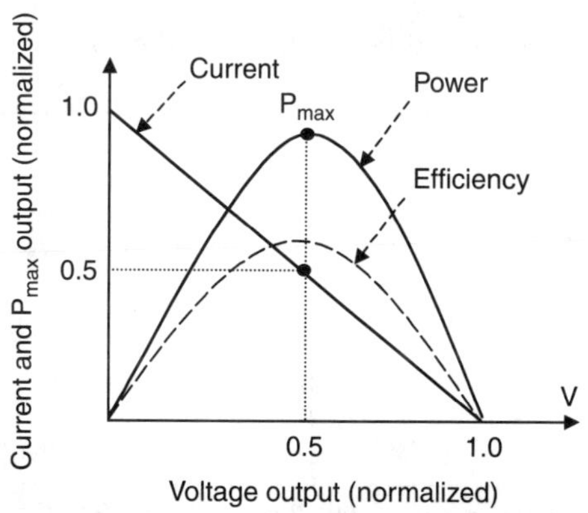

FIGURE 20.7 TEC current, power, and efficiency versus operating voltage.

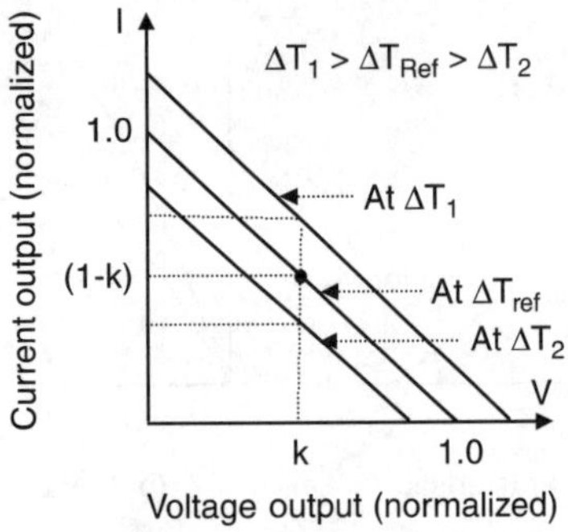

FIGURE 20.8 TEC current versus voltage at different temperature gradients.

downward for a lower $\triangle T$, as shown in Figure 20.8. The amount of shift is a characteristic of the couple material.

Aging has little effect on the RTG output, because the basic heat source has a half-life in decades. For this reason, the power generation degrades little, as shown in Table 20.2. Most power degradation is due to slow precipitation of the phosphorus doping in the n-type leg of the thermocouple. The I–V and P–V curves shifts uniformly with time and the temperature such that the maximum power point remains at the same voltage (Figure 20.9). This is a happy coincidence for the design engineer. The conversion efficiency is a function of the contact resistance and the hot and cold side temperatures as shown in Figure 20.10 and Figure 20.11, respectively.

Table 20.2 Measured power levels of the RTG in *Voyager II* versus time

Date	*Voyager II* power output, (W)	Percent of initial power generation
August 1977 (launch date)	478	100
December 1980	436	91
December 1985	398	83
December 1990	370	77
December 1995	344	72
December 1998 (21 years)	327	68% (after 21 years)

(Source: Data from Mondt, J.F., *Proc. of the 36th IECEC*, ASME, 2001, Vol. I, pp. 133–139.)

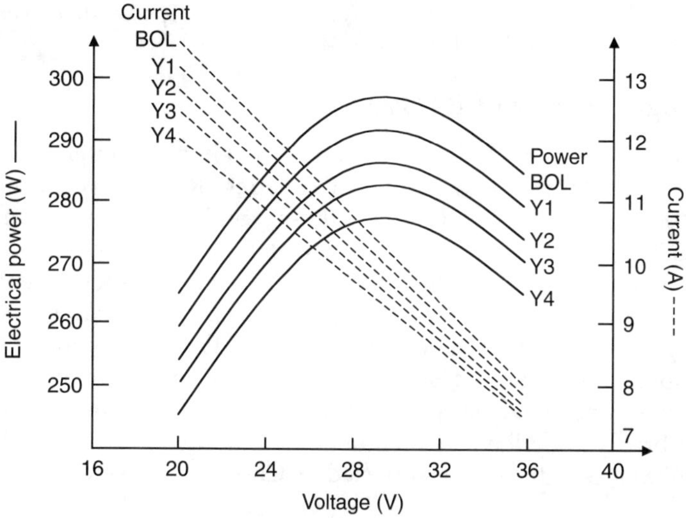

FIGURE 20.9 *I–V–P* characteristics of RTGs with age.

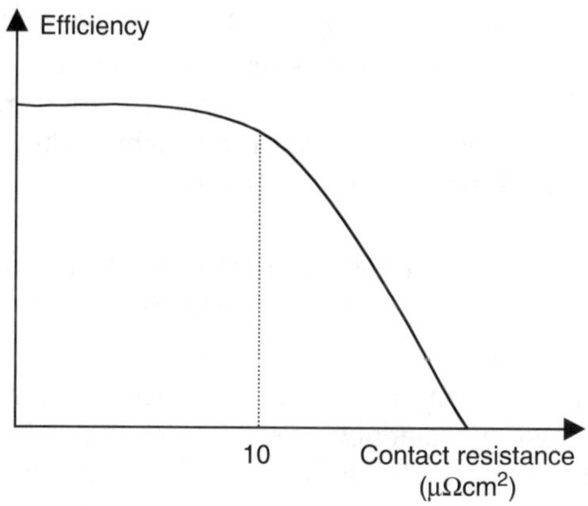

FIGURE 20.10 TEC conversion efficiency versus contact resistance.

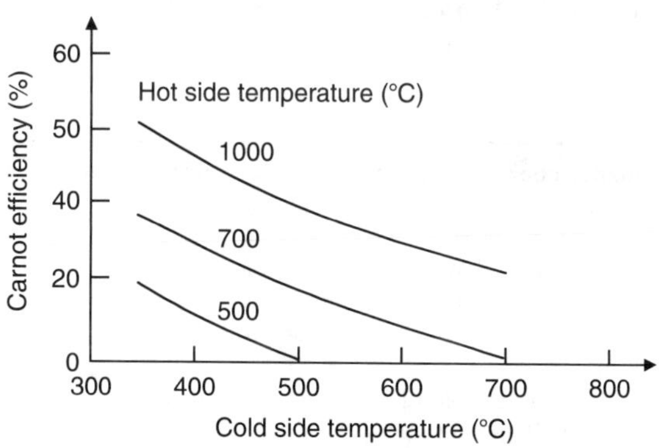

FIGURE 20.11 TEC conversion efficiency versus temperature.

20.6 Flight History of RTGs

RTG technology has been developed and flown in many interplanetary and deep space missions for a couple of decades. RTGs have been used in *Pioneer, Explorer, Voyager, Galileo, Ulysses,* and *Cassini* missions. These missions have used essentially a standard RTG assembly designed to meet mission requirements listed in Table 20.3. A number of such assemblies are used as needed for a particular mission. For example, *Galileo* spacecraft for exploring Jupiter and its moons used two RTGs (Figure 20.12), whereas *Cassini* spacecraft exploring Saturn and its rings carried three RTG (Figure 20.13). These and other outer planet missions that have used RTGs in the past are briefly described in Section 19.7. A new application of RTG in a solar probe close to the sun as reported by Choi[1] is discussed in Section 19.5.

20.7 Segmented TECs

In a unicouple TEC, the same heat and the same current is seen by each segment in each leg. High conversion efficiency can be obtained by operating the TEC material over a wide temperature range, and using different materials in each temperature range where they posses optimum performance. This can be accomplished in two ways:

- Multistage TEC, where each stage is electrically isolated but thermally in contact with each other, operates over a fixed temperature difference.
- Segmented TEC, where each of the p and n legs is made of multiple segments of different materials placed in series. Such a TEC is estimated to have a conversion efficiency around 15% versus 7% using the state-of-the art BiTe, FeSi, PbTe, and SiGe alloys. In the segmented TEC concept, the length and cross section of each segment is adjusted in accordance with its thermal and electrical conductivity to achieve the desired temperature difference across each material and the maximum

Table 20.3

Performance characteristic of the standard RTG in use
• Mission life 5 years
• BOL power 285 W
• EOL power 255 W (notice the negligible decay over mission life)
• Suitable for space shuttle launch
• RTG module 42 cm diameter × 114 cm long
• Total mass 55 kg
• TE 572 SiGe unicouple in 16 circumferential rows
• Operating voltage 28 to 30 V
• Operating temperature gradient $\triangle T = 707$ K
• Operating temperature: 1273 K hot junction, 566 K cold junction

FIGURE 20.12 *Galileo* spacecraft using two RTGs attached to two long booms. (Source: NASA.)

efficiency of the couple. The overall operating temperature drops from 1275°C to 300°C in typical segmented TEC designs.

20.8 Advanced RTGs

The DoE recently issued a directive to develop and maintain the nuclear space power program to support defense and deep space missions. The program has funded research and development for the Advanced Radioisotope Power System (ARPS) with a goal to triple the conversion efficiency of the present TECs. A key thrust is on using advance SiGe cells with improved thermal conductivity and contact resistance of less than 10 $\mu\Omega/cm^2$. The program includes building an engineering model of

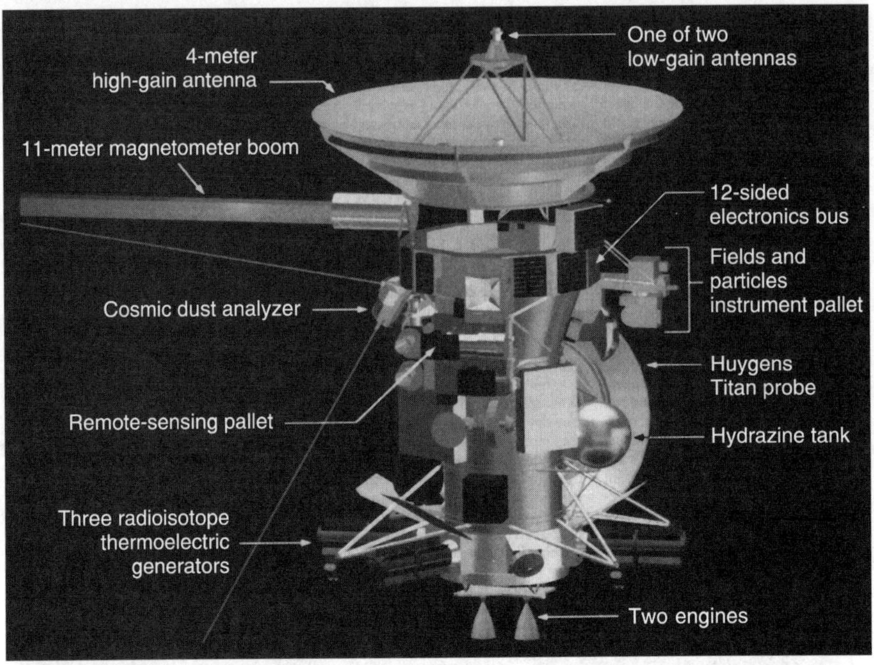

FIGURE 20.13 *Cassini* spacecraft in operational configuration with three RTGs. (From *IEEE Spectrum*, p. 62, March 1998. With permission.)

100 W output and flight qualified unit around 2005. Table 20.4 summarized the ARPS's goals.[2]

Three technologies under evaluations to achieve these goals are: (a) segmented alkali metal thermal to electric converter (AMTEC) with 15% efficiency and 7 to 9 W/kg specific power, (b) Stirling engine with 25% efficiency and 6 to 7.5 W/kg, and (c) thermo-photovoltaics (TPVs). In each case, the heat comes from the natural radioactive decay of nonweapons grade plutonium-238 pellets, clad in protective layers of graphite and iridium. The TPV was briefly described in Chapter 3, the Stirling engine is described in Chapter 21, and the AMTEC is described below.

Table 20.4 Goals of advance RTG research programs

Performance parameter	Preset state of the art RTG capabilities	Goals of the Advanced Radioisotope Power System
System efficiency	6.5%	13–25%
Specific power	5 W/kg	9–10 W/kg
Life	> 20 years	TBD
TEC junction	SiGe	Improved SiGe
Operating temperature	1275 °C to 575 °C	1000 °C to 700 °C
Isotope	Pu-238	Pu-238

In the AMTEC, the isotope heat vaporizes sodium that flows through a β-alumina solid electrolyte tube to produce electricity. β-Alumina is an alkali metal conducting ceramic: a solid electrolyte that conducts sodium ions. An electrochemical potential is generated when sodium is present at two differential pressures separated by an electrolyte. The free electrons are captured by the electrodes, which are used to power the load and then get recombined with the sodium ions. The temperature in the vapor circulation drops. The conversion efficiency is a function of the differential temperature.

The electrical characteristic of AMTEC module is similar to that of the conventional TEC. The terminal voltage drops linearly with increasing current. The power output peaks at certain voltage, approximately one half of the open circuit voltage. As in the traditional TEC, the open circuit voltage and the maximum power in AMTEC are higher at higher-temperature gradient.

20.9 Thermoelectric Cooler

The TEC junction can also be used to cool a small device locally by replacing the load by a battery and sending the current backward into the TEC. This is actually the TEC working in the reverse mode. The junction works in the reverse by taking the electrical energy and converting into a temperature drop. Such reverse operation is called the Peltier effect. The Peltier heat absorption is given by

$$Q_p = \pi_{12} J_{12} \tag{20.17}$$

where

Q_p = rate of Peltier heat absorption per unit area of the junction,

π_{12} = Peltier coefficient of the couple, J_{12} = current density taken as positive for current flowing from material 1 to material 2 at the junction.

Assuming that the cold junction is perfectly insulated thermally, it produces a temperature drop

$$\Delta T_{\max} = \tfrac{1}{2} Z T_c^2 \tag{20.18}$$

The cooling effect can also be created by using the TEC as a heat sink in the forward mode for cooling a delicate small device. In this mode, the heat near a device is converted into electrical energy, conducted by wires, and dissipated into a load away from the device to be cooled.

References

1. Choi, M.K., Power and thermal systems with thermoelectric generators at 930°C for solar probe inside 0.1 AU, *Proceedings of the 36th Intersociety Energy Conversion Engineering Conference*, ASME, 2001, Vol. II, pp. 1161–1163.
2. Mondt, J.F., Advanced Radioisotope Power System Technology development for NASA mission 2011 and beyond, *Proceedings of the 36th Intersociety Energy Conversion Engineering Conference*, ASME, 2001, Vol. I, pp. 133–139.

Chapter 21

Dynamic System with Alternator

21.1 Introduction

Solar energy can be used in systems other than the photovoltaic. A dynamic energy conversion system is an example, where the sun's energy is collected in the form of heat using a concentrator. The heat, in turn, is used to produce steam and drive a rotating turbo-generator or a reciprocating alternator to generate electricity. Such a system was a primary candidate for the space station design in the 1980s with an estimated power requirement of 300 kW. The system configuration is shown in Figure 21.1. A parabolic concentrator focuses the sun's heat on to a receiver, which boils a fluid. The fluid can be a suitable liquid metal, such as potassium chloride. A high-pressure stream of liquid metal produced in the receiver would drive a turbine based on the Rankine cycle. The fluid can also be a gas, such as a mixture of helium or xenon having the molecular weight around 40. The heated compressed gas would drive a turbine working on the Brayton cycle. A gas-based system minimizes erosion and sloshing problems in transporting the liquid metal. A functional schematic of dynamic power system for space is shown in Figure 21.2. In either a liquid metal or a gas-based system, the high-pressure high-temperature fluid drives the turbine, which in turn drives an electrical generator. Waste heat transferred to the liquid coolant is dissipated via radiator panels to space. The energy conversion efficiency is

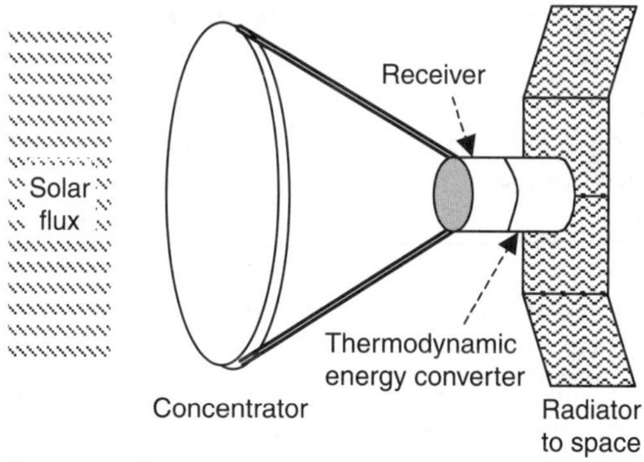

FIGURE 21.1 A solar dynamic power system schematic.

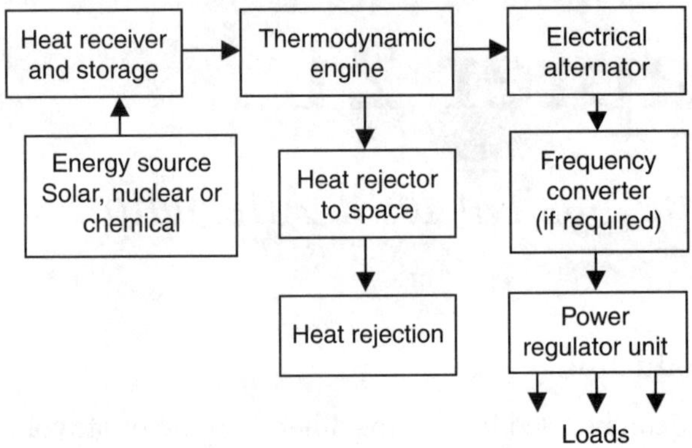

FIGURE 21.2 The functional blocks of a dynamic power system.

much higher than the photovoltaic system. This minimizes the deployed collector area and the aerodynamic drag in low Earth orbit.

The usable energy extracted during a thermodynamic cycle depends on the working temperatures. The maximum thermodynamic conversion efficiency that can be theoretically achieved with the hot side temperature, T_{hot}, and the cold side temperature, T_{cold}, is given by the Carnot cycle efficiency, which is

$$\eta_{carnot} = \frac{T_{hot} - T_{cold}}{T_{hot}} \tag{21.1}$$

where the temperatures are on the absolute scale. The higher the hot side working temperature and lower the cold side exhaust temperature, the higher the efficiency of converting the captured solar energy into electricity. The hot side temperature, however, is limited by properties of the working medium. The cold side temperature is largely determined by the cooling method and the environment available to dissipate the exhaust heat.

An indirect but major advantage of this system is that the energy storage is interwoven in the system at no extra cost. It resides in the latent heat of phase change at high temperature — around 1000 K. The systems can store thermal energy for hours with no degradation of electrical performance, or longer with some degradation. This feature makes the technology capable of meeting peak power demands with no added mass or cost of separate energy storage. It eliminates the battery requirement altogether.

Today's space power systems using well-proven PV technology provide up to a few tens of kilowatts with system specific power of 10 W/kg and life up to 15 years. Although the solar dynamic technology is not yet proven in space flights, it offers potential advantages in efficiency, weight, scalability, and the overall cost in high power spacecraft. The cost advantage comes from elimination of costly semiconductor PV cells. The systems can be extremely cost effective in a few kilowatts to hundreds of kilowatts power range. The concept is sufficiently developed for use in the near future, particularly in high-power LEO missions. It may also find applications in high-power defense spacecraft where a large solar array can make the mission nonmaneuverable and vulnerable to detection and attack by enemy. It has been considered for a 300-kW space station and for a dynamic isotopes power system (DIPS) for space defense.

The efficiency advantage in the dynamic system comes from the higher efficiency of the engine (20 to 40%) as compared to silicon solar cells (15 to 20%), and higher efficiency of thermal energy storage of the receiver (85 to 90%) as compared to the battery efficiency (70 to 75%). The greatly improved overall system efficiency as compared to the PV system translates into less solar collection area. This results in reduced drag and less concern regarding station dynamics, approach corridors, and experimental viewing angles. The reduced drag is particularly important because it allows lower flight altitudes within given constraints of drag–makeup fuel and orbit decay time. At power levels near 100 kW, such as for space-based radar, the PV solar array collector area becomes prohibitive. It is in this power range that solar dynamic power system is expected to find advantageous applications. It offers the following additional advantages:

- Low solar collection area
- High voltage a.c. power generation
- Highly scalable and mass producible
- Potentially long life components
- Inherent radiation tolerance

Recent prototype testing of a 2 kW non-optimized solar dynamic systems reported by Mason[1] demonstrated a conversion efficiency of almost 30% using 1990s component technologies. Significant improvements in efficiency can be realized for large systems in ratings above 100 kW using newer technology components and optimized design parameters.

The dynamic system uses an electromechanical energy converter in which an armature coil moves in a rotary or linear motion under an alternating magnetic field of north and south poles. The generated voltage in the armature is therefore alternating at a certain frequency. The terminal voltage output equals that generated in the armature less the internal voltage drop. The mechanical power source driving the armature is typically a steam or gas turbine or a reciprocating engine. The electromechanical energy conversion efficiency is in the 85 to 98% range depending on the system rating and configuration. The machine size and mass primarily depend on the power rating, and secondarily on the adopted thermal design to remove the internal power losses. A smaller machine generally results in lower efficiency, hence relatively higher power loss and greater thermal system mass. A rough rule of thumb in trading weight and loss over a narrow range using the same materials and design configuration is to assume that the product of the machine mass and the power loss remains constant.

21.2 Thermodynamics

On the thermodynamics side, the following two major configurations are used.

21.2.1 Stirling Engine

The Stirling concept consists of two isotope heat source modules surrounded by multifold thermal insulation, two Stirling engines driving electrical alternators, and housing that acts as the primary structure (Figure 21.3).[2] Isotope heat vaporizes a salt under pressure to drive a steam turbine and electrical generator. The working temperature is around 600 to 700°C on the hot side and 100 to 120°C on the cold side.

The Stirling technology demonstration prototype has been built and tested successfully on the ground. It used a linear alternator for rated electrical power output of 55 W at 90 V and 80 Hz frequency. The alternator

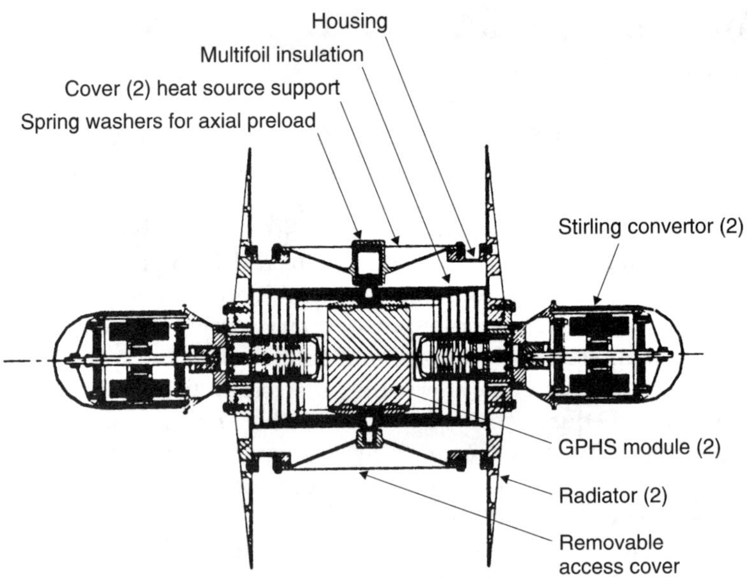

FIGURE 21.3 Stirling radioisotope power system — baseline concept. (From R. Cockfield, *Proc. 35th IECEC*, AIAA, p. 2843, 2000. With permission.)

output was electronically converted into 28 V d.c. suitable for use on the spacecraft.

The reliability can be an issue in such concepts with only two engines. If one fails, the other becomes overloaded unless both are designed for delivering full power, which would result in a heavy mass penalty. For this reason, the number of Stirling engines in practical systems designs is likely to be four. The total specific power of a few hundred watts power source with a matching converter is estimated to be around 8 to 10 W/kg in two-engine systems and 4 to 5 W/kg in four-engine systems.

As reported by Thieme, Qiu and White,[3] the baseline electrical power rating of the present Stirling converter is 55 W. Multiple units will be connected in parallel to achieve the desired power levels. Key design and operating issues are parallel operation of two thermodynamically independent free-piston Stirling converters and vibration reduction to about 2% of that of single unbalanced converters over the entire mission life even with one engine failed. An EMI compatible design for science missions is also a challenge, which is addressed by Sargent.[4]

NASA GRC and DoE with two industry partners are actively developing Stirling converters for an advanced radioisotope power system to replace the currently used RTG systems for deep space missions. A standard module of 100 W using a thin long alternator is being developed, which can be connected in parallel for higher power. Boeing is developing a multi-mission modernized 100 W RTG that will be flexible and adaptable to both the orbiter and the lander missions to a planet. On the other hand, Lockheed Martin is developing a Stirling radioisotope generator that will use plutonium-238 heat and a reciprocating alternator driven by a Stirling engine.

The multi-mission RTG is still 7% efficient, whereas the Stirling generator would have about 25% efficiency requiring less plutonium. The disadvantage of the Stirling is the moving piston that could reduce the reliability in space. The multi-mission RTG uses a lead telluride TEC, which operates at a lower temperature than the SiGe. This does not require a heat resistant refractory metal structure, which oxidizes or erodes relatively easily in any CO_2 and CO in a planet's atmosphere. One of these two new power systems is targeted for use for a Mars mission in 2010.

The Stirling system is the most developed converter option for the advanced isotope power system concepts presently being funded. It has the following advantages:

- High efficiency in excess of 20%, resulting in less heat rejection
- Reduced isotope inventory by a factor of at least 3 compared to RTG
- Specific power about 1.5 times that of RTG
- It has proven to be reliable compared to the Brayton and Rankine cycles.

21.2.2 Brayton Cycle

In this system, solar energy concentrated on to a receiver is transferred to a Brayton heat engine for conversion to electrical power. It utilizes a turbine, compressor, and rotary alternator. The working fluid is an inert gas. The gas from the receiver is expanded through a turbine, cooled through a heat exchanger where the waste heat is transferred to a liquid coolant and is pressurized in a compressor before being re-heated by the solar receiver (Figure 21.4). A recuperative heat exchanger between the turbine discharge and receiver inlet is often used to improve the thermodynamic cycle efficiency. The rotary alternator typically provides 3-phase a.c. at 1 kHz frequency for further conditioning and distribution. The closed cycle Brayton energy conversion system has significant advantage over other energy conversion system due to high efficiency (up to 40%), long life up to 30 years and low recurring cost. These advantages are important where it is necessary to supply large amount of electrical power and compressed gas. NASA Glenn has demonstrated the operation of a 2-kW closed Brayton cycle power conversion unit with a xenon ion thruster. The testing was conducted as part of a nuclear electric propulsion test bed at Glenn. The test successfully demonstrated that such power converter could supply and regulate high-voltage electrical power to an ion thruster.

21.3 Electromechanical Energy Conversion

Various types of electrical alternators that can be considered for the electromechanical energy conversion in the dynamic power system are:

- Permanent magnet (PM) synchronous or d.c. machine
- Synchronous machine with solid rotor (Rice or Lundell)
- Synchronous wound rotor machine, brushless design

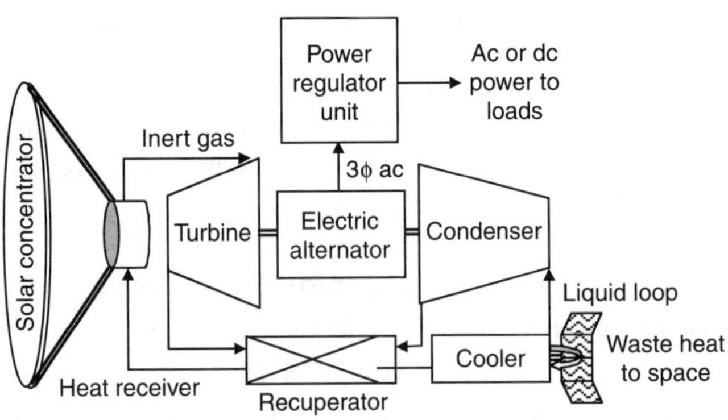

FIGURE 21.4 Brayton cycle schematic.

- Homopolar induction machine with no rotor excitation
- d.c. machine, brushless design with solid state commutator
- Homopolar d.c. machine with solid or liquid metal current collector
- Superconducting synchronous machine working at cryogenic temperature, with efficiency exceeding 99.5% and mass about 10% of the above machines

The exciting current that produces the magnetic field in these machines is always d.c., which must be controlled for regulating the voltage output. Therefore, the permanent magnet machine, which inherently gives an unregulated voltage output, is unsuitable here. The turbo-generator in the energy conversion assembly often has two field excitation coils to produce the required voltage. The series field supplies the bulk of the excitation during normal operation and provides all excitation during start up from rest. The shunt field supplies the remaining excitation for voltage control during normal operation and for power factor control during start-up.

Since power is the product of torque and speed, the power and speed ratings determine the shaft torque, and subsequently the overall machine size and mass. If the machine must operate over a range of speed, its mass will depend on the lowest speed at which the required power must be delivered. For a.c. machines, the significant parameter becomes the volt–ampere product per rpm, since the machine size depends on both the real and reactive power output. The speed in turn depends on the frequency and number of poles. Overall, the specific power improves significantly at higher power and speed ratings.

Each of the machines, except the PM machine, is capable of producing approximately twice the rated output for a short time. In addition, these machines are usually capable of delivering several times the rated current into the system fault to clear the fuse or trip the circuit breaker. This additional output capacity is extremely useful when the machine must operate over a variety of operating conditions. In a system where such added capabilities are not needed, a mass saving of 15 to 20% is possible. On the negative side, drawing twice the load for short time imposes a transient torque of 3 to 4 times the rated value on the prime-mover unless specific steps are taken to prevent it. The machine mass is also impacted by the required quality of power expressed as departures from the rated voltage and frequency, and the transient voltage deviation during load switching. Normal power quality requirements add about 10% in the machine mass.

As for the environment, the shock and vibration during launch and in orbit may not pose a severe limitation on the rotating machines design. However, the radiation damage on the insulation, bearings, and semiconductors used in construction may be of concern.

The reliability and life are limited by fatigue in the metal parts, and by thermal degradation of the insulation at operating temperature. Typical electrical machine used in the aircraft industry has an MTBF ranging from 2000 to 10,000 h. Attaining an MTBF over 100,000 h is necessary for a 10-year mission in space. Since this is an order of magnitude higher than the aircraft machines, it is apparent that the electromechanical machines for spacecraft must be design using significantly better materials and design configuration.

21.4 Electrical Generator Technologies

For a 300-kW solar dynamic power system, NASA's early space station designs screened 22 generator (alternator) concepts from which four candidates were selected for further evaluation and development in the mid-1980s. The configurations of these machines are shown in Figure 21.5.

21.4.1 Brief Description of Each Machine

21.4.1.1 Synchronous generator with PM rotor

This machine has a very simple construction. The stator has one set of output windings, and the rotor has permanent magnets to produce the magnetic field. Compared to other machines, it has a higher efficiency and lower specific volume. The output voltage is regulated by external electronics. Until recently, such machines were limited to small ratings due to limitations in the permanent magnet strength. However, new anisotropic neodymium–iron–boron permanent magnets not only have high energy exceeding 20 MGOe, but also have high heat resistance with capability of operating above 100°C. Magnequench Corporation produces the anisotropic powders, which are used to produce bonded magnets by a rapid solidification process.

21.4.1.2 Synchronous generator with solid rotor (Rice or Lundell)

This machine is a brushless synchronous generator. The field excitation coil is stationary, and the flux is carried to the rotor through two auxiliary air gaps at each end of the rotor. This improves the speed characteristics. It has a smooth solid rotor with alternating magnetic portions of the rotor separated by nonmagnetic portions, which are primarily structural elements that provide strength and pole separation in the rotor. The flux alternates under rotation of the alternating magnetic and nonmagnetic portions.

21.4.1.3 Synchronous generator with wound rotor and brushes exciter

The rotor of this machine has four sets of active elements on a common shaft, namely, the main field winding, the armature of the main exciter, the

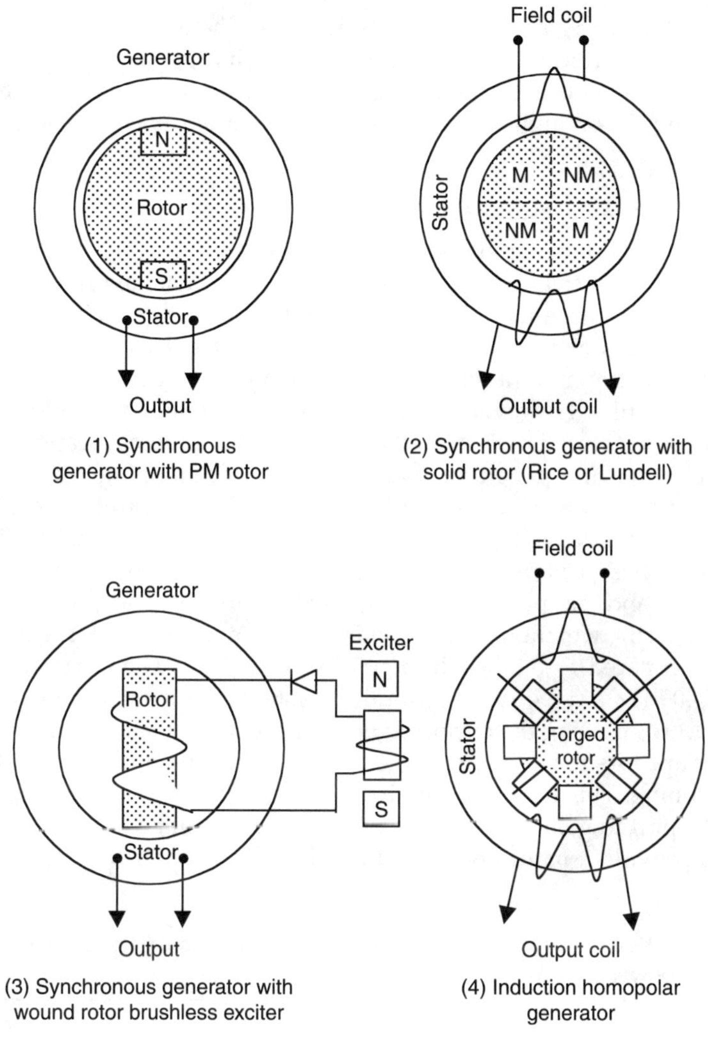

FIGURE 21.5 Constructional features of candidate electrical generators.

pilot exciter's permanent magnets, and the rectifier diodes. The d.c. power is fed to the exciter via rectifier that eliminates brushes. The voltage regulation is controlled by varying the excitation.

21.4.1.4 Induction homopolar generator

The induction homopolar machine uses a fixed toroidal field coil. When power is applied to the field, a unidirectional magnetic flux is created and it follows the path as shown in the figure. The stator output voltage of the homopolar machine is varied by varying the current of the toroidal field coil in the stator. It has a solid rotor with notches. The d.c. field winding on the stator applies variable flux to the output winding through the notched rotor, which results in a.c. power generation. This generator is mechanically simpler but has a higher specific weight and larger specific volume.

21.4.1.5 Advantages of each machine

A second level screening ranked these four candidates on a relative basis as shown in Table 21.1. The synchronous PM rotor machine ranked the first, Lundell generator ranked the second, the third was the synchronous wound rotor brushless generator, and the last one was the induction homopolar generator. Different criteria and/or different weighting factors may lead to different results. However, the synchronous generator with PM rotor and the Rice generator appear to be the most suitable generator candidates. The final selection of a generator and the power conditioning is based on the most effective power system within the technical constraints, such as d.c. or a.c., frequency, voltage level, power level, engine speed, etc. The Rice generator has several advantages such as a relatively higher speed capability and simple voltage regulation. However, it has a relatively higher specific weight, higher reactance, and more complex excitation coils when compared to the synchronous PM generator. Therefore, for a three-phase 400-Hz a.c. power distribution system, the synchronous generator with PM rotor using the cycloconverter approach similar to that used in the aircraft appears to be the preferred candidate for the power generation and conditioning. This type of generator is matched to the optimal engine speed (45,000 rpm) to generate power in the frequency range of 1000 to 3500 Hz, and to provide a naturally commutated cycloconverter to produce precise

Table 21.1 Key performance parameters of four candidate generators for dynamic power system

Generator type	Synchronous PM rotor	Synchronous Rice or Lundell	Synchronous wound rotor/brushless	Induction homopolar
Maximum rotor tip speed (m/s)	230	250	180	300
Specific mass kg/(kVA·krpm)	0.011	0.031	0.011	0.213
Specific volume liter/(kVA·krpm)	0.004	0.006	0.011	0.041
Efficiency%	88–95	87–94	82–92	80–91
Voltage regulation	Complex	simple	simple	simple
Excitation	Rotating PM	Stationary d.c. windings	Rotating diodes from PM exciter	Stationary d.c. windings
Reactance	Low	Medium	Lowest	Highest
Can operate as motor	Yes	Yes	Yes	No

400 Hz output power. In addition, the cycloconverter approach has a higher efficiency and reliability than the conventional d.c. link inverter technology.

21.4.1.6 Other criteria in machine selection

The following criteria are also important in the final generator technology selection:

- Parallel operation
- Overload capabilities
- Voltage regulation
- Speed and frequency limitations
- Start-up operations

21.4.2 Parallel Operations

Two or more identical generators are connected in parallel either to increase the power capability or to improve the reliability of the group. For this reason, only those generators which can work in parallel are suitable for reliable space applications. Moreover, they should be modular and scalable to higher ratings without designing and qualifying new designs for different missions. All of the four candidate machines in Table 21.1 can be operated in parallel.

21.4.3 Overload Capability

The ability to withstand temporary overload and to provide a high fault current for fuse clearing are desired performance features of any machine. Typical machine overload capabilities of the four machines are summarized in Table 21.2. The synchronous PM and induction homopolar machines have the greatest overload capability, and the synchronous wound rotor has the least.

21.4.4 Voltage Regulation and Field Excitation

Voltage regulation of the four machines is compared as follows.

21.4.4.1 Synchronous generator with PM rotor

The output voltage of the PM generator cannot be regulated directly since the flux is provided by the permanent magnets. The voltage regulation of this generator must be provided by external electronics such as the cycloconverter.

Table 21.2 Electrical generator overload capabilities

Generator type	Overload for 5 min	Short-time overload for 5 s	Short-circuit current for 5 s[a]
Synchronous PM rotor	150%	200%	300% minimum
Synchronous Rice or Lundell	115%	150%	300%
Synchronous wound rotor brushless	115%	150%	200%
Induction homopolar	150%	200%	300%

[a]Internal losses are different during short circuit, since the voltage output is zero.

21.4.4.2 Synchronous generator with solid rotor (Rice or Lundell)

There are two field coils, one at each end of the alternator located adjacent to the stator frame. Each coil has two sections, a series section and a control section. The series coils provide excitation from current transformers in the leads to the load. Therefore, this excitation is proportional to the load current. The control coils provide excitation from the voltage regulator (feedback control system) as necessary to maintain constant voltage regardless of the load, speed, or temperature variation in the system.

21.4.4.3 Synchronous generator with wound rotor and brushless exciter

The voltage regulation of the main generator is controlled by varying the current in the exciter field. This system has three rotors (main generator, wound exciter, and PM exciter) and rectifiers on a common shaft.

21.4.4.4 Induction homopolar generator

Voltage regulation is controlled by using a voltage regulator exciter system with magnetic amplifier feedback control. Since this type of machine has the highest internal reactance of the four machines considered, the field current from no load to full load would have the greatest variations.

21.4.5 Speed and Frequency Limitations

The output frequency (in hertz) of a generator is given by the following equation:

$$\text{frequency} = \frac{\text{rpm} \times \text{number of poles}}{120} \tag{21.2}$$

Thus, doubling the speed of a generator doubles the frequency and the electrical power output capability without requiring a change in the magnetics. However, the mechanical stress increases with increasing diameter and the speed squared, requiring increased structural material for high-speed machines. The combined effect of the diameter and the speed result in the stress proportional to the linear tip speed of the outermost fiber of the rotor. Therefore, it is advantageous to define the maximum speed as a function of peripheral tip speed in meter per second rather than in rpm, since the diameter of the rotor depends on the machine rating. The maximum rotor tip speed in meters per second for the four generator types is shown in the first row of Table 21.1, which sets the maximum frequency these machines can be designed for.

The maximum rpm is also a function of the diameter of the rotor and the kilovolt-ampere rating of the alternator. For a lower frequency application (such as 400 Hz) based on the near optimal turbine speed of the solar dynamic engine (such as 45,000 rpm), the minimum frequency of a two-pole generator would be 750 Hz. Therefore, for 400 Hz standard frequency, an effective frequency conversion using a cycloconverter or d.c. link inverter is needed. Rectifiers are needed for d.c. distribution. All of the generator types considered are capable of producing 3-phase power up to 2400 Hz. On the high frequency side, none are capable of producing 20 kHz, although it may be possible to design a 20 kHz, 1-phase, homopolar induction alternator.

References

1. Mason, L.S., A solar dynamic power option for space solar power, *Proceedings of the 34th Intersociety Energy Conversion Engineering Conference*, SAE, 1999, Paper No. 01-2601.
2. Cockfield, R., Radioisotope Stirling generator concepts for planetary mission, *Proceedings of the 35th Intersociety Energy Conversion Engineering Conference*, AIAA, 2000, Paper No. 2843.
3. Thieme, L.G., Qiu, S., and White, M.A., Technology development for a Stirling radioisotope power system for deep space missions, *Proceedings of the 34th Intersociety Energy Conversion Engineering Conference*, SAE, 1999, Paper No. 2454.
4. Sargent, N.B., The electromagnetic compatibility design challenge for scientific spacecraft powered by a Stirling power converter, *Proceedings of the 36th Intersociety Energy Conversion Engineering Conference*, ASME, 2001, Paper No. CT-37, pp. 447-451.

Chapter 22

High-Power High-Voltage Systems

22.1 Introduction

The electrical power requirement in spacecraft remained under several hundred watts for decades. It was traditionally met with the sun-regulated 28-V d.c. power system. However, the power levels for commercial, science, and military spacecraft have been rising steadily (Figure 22.1).[1] It has doubled approximately every 7 years. Because high-voltage systems are lighter and more efficient, the rising power levels have been accompanied by rising voltage levels. For example, the mass saving in just the distribution cables at various power and voltage levels are shown in Table 22.1. Today's GEO communications satellites use power in the 7 to 15 kW range at 70 to 100 V. The ISS uses 105 kW power at 120 V. Some strategic Defense Initiative weapons platforms may require steady power in several megawatts and burst power in hundreds of megawatts at voltages up to 100 kV or more. This chapter first presents high-power, high-voltage power system architectures and then discusses concerns and issues in designing such systems.

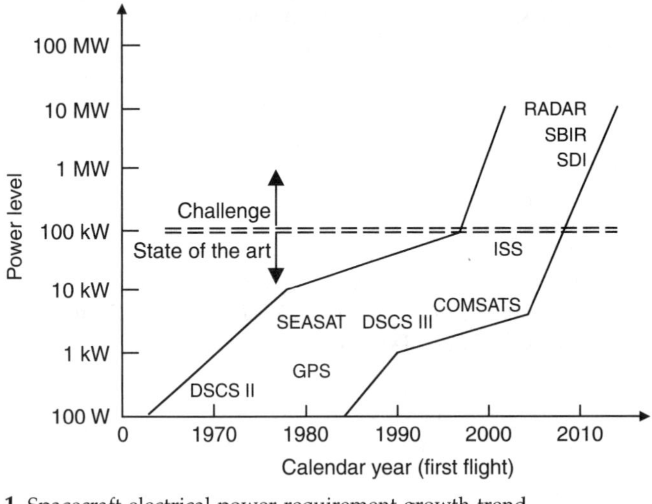

FIGURE 22.1 Spacecraft electrical power requirement growth trend.

Table 22.1 Power distribution cable mass with copper conductors versus power and voltage levels

Voltage level	20 kW over 10 m distance	100 kW over 30 m distance	Present use
28 V	155 kg	7000 kg	Heritage voltage level
70 V	20 kg	900 kg	Modern GEO satellites
120 V	7 kg	290 kg	ISS and EOS
200 V	3 kg	140 kg	SP-100

Power systems up to 100 kW have been designed at voltages up to 200 V. Distribution systems above 200 V are considered high-voltage (HV) in space due to various environmental conditions the spacecraft must undergo during the launch and operation. The space plasma, the wide variations in operating pressure and temperature, out-gassing of the materials, corona, and atomic oxygen are the key design issues.

22.2 High-Voltage PV Array

For satellites with power requirements over 25 kW, high-voltage solar arrays are needed for the system mass optimization. In LEO, the system voltage is limited to less than 200 V due to the leakage current and arcing concerns in high-density space plasma. In GEO, having low plasma density, higher operating voltages are possible after reaching the orbit. An array voltage of 1000 V has been considered for high-power solar arrays and long cables. If the system were to operate at a nominal voltage of 1000 V, a 100 kW array would require current of 100 A, thus limiting I^2R loss in the conductor. In addition, the resulting low magnetic field interacting with the Earth's would keep the drag and the re-boost cost under control. However, micrometeoroid impacts damaging the array leading to arcing and voltage breakdown between high-voltage components is a concern. Recent tests[2] on a GaAs solar array with 6-mil coverglass indicated no arcing in designs with sufficient coverglass overhang to avoid voltage creep and the wiring contacts insulated by encapsulation. The test were conducted for 7 years worth of micrometeoroids simulated by 100 μm soda lime glass spheres at velocities up to 11.6 km/s in a plasma environment typical of the GEO orbit. No arcing was detected under a bias voltage up to −438 V even where the coverglass was penetrated and cell contacts were damaged. These tests suggest that designing a solar array for voltages around 400 V may be possible in GEO orbits.

A large solar array designed for generating high power beyond a few tens of kilowatts becomes prohibitive due to the following reasons:

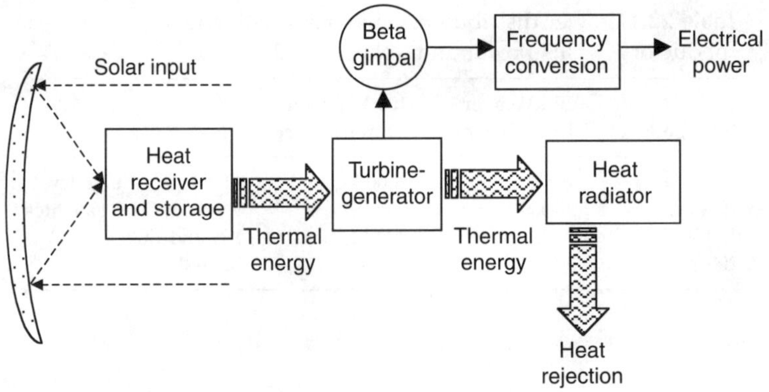

FIGURE 22.2 Solar dynamic power system schematic.

- The solar array becomes difficult to fabricate
- The large moment of inertia poses difficulty in maneuvering the spacecraft
- The spacecraft becomes highly vulnerable to offensive enemy attack, as the large exposed array would be an easy target even from a far distance

The solar dynamic power system covered in Chapter 21 is an alternative candidate for high power. Its key components are shown in Figure 22.2.[3] In such a system, the energy can be stored as thermal energy in the form of latent heat of a phase change material, thus eliminating the battery, which is generally heavier than the solar array in LEO satellites.

22.3 High-Power Nuclear TEC

The nuclear isotope with thermoelectric converter (TEC) covered in Chapter 20 is suitable for generating a few hundred watts of power. For large power in a few hundred kilowatts, the U.S. Air Force and DoE have jointly funded significant research and development since the 1980s to design, build, and qualify a nuclear reactor–TEC power system under the SP-100 program. The program was dormant in the 1990s and has been revived recently under NASA's Project Prometheus for further development and deployment for high-power interplanetary propulsion (Figure 22.3) and space defense missions with nuclear threats or with irregular orbits. The first nuclear powered space probe is planned on the *Jupiter Icy Moon Orbiter* (JIMO) for a possible launch in 2011. The static TEC offers a better alternative to the rotating turbo-generator. The effect of the high gyroscopic effect of large rotating machines in space is a major technological risk. No rotating machines of such a size have been used in space before.

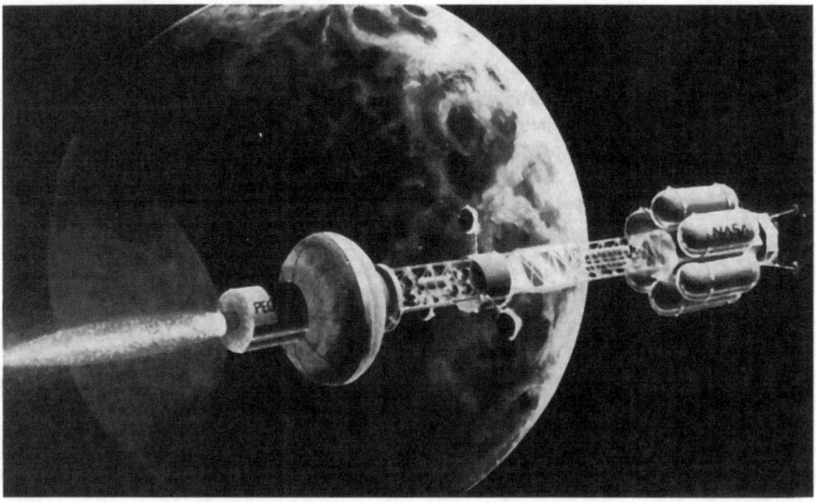

FIGURE 22.3 SP-100 power system for electrical propulsion to Mars.
(From Bell and Esch, *IEEE Spectrum*, p. 48, August 1991. With permission.)

In the SP-100 power system, the TEC modules are made of multicouple SiGe/GaP thermoelectric (TE) cells having a TE conversion efficiency of about 7%. However, the working heat losses reduce the overall system efficiency to 5%. A 100-kW_e electric power system, therefore, requires the reactor thermal power of 100 kWe/0.05 = 2000 kW_t. The remaining 95% or 1900 kW_t of the reactor heat must be radiated back to space in order to control the spacecraft temperature within the material limits. The radiator design, therefore, is a principal design driver. Figure 22.4 shows the

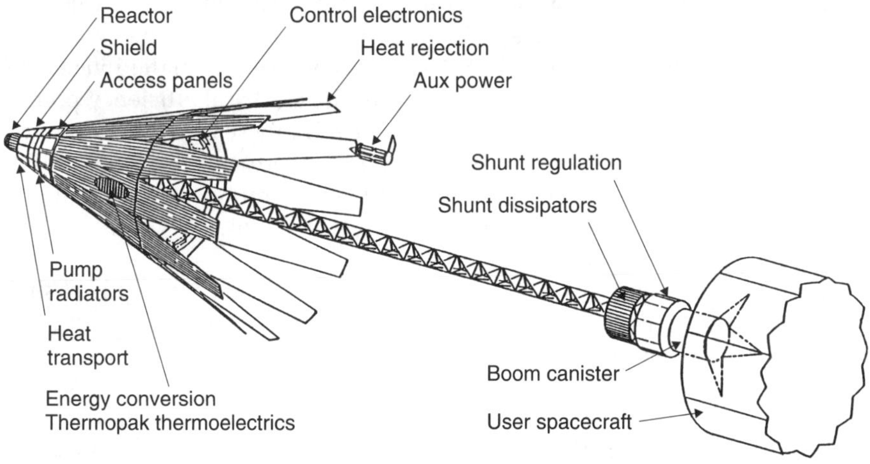

FIGURE 22.4 SP-100 power system configuration with nuclear reactor and TEC.
(Source: U.S. Department of Energy.)

radiator with panels fanning out in a conical shape for heat rejection to space. The reactor core and the radiation shield are at the tip of the cone. The user payload and sensitive equipment are located away from the heat by placing them at the other end of a long 50-m boom.

The SP-100 uses a fast spectrum liquid lithium cooled reactor and SiGe/GaP TE with $Z = 0.0007$ for the state-of-the-art design and 0.0014 for the future advanced TEC design. Each TE cell has a V_{oc} of 0.50 V, and I_{sc} of 12.8 A. Thus, the maximum power from each TE cell is $0.50 \times 12.8/4 = 1.6$ W. The specific power of the TE cell is 33 W/kg.

The nuclear reactor research and developments for space defense have been funded in the past for electrical power in the 30 kW to 300 MW range. The reactor is similar to a ground-based nuclear power plant, only much smaller in scale.

In the baseline concept, the nuclear reactor boils the liquid metal. The hot liquid is circulated on the platform by a liquid lithium pump loop, transferred through an intermediate heat exchanger and delivered to an array of heat pipes mounted to the 12-sided conical main body as shown in Figure 22.4. The TE panels receive heat radiantly from the heat pipe array. Thus, the heat is transferred from the reactor core to the TE modules, each of which generates 4.6 kW_e at 100 V. The heat not converted into electricity is wasted. It is radiated into space from the outer surface of the TE panels. The panels are deployed for effective radiation to the outer space. The lithium coolant is pumped by an electromagnetic pump driven by a dedicated TE module. The TE for the pump is powered by the temperature drop between the working fluid and the pump radiators to assure pumping as long as the reactor is at a temperature.

Twelve 4.6-kW_e TE modules are connected in parallel to produce 50 kW_e at 100 V. Two such groups are connected in series with the mid-point electrically grounded to the spacecraft structure. The voltage between two outer conductors is therefore 200 V (Figure 22.5). The usable voltage to avoid electrical breakdown in the SP-100 mission orbits is in the 150 to 200 V range. However, the optimum voltage from the electrical efficiency point of view was determined as exceeding 200 V. The finally selected bus voltage of 200 V is derived by using a center ground conductor between $\pm$100 V conductors. This, in effect, gives only 100 V operating voltage from the structure ground, while giving the load voltage of 200 V, keeping the current level and ohmic loss low.

The SP-100 power system uses a regulated voltage bus with a shunt dissipater. The TE modules continuously generate 100 kW_e whether the payload needs that much power or not. When payload power consumption is less than 100 kW_e, the excess power is absorbed by the shunt dissipaters and dissipated into space. The shunts are made of traditional high-temperature tungsten filament resistance wires. The shunts are turned on automatically in response to the bus voltage error signal. The power output of the TE panels are diode-isolated and are combined in a single 200 V d.c.

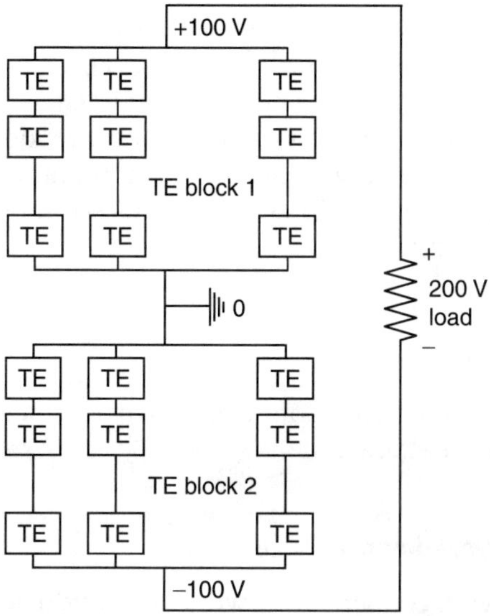

FIGURE 22.5 SP-100 thermoelectric converter assembly for 200-V system between ± 100-V lines.

bus at the power module, which is located on the opposite end of the extension boom. Such physical isolation helps reduce the amount of shielding required.

A 300-W, 28-V auxiliary power supply from a separate RTG is provided for startup, maintenance, and the shunt-down period. Under certain mission constraints, solar or battery-only options may also be considered for this purpose.

The following were the SP-100 system requirements:

- 7-year mission life
- Survival to a classified nuclear threat
- 100 kW_e EOL power, scalable to 30 kW_e to 1000 kW_e power levels
- 200-V fully regulated bus
- Reliability goal of 0.990
- Heat source upper temperature 1310 K

Although significant development work has been done in the USA, no SP-100 unit has flown. The Russians, on the other hand, have flown dozens of such units. The U.S. research funding was reduced in the early 1990s after the U.S. procurement of similar technology from Russia following the collapse of the former Soviet Union. However, from 2003 to 2008, the SP-100 budget is expected to be over one billion dollars, primarily targeted at space defense and electric propulsion. A key thrust of this research is the development of high-temperature materials. High power requires high temperature to effectively reject waste heat with small compact radiators.

Multi-megawatt burst power systems for strategic defense may consider nuclear reactors with rotating machines. For minimizing the gyroscopic disturbance, such systems would necessarily use counter-rotating machines in pair(s), as discussed in Section 22.11.

22.4 High-Voltage Design Issues

The design issues that are common to all power systems using high voltage are discussed in this section.

22.4.1 Paschen Breakdown Voltage

The breakdown voltage between two bare conductors in a given gas depends on the *p–d* product, where *p* is the gas pressure and *d* is the separation distance. It is shown in Figure 22.6, known as the Paschen curve. The spacecraft passes through a wide pressure variation, ranging from atmospheric pressure on Earth before launch to 1×10^{-10} torr in high orbits. During this transition, it always passes through the minimum breakdown voltage point on the Paschen curve. The power distribution system normally has numerous bare spots of conductors and terminals. They would spark during the spacecraft's journey from the Earth to a high orbit regardless of the separating distance between them. Out-gassing can make

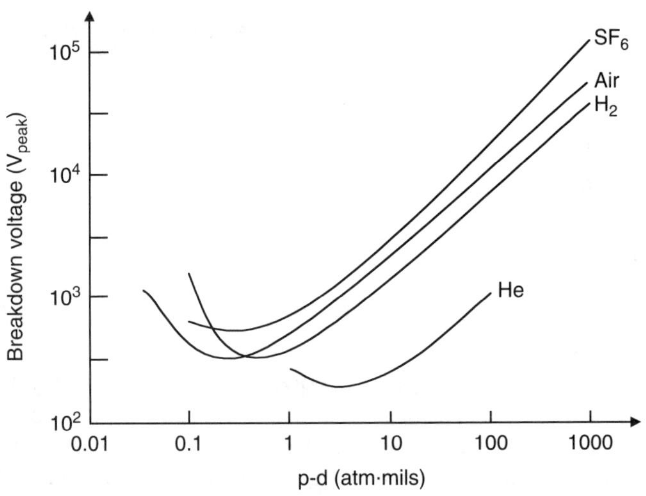

FIGURE 22.6 Pressure-spacing product versus breakdown voltage in various gases (Paschen curves).

the pressure prediction difficult. For this reason, a high-voltage system can be turned on only after reaching the orbit (0.1 Pa vacuum). Alternatively, the equipment must be filled with insulating gas, such as nitrogen, or completely encapsulated with a solid insulation.

Preliminary data have shown that material out-gassing causes the local pressure around the high-voltage component to increase to the point that it is no longer a good vacuum. This, in turn, results in lower corona inception voltage. The approach to increasing the corona voltage, therefore, is to select the material with a low out-gassing rate in the first place. Figure 22.7[4] depicts out-gassing test results on a few materials. It indicates that Teflon has a low out-gassing rate among electrical insulation materials, making it a good insulation candidate for high-voltage space applications.

22.4.2 Dielectric Stress Concentration

If a HV rectangular conductor is placed in an enclosure filled with gas, the dielectric stress at the ends is not uniform, but is concentrated near the conductor edges. If the conductor radius is r and end distance e (Figure 22.8), the maximum dielectric stress is almost[5]

$$E_{\max} = \frac{V}{e}\left(1 + 0.25\frac{e}{r}\right) \tag{22.1}$$

For $E_{\max}$ to be equal to V/d, i.e., the same as the uniform stress in the flat part with ends fully radiused, i.e., $r = a/2$, the end distance is derived by equating

$$E_{\max} = \frac{V}{e}\left(1 + 0.25\frac{2e}{a}\right) = \frac{V}{d},$$

which gives

$$e = \frac{a\,d}{a - 0.5\,d} \tag{22.2}$$

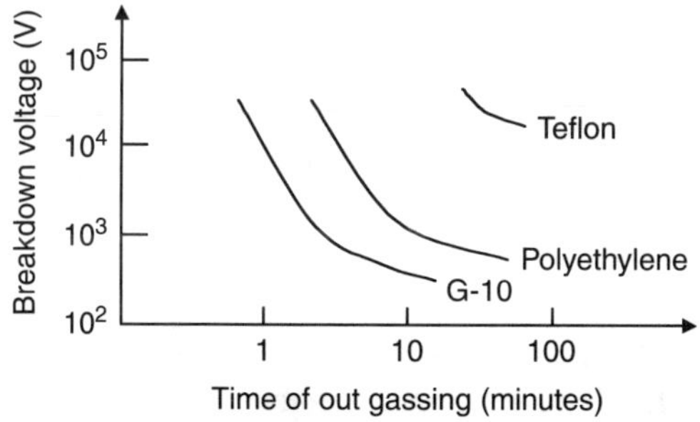

FIGURE 22.7 Breakdown voltage versus out-gassing time.

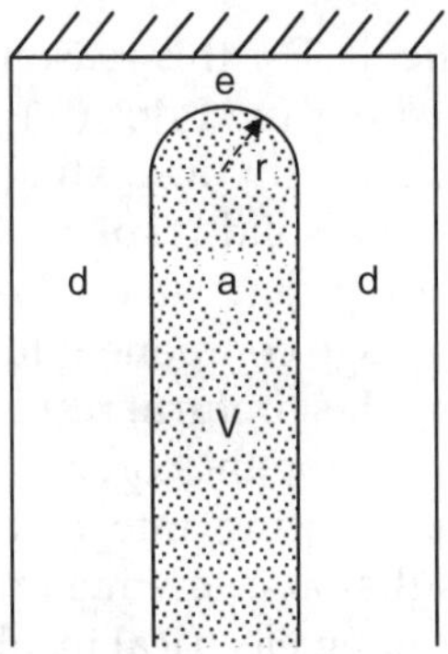

FIGURE 22.8 Voltage stress concentration at flat conductor edge with radius.

For example, with $a = 3$ mm and $d = 1$ mm, e must be 1.2 mm, i.e., 20% more than d, the insulation thickness. For a smaller radius, say $r = a/4$, $e = ad/(a - d) = 1.5$, i.e., 50% more insulation is required at the ends.

The stress concentration at sharp corners and their effects on corona voltage and dielectric breakdown in atmospheric conditions on the ground are well studied and reported in the literature. Such is not the case in the space environment. A thorough systematic study on this aspect of the insulation design is required. Until then, the available data must be used with interpolations and extrapolations using wide margins to minimize the associated design risks.

22.4.3 Corona Degradation

The insulation design should be such that partial discharges (corona) should not start. If started at some higher transient voltage, it must extinguish during the down-slope of the transient at some voltage higher than the steady state operating voltage. Corona has two degrading effects on the insulation performance. It chemically erodes the material, thus shortens the life; and it produces radio frequency EMI that may interfere with the surrounding equipment. Both of these effects are undesirable, and should be avoided.

The corona voltage is a function of the type of material and its thickness used in the design. For a 1000-V electrical component, corona tests on Earth on polypropylene insulation gave the test results reported in Table 22.2. These results must be modified to account for the Paschen minimum and plasma effects depending on the orbit in which the spacecraft is designed to operate or pass through.

A critical quantity needed for evaluating solid insulation subject to corona is E_i, which is the partial discharge inception voltage stress. This can be obtained directly by measurements, from extrapolation of the

Table 22.2 Corona test results on polypropylene insulation

Thickness (mils)	Inception(V_{rms})	Extinction (V_{rms})
5	880	835
10	1570	1420

(Data source: M.R. Patel, NASA Report No. CR-175071, 1986.)

breakdown voltage–time curve, and from careful analysis of the voltage stress at the conductor edge or in a void if necessary geometric parameters are known. Large voids must be avoided in solid insulations. The dielectric stress is εE in a laminar void and $3\varepsilon E/(1 + 2\varepsilon)$ in a spherical void much smaller than the insulation thickness. Here, ε is the relative permittivity and E is the uniform stress in the solid insulation. Partial discharge occurs in a void when the product of the pressure in the gas and the void thickness (electrode separation) exceeds the breakdown voltage according to the characteristic Paschen curve. It can be aggravated by dielectric stress concentration around sharp corners. Good HV conductor design avoids sharp corners to keep the dielectric stress concentration low and corona inception voltage high.

The following analysis establishes the largest permissible void size in space-worthy high-voltage insulation. The laminar void is assumed, since it produces more stress concentration than the spherical void. If a laminar void of size d_o exists in a solid insulation between two conductors due to any reason, the breakdown voltage across such a void will depend on the gas pressure in the void. Since this pressure may vary over a wide range in and around the spacecraft at different times of the mission, the minimum Paschen breakdown voltage must be used in the calculations, which is about 297 V_{peak} or 210 V_{rms} between copper electrodes. There are now two dielectrics in series, a gap of laminar thickness d_o and the solid insulation of thickness d. If the dielectric stresses are E_o and E in the gap and the solid, respectively, the voltage across the conductors would be $V = E_o d_o + Ed$. Polypropylene insulation with a relative dielectric constant of 2.25 would take 1/2.25 times the gap stress, i.e., $E = E_o/2.25$. With $d = 20$ mils as an example, the above voltage then becomes

$$V = E_o d_o + \left(\frac{E_o}{2.25}\right) \times 20 = E_o d_o + 8.89\, E_o = E_o d_o\left(1 + \frac{8.89}{d_o}\right) \tag{22.3}$$

where d_o is in mils. The product $E_o d_o$ is the voltage across the gap, which has the minimum Paschen value of 210 V_{rms}. The corona would start at this voltage across the gap. The corona inception voltage of a cable with polypropylene insulation is then

$$V_i = 210\left(1 + \frac{8.89}{d_o}\right) \tag{22.4}$$

Equation 22.4 is plotted in Figure 22.9, which shows that it takes about a 2-mil delamination gap for corona to start at 1200 V_{rms}. This condition would not normally lead to an immediate failure, but eventually would do so under gradual corona degradation. Should the gap completely fail, the dielectric stress in the solid insulation would be 1200/20 = 60 V/mil, which is two orders of magnitude lower than the solid insulation's short time breakdown strength.

As seen in Figure 22.9, the corona inception voltage decreases with increasing gap size and asymptotically approaches the minimum Paschen value of 210 V_{rms}. This is because the Paschen minimum is a constant value, although occurring at different pressures for different gap sizes. As the gap increases, the corona inception voltage across the gap remains constant at 210 V, while the voltage contribution of the decreasing dielectric stress in the solid insulation decreases. This voltage is based on the worst possible combination of the gap pressure and the gap size (i.e., the minimum Paschen value). It is likely that there is always some pressure in the void due to continuous out-gassing of the insulation, giving a higher corona inception voltage. The location of the void along the cable length, therefore, matters. If the void is at mid-length of the cable, out-gassing may take a long time due to the high resistance offered to the gas flow in a long narrow duct. For increased reliability, it may be desirable to bleed some gas (such as nitrogen) along the length of high-voltage cable insulation. This would assure the desired gas pressure to maintain the required dielectric strength of the cable.

There is an increasing push for mass reduction to reduce launch cost. For this reason, many HV components do operate with some corona, which

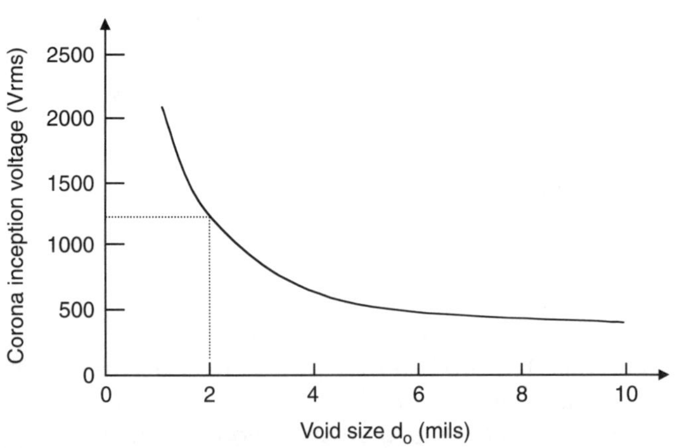

FIGURE 22.9 Corona inception voltage versus laminar void size in a flat conductor cable.

must be controlled in manned spacecraft to protect the astronauts from corona charging currents. The high value of mass minimization justifies a sacrifice in the equipment life, provided reliability and safety are not sacrificed. A life of 5 to 15 years has been typical for space components, but reliability must be high since these components must operate without servicing.

22.4.4 Atomic Oxygen

The interaction of atomic oxygen with the insulation surface is of particular concern in low Earth orbit. This interaction is a reaction and/or removal of mass from the surface during the mission life. The amount of interaction depends on the total integrated dose of atomic oxygen impingement, which is a strong function of the orbit altitude. Figure 22.10 shows such an interaction in a typical polymer. The insulation design should allow such a loss of material over the mission life in space.

Table 22.3 shows how various insulating materials degrade under radiation and atomic oxygen. It shows that Teflon® is bad under the effect of either one. An estimated radiation dose on the ISS over 5 years is about 100 krads, so silicon rubber may be used in the design, provided that the degradation under atomic oxygen is acceptable. The atomic oxygen fluence rate on polar orbits at 400-km altitude is about 3×10^{21} atoms/cm^2·year, which is less than that on the ISS.

22.4.5 Plasma and Charged Particles

The plasma in the mission orbit exerts the greatest influence on the voltage that two bare conductors can withstand in space. The plasma density in electrons/cm^3 versus orbit altitude is depicted in Figure 22.11. It peaks

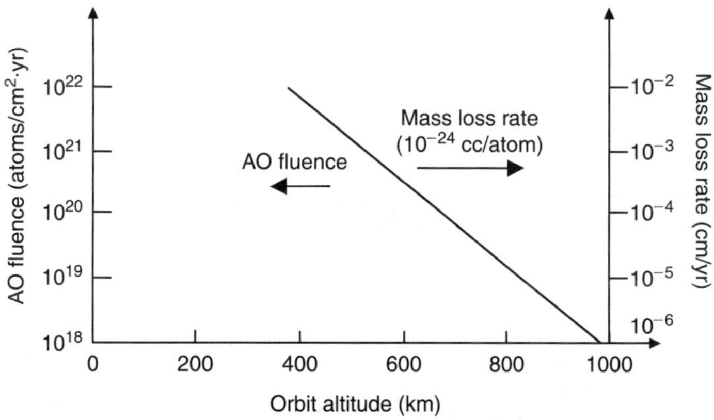

FIGURE 22.10 Mass loss due to atomic oxygen versus orbit altitude.

Table 22.3 Depletion rate of electrical insulation in space environment

Insulation	Radiation dose before adverse effect is noticed (Mrads)	Reactivity, 1×10^{-24} cm^3/atomic oxygen atom	Depletion rate in 3×10^{21} atoms/cm^2·year orbit, mil/year
Teflon	0.1	0.03	0.036
Silicon rubber	1	0.05-0.2	0.06-0.24
Kapton	100	3	3.6
Epoxies	1000	2.5	3.0

inLEO, and is negligible in GEO and beyond. The plasma acts as a shunt resistance across the power components, such as the solar array, causing power loss due to the resulting leakage current between bare conductors. This loss becomes noticeable above 160 V. A voltage much higher than 160 V leads to voltage breakdown. The breakdown voltage versus plasma density is shown in Figure 22.12. Combining Figure 22.11 and Figure 22.12 leads to the arcing voltage between bare conductors versus orbit altitude, as plotted in Figure 22.13. The arcing voltage determines the maximum solar array output voltage and also the minimum distance between printed circuit board traces to withstand the voltage over the design life of the mission. For this reason, NASA limited the ISS solar array output voltage to 160 V, with a 120-V distribution voltage and step-down converters where required for existing 28-V hardware.

In plasma-free orbits such as GEO, the conductor in a near perfect vacuum can withstand kilovolts or even higher voltages. However, the

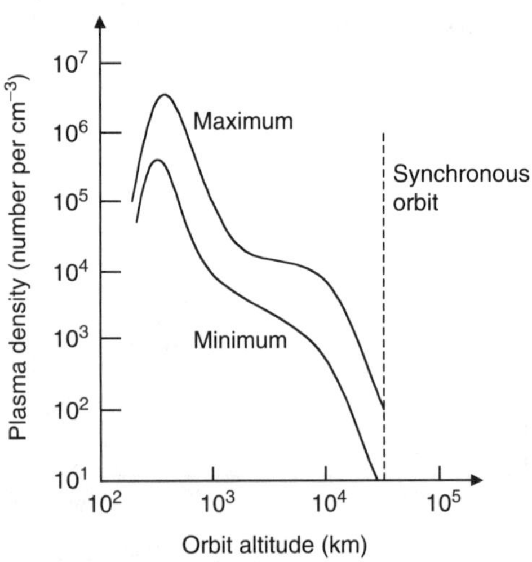

FIGURE 22.11 Plasma density versus altitude in equatorial orbit.

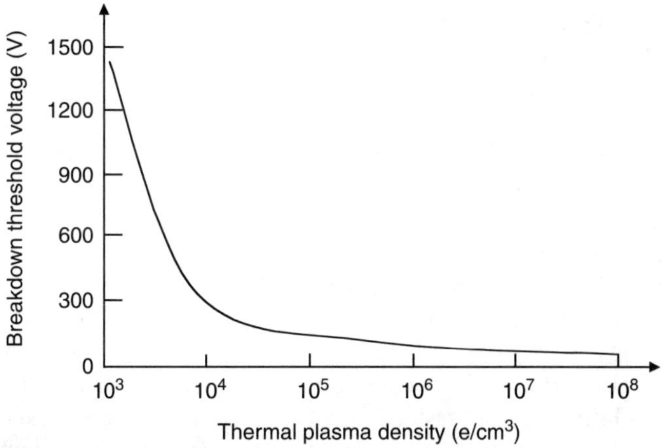

FIGURE 22.12 Breakdown voltage threshold versus plasma density between bare conductors.

Paschen minimum breakdown voltage limits its operation only in GEO and not during launch or transfer orbit. The out-gassing of the material inside the spacecraft body and the flue plumes outside create an uncontrolled gas pressure around the conductors. Moreover, the spacecraft interior during launch undergoes a decreasing air pressrun as it attains the orbit, and may take a long time after the launch to achieve a good vacuum. It is probable that at some instant the *p–d* product may achieve a value corresponding to the minimum breakdown voltage. Under such a condition, the bare conductor would cause arcing and short circuit. For this reason, the system designed to operate at higher than 160 V in GEO can be turned on only after achieving the GEO orbit.

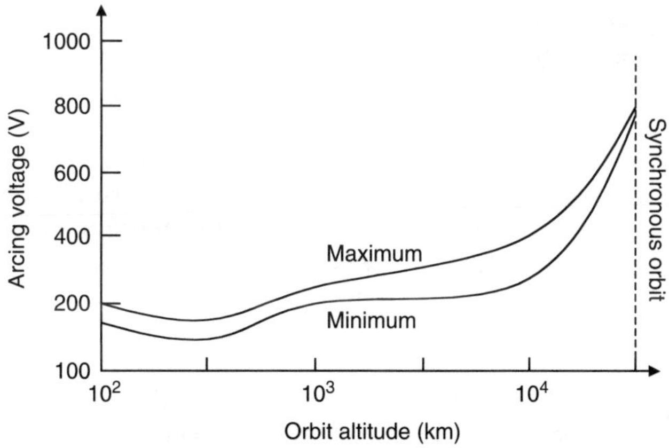

FIGURE 22.13 Arcing threshold voltage between bare conductors versus altitude.

As seen in Figure 22.13, the 300-km high orbit is the worst with the arcing on-set voltage of only 160 V. If the spacecraft must go through this orbit, the power distribution voltage must be below 160 V. If a solar array is assembled by standard construction techniques — with coverglass not completely shielding the metallic interconnects — is operated at voltage higher than 160 V in LEO, the bare interconnects can act as plasma probes and attract or repel charged particles. The generated voltages will be equal to the space plasma potential at some location on the array. Since electrons are more mobile than ions, the array will float at voltages that are mostly negative with respect to space plasma potential. Cell interconnects at voltages above this space plasma potential will collect electrons; those at voltages below this will collect ions. The voltage distribution in interconnects relative to the space plasma potential must be such that these electron and ion currents are equal (i.e., the net current collected is zero). This flow of particles can be considered to be a current loop through space that is in parallel with the operational system and hence is a power loss. In addition, the coverglass used on the solar cells must also have a zero net current collection. This interaction with space plasma forces the coverglass to a small negative potential and can produce large voltage gradients in the gap region between solar cells. This can subsequently give rise to arcing conditions or transient breakdowns to space.

The severity of such plasma interactions depends on the array operating voltage and the charged-particle environment. The operating voltage is determined from power system trades, but will probably be less than 1000 V. At the projected operating voltage only the low-energy or thermal plasma environment should be of concern since the array voltage is too low to influence the higher energy environmental particles. This plasma environment has particles with temperatures of about 1 eV and densities that vary from a maximum of about 3×10^6 cm^{-3} at 300 km to between 1 and 10 cm^{-3} at geosynchronous altitudes. Hence, plasma interactions should be more severe at the lower altitudes than at synchronous altitude.

A 100-kW space-worthy system operating at 500 V d.c. was designed by NASA. It was made of ten arrays in parallel each with 10-kW capacity. It was found that, under normal quiescent conditions, the power drain due to the electron coupling current would be negligible. However, the arcing in the negative voltage regions could seriously disrupt system operations either by introducing a ripple on the output or by terminating operations depending on the severity and location of the breakdowns. The orbital oscillations range through ram, wake, and eclipse conditions. The shuttle experiments that indicated that large space structures could create their own plasma environment tend to make plasma interactions even more critical.

Over the past decade, the advantages and disadvantages of large space power system operations at high voltages have been argued and discussed. There are obvious advantages to using high voltages in space. Possible hazards to such operations with standard array technology have been addressed. These interactions are not insurmountable but can be overcome given adequate understanding of the phenomena. What is needed is a systematic investigation to determine why discharges occur and how to prevent them. This would require test programs involving large arrays with self-generated voltages and flight experiments to prove that all of the interactions can be minimized.

Tests in ground simulation facilities in which small solar array segments were biased to positive and negative voltages in a plasma environment have shown that interactions can be detrimental. When positive voltages are applied, electron currents can be collected that become proportional to the panel area at voltages greater than 100 V. Arcing or breakdown can occur under negative bias voltages. This arcing threshold depends on the plasma density and can be as low as −300 V in simulated 300-km orbit plasma environment as shown in Figure 22.14(a) and (b).

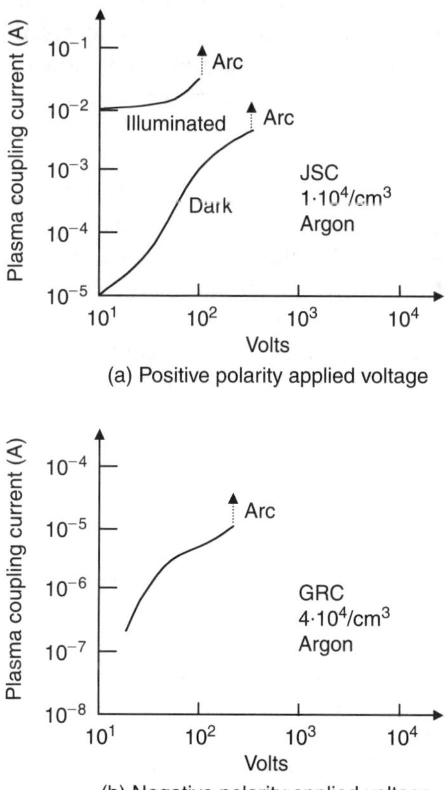

FIGURE 22.14 Plasma coupling current versus applied voltage.

Relatively few tests have been conducted in which an array capable of generating high voltages under solar simulation conditions was operated in a plasma environment. PIX-I and PIX-II space experiments conducted by NASA have shown that a 9-block array would float electrically such that one block would have an average positive voltage that would be 10% of the overall voltage, with the other eight blocks progressively more negative. This test indicated that array behavior could be approximated by considering the interaction with separate blocks at an average voltage.

As for the effects of radiation, Bouquet and Winslow[6] present their study results on the lowest doses for which radiation effects on material properties are noticeable. The report compiles data on the effects of radiation on the physical properties of polymers and synthetic organic materials. Emphasis is on materials of interest to nuclear equipment and nuclear reactor designers. The data cover five categories of polymeric materials: insulators, elastomeric seals and gaskets, lubricants, adhesives, and coatings. More than 250 materials are represented. The document reflects a continuation of the work in progress. It gives available data on physical, electrical, and optical properties for each material in terms of the lowest reported threshold dose. It indicates the radiation level at which radiation effects become apparent and the dose for a 25% change in the specified property. The literature source for the data is also indicated. Both the threshold and the 25% change dose vary widely within each of the five categories of polymer. Most of the polymeric materials have thresholds above 10^5 rads. Some materials — fluorocarbons, nylons, epoxy resins, polyethylene oxides, and polyalphamethyl chloroacrylates — have thresholds below 10^5 rads. The threshold data are useful for setting a lower limit at which changes can be expected. They may also be used as a guide for deciding when the expected radiation is low enough to be disregarded for a given polymer. The 25%-change dose is a valuable radiation statistic because changes are real at this dose level and are neither artifacts of the experiments nor results of sample variations.

22.4.6 Temperature Extremes

The spacecraft component temperature can swing over a wide range during sunlight and eclipse durations. Such temperature extremes must be considered in the insulation selection. The components exposed to space temperature (–200°C) without survival heat may experience damage. For example, most polymer insulation may shatter (crack) at temperatures below –75°C. After such an exposure, corona may start in the cracked insulation, thus shortening the component life in supporting high voltage.

22.4.7 Design Guidelines

The corona, out-gassing, plasma, attitude control exhaust, permeation from pressurized chambers, bubbles in liquid dielectrics in zero gravity, etc. make high-voltage space insulation a different science than on Earth. More experiments are needed to understand these factors.

Two types of insulation breakdown dominate in vacuum or low-pressure gaps between bare conductors. Breakdown in small gaps is dominated by the electrode surface, whereas in large gaps it is dominated by particle contamination in the gap volume. The surface breakdown generally occurs at lower stress levels than the volume breakdown. The pressure around out-gassing surfaces can be orders of magnitude higher than the ambient pressure, and can be a function of external stimuli, such as temperature and incident charged particles.

If the spacecraft must operate or pass through the worst orbit in an electrically ON state, the spacecraft power distribution voltage must be lower than the breakdown voltage in plasma and the Paschen minimum voltage. For voltages higher than that, several possible design approaches can be considered:

- Avoid the effect of plasma by placing the high-voltage component in a metal enclosure. The enclosure drains the plasma to the spacecraft ground, thus keeping it from reaching the electronic component (Figure 22.15).
- Avoid the worst pressure for the Paschen minimum by venting a gas (nitrogen) around the bare electrodes until the spacecraft reaches a safe orbit.
- Encapsulate the high-voltage component in solid insulation. The material selected for encapsulation must have low out-gassing properties and the manufacturing process must be void-free to minimize corona.
- Place the high-voltage component in liquid or gaseous insulation. The high fluid leakage rate in a hard vacuum must be accounted for in space designs.

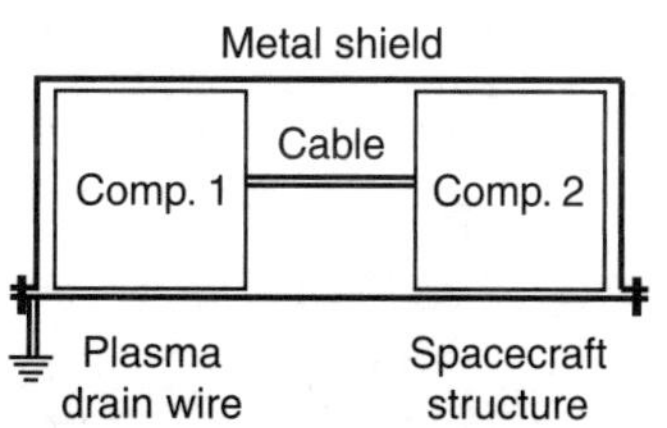

FIGURE 22.15 Space plasma shield configuration.

- Turn ON the electrical system only after reaching the altitude where the arcing onset voltage is higher than the distribution voltage. Until then, the electrical power requirement (for survival heat, navigation or controls) must be met by batteries operating at low voltage.

22.5 High-Voltage Direct Current

The automobile industry used 6-V power in the first 50 years. It has been using 12-V d.c. since the early 1950s, but has now started moving towards 42-V d.c. to meet the growing power needs in modern automobiles. Military tanks, electric guns, and turret drives in the 45 to 60-kW range use 270-V d.c. power distribution. The aircraft industry has been using 400-Hz power systems, but new standards have been developed for the industry to adopt 270 V d.c. The U.S. Navy has been using 155-V d.c. that meets MIL-STD-1399.[8]

Major factors that influence the high-voltage design in space are:

- Availability of space-qualified parts, such as tantalum capacitors, semiconductors and power devices rated for high voltage may be limited.
- High-voltage d.c. contactors present a significant challenge in designing switching relays.
- Electrical discharge may occur in the 300 V range even under space vacuum conditions.
- Plasma discharge power loss in the array, which is 1% at 400 V d.c.

The interaction of plasma environment with a high-voltage d.c. system is minimized by connecting the solar array wings at ± voltage with center ground (Figure 22.16). This results in a power system operating at twice the plasma interaction voltage. The EMI considerations in high voltage d.c. systems are similar to those in the low voltage d.c. systems. In particular, MIL-STD-461 on the conducted ripple current should be achievable.

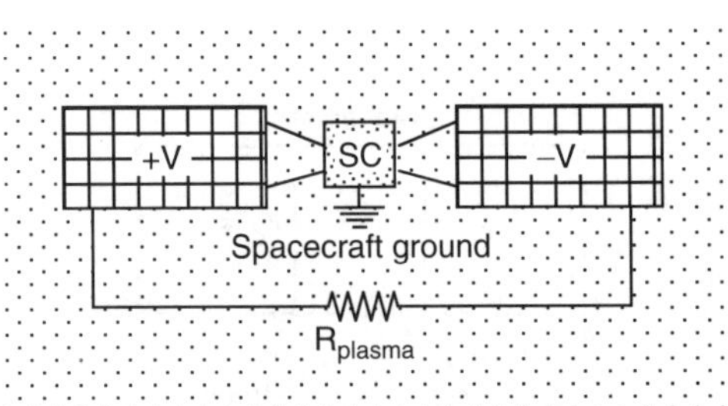

FIGURE 22.16 Plasma interaction on high-voltage solar array wings.

Direct-current systems are also expected to have low levels of radiated B- and E-fields. Power conditioners operating at high switching frequency would result in minimal EMI filters. Shielding techniques are similar to those used in low voltage systems. Solid-state switches have reduced the rate of rise and fall times in current and thereby reduced system transients.

The principal safety hazards in an electrical system are fire, shock, and audible noise. Direct-current systems are typically immune from audible noise. Fire hazards are common to all distribution systems regardless of the voltage level. Ground faults in HV d.c. are cleared by RPC and RBI to protect personnel and reduce fire hazards. No unusual safety problems with HV d.c. systems have been reported in the literature

HV d.c. components developed for aircraft applications are suitable for further development for space applications. The extensive space experience in 28 to 160 V d.c. heritage allows development of similar designs for HV d.c. systems. A number of HV d.c. components have been developed for use with HV solar array and battery assemblies. A large number of vendors have the technology and background to produce space-worthy components for HV d.c. systems competitively. Compared to HV a.c., spacecraft design engineers are much more familiar with design techniques and applications of d.c. power equipments such as d.c.–d.c. converters, d.c. power controllers, etc.

22.6 Alternating Current Versus Direct Current

The d.c. system is generally economical if power can be generated, distributed and utilized at the same voltage, or the d.c.–d.c. power conversion efficiency can match with that of the a.c. power transformer. Moreover, d.c. generates minimum EMI, but is difficult to switch off. The a.c. system, on the other hand, is generally economical where the power generation, distribution, and utilization are done at different voltages, and there are numerous users utilizing power at many voltage levels as in a ground-based utility. The high efficiency of an a.c. power transformer in converting one voltage into another is a dominant factor in favor of a.c. Moreover, switching a large a.c. current OFF when it is passing through a natural zero is easy, as is often done.

Since the d.c.–d.c. converter efficiency is rising with low-loss semiconducting devices and efficient switch mode power converters, d.c. is becoming attractive where only a.c. was considered practical a few years ago. The history of a.c. versus d.c. power systems in the aerospace industry can be examined in terms of the evolution of present day aircraft power systems, which has progressed as follows:

- 1920s: 14 V d.c., 500 W for engine ignition
- 1930s: 28 V d.c., with power in kilowatts

- 1950s to 1980s: 208 V, 400 Hz, 3-phase Y, 4-wire grounded, variable speed, constant frequency high power a.c. via mechanical gear
- 1990s: 208 V, 400 Hz, 3-phase Y, 4-wire grounded, with variable speed, constant frequency a.c. via power electronics (no mechanical gear)
- Future: 270 V d.c., or 400 Hz a.c. with variable frequency, constant voltage d.c. (no gear or power electronic inverter).

The 28-V d.c. system used in early aircraft, both military and commercial, contributed to the selection of 28 V d.c. for the first spacecraft. The 28 V d.c. remained in use until the 1980s. Modern aircraft power systems, however, have changed to a 400-Hz, 3-phase, 200 Y/115-V variable speed constant frequency system shown in Figure 22.17. The space shuttle fleet generates d.c. power using a fuel cell, but converts and distributes the power at 120 V, 400 Hz, 3-phase. An advanced system proposed for the aircraft industry centers around 270 V d.c. as shown in Figure 22.18. The spacecraft secondary power distribution can be 208 V, 3-phase Y, such that 120 V is available for single-phase utilization. The primary voltage can be any, such as 440 V 3-phase Y or △. The Japanese and European space industry preferred 120 V d.c. for secondary distribution on the space station even when NASA strongly preferred 3-phase, 208-V Y/120-V, 20-kHz system in an early phase of the ISS design. Arguments made in favor of 120 V d.c. over 120 V 20 kHz a.c. were:

- Decades-long experience
- Extensive design data base
- Low cost
- Low EMI
- High reliability
- Synergistic development using heritage

The simplicity, low parts count, and previous space qualification history of d.c. systems offer advantages over a.c. systems for high reliability in space environments.

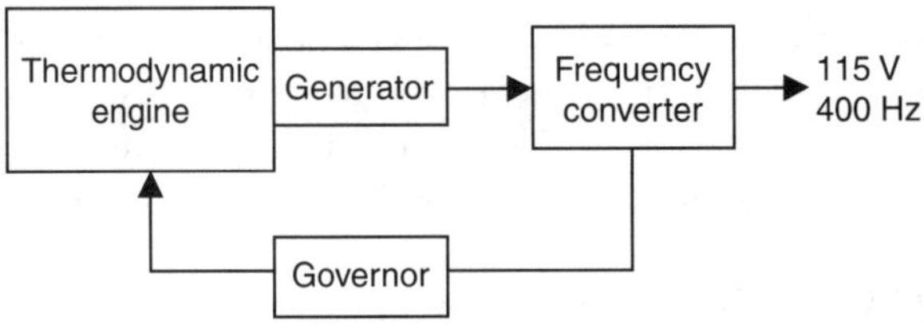

FIGURE 22.17 Variable speed constant frequency ac power system for aircraft.

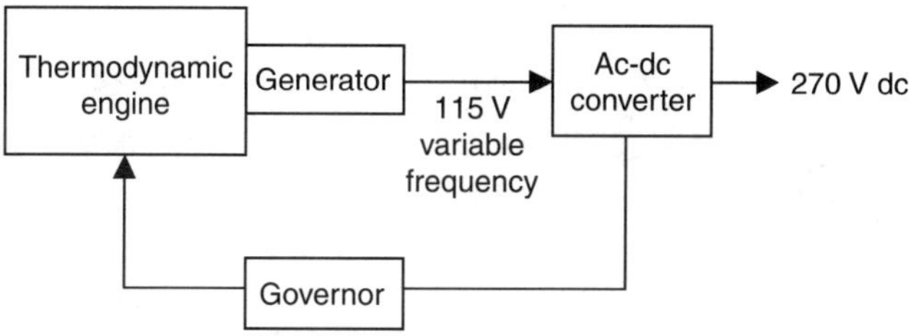

FIGURE 22.18 High-voltage d.c. power system for aircraft.

22.7 High-Frequency Alternating Current

The power system in the hundreds of kilowatts to the megawatt rating is likely to have numerous items of equipment working at different voltages much lower than the generation and transmission voltages. The distribution system for such high power platform has to be efficient, lightweight, and flexible to accommodate a variety of electrical sources and loads. It would require voltage converters for stepping down the generation voltage to suit the equipment to be powered at various locations. A very large spacecraft such as a 300- to 500-kW space station would therefore need a power system that resembles the power utility on Earth, where each user draws power at a voltage that is the optimum for a given application. The a.c. voltage conversion from one level to another can be done most efficiently using a transformer. A typical transformer efficiency is 97 to 99%, whereas that of the d.c.–d.c. converter is 80 to 95%. Also, the a.c. equipment is smaller in size and lighter in mass at high operating frequencies. For these reasons, a very high-power system is likely to be designed to operate at high-voltage and high-frequency a.c. The power system for the initial version of the ISS in the 1980s was to generate 300 kW by a combination of photovoltaic arrays generating d.c., and a solar dynamic system generating high-frequency a.c. The power requirements for the crew living areas, research projects, computers, and manufacturing facilities greatly varied, as in the ground-based utility. Studies made by NASA indicated that high-frequency a.c. distribution system would offer significant advantages in such systems. Subsequently, NASA Glenn Research Center funded development of a 440-V, 20-kHz a.c. power system that is flexible, versatile, and transparent to user equipment, while maintaining high efficiency and low mass. The output power could be either fixed or variable voltage and frequency. Any electrical source, d.c. or a.c., was to be converted to 440-V_{rms}, 20-kHz, single-phase a.c. Power was to be distributed through low-inductance cables.

At 20 kHz operating frequency, the energy transferred per cycle is reduced by factor of 50 as compared to a 400-Hz power system. The number of parts is reduced by a factor of about 5 relative to the conventional 3-phase converters. The power loss is reduced by two thirds. These factors result

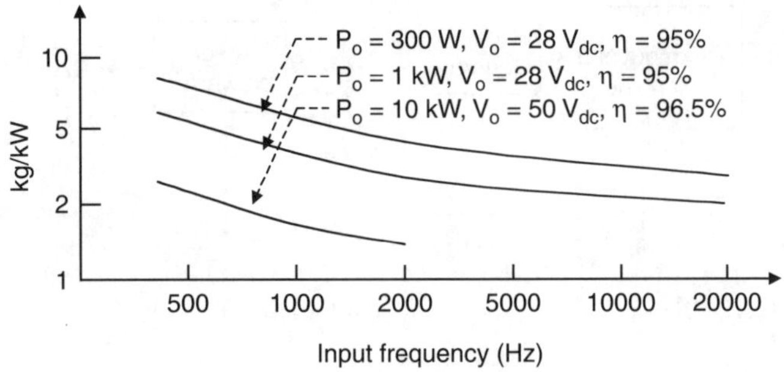

FIGURE 22.19 Load power converter mass versus input frequency of 3-phase a.c. power.

directly in mass and power savings. NASA trade studies compared the mass per kilowatt processed by various power converters. They are summarized in Figure 22.19, Figure 22.20, and Figure 22.21. Frequencies higher than 60 not only reduce the mass of magnetic component, they also offer a higher degree of safety to humans. Table 22.4 shows the relative effect of various frequencies on human cells and tissues. Frequencies above 10 kHz are safer than d.c., while 60 and 400 Hz systems cause a higher perception and paralysis effect than d.c.

As for the load power converter in high-frequency systems, the d.c. power for loads can be converted from 20 kHz a.c. using a center tap transformer with two rectifying diodes, an L-C filter and a free-wheeling diode as was shown in Figure 10.21. The free-wheeling diode carries the current when the voltage reverses, thus eliminating phase shift and a poor power factor. This approach has one serious drawback in that the reflected load current becomes a 20 kHz square wave loaded with high-frequency harmonics. The power quality requirement for the proposed NASA system required the reflected load currents to be high power factor sine waves. Therefore, additional smoothing of the output and a post-regulation may be required depending on the EMI sensitivity of the load.

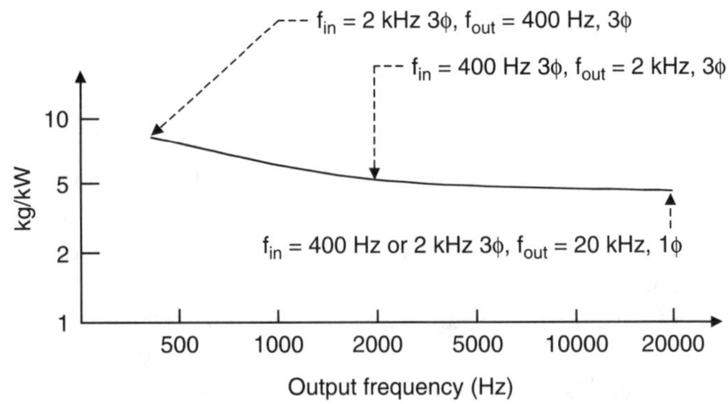

FIGURE 22.20 Central power converter mass versus output frequency.

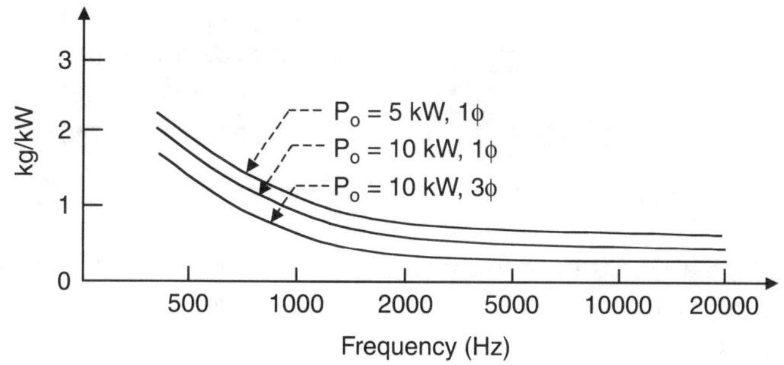

FIGURE 22.21 Power transformer mass versus operating frequency.

22.7.1 20-kHz Cable

NASA Glenn developed a flat virtually concentric (coaxial) low-inductance cable for distributing 440 V/20 kHz a.c. power in a 300-kW space station design in the late 1980s. The cable was designed and developed by Induction General Inc., initially with flat Litz braids, which were later replaced with flat solid copper conductors (Figure 22.22). The thickness of the flat conductor is less than the skin depth given by

$$\delta = \sqrt{\frac{\rho}{\pi \mu f}} \tag{22.5}$$

where ρ = resistivity of the conductor, μ = permeability of medium around the conductors, and f = frequency. For a copper conductor, δ is 9 mm at 60 Hz, 3.5 mm at 400 Hz, and 0.5 mm (20 mils) at 20 kHz. With the thickness limited by the skin depth, the ampacity of the conductor is based on the width of the conductor. In a 440-V, 20-kHz, 75-kW cable design, the center conductor is 0.5 mm thick × 5 cm wide. The two outer conductors form the return, and are 0.25 mm thick and slightly wider than the center conductor. Each conductor is insulated and the inner and outer conductors are separated by an insulating spacer to control the capacitance and inductance of the cable. The gas permeable insulating spacers incorporated by W.L. Gore & Associates allow out-gassing along the entire length of the cable, which is required to avoid voltage breakdown above the Paschen minimum.

Table 22.4 Relative effects of electric current on human cells and tissues

Frequency, Hz	0 (d.c.)	10	60	400	1 k	10 k	RF
Perception and paralyses, relative	0.2	0.9	1.0	0.7	0.6	0.2	0.01

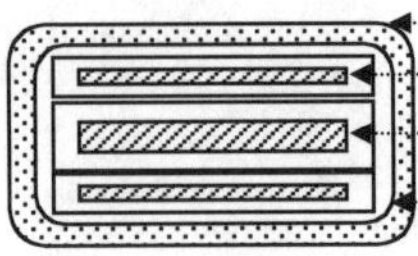

FIGURE 22.22 Virtually concentric flat cable construction.

A 30-kW, 440-V, 20-kHz flat virtually concentric cable is estimated to be, overall, 5 cm wide × 1.25 cm thick and have the following electrical parameters per meter length: $R = 2\ \text{m}\Omega$, $L = 15$ nH, $C = 2$ nF, mass = 1 kg. The 20 kHz cable connectors were also developed. One such design is shown in Figure 22.23.

22.7.2 EMI in 20-kHz Alternating Current

The EMI comparison between d.c. and 20-kHz power systems includes a study of the harness as well as the power converter noise. The d.c. information is based on actual data for conducted and radiated emissions in existing d.c. power systems. An EMI/EMC study for the ISS indicated that the d.c. power distribution is superior to 20 kHz a.c.. The following conclusions were reached against the use of a.c. on the ISS in the final analysis:

- An a.c. system is noisier than d.c., in the EMI sense.
- EMI reduction in d.c. is lighter and easier to implement than those required for a 20 kHz a.c. system.

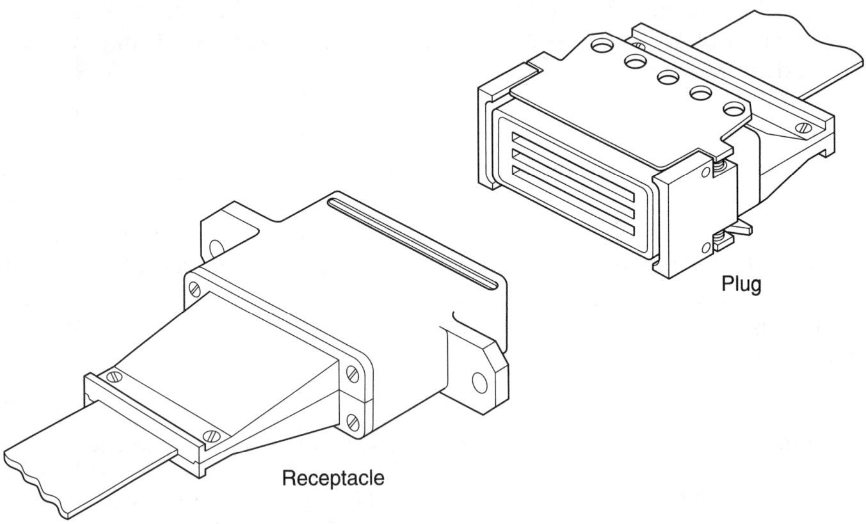

FIGURE 22.23 Power connector for a 20-kHz flat cable.
(Source: NASA Glenn Research Center.)

- Transients in a d.c. system are much less severe than those in an a.c. system.
- The grounding system in d.c. allows less structure ground current than in a comparable a.c. system.
- A d.c. harness is lighter than an a.c. harness, even without the extra shielding that an a.c. system would require to meet the space station radiated emission limits identified in JSC-30237.
- The a.c. magnetic fields in the 10 kHz to 1 MHz frequency range, which are of great concern to the plasma experiments, would be about 60 dB higher from a 20 kHz power harness than from a comparable d.c. harness.

NASA's overall conclusions from the ISS trade studies on a.c. versus d.c. in the late 1980s are summarized in Table 22.5. The NASA team debated between 400 Hz and 20 kHz for a long time, and at one time finally recommended 20 kHz. Significant developments for 20 kHz components were under way before the program priorities changed to much lower power levels for which 160 V/120 V d.c. was finally adopted for the ISS.

22.8 High-Power Components

NASA Glenn Research Center has been continuously developing high-power components for space. These components include a family of bipolar power switching transistors, fast switching power diodes, heat pipe cooled high-frequency transformers and inductors, high-frequency conduction cooled transformers, high-power high-frequency capacitors, remote power controllers, and rotary power transfer devices. Many of these components — the power switching transistors, power diodes, and the high-frequency capacitor — are commercially available for space applications. Other

Table 22.5 Frequency trade of high-power space platforms

DC	400 Hz	20 kHz
Mature power converter and switchgear in low voltage	Mature technology used in aircraft and submarines	New technology to be developed
Heavy power conversion	Competitive with 20 kHz system	High efficiency, and lower mass per kW
Fewer series component, hence more efficient	Transformer coupling to loads can improve efficiency	Loads transformer coupled to 20 kHz, simplifies LPC
Safer for crew	Simpler protection	Simpler protection
Low EMI	Manageable noise	Inaudible noise
No skin effect	Skin effect loss	Skin effect and dielectric loss

components have been developed to the prototype level. Series resonant d.c.–d.c. converters up to the 25 kW power level have been built and tested.

The power components that take advantage of both high voltage and high frequency have also been developed. These components would enable the building of large space power systems. Component technology is continuing with goals such as high-power rotary power transfer, 10-kV semiconductor switching device, high-temperature semiconductors, non-impregnant capacitors, and high-voltage a.c.–d.c. remote power controllers. The objective of this section is to summarize these new space power component technology.

22.8.1 Rotary Power Transfer

Brush and slip rings assembly is needed at the α joint of the satellite for rotary power transfer. Their design generally has the following performance limitations:

- 1 A/mm^2 with solid carbon brushes used in conventional electrical machines
- 10 A/mm^2 with copper or silver coated graphite fibers in carbon matrix
- 30 A/mm^2 with liquid metal current collectors (experimental for ground-based systems)

The roll ring is a new rotary power transfer device developed for use in the ISS. It transfers electrical power across the α joint through rotating flexures compressed between concentric conductors (Figure 22.24). Its advantages are the elimination of sliding friction and the low torque needed

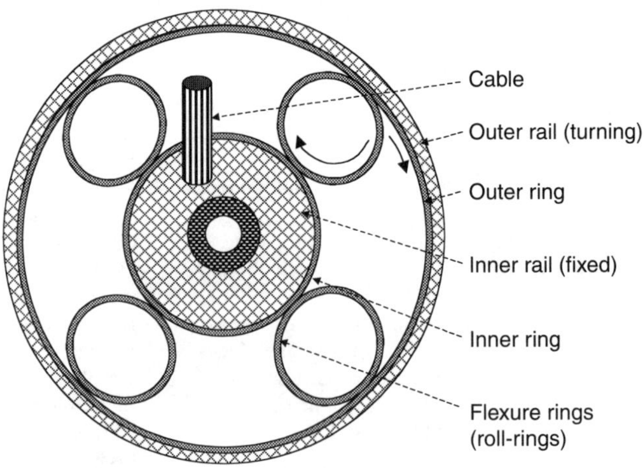

FIGURE 22.24 Roll rings for rotary power transfer in high-power d.c. or a.c. system.

to rotate the device. The roll ring can transfer both a.c. and d.c. power and is lighter and more efficient (>99.6%). It consists of two concentric conductive rings and many small rolling flexible conductive rings. The flexure is fitted and captured in the annular space between the two rings. When the rings are suitably attached to two structures on each side of the rotating axis, the flexure provides a precise electrical coupling between the structures. Performance and life advantages result from the optimum interfaces where only micro-sliding occurs. NASA Glenn with Honeywell and Sperry Flight Systems has developed a 400-kW roll ring assembly. Each power circuit is capable of conducting 200 A at 500 V d.c., i.e., 100 kW. Five power circuits may be used in parallel in a 500-kW assembly and 10 in a 1-MW assembly.

An alternative to the roll ring is the rotary power transformers at the α joint (Figure 22.25). The first design analysis performed was for 100 kW, 20 kHz unit consisting of four 25 kW modules. A 25-kW rotary transformer with 300-V input and 1000-V output operating at 20 kHz was designed based on the configuration shown in Figure 22.25. It has pancake geometry and a radial axi-symmetric gap. The study found the rotary transformer efficiency, mass and size favorable compared to those of the roll rings for the same performance requirements and input characteristics. A 2-kW, 2-kHz square wave demonstration model was built at NASA Glenn and performance characterized.

22.8.2 Switching and Protection

The ability to add and remove loads and to switch in spare units is accomplished on the ISS with remote power controllers (RPCs). In the early 1970s, NASA Glenn started an RPC program with Westinghouse to develop

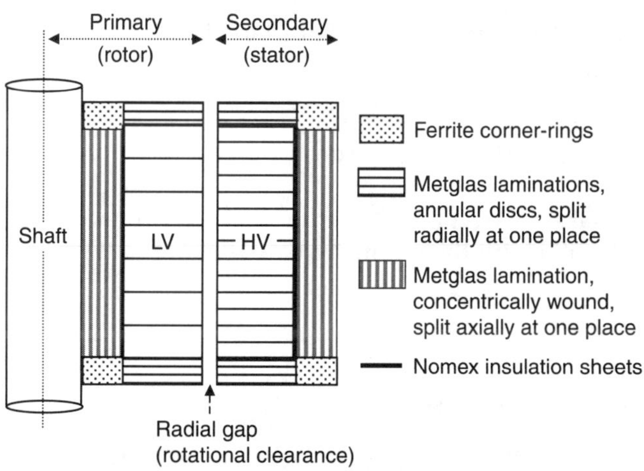

FIGURE 22.25 Rotary transformer for power transfer in high-power a.c. system.

RPCs that operate at 120 V d.c. instead of 28 V d.c. Prototypes were developed and built in 5- and 30-A ratings. A 1000 V, 25 A d.c. prototype RPC was completed with optical isolation, programmable transient protection, and ability to withstand shorts. A transistorized version of this RPC was also built. The d.c. RPC development drew from high voltage d.c. contactors used extensively and successfully in motor speed controllers, locomotion systems and submarines power switching. Large switchgear type contactors are replaced by small contactors using rare-Earth magnets to extinguish the arc generated during load interruption. This program produced prototype units that are hermetically sealed, optically isolated, computer controllable with over-current protection or I^2t trip control that are used in the 160/120 V d.c. system on the ISS.

The functions of RPCs developed for the ISS are:

- Replace electromechanical relays
- Connection or disconnection of loads and components
- Interruption of fault and overload current
- Interface with the PMAD system
- Remote operation
- Control the rate of rise and fall of current

The basic features of the ISS RPCs are:

- Controlled circuit protection
- Current limiting to prevent load transients
- Wide temperature range
- Electrical isolation of control and status signals from the power bus
- High-speed trip-out response characteristics
- Diagnostic capability
- Load capacitor charge capability
- Programmable trip characteristics

For much larger currents at the main bus level, remote bus isolator are developed for ISS with the following basic features:

- Short circuit protection
- Current limit
- Extended switching life
- Built-in diagnostics
- Status feedback

- Bi-directional current limit
- Bi-directional operation
- d*I*/d*t* limit

For a higher voltage, such as 270 V d.c., contactors are available from Hartman and Eaton Cutler for rated currents of a few hundred amperes continuous with several hundred amperes interruption capability.

22.8.3 Semiconducting Devices

High-power spacecraft require high-voltage, high-current semiconducting devices to control and condition the power. Unfortunately, the devices that are presently available lack in-field service experience, and/or substantial inventory of high-voltage d.c. components. In many cases, the available devices are not capable of providing high voltage, high current, and fast switching, while at the same time providing a low forward voltage drop. The SCRs are capable of high voltages and currents, but are slow, have large voltage drops, and require special circuitry to turn them on and off. Power MOSFETs are capable of high voltages and fast switching, but the ON resistance limits their current capability and forward voltage drop. The bipolar transistors require special drive circuitry. In the early 1980s, Glenn started developing high-voltage, high-power switching devices for either a.c. or d.c. distribution system components. Many of them are now commercially available.

For example, a 25 kW switching device program with Westinghouse developed the D60T/D62T transistor capable of handling 50 A continuous (200 A peak) collector currents with a gain of 10 at 400 to 500 V and a switching frequency of 20 to 50 kHz. It found many terrestrial applications as well as space applications. Subsequent development of the D7ST doubled the power handling capability of the D60T from 25 kW to 50 kW, with a 100 to 150 A current at a gain of 10, 400 to 500 V operating voltages and 20 to 50 kHz switching frequency. Next came the high voltage D7ST that raised the operating voltage to 1000–1200 V. The high-voltage D7ST is capable of 25 to 50 A of continuous current (200 A peak) with a gain of 10. This device has 30 kW power handling capability at switching frequencies up to 50 kHz. The next device is the augmented power transistor which raised the current capability to 70–115 A of continuous current (400 A peak) with a gain of 10. This last device has the power handling capability of 75 kW at frequencies of 20 to 50 kHz. Most of these devices can be used in 25 to 50 kW high frequency inverters, high power d.c.–d.c. converters, electric motor controllers, and power controllers.

NASA Glenn Research Center also funded, at Power Transistor Company, a fast recovery, high-voltage power diode with a recovery time compatible with the D60T and D7ST. The PTC 900 diode has a peak reverse voltage of 900 to 1200 V with an average forward current of 50 A. The diode can withstand peak surge currents of 1000 A and has a reverse recovery time of 200 ns at 50 A. A follow-on program increased the current capabilities of the diode to 150 A. These diodes are commercially available and can be used in high-voltage, high-frequency inverters, high-power d.c.–d.c. converters, high-power supplies, and motor controllers. Also, a gate-assisted turnoff thyristor (GATT) was developed to handle 200 A at 1000 V and frequencies up to 20 kHz.

For megawatts power systems, high-power semiconducting device such as the MOS Controlled Thyristor (MCT) developed at GE research center may be considered. The MCT is a power semiconductor device, which has the voltage and current capabilities of the SCR and the fast switching and low forward voltage drop of the bipolar transistor, while being controlled like the power MOSFET. These attributes make the MCT ideally suited for high-voltage, high-power applications in spacecraft power systems. The MCT is available in ratings of a few thousand volts and a few hundred amperes. If a 2500-V, 100-A MCT is qualified for space use, derated to 25% in voltage and 50% in current, it can still operate at 625 V and 50 A, thus delivering 31 kW per device.

Other components have been developed to the prototype stage that demonstrated the technology readiness. These components would enable the construction of large space power systems, high-voltage a.c. or d.c..

22.8.4 Capacitor

In the 1980s Maxwell Laboratories developed for NASA a 600-V space-worthy capacitor with a maximum current capability of 125 A at 40 kHz. The dielectric for the capacitor is polypropylene. This capacitor also has the capability to operate with a 600 V d.c. bias. The calculated loss at full load operation is 22 W, which may be compared to the 75-kVA rating of the capacitor. This capacitor has a specific mass of 0.042 kg/kVAR, which represents a decrease in size and mass by a factor of 7 and is commercially available. It can be used in high-frequency d.c.–d.c. converters, d.c.–a.c. inverters, tank circuits, and filtering and power factor correction.

22.8.5 Power Converters

The most commonly used buck converter topology is not optimal for power converters rated in tens or hundreds of kilowatts power and operating at high voltage (300 V d.c. or higher). Full utilization of the power capability of switching semiconductors in these ratings favors a topology incorporating a transformer with an appropriate turn ratio. Even where voltage differences do not require a transformer, the input-to-output isolation may still be required for system compatibility. Another potential system requirement in high power systems is to minimize EMI resulting from switching high currents. These requirements can be met by a transformer-coupled converter using Cuk integrated magnetics. The power stage topology that was shown in Figure 10.22 uses three sets of power semiconductors and integrated magnetics with common input/output filter capacitors and a shared Cuk capacitor bank. Each power stage operates at a fixed frequency of 10 kHz with the beginning of each stage's power cycle synchronized by a ring counter. Separate control circuits determine the duty ratio for each power stage. The integrated magnetics utilize a powdered iron toroid rather than a gapped core to minimize flux radiation. The assembly sequence for the integrated "lumps" requires winding the input and output inductors separately, then stacking these assemblies and having the transformer windings use the cores of both inductors. Since adjustable core gaps had already been eliminated, other ways were explored to give the zero ripple current condition in the input and output inductors. The simplest alternative is to adjust the inductor/transformer turns ratio slightly. However, the lump power requirements called for the windings to be a few turns of large wires, with the result that any one-turn adjustment has too large an effect. Another approach adjusts the ratio of the values of the input and output Cuk capacitors (normally proportional to the square of the transformer turns ratio) to give slightly different voltage waveforms on each transformer winding. This technique can be utilized to eliminate the fundamental ripple current on both input and output with only a small residual ripple, primarily at the second harmonic of the main switching frequency. However, this approach requires that each lump have its own set of Cuk capacitors. When the Cuk capacitors are shared in a stagger-phased mode, certain power stage duty ratios (33% and 67% with three phases) exist with almost no power ripple in the Cuk capacitors. The value of the capacitor bank then becomes almost irrelevant. Only in a single power stage setup can a disproportionate voltage ripple on the Cuk capacitors be utilized over a wide operating range to *steer* the ripple currents from the inductor windings into the transformer windings.

22.8.6 Transformer and Inductor

NASA Glenn with TRW Corporation developed a heat pipe cooled 2.2-kW, 20-kHz transformer with 30% reduction in mass and volume to yield a specific mass of 0.6 kg/kVA. The transformer was designed so that the decrease in transformer mass would only increase the thermal radiator and solar array mass by 50% of the decrease. Its size could replace the one used in a 30-cm ion propulsion engine power processor. Around the same time, Thermal Technology developed a space-worthy high frequency, 25-kVA, 20-kHz hypercooled aluminum transformer with efficiency greater than 99% and a specific mass of 0.13 kg/kVA. Power systems using such transformers should show a marked improvement in system mass, size, and efficiency. In conjunction with the heat pipe cooled transformer, a heat pipe cooled inductor was also developed. The mass of the inductor was reduced by 40% compared to the conventional inductor, resulting in a smaller capacitor bank in the input filter. This inductor was also designed to fit into the 30-cm ion propulsion engine power processor. As for a scaled-up design, a 1-MW transformer design using heat pipes and aluminum conductor is estimated to have a specific mass of 0.14 kg/kVA at 400 Hz, 0.080 kg/kVA at 1 kHz, and 0.04 kg/kVA at 20 kHz.

22.9 Very-High-Voltage System

Voltages ranging from 200 V to 100 kV may be needed for the future high-power civil and military space missions. The state of space power technologies is changing, and high-voltage systems are evolving rapidly. Therefore, space-qualified high-voltage power components will assume an increasingly greater role in the coming decades. Key design considerations for such components and systems are presented below.

The plasma arcing inception voltage in a space environment varies from 200 V in LEO to 1000 V in GEO. It is therefore obvious that high-voltage equipment rated in kilovolts cannot have bare conductors exposed to space plasma. This requires positive insulation systems for the equipment consisting one or more of the following design features:

- Insulating gas, typically at high pressure
- Insulating liquid
- Perfect plasma-free vacuum with minimum out-gassing materials used in the construction
- Encapsulation with solid insulation, typically an epoxy-resin compound
- The incoming and outgoing insulated cables cannot be terminated in bare exposed terminals. The insulated cable must enter inside a sealed casing containing the insulating medium

The gas insulation, being mass effective, is likely to find applications in very high voltage spacecraft. A 3-phase gas insulated transmission line concept is shown in Figure 22.26 and Table 22.6 list the relative dielectric strengths of commonly used insulating gases at atmospheric pressure. This strength increases with increasing pressure. Notice that Freon^{TM}, which is commonly used in 2-phase cooling systems, has the best electrical strength, followed by SF_6, which is inert, nonflammable, nontoxic, and odorless. Hydrogen is the least preferred electrical insulation, in addition to it being a safety hazard.

The liquid insulation, being 3 to 4 orders of magnitude denser than gases, possesses much higher dielectric strength and heat removal capacity.

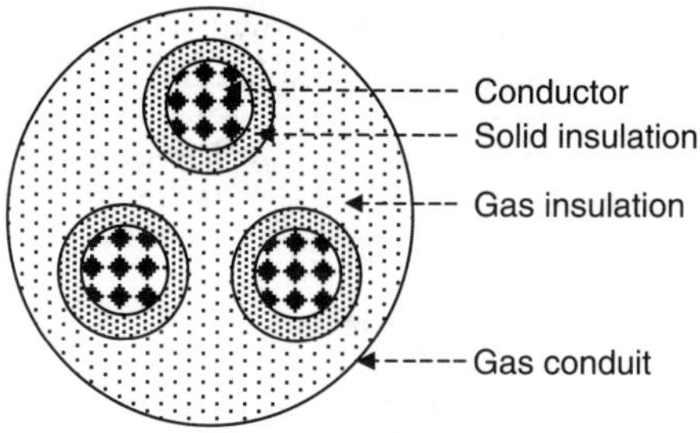

FIGURE 22.26 Gas insulated 3-phase transmission line for very high voltage.

Moreover, like gas but unlike solid insulation, liquid tends to be self-healing should a discharge occur in it. The commonly used liquid insulations with their dielectric breakdown stresses are listed in Table 22.7. The design stress margin depends on the dielectric stress concentration factors and other uncertainties in the design calculations.

Major factors that cause the insulation to degrade are the operating temperature and the corona discharge. Temperature degradation is well understood. The insulation life reduces to one half for every 7 to 10°C rise in the operating temperature. As for the corona, it degrades different materials at different rates. The material which has high dielectric withstand capability when new generally degrades faster under corona. The material selection must therefore match with the mission duration. Yet another degradation factor, applicable only to solid encapsulated components, is thermal cycling. The differential coefficient of thermal expansion between the conductor and the polymer encapsulant induces thermo-elastic stress cycles. These may lead to internal cracks in the solid insulation, thus degrading the dielectric properties of the insulation system.

As the power transmission distance increases, the voltage drop and power loss can be kept reasonable only by using high voltage. For this

Table 22.6 The relative breakdown strength of commonly used insulating gases at atmospheric pressure

Gas	Relative strength
Air	1
Nitrogen	1.05
CO_2	0.95
SF_6	2.5
C_5F_8 (Freon)	5.5
Hydrogen	0.6

Table 22.7 The breakdown strength of commonly used insulating liquids

Liquid	Design stress (kV/mm)
Petroleum oil	2–6
Chlorinated hydrocarbons	4–12
Silicone oils	4–12
Synthetic hydrocarbons	10–25

reason, several kilometer long lines would require several kilovolts if the drop were to be kept below 5%. At that voltage and distance, the use of SF_6 to reduce the line mass by as much as 50% may be considered. As an example, some kind of high dielectric strength gas insulation may be necessary for connecting the SP-100 power source with a space station by a several kilometers long tether cable operating at very high voltage. For such a system, it is envisioned that the SP-100 power output may be stepped up to about 10 kV and transmitted to a space station where it would be stepped down to the user's level. The electrical and mechanical performance and mass characteristics of such a transmission line are compared using both SF_6 and vacuum. A 5-km-long transmission line mass can be significantly reduced to about one half by substituting pressurized SF_6 gas for vacuum insulation. The improvements basically result from the higher dielectric field strength that can be allowed in SF_6. The insulating distance between the several kilovolt line conductors and the grounded outer tube substantially decreases with SF_6, thus decreasing the outer tube diameter and the support mass by a significant amount. There is an added mass of the pressurized SF_6 gas management system (reservoir, pumping, and pressure monitoring systems) in a sealed line construction. However, this added mass would be small compared to the mass of several kilometers long line. On the other hand, the insulation mass reduction potential in the 50-m long SP-100 boom cable is much smaller than the added mass that may be required for the gas management system. For comparison purposes, the mass of a 100-kW, 10-kV, 5-km line is estimated to be 1600 kg with SF_6 gas insulation versus 3200 kg with vacuum insulation, a net reduction of 1600 kg. The 200-V boom cable in the present SP-100 concept would weigh 7200 kg for a 100-kW 5-km line.

22.10 Repetitive Pulse Power

The smart bombs in the two Gulf Wars fell from U.S. fighter aircraft precisely onto laser-illuminated targets. The carbon dioxide lasers in those systems were driven by pulsed-power sources. Such specialized applications demand very high power for a very short time, i.e., pulsed power. The pulsed laser is one such application. Others include medical, particle accelerators, and even strategic defense projects, such as simulation of the electromagnetic pulses (EMPs) that result from nuclear explosions. The concept of pulsed power is not new. It was used in radar systems in World War II and its new applications are expanding. Pulsed-power sources are fully integrated systems, rather than single-power supplies. In such systems, a high-voltage constant current source (rather than a constant voltage source) is used to charge a bank of capacitors. The power supply is not directly connected to the load; its sole use is to charge the capacitor bank. A switch is added to control the discharge of the capacitor, and a resistor limits the current. Fuses protect the equipment under adverse conditions.

High-voltage power supplies can deliver up to 8000 J of energy per second, with voltage ratings up to 50,000 V. Since capacitor charging varies from zero to peak voltage, the energy measured is an average delivered to the load. Though both linear and switching supplies are used in pulsed-power systems, the trend is towards switch mode technology. Most units are designed to fit in standard 19-inch racks. The switching frequency is limited to 50 kHz, since the component cost for high-voltage applications rises with the frequency.

In pulsed-power applications, the capacitor capability is measured in terms of the specific energy. A good capacitor needs a high-capacitance, low-temperature coefficient, and relatively low inductance. The capacitor dielectric has to withstand very high voltage and current, offer temperature stability, and have a fast discharge frequency. Pulse requirements vary with the application. Rail-gun applications need a 5 to 40 kV pulse with pulse widths of a few milliseconds. EMP simulation needs multi-megawatt pulses of a few nanoseconds duration.

The trend in repetitive pulse-power system design is toward higher energies, larger average power levels, and faster pulse repetition rates. This trend results in requirements for components with higher reliability, lower losses, and longer life times. Emphasis is placed on those components and material requiring significant development efforts. Switches are by far the weakest active components in pulse-power systems. Capacitors, for example, are one of the weakest elements in high-power pulsed systems, especially when operated at high repetition frequencies for extended periods of time. Improved insulation in systems and components is essential to increasing energy density and reliability. The pulse transformer and inductor designs require special cooling and thermal time-constant considerations.

The development of long-life, fast-discharge components is essential to many future high-energy applications. Many of these systems are needed where compactness, light mass, high reliability, and low loss are prime requirements. Capacitors and switches presently available cannot meet this combination of projected requirements in high repetition rate power systems. At present, the energy density of capacitors operated at continuous high pulse rates is typically a factor of three to ten less than those used in single pulse systems.

The basic characteristics and problems associated with spark gaps and thyratron switches used in pulsed power systems need to be addressed. Gas dynamic spark gaps in the thyratron switches often require more power to furnish compressed insulating gas than the average power they can pass. The corona characteristics of liquid and solid insulation limit the permissible dielectric stresses in the materials, and consequently the energy density of the system. This following brief discussion is intended to illustrate the major areas of development necessary for pulsed power components.

Capacitors are the basic electrostatic energy storage elements in pulsed-power systems. The rationale in evaluating capacitor technology is to determine and understand the failure mechanisms and then work with manufacturers to develop improved units. Recent programs have sought to determine the capacitor's weak points, the characteristic failure mechanisms, and the required dielectric characteristics compatible with high dI/dt discharge capacitors operating at 1-kHz pulse rate frequency and 100 ns discharge time.

The performance evaluation of low-inductance type of capacitors in the 100 ns, 1-kA, 1-kHz regime has led to the following conclusions:

- Partial discharge (corona) is the main cause of failure in the pulse power capacitors and are detectable during energy discharge.
- The partial discharge inception stress level during discharge is about 257 V d.c., and is dependent on time duration of discharge and the pulse repetition rate.
- A high-quality, high-frequency diagnostic system is essential to observe transients within capacitors during discharge. The high-frequency, high-voltage probes developed for the system are required as diagnostic tools.
- Preliminary tests indicate that Teflon–silicone oil and mica-paper capacitors have longer life in the regime tested than the polypropylene-silicone oil units, but with lower energy density. The latter is attributed to the impregnation characteristic of Teflon units and fabrication methods of the mica units, and may not represent intrinsic limitations.

The major components of such power systems are the high voltage capacitor, thyratron switch and the air-core transformer. The capacitor is best suited for repetition rates in kilohertz. A battery may be used for a few hertz rate. For a still lower rate, superconducting coils may find applications.

The following are the rough guidelines for selecting the energy storage technologies for a pulse-power system:

- Capacitor for high pulse rate in kilohertz
- Battery for medium pulse rate
- Flywheel or superconducting coil for low pulse rate

22.11 Multi-Megawatt Burst-Power

The power requirements for some strategic defense spacecraft range in tens to hundreds of megawatts of burst power. Directed energy weapons, typically placed in LEO, are examples. Such a power system is required to deliver high power for a short time of engagement, after which it must recharge. Important considerations in selecting the system architecture and the subsequent design detail include the following:

- Primary energy source, energy conversion, and energy storage to meet the peak power requirement and rapid startup.
- Power conditioning and distribution
- Thermal management during engagements
- Optimum voltage selection for mass minimization
- Low vibration and low torque on the weapon platform
- Long stand-by with low quiescent power requirements
- Low mass, low cost, and high reliability

A future reusable directed energy weapon in LEO is predicted to require power over 10 MW at 270 V d.c. for several minutes occasionally. For such a short duration occasional, very high burst power, large-scale energy storage in a flywheel or superconducting coil may be considered. However, the fuel cell may be an ideal candidate. For the Air Force Research Laboratory, Allen[9] evaluated 10-300 MW burst power system design. He considered several top-level options — chemical-dynamic, nuclear-dynamic, chemical-fuel cell, and battery and flywheel which could be recharged using a solar array power.

The study concluded that a system using regenerative solid oxide fuel cell (SOFC) with hydrogen and oxygen reactants could deliver the required power with the lowest power system mass, and is the best approach with the presently available technologies. The PEM fuel cell system results in higher mass, primarily from the extra cooling fluid and the associated storage tank, which was significantly higher than the mass of the high temperature radiator used with the SOFC. The fuel cell is the lowest mass system primarily because it produces the power only when needed and at

270 V d.c., which does not require additional power conversion and condition. The flywheel generally results in heavier power system that is not used continuously. Cooling of the power system and the weapon components in the study was achieved by a circulating ammonia loop.

The study found the following system optimization at the 10 MW power level:

- Power during engagement 10 MW
- Primary energy source Chemical
- Energy conversion Fuel cells
- Energy storage None major
- Voltage level(s) 270 V d.c.

The 10-MW SOFC system features are summarized in Table 22.8, and the study results are summarized in Figure 22.27 for 5 min of the weapon run time.

The solar array-SOFC system mass breakdown in percentage of the total power system mass was reported as follows: 60% in the fuel cell stack, 10 in fuel and steam tanks, 8 in reactants, 6 in radiators and plumbing, 14 in solar array, and 2% in all other components.

Recharging the fuel cell requires electrolyzing the by-product water back into hydrogen and oxygen. The SOFC can electrolyze water using the recharging process. The electrolyzer is generally a separate unit from the stack and the two cannot operate simultaneously. Recharging the fuel cell in 1 day in Allen's study required a 213-m^2 solar array generating 70 kW. Extending the recharge time to 5 days cuts the array to 55 m^2 generating 18 kW power (Figure 22.28). However, this may conflict with some mission scenarios due to condensed periods of conflict.

The dynamic system may be required for sustained high power. Some strategic defense weapons platforms may require burst powers up to 300

Table 22.8 Ten megawatt solid oxide fuel cell burst mode power systems features

Burst power	10 MWe, scalable to 300 MWe
Burst mode duration	100–1000 s
Test duration	60–240 s/year
Mission life	10–15 years
Fuel cell efficiency	70%
Operating temperature	1250°C
Working pressure	135 atmosphere
Oxygen consumption rate	1.75 kg per MW·sHydrogen consumption rate 0.015 kg per MW·s
Effluents	Negligible

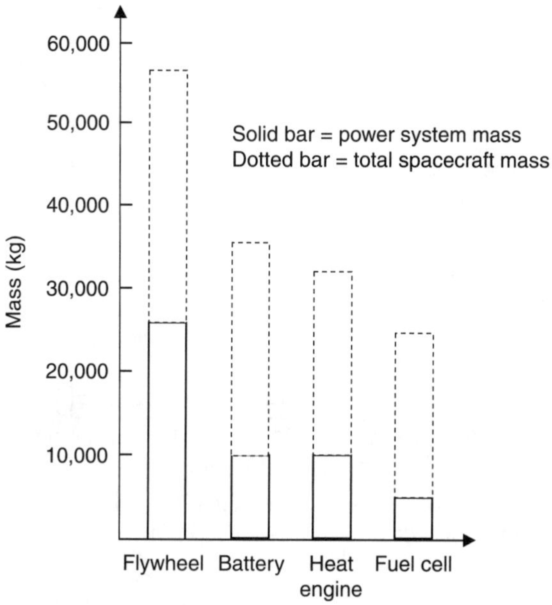

FIGURE 22.27 10-MW_e power system mass with alternate design concepts. (Data source: D.M. Allen, *Proc. 36th IECEC*, pp. 243–249, ASME, 2001.)

MW. Such high power requirements are expected to consider nuclear or chemical sources for the primary energy, and rotating electrical machines for energy conversion. In a baseline concept, the reactor heats a fluid, typically liquid metal such as mercury or sodium potassium. The vapor then drives a turbine-generator using the Brayton or Rankine cycle. The space nuclear system is similar to a ground-based nuclear power station. The major difference in space is that the waste heat must be rejected only by radiation back to space. Both systems use nuclear fission of a heavy metal atom, which releases a large amount of energy, primarily in the kinetic energy of the fission fragmentation. With uranium-238, the energy released per fission amounts to the following:

- 166 MeV in kinetic energy of fission fragments
- 10 MeV in gamma radiation
- 17 MeV in neutrons and neutrinos
- 10 MeV in further fragmentation

Thus a total of 200 MeV thermal energy is released per fission. The conversion between million electron volts and watts gives 3.1×10^{10} fissions per second needed to release 1 thermal watt. This energy is used to generate steam and drive a turbine-generator to convert to electrical power. In a

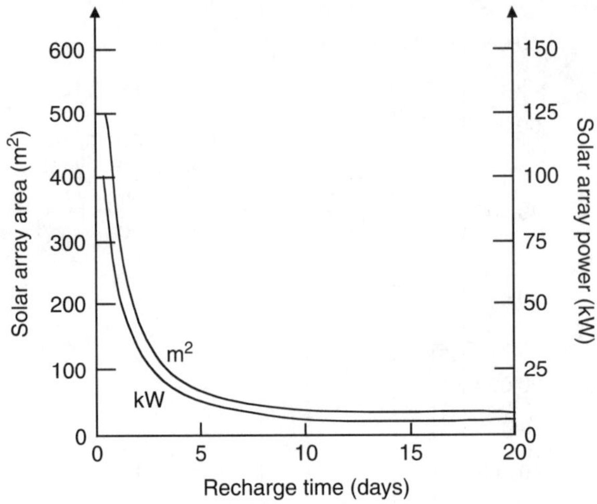

FIGURE 22.28 Solar array versus recharge time after 5 min of weapons run with 10 MW_e load. (Data source: D.M. Allen, *Proc. 36th IECEC*, pp. 243–249, ASME, 2001.)

large-scale power system in the megawatt range, the overall thermal to electrical conversion efficiency is 40 to 50%.

The nuclear reaction must be contained in a well-shielded enclosure. In principle, the chain reaction of the fission, once started, continues as long as the critical mass is maintained, and the energy can be released at any desired rate. However, in practice, the release rate is limited by temperature problems, so that a given reactor has a maximum achievable release rate.

In using the synchronous generator, the steady-state torque on the platform is a nonpulsating constant torque (of d.c. nature), and does not affect the platform pointing accuracy. However, the transient torque during a fault (e.g., short circuit on the output terminals of the generator) is of a pulsating nature, and constitutes a major dynamic disturbance on the platform. Such a torque is the product of the alternating short-circuit current and the constant magnetic flux in the air gap, which becomes frozen on the instant of a short circuit, in accordance with the constant flux-linkage theorem.

Spacecraft momentum and vibration issues are important integration considerations in any system using large rotating machines. However, as cited above, rotating machines produce large dynamic torque on the platform, particularly during faults, which may hinder the pointing accuracy of the spacecraft. The disturbing torque is minimized by using such machines in one or more pairs of counter-rotating machines to cancel the torsion effects of each other. Figure 22.29[10] shows one such arrangement of three pairs of counter-rotating turbo-generators on a space platform. There always remains a net torque on the platform, as two machines in a pair cannot be located exactly at the same place. However, the momentum

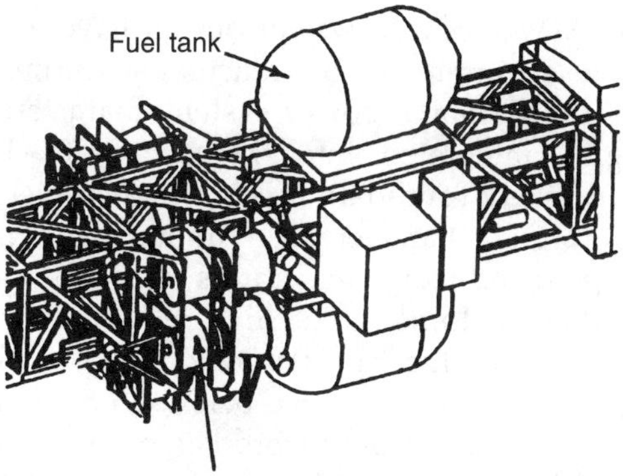

FIGURE 22.29 The placement of multiple pairs of large rotating machines on a space platform.

cancellation of 0.999999 by using counter-rotating machines is required to minimize the spacecraft uncontrolled spin and/or tumble. In a MMW power system, a mismatch of even less than 0.01% could gyrate the spacecraft out of control, and a mismatch in the order of 0.001% could introduce significant pointing errors.

In the case of a chemical-dynamic system, the Brayton engine is lighter than the Stirling. Both engines may be used in counter-rotating pairs, although the concern of mismatch is less severe due to their inertial energy being lower by several orders of magnitude than in high-speed flywheel energy storage. However, the vibration concerns remain. Here, a more important integration issue is the fluid loop, which can reverse direction under a high slew rate. Both heat engine and fuel cell reactants can be stored at high pressure (3500 psi) or as a liquid at cryogenic temperatures. Cryogenic storage results in lower system mass.

A major design challenge in the MMW weapons spacecraft is the thermal management to remove heat from the electrical power source and then remove it from the spacecraft while maintaining a narrow thermal gradient between the components. A 100-MW system with 35 to 40% thermal-to-electrical conversion efficiency engaged for 10 min needs 36 GJ of heat to be dissipated eventually into space. The heat may be removed from the spacecraft as generated, or stored on the spacecraft for gradual removal. The thermal system options are direct radiation, liquid heat capacity storage, and latent heat phase change storage with spray cooling. The former would have the greatest mass, whereas the latter would result in the least mass system. However, a major concern with spray cooling is the large pressure differential required to pump the liquid and the vapor. A second concern is the momentum of the fluid caused by high mass flow rate. A thermal system split into at least two counter-directional fluid loops can reduce the net momentum to a tolerable level.

Many aspects of MMW power system design have to be considered in conjunction with the requirements of an actual space mission. In order to take advantage of the dynamic power system characteristics, the system must be thoroughly integrated into the spacecraft. The synchronous generator is a more likely candidate in such high-power missions due to the high air-gap torque characteristic. Among such machines, superconducting or hyper-conducting generators could be advantageous for their low mass, low volume, and high efficiency. Superconducting machines use a superconducting d.c. field coil on the rotor, which completely loses its electrical resistance below a critical temperature, T_{cr}. Until recently, the T_{cr} of superconductors has been around that of liquid helium, but is now closer to that of liquid nitrogen, with the discovery of some higher temperature superconductors. The a.c. stator winding in this machine still uses conventional conductors, as a.c. superconductors do not exist — not yet, at least. The hyper-conducting machine, on the other hand, uses a high-purity aluminum conductor in both the rotor and the stator at liquid nitrogen temperature. It exhibits extremely low electrical resistance, but not precisely zero as in the superconducting machine. Both the superconducting and the hyper-conducting generators are likely candidates for high-power space missions.[11]

The design characteristics of a 300-MVA (270 MW) superconducting generator are summarized in Table 22.9. Although a higher frequency machine may be advantageous on space platforms for their lower mass and higher efficiency, a two-pole, 60-Hz machine is considered here for two reasons:

Table 22.9 Three-hundred megavolt ampere superconducting generator design parameters

Rating	300 MVA. 0.9 power factor
Voltage	24 kV, 60 Hz, 3-phase
Rotor speed	3600 rpm
Rotor cooling	Liquid helium
Stator cooling	gaseous hydrogen
Efficiency	99.5%
Active length	78 inches
Rotor outer radius	18.6 inches
Armature inner radius	23.5 inches
Armature outer radius	32.5 inches
Core inner radius	35.5 inches
Core outer radius	52.0 inches
Rotor inertia constant	0.62 s

(Data source: M.R. Patel, *Proc. 25th IECEC*, p. 990-X, AIAA, 1990.)

- The centrifugal stress in the rotor of high-power machine limits the rotor speed to around 3600 rpm, and
- Some applicable design data are available in published literature on the 300 MVA superconducting generator development program funded by the Electric Power Research Institute[12]

The platform's dynamic-control and the pointing system must account for the net effect of all the rotating machines (generators, turbines and pumps) on the platform, which depends on their exact placement. The net effect of all such machines is platform-specific and varies from one platform to another.

22.12 High-Temperature Components

High-temperature components are required in high-power solar dynamic, SP-100, pulse-power, and electromagnetic pumps for liquid metal. For energy storage at high temperature, a sodium–nickel chloride secondary battery may be used up to 275°C, sodiumsulfur up to 350°C, and lithium-ion sulfide up to 450°C. A very large sodium–sulfur battery with cylindrical cells of total capacity 48 MWh has been built in Japan for terrestrial use.

Premium grade silicone rubber or polyamide fibers such as Nomex and Kevlar are widely used insulation for service temperature up to 250°C and voltages of 10 kV a.c. or 25 kV d.c.

High-voltage, high-temperature wires rated up to 600 V and 450°C are commercially available for use in furnace ignition leads, thermocouples, aircraft, and missiles. They are generally made from nickel-clad copper wire insulated with glass reinforced mica tape or coated with a glass–ceramic compound. Tungsten or nichrome wires with 3M Nextel™ ceramic fiber or Rockbestos Phosroc™ insulation can operate up to 1200°C continuously and 1400°C in the short term. Such wires have also been considered for use in submarine fireproof areas. Their radiation resistance has been tested up to 10^7 rads.

The life of high-temperature conductors commonly used in heating elements is shown in Figure 22.30. On the insulation side, it is always important to choose a dielectric with thermal stability. One high-temperature wire candidate is cuprous oxide thin film insulation on a copper substrate. It is available from Permalustetr Inc. in rectangular wires and thin strips. The insulation is impervious to nuclear radiation, gamma rays, corrosion, corona, high vacuum, and cryogenic temperatures. It can operate at 1000°C without additional corrosion or oxidation of the copper. The insulation film thickness is 3–4 micron for 100-V insulation. It is highly flexible, does not out-gas or craze, and allows other over-coatings. It is ceramic in nature. It is not a deposition, only a conversion coating. It would allow compact coils in inductors and transformers.

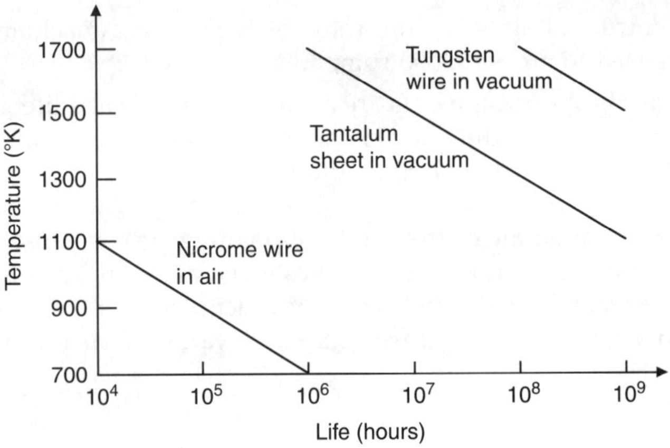

FIGURE 22.30 High-temperature heater element life characteristics.

Nickel connectors are good up to 650°C. The bare connectors may be coated with some sort of insulation, such as ceramic sprays, potting compound, and ceramic fiber tapes.

The following is a brief summary of NASA/Westinghouse Lima research on the development and evaluation of magnetic and electrical materials capable of operating in the temperature range of 800 to 1600°F. Advanced space power systems are the area of eventual application of such materials.

High-temperature magnetic materials with satisfactory strength are needed for use in solid rotors of electrical generators. Screening tests have been conducted on martensitic alloys, using additions of tantalum plus tungsten and tantalum plus silicon to a base composition of iron with 15% nickel and 25% cobalt. They indicate that these alloys show promise for meeting the desired creep stress exceeding 60,000 psi at 0.4% creep strain, and magnetic saturation exceeding 13 kilogauss at 1100°F. In the 1200 to 1600°F range where dispersion strengthening is applied to cobalt and cobalt–iron base alloys, results show the saturation magnetization goal of 12 kilogauss can be met by 27 cobalt–iron base alloys containing as much as 20 volumepercent dispersoid. No problem appears to exist in meeting the goal of a coercive force of less than 25 oersteds at 1200 to 1600°F in either 27 cobalt–iron base or cobalt base compositions. Creep tests on Nivco alloy at 1100°F were conducted at pressures in the 3 to 5×10^{-9} torr range at 1100°F. Design analysis suggests creep strains in excess of 1% should be run above 1050°F. This would provide more reliable design data for rotors in inductor alternators even though the design is based on only a total creep strain of 0.4%, because this represents the integrated value of a wide range of incremental creep strains.

The tests also determined the feasibility of high-temperature capacitors using high-quality dielectric materials. Comparison of electrical data on boron nitride (boralloy), polycrystalline alumina (Lucalox), polycrystalline beryllia, and single-crystal alumina (Linde sapphire) showed that pyrolytic boron nitride has significantly lower a.c. power loss, higher d.c. resistance, and exhibits less change in capacitance with temperature than any other material investigated in the temperature range from room temperature to 1100°F. In addition, the d.c. breakdown strength of pyrolytic boron nitride is 7000 v/mil at 1100°F, which is several times greater than that obtained with sapphire or beryllia.

Alkali-metal compatible ceramic-to-metal seals and combinations of material designed into generator stator with a bore seal, a transformer, and a solenoid were investigated of compatibility under electrical and magnetic stresses at elevated temperature and under high vacuum. Ceramic out-gassing studies indicated that the firing cycle applied to a ceramic using nitrogen–hydrogen atmosphere followed by the vacuum treatment, which they undergo during brazing is sufficient for minimizing oxygen bearing out-gassing products. These would add only a few parts of oxygen to the alkali-metal charge in a typical advanced electric power system. The stator, transformer, and solenoid to provide data on a stability of materials applied to typical electrical designs were installed in the thermal vacuum chamber and stability testing was done at 1100°F hot spot temperature. The chamber pressures were approximately 4×10^{-7} torr at the beginning of the 5000-h tests and decreasing. The minimum cold chamber pressure obtained after system bake-out was 1×10^{-10} torr.

Magnetics for use at temperatures up to 600°C and voltages up to 600 V d.c. may consider nickel clad copper coated with ceramic insulation or electrochemically formed aluminum oxide film. Such a coating is primarily a refractory-glass compound that can withstand flexibility and braiding requirements consistent with normal coil winding operation.

To keep the surface from oxidation films and to enhance the mechanical strength of copper at high temperature, nickel cladding is required. The word *cladding*, as used by the copper wire industry, actually means electroplating 27% of pure nickel by mass (or area, as both nickel and copper have about the same mass density). The cladding reduces the effective conductivity of the composite wire to about 75% of the core copper because of the lower conductivity of nickel. With the core copper conductivity of 85%, the composite wire would then have the conductivity of $0.85 \times 0.75 = 0.64$ times that of pure copper at room temperature. At higher temperature, this value may be different. In magnetic environment, nickel is slightly ferromagnetic at room temperature, but loses its magnetism at about 700°F (370°C).

The SP-100 boom cable design requires electrical conductor and insulating materials that can continuously operate in 570°C environment. If we add approximately 30°C for the internal temperature gradient, we need conductor and insulating materials that can withstand about 600°C continuous temperature at the reactor end. The cable temperature at 30 m away from the reactor would be around 100 to 200°C depending on the wire gage used, i.e., on the internal power loss in the wire. The total insulation mass of a 200-V cable is less than 1.5 kg per 100 m. The 48 pairs of 50-m boom cable have a total cable length of 4800 m with total insulation mass of about 72 kg.

The conductor grade, which has a high electrical conductivity (close to copper) and at the same time retains the necessary mechanical strength at 600°C suitable for the SP-100 boom cable application is CdCr copper. This precipitation hardenable alloy has high mechanical strength, resistance to softening at elevated temperatures, high flex life, and freedom from embattlement. The boom cable is required to be flexible for ease in deployment. Such flexibility is achieved by making the cable from numerous thin strands (10 to 20 mils diameter). However, the thinner the strands, the quicker they melt, should the temperature approach the melting point even for a brief period (such as under a laser beam attack). The threats and the resulting over-temperatures determine the final conductor strand size. In the absence of a specific threat, adopting the industry and military standards appears to be reasonable, which calls for 133 strands of 11 mils diameter in AWG No. 8 wire and 133 strands of 14 mils diameter in No. 6 wire. From the high-temperature insulation candidates shown in Figure 22.31, glass ceramic or asbestos fibers are potential candidates, although Nextel of 3M or Phosroc would surpass the requirement.

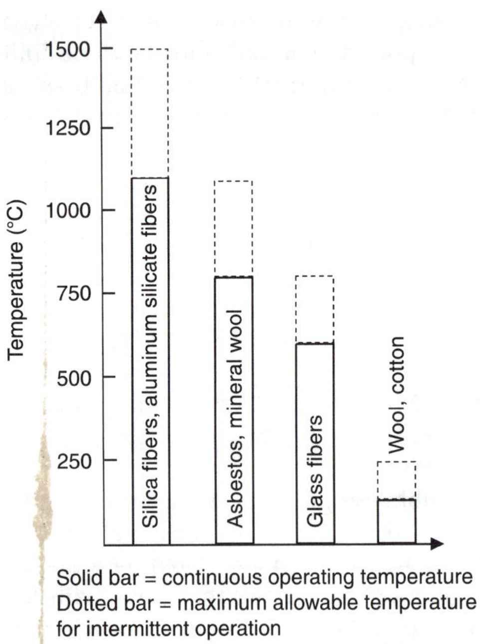

FIGURE 22.31 High-temperature insulating material operating characteristics.

References

1. Barthelemy, R.R., Messie, L.D., and Borger, W., Military space power system technology for the 21st century, *Proceedings of the Intersociety Energy Conversion Engineering Conference*, 1986, Vol. Ill, pp. 1401–1410.
2. Brandhorst, H.W. and Best, S.R., Effects of hypervelocity impacts on solar cell modules at high voltage, *Proceedings of the 36th Intersociety Energy Conversion Engineering Conference*, ASME, 2001, Vol. I, pp. 353–358.
3. Binz, E. and Hartung, J., Solar dynamic power for a space station, *Proceedings of the 21st Intersociety Energy Conversion Engineering Conference*, pp. 2072–2076.
4. Merryman, S.A., Bandy, A.J., and Gordon, L.B., The breakdown characteristics of an out-gassing dominated vacuum region, *Proceedings of the Intersociety Energy Conversion Engineering Conference*, 1987, Vol. Ill, pp. 763–765.
5. Patel, M.R., *High Frequency Power Distribution System*, NASA Report No. CR-175071, 1986.
6. Bouquet, F.L. and Winslow, J.W., *Radiation Data Definitions and Compilation for Equipment Qualification Data Bank*, NASA Tech Briefs, January 1987.
7. SAE STD-1031, *Aircraft 270 V dc power distribution bus, electrical power, HV dc, aircraft, characteristics and utilization of.*
8. MIL-STD-1399, *Submarines 155 Vdc power distribution (100 kW range), interface standards for shipboard systems.*
9. Allen, D.M., Multi megawatt space powers technology program, *Proceedings of the 36th Intersociety Energy Conversion Engineering Conference*, ASME, 2001, Vol. I, pp. 243–249.
10. Patel, M.R., Dynamics of high power rotating machines on space platforms, *Proceedings of the 25th Intersociety Energy Conversion Engineering Conference*, AIAA, 1990, Paper No. 990-X.
11. Gamota, G., One two three - zero resistance in defense applications, *IEEE Transactions on Magnetics*, Vol. MAG-17, January 1981.
12. Patel, M.R., Nathenson, R.D., and Ahmed, M.E., "Designing pancake coils of large 300 MW superconducting generator for electromagnetic and thermal loads, *IEEE Transactions on Power Apparatus*, Vol. PAS-102, August 1983, pp. 2710–2716.

Chapter 23

Electric Propulsion

23.1 Introduction

Electric propulsion continues its expansion in space. More than 160 spacecraft in the Earth orbit and interplanetary space now use it in some form. Many new programs have begun to develop the next generation of electric propulsion for solar and nuclear powered spacecraft. Although it is another load on the bus, some of its unique features that influence the power system design and operation are discussed in this chapter.

In the geosynchronous orbit, the satellite must maintain the assigned orbit altitude and inclination to stay locked with a fixed ground station. But the orbit slowly degrades due to various drags in the space. The propulsion system maintains the satellite in the desired orbit and controls its attitude towards the Earth. Propulsion is also needed for orbit transfer and to deorbit the satellite for disposal at the end of a mission. The requirements for these functions are listed in Table 23.1.

The propulsion thrust in a typical satellite is produced by chemical fuels carried on board. In a geosynchronous satellite, over 90% of the on-board fuel is used for N–S station keeping, and the rest for E–W station keeping, attitude control, and orbit transfer. The resultant thrust vector of all station-keeping thrusters must pass through the center of the satellite mass. The thrust vector for the attitude control, on the other hand, must be at a right angle to the center of mass in order to generate the maximum moment.

Chemical propulsion typically uses a catalytic monopropellant hydrazine (N_2H_4) fuel. No oxidizer is needed. Heat or a catalyst decomposes the fuel, and the chemical reaction produces heat and thrust of the exhaust gas. Some satellites use fuel and oxidizer in separate tanks to achieve higher

Table 23.1 Propulsion requirements for three key functions

Function	Propulsion requirement
Orbit achievement	High thrust, high efficiency, low precision
Orbit adjustments	High precision, efficiency not an issue
Disposal	High efficiency, high dormant reliability, low precision

thrust levels. Such bi-propellant fuel systems generally uses hydrazine and nitrogen tetroxide (N_2O_4). This yields specific impulse around 300 s.

At present, electric propulsion is used in low-thrust maneuvers. However, attractive mass saving results in applications requiring high-velocity increments, such as for N–S station keeping (50 m/s per year), and for orbit raising (1 to 5 km/s). Full orbit transfer from low to geosynchronous orbit requires about 6 km/s velocity. Electric propulsion has several times the efficiency of a chemical system, and is attractive for attitude control in LEO satellites. In GEO satellites, it is often used for N–S station keeping, and sometimes for orbit raising, in which the transfer time can be long. Although in theory, the thrust-to-drag ratio must be greater than 1 to raise an orbit, it must be greater than 3 for practical control stability requirements.

For the N–S station keeping (NSSK), electric propulsion can be powered by the satellite battery without additional mass except for a dedicated load power converter. Orbit raising requires much higher power, which must be derived from the solar array. Small orbit raising using low specific impulse devices may be considered using the existing solar array and battery. The propulsion operation is performed during sunlight in the non-eclipse season. During this period, the extra power available from the solar array after meeting the load power requirement is used to fire the propulsion. Additional power, if needed, can come from the battery up to the maximum allowable DOD at the end of station keeping. Electric propulsion that requires large amounts of power significantly impacts the power system architecture and design. For example, full orbit raising would definitely need an additional solar array or some other power source, which would add mass.

Many electric propulsion technologies require a high-voltage power source to generate arcs and plasma that is electrically conductive. The power system control circuit must provide protection against damage to either the thruster or power system components in the event of inevitable arcing between high-voltage elements. The most frequent place for an arc to occur is in the thruster. The high-voltage power source is designed in two ways to provide protection against such arcing. One way is to incorporate an over-current sensing circuit, which turns off the high voltage in case of an arcing fault. Alternatively, the source can be designed with high output impedance so that an arc merely results in deep voltage sag. In the cesium thruster, discussed later in this chapter, the thruster electrodes are subjected to excessive cesium atom deposition when the high voltage is turned off. The cesium continues to flow from the ion source, and the evaporation from the ionizer is neutral cesium since no electrostatic field exists to draw off the ions. The voltage should therefore be reapplied quickly. Under a loss of high voltage, electrode and insulator contamination leading to excessive drain current and arcing is possible at any time the cesium is flowing from the ion source. The power system design must provide a method for handling an arc or loss of voltage for a specified period of time.

23.2 Specific Impulse

The function of propulsion system is to apply impulse to change velocity of the spacecraft. When the mass is ejected through nozzles, the reaction propels the spacecraft in the opposite direction. Since the ejected mass must be carried on board at the time of launch, the impulse produced per kilogram of the ejected mass is taken as a figure of merit of the propulsion system. The impulse is defined as the time integral of force. A constant mass flow rate of $\mathrm{d}m/\mathrm{d}t$ exhausting at an effective velocity V_e through a nozzle creates a steady force $F = (\mathrm{d}m/\mathrm{d}t)\ V_e$. Such a force applied for $\triangle t$ seconds results in impulse $I = F{\cdot}\triangle t$. The impulse created per kilogram mass of propellant, called the specific impulse, is then given by

$$I_{sp} = I/\Delta m = F/(\mathrm{d}m/\mathrm{d}t) \tag{23.1}$$

The specific impulse is also known as the propellant efficiency or the thrust to mass flow rate. Its unit is second. The total impulse a fuel of mass $\triangle m$ can create is then $I_{sp}{\cdot}\triangle m$. This figure of merit, traditionally used for comparing chemical propulsion systems, is less meaningful for comparing electric propulsions. It is therefore modified to be the impulse created per kilogram of total additional mass that must be carried on-board for electric propulsion. The additional mass is mostly in the power system, and to a lesser extent in the propellant. Most of the added mass is in the solar array, and some in the power conditioning equipment. A modified figure of merit for electric propulsion is developed below using the following notations:

P_e = additional power needed for electric propulsion, watts

M_p = mass to generate additional power for propulsion, kg

P_{sp} = specific power of the power source = P_e/M_p, watts per kg

M_e = mass of the expellant needed on board, kg

M_{sc} = spacecraft mass without additional masses, kg

M_t = total spacecraft mass = $M_{sc} + M_p + M_e$, kg

V_e = expellant exhaust velocity, m/s

T_b = propulsion burn time, seconds

Assuming that the expellant mass flow rate is constant during the propulsion burn time, it is M_e/T_b. We also assume a constant expellant velocity during the burn. Then, the energy conservation requires the balance of power, i.e.,

$$P_e = \frac{1}{2}(M_e/T_b)V_e^2$$

which, by using classical rocket equations, leads to the following two expressions:

- expellant mass,

$$M_e = \frac{M_t - M_{sc}}{1 + \left(\frac{V_e^2}{2T_b P_{sp}}\right)} \tag{23.2}$$

- power system mass,

$$M_p = \frac{M_t - M_{sc}}{1 + \left(\frac{2T_b P_{sp}}{V_e^2}\right)} \tag{23.3}$$

Both M_e and M_p rise and then fall with changing parameters. An optimization study is therefore needed to minimize M_e for a given mass ratio of M_{sc}/M_t. The study would also show the power related mass M_p is lower at higher exhaust velocity up to a certain point, beyond which M_p rises.

For electric propulsion, one figure of merit is the impulse per kilogram of total additional mass ($M_e + M_p$), and another is the thrust-to-power ratio (TPR) given by

$$\text{TPR} = \frac{\frac{dM_e}{dt} \cdot V_e}{P_e} \tag{23.4}$$

With specific power achievable with the present power system technologies, a long burn time and low exhaust velocity are inherent characteristics of electric prolusion. Significant power improvements of several fold are needed to make electric propulsion viable for lift-off from a planetary surface or orbit raising. However, electric propulsion for on-orbit maneuvering and interplanetary missions is practical and is being used at present.

In designing any type of electric propulsion, the exhaust velocity must be selected to minimize the total mass added in the power system and the propellant. That would maximize the available payload mass in a given spacecraft with a given launch vehicle capacity. This trade is depicted in Figure 23.1.

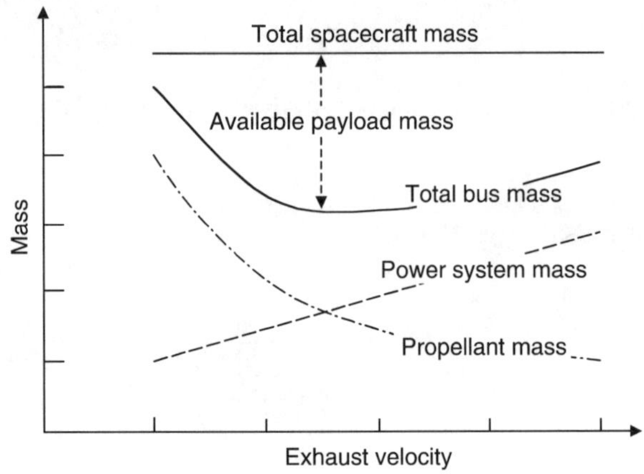

FIGURE 23.1 Payload maximization with electric propulsion.

23.3 Types of Electric Propulsion

Various types of electric propulsion are described below.

23.3.1 Electrothermal Hydrazine Thruster

The EHT, also known as Resistojet is the simplest of all in concept and is depicted in Figure 23.2. A propellant is heated by a tungsten heating element. The energy (enthalpy) of the propellant gas is increased by heating, which is then converted into kinetic energy. The accelerated gas is directed through nozzles and propels the spacecraft. The exhaust velocity is a function of temperature, and is generally kept below 10 km/s from the material life considerations. The achievable thrust level is in a fraction of newton. The propulsion efficiency is around 75%. The propellant can be hydrogen, nitrogen, or ammonia. Hydrogen, having high specific heat and thermal conductivity, works better than nitrogen or ammonia. However, it is corrosive, and requires cryogenic temperature for storage. Ammonia, on the other hand, is easy to store without refrigeration, but is also corrosive, limiting the life of both the heater and the nozzle.

23.3.2 Hydrazine Arcjet

The HAJ, in principle, is similar to EHT, except that the propellant is heated by its own ohmic heating by passing through arcing electrodes, instead of using a separate heating element. Thus, the propellant is directly heated. It uses battery power during non-eclipse seasons. The specific impulse is high in excess of 1500 s, but the excessive electrode wear is life limiting. A typical GEO communications satellite carries four arcjet thrusters for N–S station keeping maneuvers. Two arcjets are fired at a time for NSSK, one from each

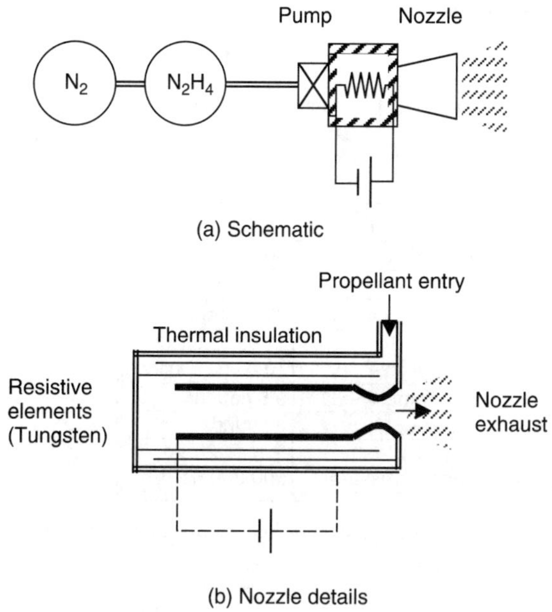

FIGURE 23.2 Electrothermal hydrazine thruster.

battery, 2 weeks before or after an eclipse season. The arcjets are always fired during sunlight, nominally at 1200 and 2400 hours with respect to local midnight. Each HAJ consumes 1 to 2 kW power depending on the spacecraft size. The arcjet power circuits are always fused. Since this is usually the heaviest single load on the spacecraft, the arcjet fuse-blow voltage transient is generally the most severe on the bus. The computer simulation of this transient is therefore required for all new designs. The hydrazine arcjet thruster typical envelope and electrical performance are shown in Figure 23.3.

23.3.3 Ion Thruster

In the ion propulsion system (IPS), charged particles are accelerated in an electrostatic field. The thrusters are classified according to how charged ions are created from the neutral propellant. In a simple method, vapor of a material having low ionization potential is ionized by passing through a heated porous tungsten plate. Cesium is often used for the vapor material. High heat loss due to thermal radiation in the process limits the overall efficiency.

The schematic of ion thruster in *Deepspace-1* spacecraft is depicted in Figure 23.4. In this scheme, electrons emitted from thermionic cathode pass through concentric cylindrical anodes. An externally applied weak magnetic field sets the electrons in a spiral motion. The collision between

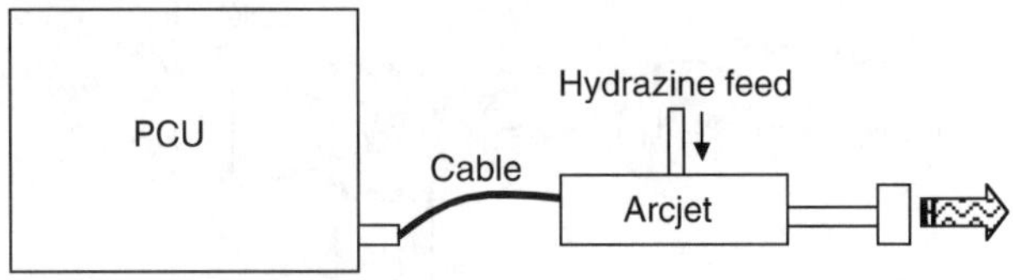

Input power	1800 watts
Rated voltage	80 volts
Input voltage	95-65 volts
PCU efficiency	90 percent
PCU size	11×9×3.5 inches
Arcjet size	10×3.5×2 inches
PCU mass	9.5 pounds
Arcjet	3 pounds
Valve power	9 W @ 28 Vdc
Feed pressure	300-200 psi

FIGURE 23.3 Hydrazine arcjet thruster typical envelope and electrical performance.

these accelerated electrons and the propellant ionizes the propellant. An inert gas, such as argon or xenon, is used as the propellant. Thrust levels in excess of 1 N are possible in large units. Good mass saving is achieved using such propulsion for the N–S station keeping in geosynchronous satellites. A typical chemical thruster emits particles at 3 km/s. A typical ion thruster emits ions at 30 km/s, and needs only 10% propellant. The ion propulsion thus saves launch cost significantly.

In the working principles, a strong electrostatic field is created between two plates held at high voltage, generally at several kilovolts. A charged particle injected into the E-field is accelerated along the electric field lines with force equal to qE, where q is the particle charge and E is the electric

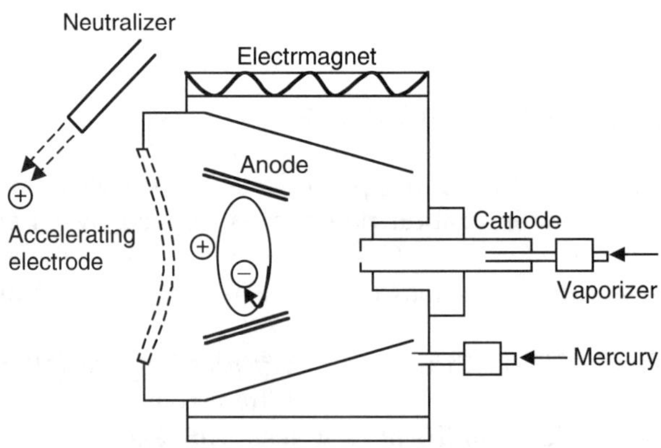

FIGURE 23.4 Ion thruster used in a *Deepspace-1* spacecraft.

field intensity. Under this force, negatively charged particles move along the rising field, while positively charged particles move along the falling field. If the expellant mass M_e having a total charge of Q is exhausted at velocity V_e through the electric potential difference of U volts (not to be confused with velocity V), the energy balance (in joules) gives

$$\tfrac{1}{2}M_eV_e^2 = QU \tag{23.5}$$

The propulsion power (in watts) required to create the propulsion thrust F then follows the balance of power, i.e.,

$$P_e = \tfrac{1}{2}FV_e \tag{23.6}$$

The above two equations lead to the propulsion thrust to electric power ratio

$$\frac{F}{P_e} = \sqrt{\frac{2M_e}{QU}} \tag{23.7}$$

The thrust-to-power ratio, a figure of merit, is maximized by using a propellant having large mass to charge ratio (M_e/Q). Argon, xenon, and cesium fall in this category.

A version of this propulsion is being developed in Europe, and uses ions created directly from a liquid metal surface by means of strong electric field. Such propulsion gives a high force over short pulse time, and has high efficiency. The gridded (conventional) ion thruster has a high exhaust velocity. It has been developed extensively in the United States, Japan, Germany, England, and France. The preferred version in this family is the xenon ion propulsion system (XIPS) developed and tested at NASA/JPL. The unique feature of the gridded ion thruster is the separation of the ionization process from the accelerated beam by a series of electrostatic grids. Neutral xenon is fed into the ionization chamber, where a low voltage discharge is struck. A high ionization efficiency is achieved by hindering the electrons from reaching the anode by a peripheral magnetic field. Ions are extracted from the plasma, accelerated, and focused into tight beamlets by the grid system. The negative accelerating grid prevents transmission of electrons in either direction. Electron-rich plasma is emitted from the neutralizer downstream of the grids, thus neutralizing space charge of the ion beam.

The gridded xenon ion thruster differs significantly from the Hall thruster discussed latter. XIPS has by far the highest exhaust velocity of all the electric propulsion devices, but also gives a low thrust per unit power and longer trip time. The ion flow through the grid is limited by space charge buildup and the thrust density is low. On the positive side, the exhaust beam is very well focused. A larger model of XIPS is being used in a four-thruster system designed to do both the N–S station keeping and a significant amount of orbit raising in large commercial satellites. Boeing's BSS-702 and BSS-601HP busses are such examples. The mass savings could be up to three times as great as that associated with NSSK alone. For low- and mid-Earth orbits, the entire orbit raising and de-orbit functions could be performed by an ion propulsion system.

NASA's *Deepspace-1* interplanetary probe uses a gridded ion propulsion system using PV power. It produces thrust with specific impulse (propellant efficiency) 10 times greater than a typical chemical propulsion system used during interplanetary missions. It uses a hollow cathode to produce electrons that ionize xenon gas through collisions, yielding positively charged atoms of xenon. The xenon ions are then electrostatically accelerated by a potential of more than 1000 V and emitted from a 30-cm thruster nozzle through an electrically charged grid. The power available to the IPS during the *Deepspace-1* mission comes from solar arrays.

The ion propulsion degrades the solar array in one way. It generates ions that are discharged into space at high velocity. A large number of low-energy electrons are also emitted to prevent the satellite from building up a negative ion slowing down the charge. A few non-ionized atoms drift out of the thruster and get ionized in collision with ions in the high-velocity stream and form plasma. Through the drifted plasma, the low-energy electrons can flow as the leakage current, draining power from the solar array.

23.3.4 Stationary Plasma Thruster

The stationary plasma thruster (SPT), also known as the Hall-effect thruster, falls into the mid-range of the exhaust velocity spectrum. Plasma (an ionized gas that is electrically conductive) is made by stripping atoms of their electrons, confining the resulting products, and preventing them from recombining. Usually this means heating an appropriate gas mixture to a temperature at which collisions break up the atoms and then keeping the resulting product in equilibrium as long as needed. Forcing a large electric current through a gas mixture can create a plasma. The SPT was originally developed in the former Soviet Union, but recently has been improved and tested extensively by a Russian, French, and American consortium. The SPT is essentially an electrostatic ion thruster without grids. An axial electrostatic field and a radial magnetic field are established in the annular flow chamber. An equilibrium discharge is set up when electron-rich plasma is fed from the cathode while neutral xenon atoms are emitted through the anode. Some of the electrons are trapped in the crossed electrostatic–magnetic field and forms a virtual negative electrode, which accelerates the ions. They ionize the xenon propellant by electron impact in the discharge region and also neutralize the space charge of the accelerated ion beam. The magnetic field produces a small azimuthal torque to the ions, which does not change the thrust.

SPTs with improved performance are in various stages of development in the 3 to 5 kW power range. A high power model SPT-100 has been used for N–S and E–W station keeping operations on many satellites by Russia, and is being used by Astrium, the French–German–British satellite builder on *Inmarsat* and *Intelsat* communications satellites. It gives about 16 km/s and generates about 80 mN of thrust using 1.5 kW of input power. This is a good combination of thrust to power ratio of about 50 mN/kW and a fairly high specific impulse. Its main advantage over the gridded ion thruster is the high thrust density, which is possible with gridless acceleration. The thrust output to power input ratio for the SPT is in the 50 to 70 mN/kW range. Its drawback is a wide beam angle due to not having the focusing grids. This may interfere with nearby spacecraft components. However, various other alternatives for narrowing the exhaust plume are being developed.

23.3.5 Magneto-Hydrodynamic

The magneto-hydrodynamic (MHD) propulsion — also known as the electro-hydrodynamic (EHD) — uses plasma. Neutral plasma is accelerated under both the ohmic heating and the electrodynamic force generated by the interaction of the electric field $\boldsymbol{E}$ and the magnetic field $\boldsymbol{B}$. The thrust that propels the spacecraft in MHD comes from the Lorentz force $F = \sigma E \times B$, where σ = electric conductivity of the plasma. The force is perpendicular to both the $\boldsymbol{E}$ and $\boldsymbol{B}$ fields. The accelerating force is imparted to neutral particles in the gas by collision with ions, thus providing a direct and efficient increase in the propellant velocity that propels the vehicle. The electric conductivity of the plasma-mixed exhaust is an important performance factor. It varies directly with the temperature and inversely with the pressure. Its value of 0.10 Ωm is considered adequate for acceptable performance in the MHD accelerator. Since the self-induced magnetic field accelerates ions, the thrust is proportional to the current squared. The overall efficiency ranges from 10% in the kilowatt range to 40% in the megawatt range. A pulse mode of operation is essential in high power ion thruster. Significant development in this area has been under way in Japan. A thrust power of a few newtons with a specific impulse of 2000 s is achievable. Cathode erosion is a major life-limiting factor.

A constant E-field applied to plasma is not of interest here, as it adjusts itself by developing a thin sheet of space charge, which shields the main body of plasma from the E-field (as in a conductor body). On the other hand, a constant B-field causes a Lorentz force on the charged particle in plasma. The force is given by $F = Q\ (\boldsymbol{E} + V \times \boldsymbol{B})$, which is always perpendicular to the velocity, V, of the particle. It causes the charged particle to gyrate about the field lines without altering the internal space-charge neutrality of the plasma. This force provides acceleration, keeping the kinetic energy constant, as Lorentz force can do no work.

In the MHD propulsion, a trace of conducting fluid such as potassium or cesium is added in a chemical rocket engine exhaust to ionize it. The thermal plasma in the exhaust is propelled under a strong magnetic field created by superconducting electromagnets. MHD accelerators have been built in the USA and Russia, although not much work is currently under way.

23.3.6 Pulsed Plasma Thruster

The pulsed plasma thruster (PPT) is a new system that has been successfully tested for the first time on NASA's *Earth Observing Spacecraft* managed by Goddard Space Flight Center. Launched in November 2000, it is a testbed for new instruments and technologies. In January 2002, the reaction wheel that controls the craft's pitch (up and down motion), was turned off and the new system took over for a 4-h test. Millisecond pulses of electricity fired across solid bars of Teflon turned portions of the Teflon into an electrically charged gas that is forced out of the nozzle to generate thrust. Tests showed that the system was about three times more efficient than conventional chemical thrusters. This first test also eased concerns that the system would produce electromagnetic interference that could affect other systems on the spacecraft.

23.4 Performance Comparison

EHT and HAJ are both electrothermal hydrazine-based thrusters in which a monopropellant is fed through a conventional thrust chamber and nozzle, usually via a preliminary catalytic decomposition chamber. The gain in exhaust velocity is derived from the electrical power added to the thrust chamber. The propellant exhaust velocity in these two devices is relatively low.

In EHT, the electric power is added using a resistance element, which is the hottest component in the system. The maximum power that can be added to the propellant is restricted by the melting point of the resistance element. This, in turn, limits the rise in exhaust velocity to 2 to 3 km/s, which is competitive with storable bipropellants giving 3 km/s. The EHT has been used for NSSK on many satellites, and to raise small satellites from the parking orbit to the final orbit.

In HAJ, the electric power is added directly to the propellant in the form of a co-axial arc. The temperature in the arc region is several tens of thousand kelvin, but most of the propellant near the walls is much cooler. The HAJ has been used for NSSK on many large commercial satellites since the early 1990s. It produces an average exhaust velocity of 5 to 6 km/s;

Table 23.2 Specific impulse of various propulsion systems

Propulsion class	Specific Impulse
Chemical (liquid and solid)	200—400 s
Electrothermal hydrazine (EHT)	300–400 s
Hydrazine arcjet thruster (HAJ)	500—1500 s
Pulsed plasma thruster (PPT)	1000–4000 s
Ion propulsion system (IPS)	2000–8000 s

about twice that of storable bipropellant, and the thrust of about 200 mN using an input power just under 2 kW. Its thrust-to-power ratio of about 100 enables relatively fast maneuvers. However, its low exhaust velocity limits the mass benefit relative to chemical propulsion. HAJ lends itself well to spacecraft that already use hydrazine as a chemical propellant. HAJ used for NSSK in many large commercial communication satellites gives an I_{sp} of about 500 s, compared to 200 s in traditional monopropellant thrusters, and 300 s in bi-propellant and EHT thrusters. This reduces the propellant mass required for station keeping.

A gridless electrostatic ion thruster, such as the Russian SPT, has been used on short duration missions with a specific impulse approaching 2000 s. The gridded ion thruster can achieve a specific impulse in the 3000 to 4000 s range.

The specific impulses of major types of propulsion systems are compared in Table 23.2, and the power-to-thrust ratio versus exhaust velocity is plotted in Figure 23.5. As for the voltage requirement, the HAJ generally

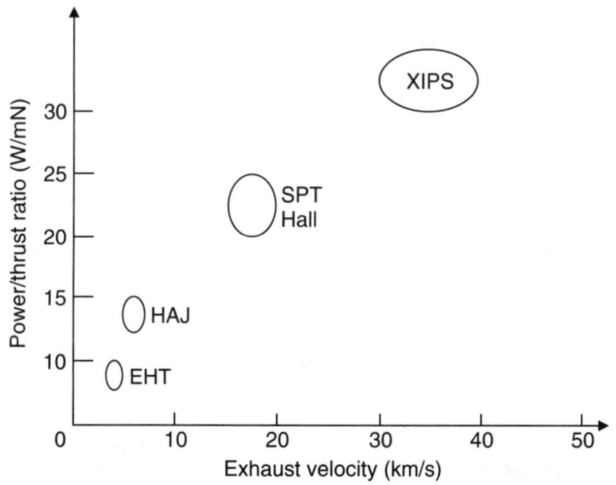

FIGURE 23.5 Power-to-thrust ratio versus exhaust velocity.

works at the bus voltage, the SPT thruster works around 300 V, and the IPS works at voltages exceeding 1000 V.

The HAJ, SPT, and XIPS, each have their own inherent advantages and disadvantages. The development continues in order to incorporate these systems in full orbit raising. When achieved, it would significantly change the Earth-orbiting satellite designs.

23.5 Solar PV Propulsion

Many new developments are under way in electric propulsion, which will require high power to reach new frontiers. In particular, electric propulsion is being studied for interplanetary missions since it gives mass efficient architecture. However, multi-megawatt power levels are needed to obtain reasonable transfer times. The power requirement drops to less than 1 MW if cryogenic upper stage transportation technology is utilized with electric propulsion. Solar electric propulsion becomes an attractive alternative to the nuclear dynamic power system for a 1-MW system. It is applicable to a wide range of NASA missions, such as human Mars exploration and robotic exploration of the outer planets. For human Mars exploration, Kerslake and Gefert[1] have developed a 1-MW class conceptual design using various PV technologies (Figure 23.6). The time required to travel to Mars, as estimated by a NASA GRC computer code, by using a xenon Hall thruster, is about 230 days.[2] The Kerslake study summarized in Table 23.3 shows that using the present $CuInS_2$ thin-film PV technology, it is feasible to design a 1-MW power system with a mass under 10 metric tons and deployed solar array of 5800 m^2. The EPS in this study consists of a 500-V d.c. distribution system with 16 channels feeding eight 100-kW_e thrusters. Since thrusters do not operate during eclipse, the battery requirement is modest, about 13 kWh for housekeeping and payloads. The 500 V d.c. permits *direct drive* thrusters operation that greatly reduces mass of both the harness and the power-processing unit, thus increasing the system efficiency. Copper wires with

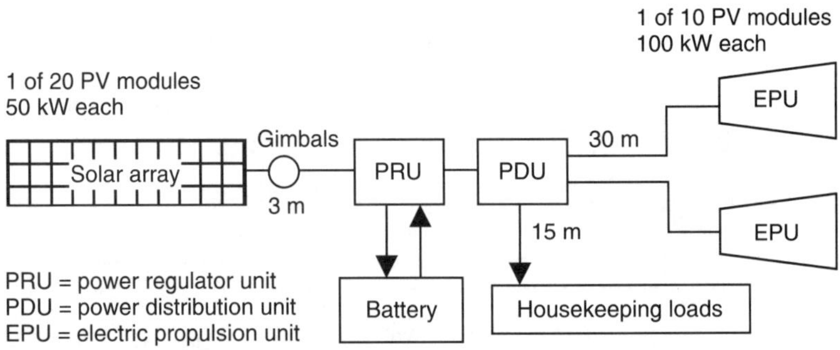

FIGURE 23.6 Power system block diagram for 1-MW class electric propulsion.

Table 23.3 One-megawatt class solar electric propulsion system for Mars exploration

Parameter	ISS cr-Si	Linear concen.	a-SiGe thin film	$CuInS_2$ thin film
BOL power	1270 kW	950 kW	1150 kW	1050 kW
EPS Mass	46 MT	21 MT	12 MT	10 MT
PV array	14,000 m^2	3500 m^2	9000 m^2	5800 m^2
Cells/string	1450	270	340	880
Strings	990	910	9500	2370
Fluence MeV/cm^2				
for current	1.0×10^{16}	2.2×10^{14}	5.9×10^{14}	5.9×10^{14}
for voltage	2.6×10^{16}	3.7×10^{14}	1.3×10^{15}	1.3×10^{14}
Maximum leakage	260 A	1 A	0	0

MT = metric ton (1000 kg).
(Data source: T.W. Kerslake and L.P. Gefert, in *Proc. 34th IECEC*, p. 2449. SAE, 1999.)

Teflon insulation are used. NASA data on the space plasma effect at ±500 V d.c. operating voltage were used to estimate the parasitic leakage current. The necessary 600-V silicon and silicon carbide based technology developments are well under way at NASA for switchgear components and remote power controllers at this voltage.

The PV cell technologies considered by Kerslake and Gefert were:

- The ISS PV array with an 8 × 8 cm crystalline silicon PV array design
- Linear concentrator array based on SCARLET array design, using a 2 × 8 cm crystalline triple-junction $GaInP_2/GaAs/Ge$ cells operating at 7.5 × sun
- Thin film triple junction 5 × 5 cm amorphous aSiGe cells on folded thin 2-mil polymer membranes, and
- 5 × 5 cm $CuInS_2$ thin film cells on thin 2-mil polymer membranes

In the conceptual design cited, the thin film cells are encapsulated with 1.5-mil-thick FEP Teflon for isolation from the space plasma. The PV array membranes are deployed using an inflatable rigidified longitudinal column and a flexible composite lateral member. The array strings are negatively grounded, and bypass diodes are incorporated every 10 cells to reduce long-term degradation. The harness is multi-ribbon flat copper sized for 3% voltage drop. The PV array surface is coated with transparent indium tin oxide to prevent space charge buildup and arcing at high orbit altitudes.

In another study, Bailey, Hepp, and Raffaelle[3] discuss the current state of the art and address some of the technology development issues required to make thin film PV viable for future solar–electric propulsion from Earth to Moon or Mars. They have developed a conceptual design for a Mars solar electric propulsion vehicle using thin film multi-junction PV array that fans out (Figure 23.7). The study has concluded that the specific power around

FIGURE 23.7 The solar electric propulsion vehicle concept for human Mars mission. (Source: NASA Glenn/Power and Propulsion/J. Hojnicki.)

1 kW/kg would be required, which is almost 40 to 50 times the present state of the art.

23.6 Solar Thermal Propulsion

This propulsion concept uses concentrated solar heat. The concentrator is generated geometrically from a right circular cone. The circular area symmetric about the axis of rotation gives an on-axis paraboloid. A primary off-axis concentrator is generated from a cone axis away from the axis of rotation. In Figure 23.8, the solid line is the on-axis paraboloid and the vertical line passes through point B, the bottom of the bowl. If a thruster were close to point B near the paraboloid focus, the rocket engine would

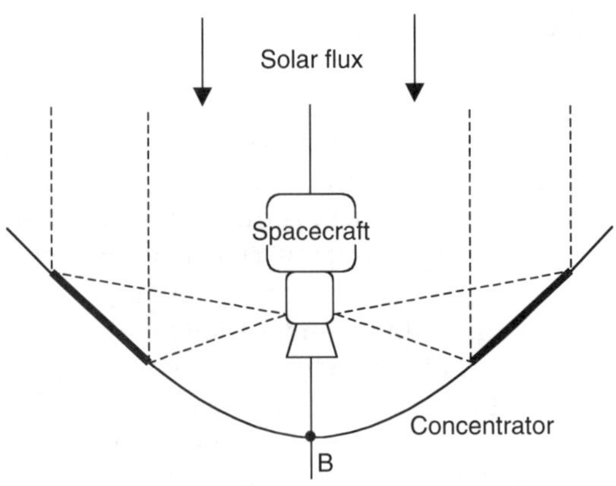

FIGURE 23.8 Solar concentrator propulsion with paraboloid collector.

block the sunlight if it were in front or behind the direction of travel. Also, the exhaust plume would impinge on the concentrator and degrade its performance. Laug and Holmes[4] have proposed a solution to this in using an on-axis concentrator, which takes two elliptical sections of the paraboloid away from the axis of symmetry. The required concentration ratio for solar thermal propulsion is 10,000:1, which is much higher than that for solar PV propulsion. This means a very small surface accuracy error tolerance, about 1 mm rms, which is comparable with that for a 15-GHz antenna.

23.7 Nuclear Power Propulsion

Humans cannot reach the outer limits of the solar system without nuclear power. Most current spacecraft rely on chemical rockets for reaching an orbit and on solar panels to generate power for on-board equipment. But solar panels have a limited capability and their strength is weakened when the spacecraft is far from the sun. For almost 50 years, NASA has been doing planetary science in the same way. That is, to accelerate for 5 to 10 min and then stop and coast along. That is an ineffective way of exploring deep space. We have reached a limit of what we can do with chemical propulsion. Although a couple of probes have left our solar system, it takes a very long time to get there. For making space travel faster, NASA is developing nuclear-powered rockets that could halve the time it takes to reach planets such as Pluto.

NASA has studied the use of nuclear-powered spacecraft in the past. During the 1990s, nuclear energy was, politically, an unsound topic and new developments on radioisotope generators and nuclear-fission reactors were shelved. However, they have been revived and now receive funding. The nuclear system initiative proposed by NASA in 2002 amounts to over a billion dollars worth of new developments over 5 years.

Radioisotope propulsion relies on heat produced by the natural radioactive decay of plutonium to generate electricity or to create a propulsive thrust using nonnuclear xenon fuel. NASA has only one basic model of a radioisotope power generator, but plans to use the new funding to develop a more advanced system using nuclear reactor. The proposed system would use a conventional chemical rocket launching a spacecraft equipped with nuclear components. At a safe distance from Earth, the spacecraft would detach from the rocket, the reactor would go online and the mission would begin. Undoubtedly, the major obstacle is the issue of safety.

All space systems we build must safely survive the worse possible scenario — a rocket blowing up on the pad. To aid safety, the nuclear hardware on board the spacecraft would be launched in a *cold* non-operating state. The reactor would be switched on when the vehicle reaches 2500 km from Earth, a distance chosen in accordance with NASA's orbital debris guidelines for preventing hazardous materials ending up in orbits near the Earth.

This may make a nuclear power propelled craft possible in near future. It is likely that the first beneficiary of the new system could be *Mars Smart Lander* to be launched in 2009. With nuclear power, it may be able to operate for years, rather than for months. Solar-powered sources last only for months on Mars due to dust in the atmosphere that coats the solar panels.

23.8 Microwave Beam Propulsion

Electric power is traditionally generated on board the spacecraft. An alternative approach studied by NASA is to generate a large amount of power on another satellite and transmit it by a microwave beam, eliminating the mass of the power generation system on the user spacecraft. NASA's interest in such a system is to reduce deep space mission time, such as manned missions to Mars.

A study by Lindberry and Chapman[5] focused on using an MHD accelerator powered by a microwave beam for augmenting the specific thrust of a chemical rocket engine from 450 s to 2500–4000 s with a corresponding thrust of about 300 to 400 N and acceleration of 0.002 to 0.02 m/s^2. No other all-electric or all-chemical propulsion can achieve such performance alone, but it would require a strong magnetic field. Superconducting magnet technology is well established and is commercially used in numerous ground-based applications for generating strong magnetic fields in the 10 to 20 T range. The old liquid helium-based superconductors (<10 K) are being replaced by newer high-temperature superconductors (around 100 K). Room temperature superconducting technology is on the horizon, which is assumed to be available in the NASA study of this concept. The major design challenge would be a magnet coil with a current density of 1000 MA/m^2, and corresponding electrode erosion, and a mechanical force containment structure. Thermal management of the equipment that handles such high currents is also a challenge.

It is estimated that the MHD accelerator would have an efficiency in the 80 to 90% range, and only a couple of percent of the system mass. The remaining 98% mass would be primarily in the receiving rectenna. A low mass rectenna is therefore a critical technology for deploying this concept.

The frequency considered for such a system is from 3 GHz to 300 GHz with the corresponding wavelengths from 0.10 m to 0.001 m. Brandhorst, in a previously cited study, considered two power beaming options, microwave and laser beams. Because of on-orbit aperture limitation, the microwave option was dismissed. For the laser option, 10.6 and 1.06 μm wavelengths were considered. The energy conversion system for a 10.6-μm laser was either a thermal engine or a thermo–PV. For the 1.06-μm laser, suitable energy conversion would be direct bandgap PV cells giving excellent conversion efficiency around 50%.

Electrothermal propulsion using microwave power is another possible concept that has been studied.[6]

The microwave power beam and superconducting energy storage are further covered in Chapter 26 and Chapter 27, respectively.

23.9 Tether Power Propulsion

Space tether is being developed for a wide range of applications, such as electrodynamic propulsion, momentum transfer, formation flying satellites, remote sensing, and space structures. In 2004, NASA Marshall Space Center plans to test a tether that will passively de-orbit a satellite back to Earth. The proposed system will deploy 5 km of bare wire plus 10 km of special insulated tether from a Delta II upper stage to achieve 0.4 N of drag thrust, for de-orbiting. The experiment will use a bare tether for current collection in lieu of the end-mass collector and insulated tether used on the TSS-1 mission.

The tether is a distinctly different propulsion system that is propellant-free. It is electrodynamic in principle, and uses a long conducting wire moving in the Earth's magnetic field to generate electrical power for propulsion. NASA has experimented with this concept a few times in space. If future development proves to be successful, it can, one day, transfer satellites from one orbit to another, and even power missions to the outer planets.

In its working principle,[7,8] electrons floating in space are made to flow in a long conductor. This constitutes an electric current. A conductor carrying current I moving in a magnetic field of density $\mathbf{B}$ generates mechanical force per unit length equal to $I \times \mathbf{B}$ in the direction that is perpendicular to both $\mathbf{B}$ and I. This force, known as the Lorentz force, works as a motoring force which can raise or lower the payload orbit depending on the current flowing towards or away from the Earth (Figure 23.9).

The tether can also be used for power generation. Experimental and theoretical work using very large high-voltage structures in a simulated LEO plasma has shown that substantial currents may couple between such structures and the surrounding ionospheric plasma. A long tethered satellite system would behave as a power generator if the induced $V \times \mathbf{B}$ voltage in a long conducting tether moving in the Earth's magnetic field is allowed to drive a negative current through an on-board load, providing electrical power at the expense of orbital energy. NASA's tether power generation concept utilizes such electrodynamic characteristic of a conducting tether in the Earth's orbit to generate useful electric power for a number of applications. It has several advantages over conventional power sources. It is 2.5 to 3 times as efficient as fuel cells in terms of fuel utilization,

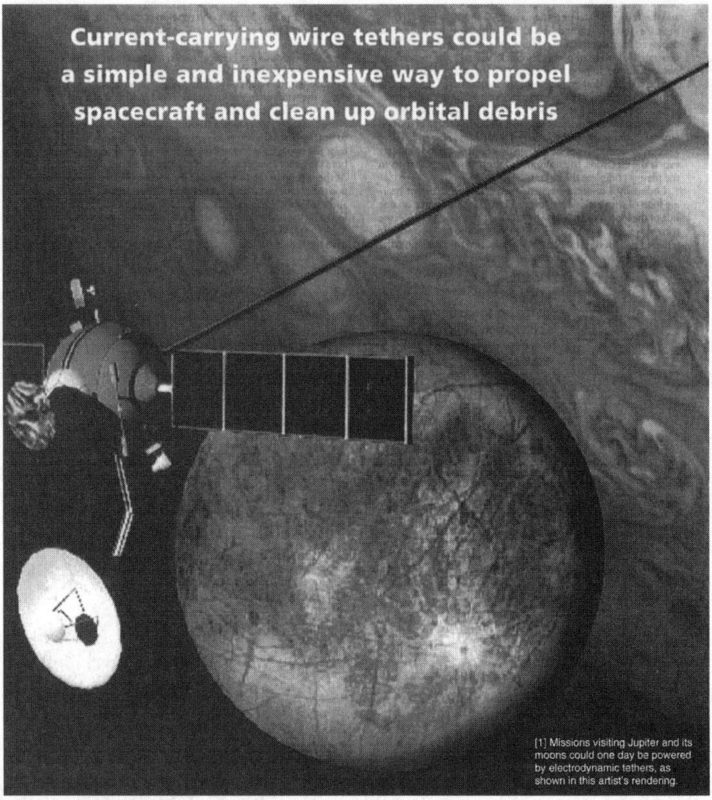

FIGURE 23.9 Tether propulsion concept.
(From L. Johnson, *IEEE Spectrum*, p. 39, July 2000. With permission.)

and is not subject to day/night cycles as are solar arrays. The magnitude of the power generated depends on the tether length. As an example, it has been estimated that a current of 5 A through a 100-km tether can produce upwards of 80 kW of power. Accounting for drag and other losses, the system could operate at a net efficiency of 75%.

The concept has been around since the 1960s. In 1993, NASA's Plasma Motor Generator Mission used a tether equipped with a hollow cathode end mass to collect electrons. In 1996, NASA flew another 20-km long, 2.5-mm-diameter conducting tether between the space shuttle *Orbiter* and a 500-kg satellite with scientific instruments provided by the Italian Space Agency. This tethered satellite system (TSS) generated over 2 kW of power and 1 A of current. While not a complete success, it provided valuable data for future research and development on this unique concept of bleeding energy from space charge.

The most critical requirement for the tether power system is the availability of tether materials with the required strength, conductivity, insulation, and protection from adverse effects of the space environment such as radiation. Even with the availability of such materials, a number of design issues need to be studied in the near term for gaining a clear understanding of the technologies necessary for power generation. These issues include advances in high-voltage technology and power processing equipment. In addition, there are basic questions on the phenomenology and design, which must be investigated. They are:

- The plasma–electrodynamic interactions affecting return current losses
- High-voltage electronic components and insulation materials
- Hollow cathode current emission technology
- D.c. tether impedance controller technology
- Passive tether control techniques

The primary focus of the tether research is the plasma interactions for power generation. It also considers plasma interactions related to the system's use as a source of thrust and as an extremely low-frequency (ELF) antenna system. The concept of the tethered satellite system as a generator or motor envisions the ejection of charge from one end of the system, and the collection of a compensating current at the other end. The effect of the magnetic field on charged particle collection is considered and is shown to be substantial. Specific calculations of charged particle collection utilize computer codes for geometrically complex objects and well-developed analytical theories of spherical probes, both for the case of unmagnetized plasmas. Without the aid of plasma contactors or large area collectors, quiescent plasma theory indicates that the effect of the magnetic field would limit the useable power developed in an initial electrodynamic tether mission to about 1 kW. Since the plasma may be turbulent, however, this estimate may be pessimistic. In any case, uncertainties in present knowledge clearly point to a need for a better understanding of probes in magnetized plasmas, including plasmas that are strongly perturbed by the passage of electron beams.

The first electrodynamic tether experiment will involve electron beams. The electromotive force (EMF) generated by motion across the geomagnetic field will be used to operate an electron gun, which ejects a beam from the shuttle into the surrounding plasma. Some fraction of the electrons generated at the gun's cathode may not escape the vehicle, instead leaking current to the anode (shuttle ground) and causing a negative potential buildup. This potential buildup, a sensitive function of the ratio of leakage current to neutralizing current, is studied for the case where the neutralization is affected by ram ions from the ambient plasma.

Estimates of the effect of leakage currents on shuttle potential ignore the effect of interactions between the beam and ambient plasma. Strong perturbations of the ambient plasma are frequently observed, but experimental results do not yet permit firm statements on the magnitudes of vehicle potentials that develop during beam ejection. One experiment, conducted on the Space Shuttle/*Spacelab-I* showed that the vehicle charge buildup is neutralized by injecting a plasma plume simultaneously with an electron beam. This tended credibility to the effectiveness of the plume as a *plasma contactor*, the term that rests on the expectation that a spacecraft generated plasma acts as a conductor contacting the ambient plasma over an effective area much greater than the geometrical area of the satellite. A theory developed by NASA indicates not only that plasma contactors substantially enhance charged particle collection, but also act as an efficient source for emission of current into the space plasma. However, the theoretical basis for these assertions, like the experimental one, is not adequate for confident predictions. Both require further development.

The electrodynamic tether system is subject to time variations induced by load current pulsation and by fluctuations in induced voltage and plasma properties as the system orbits the Earth. The magnitudes of these variations have been examined and their effect determined by considering the system as an end-loaded and distributed RC line. Although the variations are significant, the relevant time constants of the system are short, permitting it to react in a quasi-steady manner. This latter conclusion pertains also to the electron gun current pulsation that enables the system to function as an ELF antenna.

While the results of previous research indicate that the tether is an efficient generator of ELF waves compared to a tuned antenna in vacuum, and that the waves couple only weakly into the Earth's ionosphere cavity, the tether may, nevertheless, prove useful as an electrical source.

Compared to the ohmic resistance of the tether, the radiation resistance is small, not only for wave modes with frequencies in the ELF range, but also for other wave modes excited by the motion of the tether. The implication is that radiation resistance is not an important factor when computing time dependent currents in the tether. While many questions relating to wave generation and current closure through the ionosphere are not fully answered, there is yet definitive evidence that these issues will profoundly affect the electrical response of the system.

Other issues are the drag produced by deflection of particles by the tether's magnetic field and the electro-mechanical stability of the tether against kink modes. In both cases the associated physical effects are found to be insignificant.

With reference to Figure 23.10, Johnson[8] describes the tether's working as follows:

1. The conducting tether attached to a space object moving in the Earth magnetic field generates voltage across the wire given by $e = BLV$.
2. Space electrons are attracted to the positively biased far end of the tether.

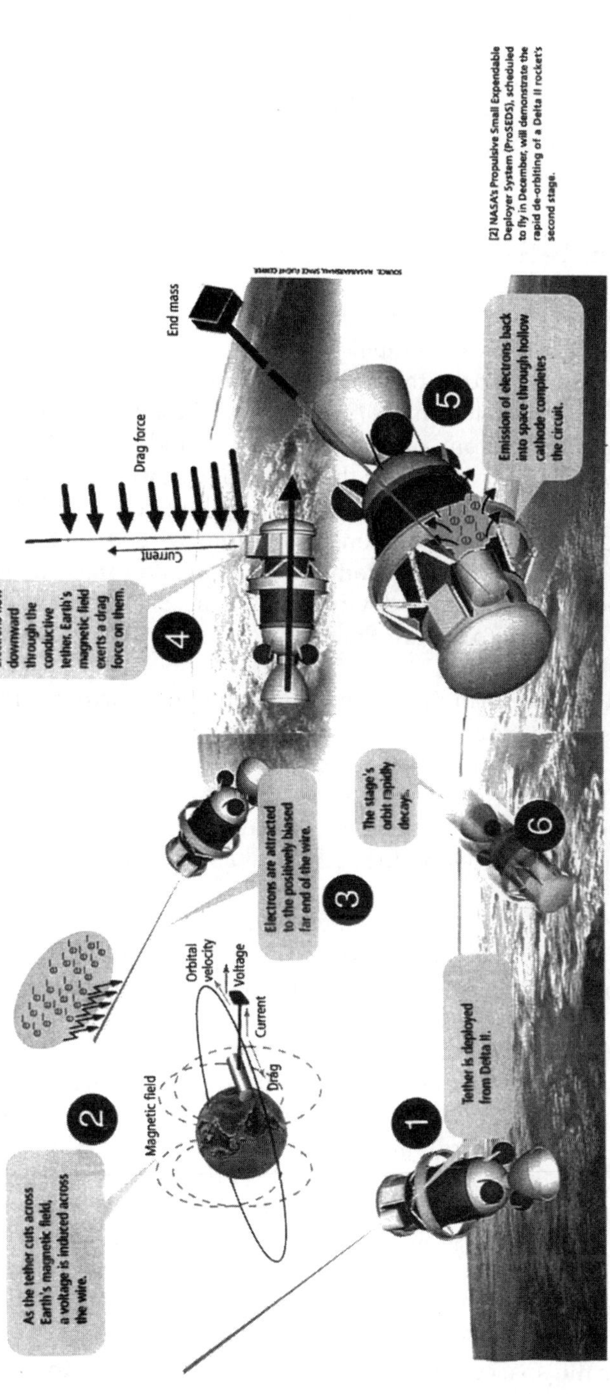

FIGURE 23.10 Tether working principle. (From L. Johnson, *IEEE Spectrum*, p. 40, July 2000. With permission.)

3. The electrons flowing downwards through the tether constitute a current.
4. The current-carrying tether moving in the Earth's magnetic field exerts a drag force given by $F = BIL$, causing de-orbiting of the space object.
5. A hollow cathode at the near end of the tether emits electrons back into space, completing the electric circuit.

The force will be reversed if a reverse current is applied by a higher voltage than that self-generated in the tether, such as from a solar array. In this case, the orbit of the space object attached to the tether would rise.

The tether forms a unique type of electric circuit, which NASA demonstrated in space with the Plasma Motor Generator in 1993 and the Tethered Satellite System in 1992 and 1996. These missions deployed long conducting tethers from orbiting spacecraft and generated kilowatts of power. The tether is dragged through the ionospheric plasma through which the spacecraft and the tether are traveling at a speed of 7 to 8 km/s. In so doing, a few kilometers long aluminum wire extracts electrons from the plasma at the end farthest from the payload and carries them to the near end. A specially designed device, known as a hollow-cathode emitter, expels the electrons to ensure their return to space. Currents in the plasma complete the circuit.

Ordinarily, a uniform magnetic field acting on a current-carrying loop of wire yields a net force of zero, since the force on one side of the loop is canceled by that on the other side, in which the current is flowing in the opposite direction. However, since the tethered system is not mechanically attached to the plasma, the magnetic forces on the plasma currents in space do not cancel the forces on the tether, and so the tether experiences a net force.

As the tether cuts across the magnetic field, its bias voltage is positive at the end farthest from the Earth and negative at the near end. This polarization is due to the action of the Lorentz force on the electrons in the tether. Thus, the *natural* upward current flow is due to the (negatively charged) electrons in the ionosphere being attracted to the tether's far end and then returned to the plasma at the near end, aided by the hollow cathode emitter (steps 3 to 5 above). The hollow cathode is vital. Without it, the wire's charge distribution would quickly reach equilibrium and no current would flow. The Earth's magnetic field exerts a drag force on the current-carrying tether, decelerating it and the payload, and rapidly lowering their orbit. Eventually, they re-enter the Earth's atmosphere (de-orbit).

A key advantage of the electrodynamic tether is that it does not need the spacecraft's on-board power. If the tether were being used to raise rather than lower the spacecraft orbit, the current would have to be forced in the other direction. An addition of a small PV power source would do it.

Some of the near term applications of tether propulsion systems are as follows.

23.9.1 Space Debris Cleanup

Over the past few decades, the space around Earth has become cluttered with debris, which could take years, and in some cases a century, to fall from orbit. The danger is that old satellites, rocket stages, and trash thrown overboard by early space travelers could collide with working satellites, space shuttles, and space stations. The International Space Station, for example, has to maneuver several times a year to avoid hitting debris, burning up precious fuel each time. NASA and ESA have recommended that national governments require future spacecraft to be self-removed from the orbit at the end of their life spans. To do that, a satellite must be loaded with extra propellant that could add as much as 25% of the satellite's weight. Using a tether to de-orbit would be inherently more mass efficient and reliable. A tether is much lighter and more compact than conventional thrusters, and can account for as little as 2% of a satellite's total weight. It has no complex valves, plumbing or circuitry that must stay operational and leak-free for years. At the end of a satellite's useful life, the tether would unreel and tether-driven orbital decay would begin. A satellite's orbit, which would otherwise take 1 year to decay naturally, could be de-orbited in a few days using a tether.

23.9.2 Orbit Transfer Vehicle

The electrodynamic tether can be attached to an unmanned space tugboat that would carry satellites to higher orbit. After being launched into low Earth orbit, the orbital transfer vehicle would grapple the satellite and maneuver it to a new altitude or inclination. The tug could then lower its own orbit to rendezvous with another payload and repeat the process. Conceivably, several such orbital reassignments could be performed without the need for propellant, making the tug relatively inexpensive to operate. Because plasma density diminishes rapidly with distance from the Earth, the tug could only operate below 2500 km. The lowest usable altitude would be about 250 km, below which the atmosphere would begin to exert too much aerodynamic drag.

23.9.3 Keeping the Space Station Afloat

The International Space Station is estimated to require 0.5 to 1.5 tons of propulsion fuel replenished from the Earth for every year of operation. This is to overcome the atmospheric drag, which would cause its altitude to drop by several kilometers per week. ISS with an electrodynamic tether could drastically cut its dependency on fuel. Atmospheric drag exerts a force of up to 1 N on the station. NASA has designed a 10-km long, 0.6 mm × 10 mm aluminum ribbon tether capable of delivering 0.5 to 0.8 N of thrust with less than 10 kW of solar array power needed to reverse the direction of the current in the tether. More research is needed before such a tether can be actually deployed on the space station. For example, if micrometeoroid or orbital debris accidentally severs the tether, a mechanism is needed to keep it from recoiling around the station and damaging key components.

23.9.4 Exploring the Outer Planets

Electrodynamic tether technology may be used to propel and power spacecraft exploring the outer planets. Jupiter has an environment particularly favorable to the electrodynamic tether. It has a strong magnetic field, and its mass dictates high orbital velocities. Present Jupiter missions rely on RTGs, but the power available is typically less than 300 W. In theory, the tether could power the craft instruments and generate thrust at the same time. For a circular orbit close to the planet, tether propulsive forces have been calculated to be as high as 50 N and power levels as high as 1 MW. This level of power can enable high-power instruments and radar. But it also means resolving the high-power conversion, energy dissipation, and tether overheating problems.

References

1. Kerslake, T.W. and Gefert, L.P., Solar power system analysis for electric propulsion missions, in *Proceedings of the 34th Intersociety Energy Conversion Engineering Conference*, SAE, 1999, Paper No. 2449.
2. Brandhorst, H.W., Power without wires, a solar electric propulsion concept for space exploration, *IEEE Aerospace and Electronic System*, Voltage. 16, No. 2, February 2001, pp. 3–7.
3. Bailey, S.G., Hepp, A.F., and Raffaelle, R.P. Thin film photovoltaic for space applications, in *Proceedings of the 36th Intersociety Energy Conversion Engineering Conference*, ASME, 2001, Paper No. AT-36.
4. Laug, K.K. and Holmes, M.R., Paraboloidal thin film inflatable concentrators and their use for power applications, in *Proceedings of the 34th Intersociety Energy Conversion Engineering Conference*, SAE, 1999, Paper No. 2552.
5. Lineberry, J.T. and Chapman, J.N., MHD Augmentation of rocket engines for space propulsion, in *Proceedings of the 35th Intersociety Energy Conversion Engineering Conference*, AIAA, 2000, Paper No. 3056.
6. Power, J.L., Microwave electrothermal propulsion for space, *IEEE Transactions on Propulsion in Space*, Vol. 40, No. 6, June 1992.
7. Cosmo, M.L. and Lorenzini, E.G., *Tethers in Space Handbook*, NASA Handbook, December 1997.
8. Johnson, L., The tether solution, *IEEE Spectrum*, July 2000, pp 38–43.

Chapter 24

Fuel Cell Power

24.1 Introduction

The fuel cell was developed as an intermediate-term power source for space applications. It was first used in a moon buggy. At present, it is routinely used to power NASA's fleet of space shuttles (STS Orbiter) that carries components and crew to the International Space Station and other space service missions.[1] The fuel cell resembles a battery in that it converts the chemical energy of a fuel directly into d.c. electricity. However, unlike a battery, it does not run down in energy and have to be recharged. It keeps producing electricity as long as the fuel is supplied. One can call the fuel cell 'the gas battery.' A typical fuel gas is hydrogen or a hydrogen-rich mixture and an oxidant.

A fuel cell finds applications in space missions lasting for a few days to a few weeks where a battery is not practical. It also has a potential use as an auxiliary power source for orbit transfer vehicles. The regenerative fuel cell integrated with an electrolyzer unit presents an attractive mass saving for LEO satellites requiring large energy storage. It was a serious candidate in place of the battery for the ISS.

The working of a fuel cell is the reverse of the electrolysis of water. In the electrolysis, electricity is passed between two electrodes in water to produce hydrogen and oxygen. In the fuel cell, hydrogen and oxygen are combined to produce electricity and water. The energy conversion is direct from chemical to electrical. Since the process is isothermal, the conversion efficiency is not limited by the Carnot efficiency. This is unlike many chemical-to-thermal-to-mechanical-to-electrical energy converters using steam or internal combustion engines. It skips the usual combustion step of conventional power systems and converts a high percentage of the fuel's available free chemical energy directly into electricity. The fuel cell efficiency, therefore, can be about twice that of the thermodynamic converter. It is as high as 65% in some designs, and 75 to 80% in solid metal oxide fuel cells developed for ground base power plants. Its superior reliability with no moving parts is an additional benefit over thermodynamic power generators.

24.2 Electrochemistry of a Fuel Cell

A fuel cell consists of anode and cathode electrodes separated by a liquid or a solid electrolyte. The electrodes are electrically connected through an external load circuit as shown in Figure 24.1. Hydrogen or a hydrogen-rich mixture is fed to the anode. The hydrogen fuel is combined with oxygen of the oxidant entering from the cathode port. The hydrogen, however, does not burn as in the internal combustion engine. It splits into hydrogen ions (H^+) and electrons (e^-), and produces electricity by an electrochemical reaction. Water and heat are the byproducts of this reaction if the fuel is pure hydrogen. With natural gas, ethanol or methanol as the source of hydrogen, as in some ground based fuel cells; the byproducts include carbon dioxide, and traces of carbon monoxide, hydrocarbons, and nitrogen oxides, all of which are negligible.

The hydrogen–oxygen fuel cell consumes hydrogen as fuel, oxygen as oxidant, and an aqueous acid solution as electrolyte. Their net reaction is $2H_2 + O_2 = 2H_2O$, and, of course, energy is released in the process. In one type of fuel cell, the electrons flow from the anode and travel through the external circuit to the cathode, powering the electrical load en route. Hydrogen ions migrate through the electrolyte to the cathode, closing the loop. At the cathode, they combine with the oxygen and the incoming electrons from the external circuit to produce water. The kind of ions and the direction in which they migrate vary, depending on the type of electrolyte.

Major components and their functions in the fuel cell are:

- The anode (fuel electrode) provides a common interface for the fuel and electrolyte, catalyzes the fuel oxidation reaction, and conducts electrons from the reaction site to the external circuit.

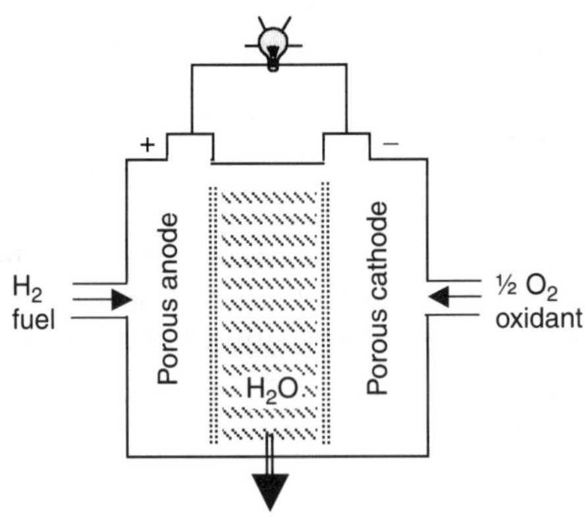

FIGURE 24.1 Fuel cell construction.

- The cathode (oxygen electrode) provides a common interface for the oxygen and the electrolyte, catalyzes the oxygen reduction reaction, and conducts electrons from the external circuit.
- The electrolyte transports one of the ionic species involved in the fuel and oxygen electrode reactions, while preventing the conduction of electrons to avoid a short circuit.
- Other components may also be necessary to seal the cell, to provide for gas compartments, and separate one cell from the next in a fuel cell stack.

The fuel cell is thus a static electrochemical device that generates electricity by using a chemical reaction without altering the electrodes or the electrolyte materials. This distinguishes the fuel cell from the electrochemical battery. Unlike the conventional battery, the fuel cell has no electrical energy storage capacity. Hence, it must continuously supply the reactant and withdraw the reaction products during operation.

24.3 Electrical Performance

The fuel cell basically works as a voltage source with an internal resistance. The electrical potential appears at the terminals of two electrodes involved in the process. The theoretical value of the fuel cell potential is 1.25 V, which matches that of the NiCd and NiH_2 batteries. Multiple fuel cells stack up in series–parallel combinations using heavy graphite pallets for the required voltage and current, just as the electrochemical cells do in the battery. However, as soon as the current is drawn, the voltage drops significantly due to various losses. Because the primary loss mechanism is ohmic loss in the electrodes, the voltage continues to drop with increasing current. The voltage drop is given by

$$V_{drop} = \alpha + \beta \log_e J \tag{24.1}$$

where J is current density at the electrode surface, and α and β are constants which depend on temperature and the electrode surface. The theoretical potential difference of 1.25 V between the anode and cathode in the hydrogen–oxygen fuel cell is determined by the difference of the free energy of the reaction product and the fuel and oxidant. This potential is different in different fuel cells depending on the reactions involved. The electrical performance of a fuel cell is represented by the electrode voltage versus surface current density, commonly known as the polarization curve or the V–I curve as shown in Figure 24.2. Ideally, a single H_2–O_2 fuel cell could produce 1.25 V d.c. under ambient conditions. Undesirable ions and products of the intermediate irreversible reactions decrease the cell

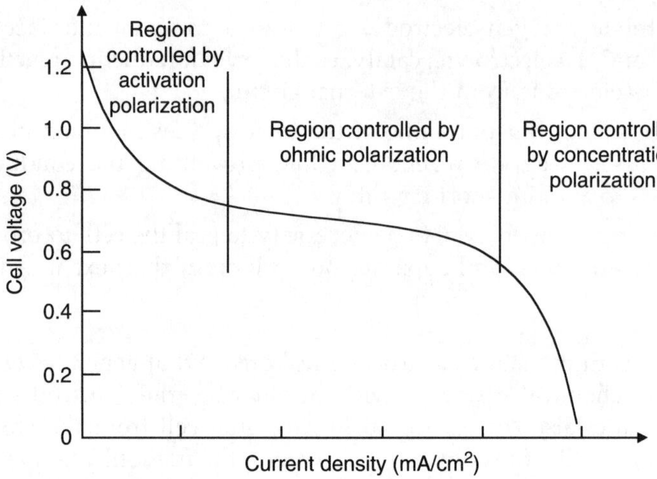

FIGURE 24.2 Fuel cell polarization curve — voltage versus current.

potential, even at open circuit. Further voltage drop under load results from various irreversible polarizations in the cell, such as:

- Resistance polarization, i.e., the electrical resistance of the electrolyte and electrodes
- Concentration polarization, i.e., the accumulation of ions and reaction products and the depletion of consumed ions and reactants in the electrolyte near the electrode surfaces
- Activation polarization, i.e., the reluctance of the fuel and oxidant to undergo reaction at each electrode
- Activation polarization causing energy losses associated with reactions
- All ohmic losses within the cell, i.e., electrodes, current collectors, contacts, and ionic impedance of the electrolyte
- Concentration polarization causing the energy losses associated with mass transport

The net result of these polarizations is that the practical fuel cell produces between 0.5 and 1.0 V d.c. at currents of 100 to 400 mA/cm^2 of cell area. Fuel cell performances can increase by increasing cell temperature and reactant partial pressure. A trade-off exists between achieving higher performance by operating at higher temperature or pressure and confronting the materials and hardware problems imposed at more severe conditions.

The practical operating range of the fuel cell is controlled by ohmic loss. The *V–I* characteristic in this region is very similar to that of the battery, except that the average discharge voltage is lower. The voltage drops approximately linearly with increasing current and also with time, a shown in Figure 24.3.[2] At any given time, the terminal *V–I* relationship can be expressed as

$$V = V_o - kI \tag{24.2}$$

where V_o = open circuit voltage and k = constant. The value of k increases and V_o decreases with time. The power at any operating point is given by

$$P = VI = (V_o - kI)(V_o - V)/k \tag{24.3}$$

The maximum power is when $dP/dI = 0$, which occurs at $V = \frac{1}{2}V_o$, leading to

$$P_{max} = \frac{1}{4}V_o^2/k \tag{24.4}$$

Unlike the PV cell, the fuel cell does not work in the *use the input energy or lose it* mode. It uses the on-board fuel to generate power. For this reason, the fuel cell is not operated at P_{max} until it approaches the end of life. It is rather operated at the maximum fuel efficiency until the EOL.

Since V_o degrades with time, so does the P_{max}. The open circuit voltage can be expressed as a function of time as

$$V_o(t) = V_o(0) - K_o h \tag{24.5}$$

where h = number of hours since the fuel cell is placed in operation. With a voltage regulating converter between the fuel cell and the load, the life of the fuel cell can be defined as the time it takes for the voltage to decay below the required input voltage, or the P_{max} to fall below the required output

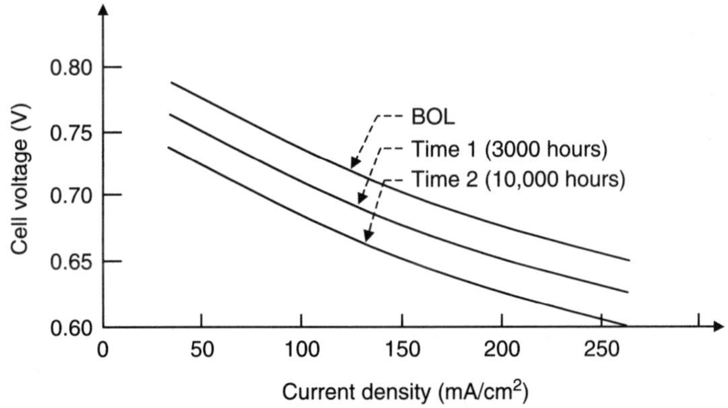

FIGURE 24.3 Fuel cell operating characteristic degradation with time.

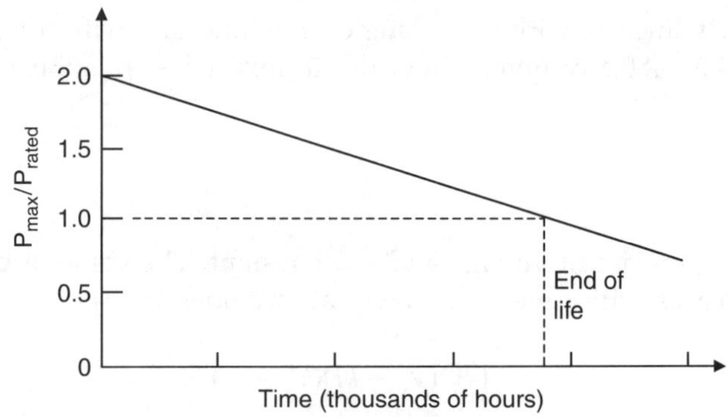

FIGURE 24.4 P_{max}/P_{rated} ratio versus time determines the life.

power. It can be predicted from the V_o versus time relation. The expected life of the fuel cell is determined as shown in Figure 24.4.[2]

The transient electrical performance model of the fuel cell include electrochemical, thermal and mass flow elements that affects the electrical output.[3] Of primary interest is the electrical response of the cell to a load change. To design for the worst case, the performance is calculated under both the constant reactant flow and the constant inlet temperature.

24.4 Types of Fuel Cells

A fuel cell is classified by the types of fuel and electrode materials used in the construction. The space-qualified fuel cell in the future is likely to take advantage of recent advances in commercial fuel cells using extensive ground experience on advanced technologies. Various types available at present are described below.

Solid polymer electrolyte (SPE) was used in the early manned missions lasting less than 1 week. Since then, two parallel fuel cell technologies have been developed: proton exchange membrane (PEM) for *Gemini* and alkaline for *Apollo* missions to the moon. An improved aqueous alkaline technology was also selected for NASA's space shuttle fleet. A lightweight alkaline fuel cell stack has been recently developed by United Technologies for NASA. It is designed to use propellant grade H_2 and O_2 to generate a peak power of 3.5 kW for an average load of 2.2 kW. Specific powers of these units were approximately 110 W/kg for the standard space shuttle fuel cell and approximately 150 W/kg for the lightweight stack.

A high-power shuttle fuel cell stack operating on dedicated cryogenic reactants H_2 and O_2 is designed for a peak power of 12 kW with brief overloads of 15 kW and an average load limited to 4.5 kW for the life consideration. The average load has now been increased to 7 kW. The space shuttle does not have batteries. It obtains starting power from the ground before launch, and then uses fuel cells after the launch. Each shuttle has three fuel cells generating 30 to 36 V d.c. power. Some of this power is used in d.c. form, and some is converted in 400 Hz a.c. for aircraft-type equipment used in the shuttle. It has 118 kg of dry mass. The output power capability of 12 to 15 kW gives a pecific power around 100 to 120 W/kg based on the stack and accessories without power conditioning and gas supplies. Although the electrical startup time is 10 to 20 min and the shutdown is instantaneous, the whole system requires about 2 h of warm-up time before it can be loaded. The design life for the space shuttle power plant is 5000 h at rated power. It is likely that the life of the shuttle power plant could be increased to 20,000 h (along with higher power capability) with the incorporation of advanced technology items.

The alkaline fuel cell used in the space shuttle combines hydrogen and oxygen to generate electricity, with water as a byproduct. Oxygen enters the cell through the sintered-nickel cathode where a catalyst produces OH ions and delivers them to the cell's potassium hydroxide electrolyte. The ions drift through the alkaline electrolyte to the anode, where they combine with hydrogen atoms to form water molecules and release electrons, and the electrical power is delivered to the external load. The energy released by 1 kg of cryogenic hydrogen combines with 8 kg of oxygen is about 34,000 Wh. That is 3800 Wh/kg of fuel, excluding the fuel tank and the supporting systems. This amounts to the energy conversion efficiency of about 50%. The byproduct water is used for the crew.

The PEM technology is found more suited for spacecraft applications at present. NASA Glenn Research Center continues developing the PEM and regenerative fuel cells for a wide variety of applications, including launch vehicles, planetary and interplanetary spacecraft, planetary surface power, aircraft propulsion, and terrestrial use. For spacecraft applications, the PEM technology offers enhanced safety, longer life, lower weight, higher reliability, higher peak power capability, compatibility with propulsion fuel, and potential for lower cost. A prototype 5.25 kW modular PEM fuel cell was tested in 2000.[4] The goal of the current PEM programs at NASA[5,6] are three times the power and four times the life compared with the alkaline fuels cell of the same mass and volume; with the peak to average nominal power ratio of six. Further developments will be needed to qualify for manned flights.

The performance of various types of fuel cell for space applications is compared in Table 24.1. Compared to the PV array, the fuel cell gives much higher power per kilogram. Flexibility is another major advantage, as it does not need sun pointing and provides the same power during both day and night. Its disadvantage is that it needs to carry fuel on board.

Table 24.1 Various fuel cell performance compared

Fuel cell technology	Specific power (W/kg)	Life, in hours
Alkaline	100–150	~ 50,000
Solid polymer	100–150	~ 50,000
Alkaline (space shuttle)	300–400	3000–5000
Lightweight cell under development	600–700	TBD

24.5 Regenerative Fuel Cells

The fuel cells presented above convert energy one way: from fuel to electricity. They are not designed for the reverse recharging operation. Recharging the fuel cell requires an electrolyzer to decompose the water back into hydrogen and oxygen. The electrolyzer is generally a separate unit from the fuel stack and the two cannot operate simultaneously. The hydrogen–oxygen regenerative fuel cell (RFC) has been developed for energy storage in short-term space missions, but not used in space as yet. The RFC was considered for the ISS but dropped in favor of the NiH_2 battery due to its poor round trip energy efficiency. However, a high-power long-term manned mission to the moon or Mars may find the RFC practical. It would use solar energy to produce power and independent electrolysis units for hydrogen and oxygen production. This is advantageous not only for the energy storage, but also for life support and for auxiliary thrusters as well.

The RFC is based on a hydrogen–oxygen proton exchange membrane fuel cell and electrolyzer technology. The fuel stacks consists of multiple cells, each with certain active area operated at 60 psi, 80°C and nominal 50 A/m^2 current density. Cells are series–parallel connected and include bypass diodes. The electrolyzer stack consists of multiple cells with certain active areas, operated at 315 psi and a nominal current density of 20 A/m^2. The hydrogen and oxygen reactants are stored in gaseous form at 3000 psi, while water is stored at 14.7 psi. Hydrogen is contained in Kevlar tanks with 10-mil titanium liners to minimize gas diffusion. Similar tanks are used to contain oxygen and water. The fuel cell operating temperature is maintained by a radiator using water as the working fluid. The electrical energy efficiency for the electrolyzer is 90% and that of the fuel cell is 60%, thus the overall round trip efficiency of the systems is about 54%. The life is estimated to be 10,000 h.

The RFC has a good peak power capability, about 10 times the base power. This may be useful for providing power pulses. The primary disadvantage of RFC is the peripheral pumps and plumbing that can handle fluids reliably over a long mission life in GEO. Providing redundancy in such a system may incur significant mass penalty. Other than the reliability issue, the fuel cell is a well-developed technology. It may be possible to service the RFC system for a LEO mission of 5 years. In LEO, it has a reasonable specific energy around 5 to 8 Wh/kg and round trip energy efficiency of 60 to 70%. Because of the lower efficiency compared to the battery, both the solar array charging power and the thermal cooling requirement would be greater. In high power GEO applications, mass-optimized RFC units may provide 20 to 35 Wh/kg. The high specific energy is largely due to the need for only a small electrolyzer unit because of the relatively long recharge time available in GEO. The specific energy, however, is sensitive to various thermal and electrical design requirements.

Proton Energy Systems[7] is currently developing 1 to 100 kW RFC for LEO and GEO applications. The company has tested 25 to 35% d.c.–d.c. round trip efficiency. The efficiency is low, but the specific energy and specific power are high compared to the battery. At stack level, 300 to 400 Wh/kg and 70 to 100 W/kg are possible, including 10% for hydrogen storage. This compares with about 200 Wh/kg with the Li-ion battery.

24.6 RFCs for Space Colonies

When human exploration of space grows into colonies on the moon and Mars, power systems that last for decades will be required. Small nuclear-based power systems can provide long-term sustainable power for interplanetary missions and for human habitation in space colonies. However, the systems issues related to the use of large nuclear power plants needed to provide large amounts of power for life support, are yet to be resolved. These issues include the system integration of all functions of the habitat, as well as safety concerns. The regenerative fuel cell provides an alternative with significant advantage. It coincidentally furnishes a backup supply of oxygen and water valuable to the inhabitants in the event of emergencies and unforeseen circumstances. A simple architecture of such a power system is shown in Figure 24.5. It includes the following:

- PV array for power generation during the cyclic day time periods of approximately 14 Earth days
- Electrolyzer that uses d.c. electricity from the array and electrochemically converts water into hydrogen, oxygen, and low-grade heat
- Fuel cell stack that electrochemically converts hydrogen and oxygen into d.c. electricity, water, and low-grade heat during the cyclic night period of approximately 14 Earth days
- Storage tanks for pure water
- High-pressure hydrogen and oxygen tanks
- Automated controls

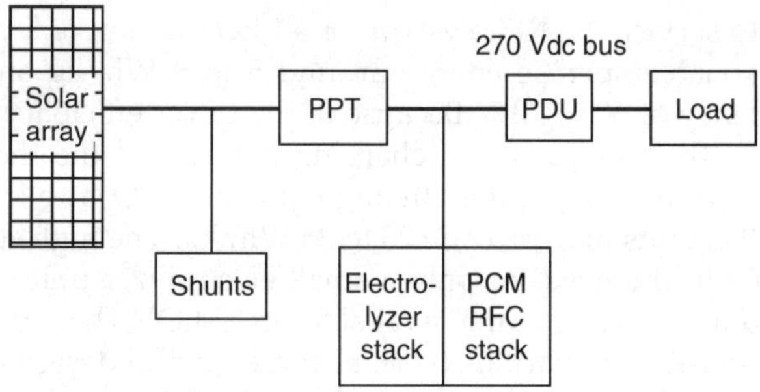

FIGURE 24.5 Power system architecture for human colony in space (Moon or Mars.)

A preliminary study of a 25-kW power system for a lunar habitat by Voecks[8] compared the KOH alkaline technology with the PEM technology for use in both electrolysis and fuel cell units, each with 65% efficiency. The study identified significant advantages associated with the PEM technology for the following reasons:

- PEM fuel cell and electrolyzer requires less pressure control complexity.
- Containment of the aqueous KOH electrolyte is much more difficult than polymeric membrane.

The environment is very demanding, while some aspects are simpler for operating an RFC on the moon compared to on the Earth. The moon operation requires storage of hydrogen, oxygen, and water, whereas the terrestrial operation does not need to store oxygen, since air can be fed through the cathode on a one-pass basis. It is easy to maintain the PV power generation schedules on the moon, as the sun time and dark are nearly constant in the lunar month (also day, which is 29 days, 12 h, 24 min and 2.9 s of the Earth.) On the Earth, the seasonal variations and the clouds can make the sunlight vary over a wide range.

24.7 RFCs for Satellites

The RFC may perhaps replace the battery in satellites, although not in very near future, because even the efficiency-optimized RFC design has a much lower round trip energy efficiency than the battery (Table 24.2). Moreover, the radiators and heat pipes required in the RFC system are significant in terms of both the mass and integration into the spacecraft design. The protection against single-cell failure using a bypass diode is not practical in case of the fuel cell. For a space-based radar power system as a hypothetical example, the mass and volume of the fuel cell energy storage alone — not counting the systems integration related mass — are roughly estimated and

Table 24.2 Round trip energy efficiency of RFC and battery

Energy technology	Round trip energy efficiency
Nickel cadmium battery	70–80
Nickel hydrogen battery	75–85
Hydrogen–oxygen RFC:	
Mass optimized design	55–60
Efficiency optimized design	65–70

compared with that using two batteries in Figure 24.6 and Figure 24.7, respectively. This study assumed a mission life of 5 years, orbit inclination 90° at various altitudes, power 20 kW, discharge time equal to charge time giving the duty ratio of 50%, and the battery DOD 25% for NiCd and 35% for NiH_2. It is seen in the figures that the RFC power system results in low mass in low orbits, comparable mass in mid orbits, and a very large volume in both orbits. In GEO, the RFC would certainly result in much greater mass in almost all cases.

However, when the designs are further developed, the RFC may find applications in high-power spacecraft. For a 10-MW pulse power directed energy weapon platforms in LEO as an example mission, Allen[9] reports a conceptual design study comparing the alternative systems, such as the RFC, heat engines, flywheel, and batteries that could be recharged using PV array. The study concluded that the solid-oxide RFC with PV charge power could deliver the required power with the lowest mass. The details of the Allen study are given in Chapter 22.

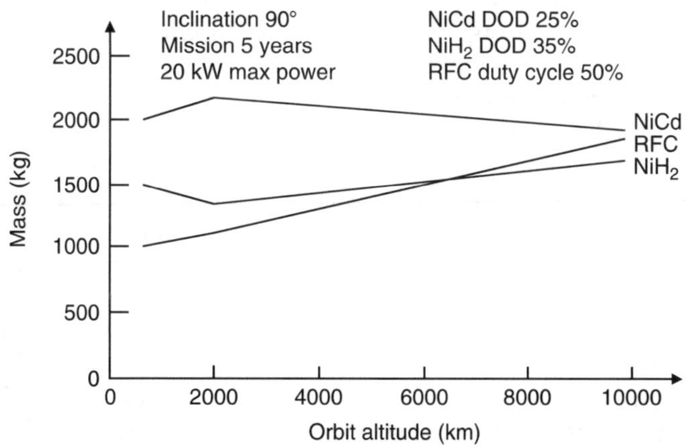

FIGURE 24.6 Alternative energy storage mass for 20-kW space-based radar.

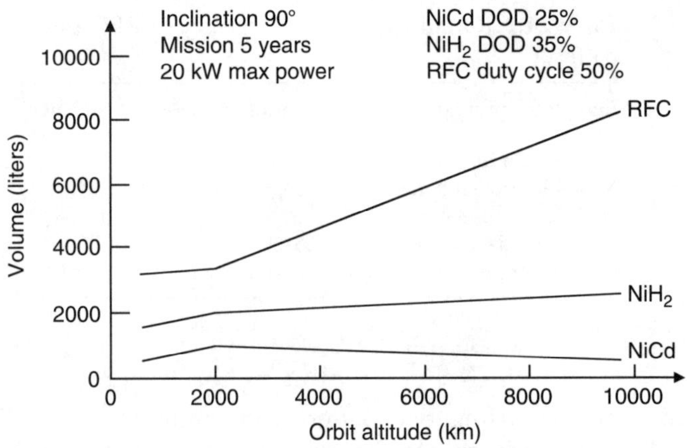

FIGURE 24.7 An alternative energy storage volume for 20-kW space-based radar.

24.8 Commercial Fuel Cells

In terrestrial applications, the fuel cell has the potential of replacing the combustion engines in automobiles, downtown buses, and hospital standby power generators. Large cells for off-grid power market include some military applications. The U.S. energy policy has placed significant emphasis on the hydrogen-based economy. Some of these terrestrial fuel cells research may be useful for future space applications.

The low temperature (250°C) fuel cell is now commercially available from several sources. It uses phosphoric acid as the electrolytic solution between the electrode plates. The high temperature fuel cell has higher energy conversion capability per kilogram of weight, but cost relatively more. Solid oxide, solid polymer, molten carbonate, and proton membrane exchange fuel cells in this category are being developed. The Fuel Cell Commercialization Group in the USA has recently field-tested a molten carbonate direct-fuel cell for a 2 MW utility-scale power plant. The German car manufacture Daimler-Benz and Ballard Power Systems of Canada are developing solid polymer fuel cells for automobiles as an alternative to battery-powered vehicles. Their target is to sell the first commercial fuel cell powered car before the year 2005. Major automakers around the world have accelerated research funding to be ready for consumers by 2010.

Howaldtswerke-Deutsche Werft AG of Germany has reported having built four 212A-class submarines that are fuel cell powered. The design is air-independent, and produces no noise or exhaust heat. It carries a crew of 27 with displacement of about 1450 tons. It is scheduled of commissioning in the German Navy in 2004. Large alkaline fuel cells have been built for deep submergence rescue vehicles (DSRVs) with 20 to 35 kW power levels.

Solid-oxide fuel cells of several different designs, consisting of essentially similar materials for the electrolyte, the electrodes, and the interconnections, are being investigated worldwide. Most success to date has been achieved with the tubular geometry being developed by the Westinghouse-Siemens Corporation in the USA and Mitsubishi Heavy Industries in Japan. The cell element in this geometry consists of two porous electrodes separated by a dense oxygen ion-conducting electrolyte. It uses ceramic tube operating at 1000°C. The fuel cell is an assembly of numerous such tubes. It is developed for multi-megawatt combined cycle gas turbine and fuel cell plants up to 60 MW capacity. The cell voltage degrades less then 0.1% per 1000 h of operation, and has life in tens of thousand hours of operation. The prototype has been tested for over 1000 thermal cycles with zero performance degradation, and 12,000 h of operation with less then 1% performance degradation.

References

1. Oman, H., Fuel cells power for aerospace vehicles, *IEEE Aerospace and Electronics System Magazine,* Vol. 17, No. 2, February 2002, pp. 35–41.
2. Babasaki, T., Take, T., and Yamashita, T., Diagnosis of fuel cell deterioration using fuel cell current–voltage characteristics, in *Proceedings of the 34th Intersociety Energy Conversion Engineering Conference,* SAE, 1999, Paper No. 01-2575.
3. Hall, D.J. and Colclaser, R.G., Transient modeling and simulation of tubular solid oxide fuel cell, *IEEE Power Engineering Review,* Paper No. PE-100-EC-004, July 1998.
4. Perez-Davis, M.E. et al. Energy storage for aerospace applications, in *Proceedings of the 36th Intersociety Energy Conversion Engineering Conference,* ASME, 2001, Vol. I, pp. 85–89.
5. Hoberecht, M. and Reaves, W., PEM fuel cell status and remaining challenges for manned space flight applications, in *Proceedings of the 1st International Energy Conversion Engineering Conference,* AIAA, 2003, Paper No. 5963.
6. Burke, K., Fuel cells for space science applications, in *Proceedings of the 1st International Energy Conversion Engineering Conference,* AIAA, 2003, Paper No. 5938.
7. Barbir, F., Dalton, L., and Molter, T., Regenerative fuel cells for energy storage – Efficiency and weight trade off, in *Proceedings of the 1st International Energy Conversion Engineering Conference,* AIAA, 2003, Paper No. 5937.
8. Voecks, G.E. et al. *Operation of the 25 kW NASA Lewis Research Center Solar Regenerative Fuel Cell Testbed Facility,* NASA Technical Report No. 97295, 1997.
9. Allen, D.M., Multi-megawatts specifications power technology comparison, in *Proceedings of the 36th Intersociety Energy Conversion Engineering Conference,* ASME, 2001, Vol. I, pp. 243–249.

Chapter 25

Flywheel Energy Storage

25.1 Introduction

A flywheel stores kinetic energy in rotating inertia. The stored energy can be converted into electricity using an electromagnetic machine in the generating mode. Recharging — spinning the wheel — is done using the same machine in the motoring mode. The round trip energy efficiency of the flywheel system is high — 85 to 90% compared to 70 to 75% with the electrochemical battery. Flywheel energy storage is an old concept that has now been developed for a variety of commercial applications. Automobiles, utility load leveling, and uninterrupted power supplies (UPS) are a few examples. At present, commercial UPS systems using flywheel–battery cost \$300 to \$500 per kWh. The space-qualified NiH_2 cells cost \$80,000 to \$100,000 per kWh. The flywheel system is being developed by NASA and industry partners for replacing the battery in the spacecraft power system.[1–4] One is scheduled to fly for flight tests on the International Space Station (ISS) in 2006. It is estimated that flywheel energy storage can improve the depth of discharge (DOD) and life by a factor of 2 compared to those of the battery. An integrated energy and momentum storage flywheel system can be even more advantageous.

The following developments have made flywheel technology more attractive now than ever before:

- Advances made in high-strength lightweight fibers. For example, fibers with ultimate tensile strength of one million pounds per square inch are commercially available now.
- Improvements in the fiber–resin composite rotor design and manufacturing process[5] and in understanding their performance characteristics.
- Availability of high-speed magnetic bearings,[6–8] eliminating the mechanical contact between the stator and the rotor, thus making the friction zero and the self-decay rate negligible.

The above technological advances when combined make it feasible to design a flywheel to operate at up to 100,000 revolutions per minute, yielding much higher specific energy compared to that with electrochemical batteries.[9]

Battery performance is sensitive to many factors having highly nonlinear relationships. Therefore, the design, analyses, and manufacturing, are time consuming. Moreover, the battery depth of discharge is limited to a value far below full capacity. The DOD is 30 to 35% in LEO satellites, where the battery is often the heaviest components of the power system. The flywheel can replace the battery with the following benefits:

- Much higher specific energy, hence lower mass and volume
- Deeper depth of discharge, up to 90% even in low Earth orbits, resulting in lower Wh rating and mass
- Lower hardware manufacturing and launch costs
- Insensitive to temperature, hence lower thermal system mass and cost
- Easy charge/discharge control as the stored energy depends on the wheel speed alone
- The charge/discharge ratio has no impact on life
- No capacity degradation over life
- No voltage sag over life
- No trickle charging required
- Shorter engineering design and analysis time
- Longer cycle life under charge/discharge cycles
- Flexibility in designing for a given voltage and current
- Improved quality of power as the electrical machine design can be made stiffer than the battery
- Matches well with peak power tracking solar array
- High peak to average power ratio (LEO communications satellites need more power while traveling over highly populated areas)

These benefits have the potential of making the flywheel the least-cost energy storage alternative per Wh delivered over the operating life on ground or in space. However, it has the following negative aspects:

- Higher output impedance because of the electrical machine (about the same as PRU converters in a fully regulated bus)
- Lower reliability, as a single failure would cause loss of significant power capability
- Most important issue is safety at speeds > 50,000 rpm
- Vibration and transient torque tolerance on the spacecraft platform for attitude control

- No launch and ascent power, as the flywheel cannot be spinning during launch
- Gyroscopic torque tolerance on attitude control

25.2 Photovoltaic–Flywheel Power System

A functional block diagram of the spacecraft power system using a solar array for power generation and a flywheel for energy storage is shown in Figure 25.1. The configuration represents a direct energy transfer fully regulated bus, which is used in most high-power satellites, except that the flywheel now replaces the battery. Attached on the flywheel shaft is an electromagnetic machine, which works in either mode, as a generator or as a motor. The machine converts mechanical energy into electrical energy and vice versa. Among several types of motor–generator designs available, the brushless d.c. machine is the most suitable for use with the flywheel. In operation, the PV array power during sunlight supplies the load and charges (spins up) the flywheel using the electrical machine in the motoring mode. During the eclipse, the flywheel is despun and the energy is fed to the load via the electrical machine in the generating mode. The bus-sense electronics senses the bus voltage, compares it with the reference voltage, and feeds the amplified error signal to the mode controller. The mode controller activates the buck or boost converter in the power regulator unit (PRU). It also controls the charge (motor) mode, the discharge (generator) mode and the shunt mode as required during the eclipse and sunlight periods.

The machine voltage, being a linear function of the flywheel speed varies over a wide range. The charge and discharge converters in the PRU provide interface between a widely varying machine voltage and the closely regulated bus voltage. It is possible to design PRU converters with input

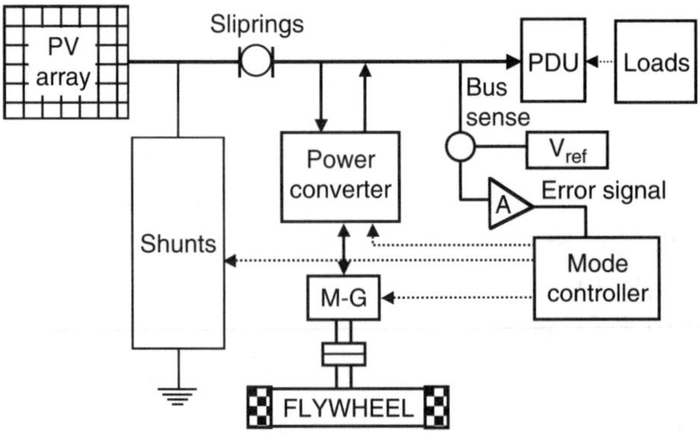

FIGURE 25.1 Spacecraft power system architecture with flywheel energy storage.

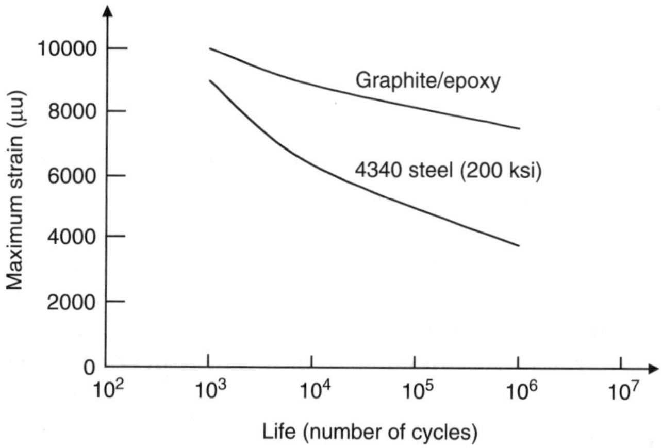

FIGURE 25.2 Fatigue life of graphite fiber–epoxy composite compared with stainless steel.

voltages varying over the 1 to 3 range. This allows the machine speed to vary over the same range. That is, a low rotor speed can be one third of the full speed. Since the energy storage is proportional to the speed squared, the flywheel state of charge at low speed can be as low as 10%. This means 90% of the flywheel energy can be discharged with no design difficulty in the power electronics, or in any other components of the system.

The fatigue life of the composite rotor may limit the number of charge/discharge cycles the flywheel can withstand. Experience indicates that polymer fiber composites generally have longer fatigue life than solid metals (Figure 25.2). A properly designed flywheel, therefore, can last much longer than the battery and can discharge to a much deeper level. Flywheels made of composite rotors have been fabricated and tested to demonstrate more than 10,000 cycles of full charge and discharge.[10]

25.3 Flywheel System Components

The major components of the flywheel assembly are:

- Rotor rim, where the energy is stored
- Hub interfacing the rim with the shaft
- Magnetic bearings for on-orbit operation
- Mechanical bearings for emergency, launch, and high slew rates
- Electrical machine (d.c. brushless motor–generator)
- Sensors embedded in the wheel to sense physical variables, such as position, vibration, temperature, and the overall health of the wheel during all phase of the mission operations

Each component is briefly described below.

25.3.1 Rotor Rim

The rotor rim is the primary component of the flywheel that stores the required energy and momentum. It is made of a high-strength fiber–resin composite matrix, with fibers principally oriented in the hoop direction to withstand high hoop stress under the rotational centrifugal force. The hoop stress is a maximum at the outermost tip of the wheel. For economical two-rim designs (Figure 25.3), the outer rim is made of high-strength fiber composite and the inner rim is made of low-cost fiber composite.

25.3.2 Hub

The hub supports the rotor and allows it to adjust radially under mechanical and/or thermo-elastic stress. It also is made of fiber–resin composites, with fibers oriented at suitable angle(s) to support the multi-dimensional stresses.

25.3.3 Magnetic Bearings

Conventional mechanical bearings are used up to speeds below 50,000 rpm. At high speed, they have high friction and vibration. Speeds approaching 100,000 rpm can be possible only by using magnetic bearings, in which the mechanical contact is eliminated, thus eliminating the friction. The aerodynamic windage is eliminated by running the rotor in vacuum. Low-flux pulsation as the rotor rotates around the discrete actuator coils produce small electromagnetic loss in the metallic parts. This loss, however, is negligible compared to the loss due to friction in mechanical bearings.

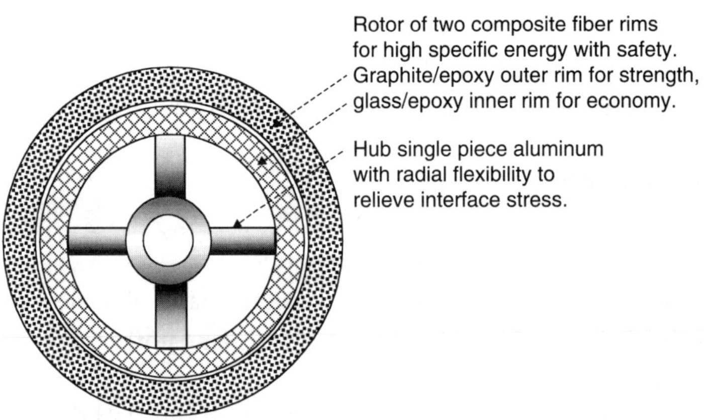

FIGURE 25.3 Two-rim flywheel rotor design.

The magnetic bearing raises and controls the shaft position using passive magnetic and/or active electrodynamic forces. The passive design uses permanent magnets of like polarities to keep them separate under the repulsive force between them. The active magnetic bearings come in a variety of configurations using permanent magnets and dynamic current actuators to achieve the required restraints. A rigid body can have six degrees of freedom. The bearings retain the rotor in five degrees of freedom, leaving one for rotation. Among various alternatives, one configuration is depicted in Figure 25.4. The dynamic actuator coils provide the horizontal restraint. The currents in the coils are controlled in response to the feedback loop controlling the rotor position. The permanent magnets are used to provide free levitation support of the shaft. The permanent magnet bias also helps stabilize the shaft under a rotor drop. The electromagnet coils provide stabilization and control. The control coils operate at low duty cycle, and only one servo controller loop is needed for each axis. The servo control coils provide active control to maintain the shaft stability by providing restoring forces as needed to maintain the shaft in the centered position. The position and velocity sensors are used in the active feedback loop. The electric current variation in the actuator coils compels the shaft to remain centered in the air gaps.

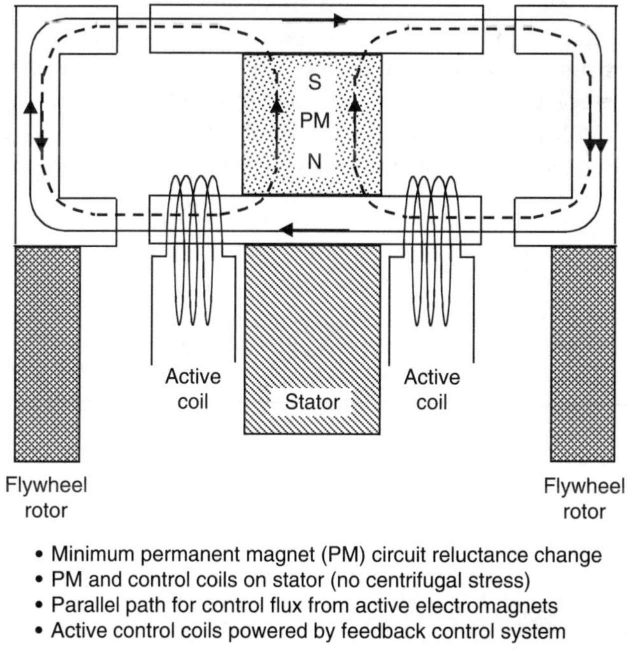

FIGURE 25.4 Magnetic bearing in one configuration (U.S. Patent 4000929.)

25.3.4 Touchdown Mechanical Bearings

These bearings support the shaft before the magnetic bearings are activated, or during high-*g* maneuvers and shocks, or in case of a magnetic bearings failure. These bearings can be ball, roller, or any combination. They last a long time for speeds up to 30,000 rpm (Figure 25.5), but can support occasional use at higher speeds in case of magnetic bearings failure.

25.3.5 Electrical Machine

The electrical machine works as a generator or motor and converts mechanical energy into electrical energy and vice versa. Among several types of moto—generator design available, the brushless d.c. type appears to be the most suitable with the flywheel system for space applications.

25.3.6 Sensors

Various sensors, some embedded in the wheel, sense physical variables, such as position, vibration, temperature, etc. They monitor the health of the wheel during all phases of operation. They also detect failures to command shutdown of the system if necessary.

Various sensors and bearing locations in one configuration of the flywheel assembly are shown in Figure 25.6. In another configuration, the rotor can be located radially outward, as shown in Figure 25.7, which forms a volume-efficient packaging. Moreover, it minimizes the thermo-elastic or rotational stress at the rotor–hub interface by providing radial resilience. Figure 25.8 depicts a NASA concept for the flywheel system packaging. A pre-stressed rotor, a strain-matched rotor, and mass-loaded designs are other configurations that can minimize stresses.[11]

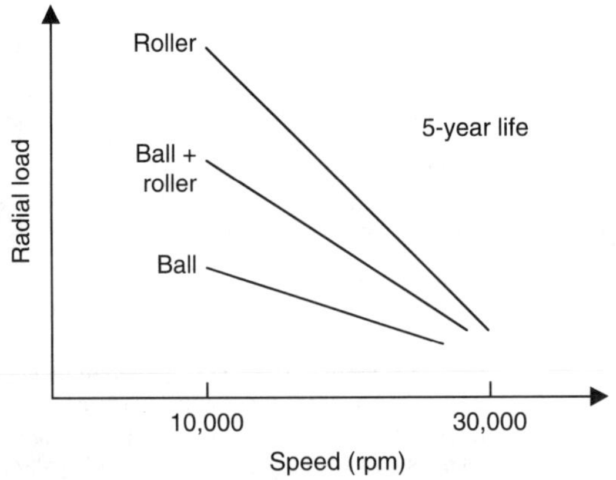

FIGURE 25.5 Radial load and speed limitations of ball and roller bearings.

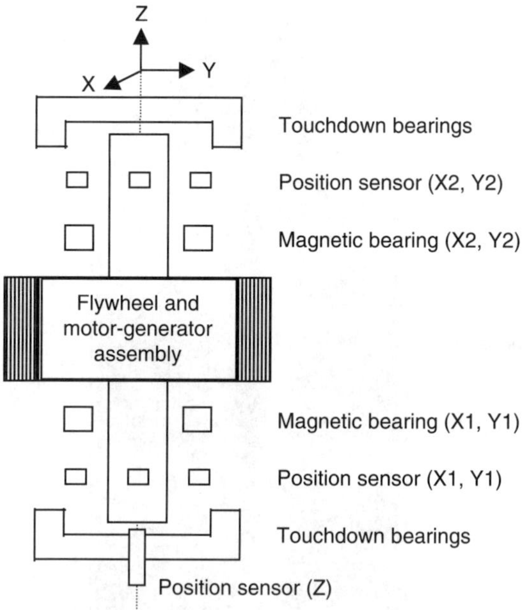

FIGURE 25.6 Flywheel sensor and bearing locations.

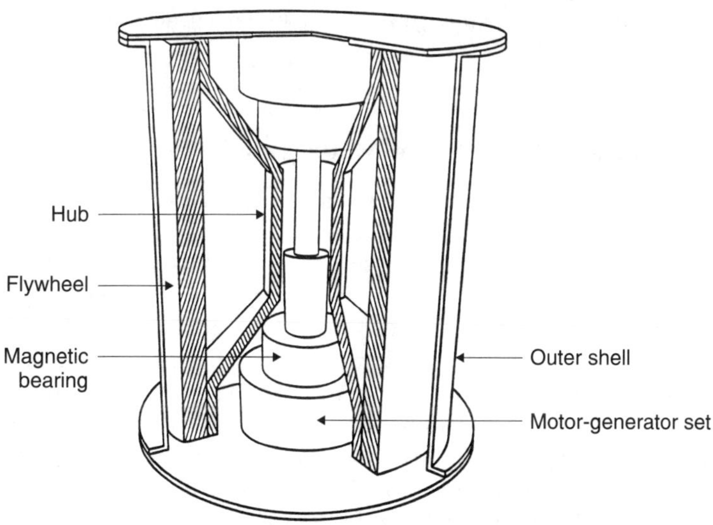

FIGURE 25.7 Flywheel components and assembly packaging with stress-relieving hub.

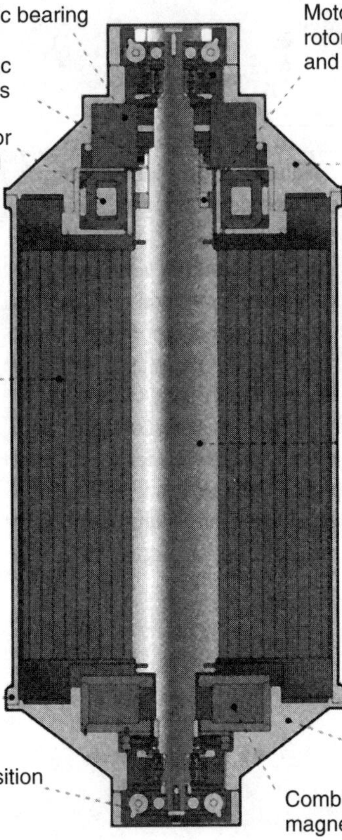

FIGURE 25.8 NASA concept of flywheel packaging.
(From S. Stankiewicz, *IEEE Spectrum*, p. 49, April 2002. With permission.)

25.4 Power and Momentum Management Electronics

Although not a part of the wheel assembly, this is an important part of the flywheel system. It is an independent box, separate from the wheel. It senses the bus voltage and bucks or boosts the bus voltage to match with the motor–generator voltage depending on the wheel speed at a given time. It provides a total interface between the motor–generator and the regulated voltage bus. It also contains the charge and discharge converters, the power management function, the momentum management function and the associated telemetry and commands. It controls the charge (motor) mode, the discharge (generator) mode, and the shunt mode as required during an eclipse and sunlight periods. Additionally, it controls the torque as commanded by the attitude control system.

25.5 Energy and Stress Relations

The kinetic energy stored in a flywheel having the mass moment of inertia J rotating at an angular speed Ω is given by

$$E = \tfrac{1}{2}J\omega^2 \tag{25.1}$$

The centrifugal body force is resisted by the hoop stress. The outermost fibers on the wheel are stressed the most. This stress is kept below the permissible limit. Stress analysis would show that the maximum hoop stress is directly proportional to the linear speed of the outermost fiber (tip speed of the wheel). For this reason, among wheels made from the same material, a wheel of twice the diameter can be spun only at one half the speed to keep the linear tip speed the same. Denoting this speed limit as V_{tipmax}, the maximum energy that can be stored per unit mass in a wheel of inner radius R_i and outer radius R_o can be shown to be

$$E_{\text{max}} = \frac{1}{4}V_{\text{tipmax}}^2\left[1 + \left(\frac{R_i}{R_o}\right)^2\right] \tag{25.2}$$

Since the ratio R_i/R_o is always less than unity, the above equation clearly shows that the thin rim flywheel design with R_i/R_o approaching unity maximizes the specific energy storage for a given tip speed limitation. The maximum design tip speed achievable in the modern fiber composite rotor is 1150 m/s with the burst speed of 1500 m/s without preloaded design. A 30-cm (12-inch) rotor operating at 100,000 rpm would produce 1500 m/s tip speed. Thus, a 100,000-rpm wheel has to be less than 25 cm in diameter.

For a thin rim flywheel of mass M and average radius R, the mass moment of inertia $J = MR^2$. Therefore, Equation 25.1 can be written as

$$E = \tfrac{1}{2}MR^2\omega^2 \tag{25.3}$$

It is the specific energy that matters the most in spacecraft power system designs. The specific energy in Wh per unit mass, denoted by E_1, is E divided by M. That is, from Equation 25.3,

$$E_1 = \tfrac{1}{2}R^2\omega^2 \tag{25.4}$$

The hoop stress σ at the outermost fiber of wheel made with material of mass density ρ is given by

$$\sigma = \rho R^2\omega^2 \tag{25.5}$$

Combining Equations 25.4 and 25.5 leads to the specific energy

$$E_1 = \tfrac{1}{2}(\sigma_{max}/\rho) \tag{25.6}$$

Thus, the specific energy of the flywheel is proportional to the maximum hoop stress allowed in the outermost fiber of the rim divided by the mass density of the material. Obviously, high specific energy can be achieved by using high ultimate tensile strength materials, which are also light in mass density, such that the σ_{ult}/ρ ratio is high. Fiber composites far exceed most metals in this respect. Therefore, they are used almost universally in all flywheel designs being tested for potential applications. Figure 25.9 shows the specific energy in rim flywheels of various materials. The figure shows that the specific energy of composite fiber wheels is much higher than the metal wheels, and the fused silica fibers have the theoretical maximum specific energy 20 times that of the state-of-the-art NiH_2 battery. Although flywheel system overheads are much higher than those in a battery, such high specific energy makes flywheel energy storage development attractive for space applications.

A preliminary design of a 10-kWh flywheel using various fiber composite materials is depicted in Figure 25.10. It shows what Equation 25.6 indicates. That is, the specific energy increases linearly with the (σ_{ult}/ρ) ratio of the material. The maximum hoop stress σ_{max} has been kept two thirds of the ultimate tensile strength σ_{ult} in this 10-kWh example design. The specific

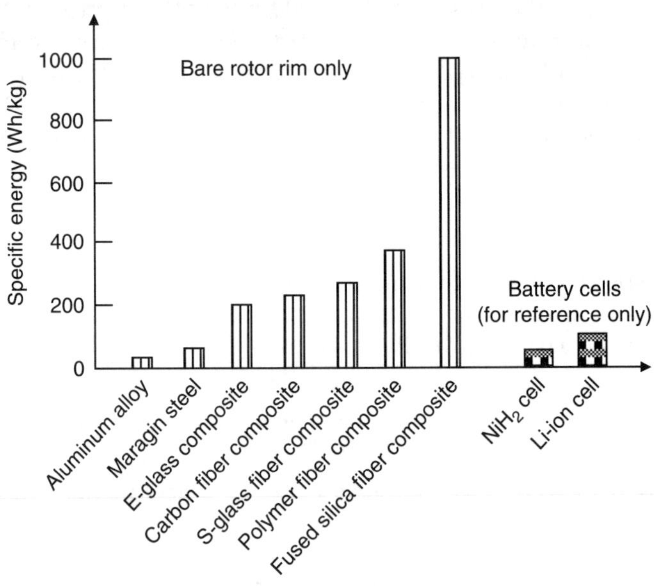

FIGURE 25.9 Theoretical maximum specific energy of rim flywheel of various materials.

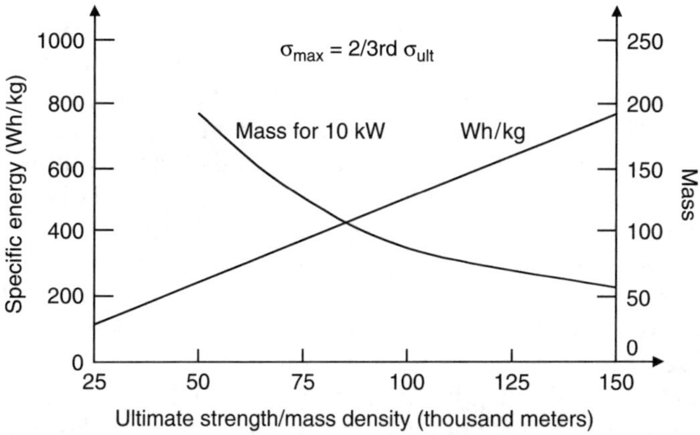

FIGURE 25.10 Specific energy and mass of 10-kWh flywheel versus σ_{ult}/ρ ratio.

energy of 90 to 120 Wh/kg has been demonstrated at the rotor (rim + hub) level in a commercial design running at 60,000 rpm.

25.6 Flywheel Application Example

An LEO satellite requires a heavy battery due to low depth of discharge dictated by the cycle life considerations. It would therefore benefit the most by using the flywheel energy storage. Patel[12] evaluated the benefits of replacing a battery with an equivalent flywheel in NASA GSFC's EOS-AM satellite as an example. The study was a point design based on reasonable possibilities. Table 25.1 gives the power system performance requirements and key design parameters using the NiH_2 battery.

Table 25.1 Power system requirements and key design parameters for an EOS-AM mission using an NiH_2 battery

Parameter	Requirement
Spacecraft life	5 years
Load power during eclipse	2530 W
Load power during sunlight	2530 W
Orbit period	1.65 h
Eclipse duration	38 min
Bus voltage	120 V
Energy storage batteries	Two 54 cells 50-Ah NiH_2
Battery depth of discharge	30% (max)
Photovoltaic array output after sliprings	5000 W

The first task for the system sizing is to perform an energy balance analysis over one orbit period. Such analysis using a flywheel in the example mission was performed using a spreadsheet program shown in Figure 25.11. The analysis determines the component ratings, as seen in the bottom half of the figure. It is noteworthy that the PV array output required with the PV–flywheel system is lower, 4662 W versus 5000 W in the PV–battery system. This reduction is because the round trip energy efficiency in the flywheel system with the motor–generator machine is higher than that in the battery system. The result is a 6.75% smaller PV array, which is a significant side benefit of the flywheel system, in addition to its primary benefit of reducing the energy storage mass.

The ratings derived in Figure 25.11 are for the total power system. A 20% margin was added in all component ratings on and above those derived in

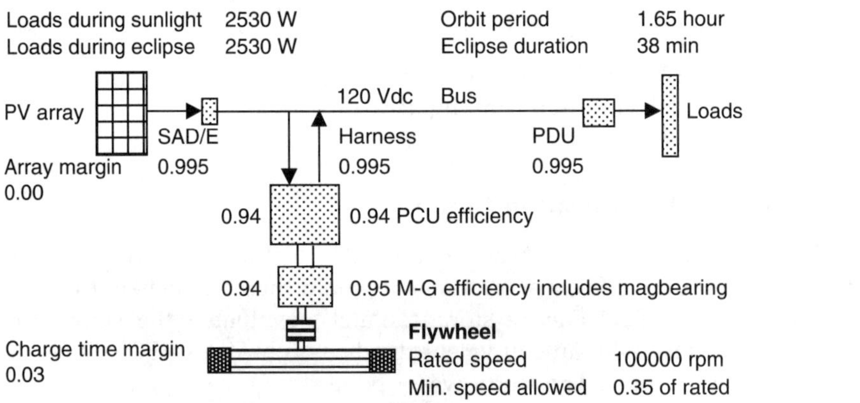

Notes: Bearing loses are included in the M-G losses
Constant power charging is used in the system design

Energy balance analysis key results:

Energy discharged from flywheel in eclipse	1821.5 Wh
Charging power required from the bus	2078.8 W
Load at PCU during sunlight (before charging power)	2555.5 W
Load at PV array during sun (including charge + diode)	4657.6 W

Required minimum steady state ratings of PV/FLYWHEEL components:

PV array EOL capability including margin	4662.2 W
PCU charge converter rating	1944.3 W
PCU discharge converter rating	2568.3 W
M-G motor rating	1847.1 W
M-G generator rating	2732.3 W
Depth of discharge of energy during longest eclipse	0.878 pu
Energy storage rating of the flywheel	2075.8 Wh
Max accelerating (motoring) torque on flywheel shaft	0.5305 nm
Max decelerating (generating) torque on flywheel shaft	0.7847 nm
Angular moment of inertia of the flywheel	0.1363 nms^2
Angular momentum of the flywheel at rated speed	1427.2 nms

FIGURE 25.11 Power system configuration and energy balance analysis for determining power system component ratings.

the figure. The mass and volume estimates were then developed for the flywheel system assuming the following performance parameters:

- Three equal wheels, two counter-rotating and one extra for reliability. The number of wheels, however, may be greater than three depending on the overall mission design
- Rotor design speed 100,000 rpm at full spin-up (operating tip speed <1.0 to 1.2 km/s)
- Ultimate tensile strength of the fiber–resin composite 300,000 psi
- Maximum allowable hoop stress in the outermost fiber of the wheel 200,000 psi
- Rotor length-to-diameter ratio 0.75. This ratio influences the vibration modes and the critical speeds
- Minimum speed at the end of eclipse 35% of the full spin-up speed. This gives the maximum depth of discharge of 88%
- Efficiency of the charge and discharge converters in the PRU is 94% each
- Efficiency of the motor-generator machine is 95% in either direction

The mass estimate of various components using the NiH_2 battery and also the flywheel for this example spacecraft are given in Table 25.2. Table 25.3 summarizes the total mass and volume of the power system using the NiH_2 battery and also the flywheel. Table 25.4 highlights differences between the two. It indicates that for LEO satellites with a 2500-W load, the flywheel system can save 458 pounds mass (35%), 13.8 cubic feet volume (55%), and 32 square feet solar array area (6.7%). Although the cost benefit of using flywheel energy storage is not predictable at this stage, the reduction in hardware and launch costs are expected to be several million dollars per spacecraft.

25.7 Integrated Energy and Momentum Storage

The momentum storage and control requirements of the attitude control system are traditionally met by using flywheels such as momentum wheels and reaction wheels. Significant savings are possible by using dual function flywheels for storing energy for the power system and momentum for the attitude control system of the satellite.[13–15] In the integrated energy and momentum storage system (IEMS), at least one energy-momentum-wheel on each axis controls the momentum in that axis and stores the corresponding energy at the same time (Figure 25.12). Placing an additional wheel on each axis would give a degree of freedom by simultaneously storing the required energy and the required momentum. One more skewed wheel on multi-axis gimbals is needed for redundancy. Thus, a

Table 25.2 Mass of EPS components using NiH_2 battery and flywheel (pounds)

Component	EPS with NiH_2 battery	EPS with flywheel
Solar array assembly	396	370
Array drive electronics	15	15
Power equip battery assembly	310	
Hex bay battery assembly	316	
Battery power conditioner	47	
Sequential shunt unit	98	90
Power distribution unit	102	102
Power switching unit	14	14
Rotor rim		45
Rotor hub		10
Motor–generator		10
Power management electronics		80
Magnetic bearings		18
Magnetic bearing electronics		10
Magnetic bearing sensors		5
Touchdown bearings		10
Auxiliary power supply		5
Shaft (steel)		15
Housing		20
Launch battery (primary)		21
Total EPS mass (pounds)	1298	840

(Data source: M.R. Patel, *Proc. IECEC*, SAE, 1999.)

Table 25.3 Total EPS mass and volumes using NiH_2 battery and flywheel in the example mission

Design parameter	EPS using NiH_2 battery	EPS using flywheel
EPS mass	1298 lb	840 lb
EPS volume	23.4 cu ft	10.8 cu ft
Solar array area	483 sq ft	449 sq ft

Table 25.4 Overall benefits of using flywheel in the example mission

Benefit	EPS using flywheel versus battery
Mass reduction	458 lb (35%)
Volume reduction	13.8 cu ft (55%)
Solar array area reduction	32 sq ft (6.7%)

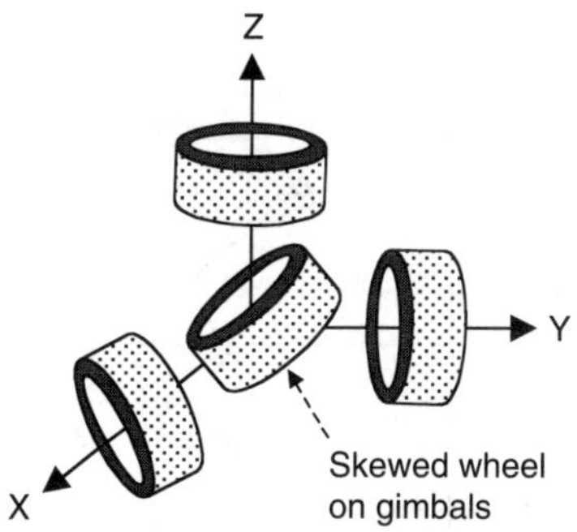

FIGURE 25.12 Energy–momentum wheel placement on three axes.

typical single fault-tolerant IEMS would require at least seven energy-momentum-wheels.

25.7.1 Energy and Momentum Relations

A flywheel having a mass moment of inertia J spinning at speed ω stores the energy $E = ½J\omega^2$, which is a scalar quantity. It also stores the momentum $\boldsymbol{M} = J\omega$, which is a vector quantity. If two wheels spinning on a given axis have mass moments of inertia J_1 and J_2 and angular speeds ω_1 and ω_2, respectively, then the total energy storage is

$$E = ½J_1\omega_1^2 + J_2\omega_2^2 \tag{25.7}$$

which is a scalar quantity describing an equation of ellipse in the ω_1–ω_2 plane. The total momentum of such two wheels is

$$\mathbf{M} = J_1\omega_1 + J_2\omega_2$$

Rearranging the above, we have

$$\omega_2 = \mathbf{M}/J_2 - (J_1 J_2)\omega_1 \tag{25.8}$$

which is a vector quantity describing an equation of straight line with slope $-(J_1/J_2)$ and the ω_2-axis intercept of $\boldsymbol{M}/J_2$. Figure 25.13 shows a constant energy ellipse and a constant momentum line of a pair of two unequal wheels. Two unequal wheels required to store a given level of energy while simultaneously maintaining a certain momentum level would operate at either of the two intersection points, P_1 or P_2. Both points of operation are dynamically stable; hence the system can indeed operate at any one point, which is closest to the prior operating point.

If two wheels on each axis are equal, each with inertia J, then the total energy stored is

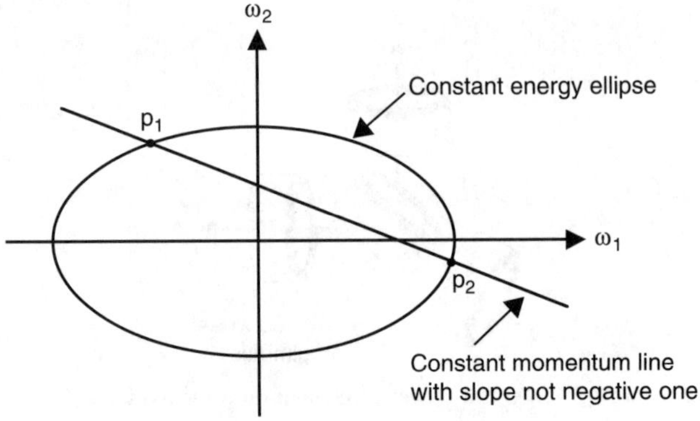

FIGURE 25.13 Energy–momentum wheel operation with two unequal wheels on each axis.

$$E = \tfrac{1}{2} J\omega_1^2 + \tfrac{1}{2} J\omega_2^2 \tag{25.9}$$

Rearranging the above gives

$$\omega_1^2 + \omega_2^2 = 2E/J \tag{25.10}$$

This is the equation of a circle with a radius of

$$\sqrt{\frac{2E}{J}}$$

The total momentum of two equal wheels on each axis is

$$\mathbf{M} = J\omega_1 + J\omega_2 = J(\omega_1 + \omega_2) \tag{25.11}$$

Rearranging the above leads to

$$\omega_2 = (\mathbf{M}/J) - \omega_1 \tag{25.12}$$

which is a straight line with slope of -1 and ω_2-axis intercept of $\boldsymbol{M}/J$. Thus, with two equal wheels on each axis, the constant energy ellipse becomes a circle and the momentum line takes the slope of -1.

25.7.2 Energy Depletion Constraints

As the spacecraft cannot discharge energy without discharging the momentum, the maximum depth of energy discharge is limited by the minimum momentum storage requirement on a given axis. Such an operating constraint is analyzed below.[14]

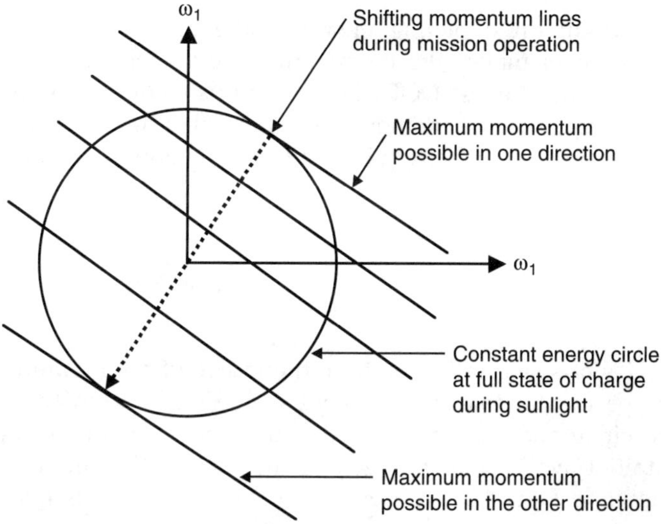

FIGURE 25.14 Energy–momentum wheel operation during sunlight with two equal wheels on each axis.

A pair of equal wheels that must maintain a given energy level, but change its momentum level to meet the attitude control requirement, must operate at any point on the circle, while changing the momentum line closer or farther from the origin. This is shown in Figure 25.14. On the other hand, if the system is required to deplete the energy without depleting the momentum, the speed of the wheels must change to contract the energy circles while maintaining the momentum line as shown in Figure 25.15. A

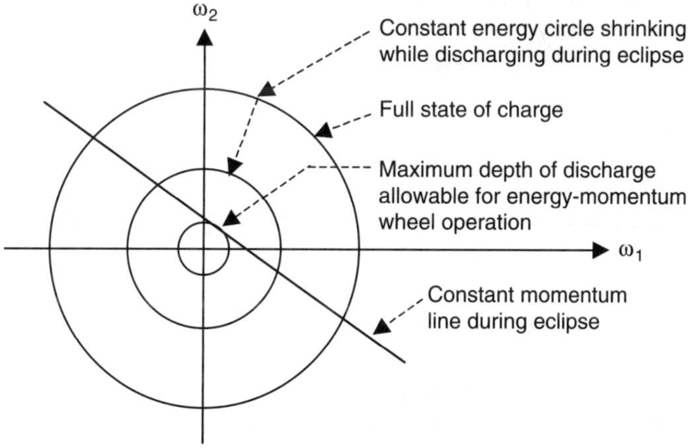

FIGURE 25.15 Energy–momentum wheel operation during eclipse with two equal wheels on each axis.

mission constraint on depleting energy while meeting the momentum storage requirement makes the momentum line tangent to the minimum energy circle at the deepest DOD. Or, the radius of the minimum energy circle must be equal to the momentum requirement at the end of eclipse. This gives the necessary requirement for energy-momentum-wheel operation as

$$\sqrt{\frac{2E_{\text{min}}}{J}} > \mathbf{M} \quad \text{or} \quad E_{\text{min}} > \frac{1}{2} J \mathbf{M}^2 \tag{25.13}$$

In all modes of spacecraft operation, the radius of the minimum energy circle must be greater than the distance of the momentum line from the origin. The circle and the line have two intersecting points of operation, with the limit of the line being tangent to the circle of the minimum energy level as shown in Figure 25.15. Hence, the maximum permissible DOD is

$$\text{DOD}_{\text{max}} = 1 - \frac{E_{\text{min}}}{E_{\text{max}}} \tag{25.14}$$

This is the upper limit on the DOD from the energy-momentum-wheel operational point of view. It is necessary that the minimum energy circle at the end of the eclipse must be no smaller than being tangent to the required momentum line on each axis. This establishes the allowable depth of energy discharge in the energy-momentum-wheel system.

25.8 Energy–Momentum Wheel Example

The EOS-AM power system is again taken as an example to estimate potential benefit for replacing the battery and the momentum wheel with IEMS.[15] The primary design parameters, which influenced the IEMS estimates summarized here are listed below.

1. The power system ratings derived in Figure 25.11 are for the total EPS, i.e., for all operating wheels combined. The ratings of each wheel, the motor-generator and the power and momentum management electronics (PMME) would depend on one of the configuration options applicable in the satellite under consideration.
2. If the flywheel is used only for the energy storage, the net zero momentum requires two wheels counter-rotating on each axis. One extra wheel is required for redundancy. Hence three wheels, each with 50% of system rating, are required.
3. In the IEMS configuration, however, one wheel is needed for the momentum control on each of the three axes, which also stores energy within its momentum limitation. One additional wheel is required for

the remaining energy storage requirement, and one more for redundancy. Hence, five wheels are the theoretical minimum number of wheels needed. Each wheel with 25% of system rating is used in the IEMS mass and size estimates. The momentum constraint is not applied in this study.

4. All wheels are assumed to be equal in size. However, it is recognized that operational considerations in some satellites may require wheels of different sizes.
5. A 20% margin is added in all component ratings on and above those derived in Figure 25.11.
6. The rotor design was considered around three major design variables: (a) the hoop stress, (b) the ultimate strength of the fiber–resin composite, and (c) the length-to-diameter ratio, which influences the vibration modes and the critical speeds. With the maximum hoop stress to ultimate strength ratio of 0.7 and the length to diameter ratio of 0.75, the rotor design speed of 100 krpm is used. This speed is considered achievable. Chen et al.[16] have reported a 100-Wh, 200 krpm flywheel on magnetic bearings, where the rotor was made of fiber–epoxy composite with the maximum operating hoop stress of 150 kpsi in the outermost fibers.
7. The hub mass is taken as 20% of the rim mass. This is the best estimate available at present.
8. The PMME provides a total interface between the motor–generator and the regulated voltage bus. It contains the charge and discharge converters, the power management (mode control) function, the momentum management function and the associated telemetry and commands.

25.8.1 Providing Launch and Ascent Power

One potential design issue with the IEMS is that the wheels may or may not be spinning at the time of launch to provide the required power during launch and accent. This issue must be resolved with the launch vehicle prime contractor. As a conservative approach here, it is assumed that the wheels cannot be spun until the PV array is exposed to the sun. At that time, the first task would be to spin-up the wheels to full speed (100% state of charge). Then onward, the system must be in energy balance in each orbit.

An effective alternative for providing the launch and ascend power is the primary battery. Among available options, the lithium carbon monofluoride (LiCFx) has high specific energy and a long flight history in launch vehicles. LiCFx cells are available in several ampere-hour ratings. Two primary batteries, each with 24 LiCFx cells of 40 Ah capacity, would provide an equivalent of full charge of the EOS-AM batteries, and at the same time be compatible with the 120-V bus. The mass of such batteries (inclusive of all battery hardware) has been estimated and included in the IEMS designs for comparison with the existing EPS.

25.8.2 Replacing Batteries and Reaction Wheels with IEMS

Based on the design parameters and assumptions described in above, the IEMS power system component mass and volumes have been estimated by Patel[15] for EOS-AM satellite and are compared with those using the existing and future batteries. Table 25.5 summarizes these results.

Various new secondary battery technologies are under development. The Jet Propulsion Laboratory recently evaluated about 45 major and minor technology development plans to increase the specific energy of batteries. The study selected lithium-ion as the most promising technology for future

Table 25.5 IEMS vs. battery mass (kg) in an EOS-AM satellite

Component	EPS with existing NiH_2 battery	EPS with future battery	EPS with IEMS
Solar array	180	180	167
Array drive electronics	6.7	6.7	6.7
Power equip battery assembly	140	76	
Hex bay battery assembly	144	77.5	
Battery power conditioner	21.5	21.5	
Sequential shunt unit	44.6	44.6	41.5
PRU[a]	46.5	46.5	46.5
Switching and fusing unit	6.5	6.5	6.5
Reaction wheel	57.5	51.7	Incl.[b]
Rotor rim			16.8
Rotor hub			3.6
Motor–generator			3.8
PMME [c]			37.3
Magnetic bearings			11.5
Magbearing electronics			5.7
Magbearing sensors			4.5
Touchdown bearings			5.7
Auxiliary power supply			4.5
Shaft (steel)			7.6
Housing			11.2
Launch battery (primary)			9.6
TOTAL MASS	647.3 kg	511 kg	390 kg
TOTAL VOLUME[d]	780 liters	550 liters	315 liters
SOLAR ARRAY AREA[d]	44.8 m^2	44.8 m^2	41.6 m^2

[a]Power regulator unit.
[b]Function included in flywheel.
[c]Power and momentum management electronics.
[d]Summary results of size and volume are given here for brevity.
(Data source: M.R. Patel, *Proc. IECEC*, p. 200-x, AIAA, 2002.)

spacecraft. The on-going NASA/JPL battery development programs are targeted to achieve the following goals in 5 to 7 years:

- Improve the specific energy from the present 50 Wh/kg to as high as 100 W/kg at the cell level.
- Increase the permissible depth of discharge in LEO satellites from the present 30% to as high as 50%.

Several parallel battery development programs are under way to meet these goals. It is difficult to predict which particular one would come close to the stated goals. Even after the predictions are achieved in the laboratory over the next 5 to 7 years, electrical power systems engineers would incorporate the results in their designs only after completing the time-consuming life tests and accumulating adequate performance data on actual flights. Experience indicates that this may take an additional 5 years.

Without predicting which particular technology would become flight-qualified in the next 5 to 7 years, it is assumed here that the batteries coming to the design stage after 5 to 7 years would have specific energy of 75 Wh/kg at the cell level, and can be operated for 10 years in LEO at 40% depth of discharge. With these improvements, the future battery mass and volume would reduce to almost one-half of their present values. Since IEMS would have to complete with these improved batteries, the EOS-AM power system estimates with the future battery technology is also included in the present study. Improvements in the reaction wheel assembly, however, are likely to be small in view of its technical maturity. Therefore, it is assumed that the mass and volume of the existing reaction wheel will shrink by 10% over the next 5 years.

25.8.3 Summary of Point Design Example

The bottom of Table 25.5 shows the benefits of using IEMS in an LEO satellite having 2500 W payload power requirement. As can be seen, the IEMS could save 40% mass, 60% volume, and 7% solar array area when compared to the present batteries. With future batteries, the IEMS could save 24% mass, 43% volume, and 7% solar array area. Although the cost benefits of using IEMS are not addressed in this study, the potential reduction in the hardware and the launch costs could be in several million dollars per spacecraft.

25.9 Electrical Machine Options

Various types of electrical machines and their suitability for use in the flywheel energy and momentum storage system are examined.

25.9.1 Synchronous Machines

Synchronous machines are further classified as: (a) excited machine (brushless version), (b) PM machines (always brushless), and (c) reluctance machines (no excitation needed). Out of these three, the reluctance machine, although the simplest and the least expensive of all, is unsuitable with a flywheel, as it works only as a motor. The flywheel system needs bi-directional energy conversion (both in the motor and the generator modes). The excited synchronous machine with slip rings and brushes in any configuration is not suitable either. The PM synchronous machine is widely used in many applications. Chen et al.[16] have reported a small 100-W, 2-pole, 15 kHz, radiation cooled, brushless, PM synchronous machine design for 200-krpm flywheel on magnetic bearings. The neodymium–iron–boron PM is common in such designs, since it offers high magnetic strength. It is also strong in compression, although weak in tension which is minimized in the design.

With f = frequency and P = number of poles, the machine speed (in rpm) is given by

$$\text{speed} = \frac{120f}{P} \tag{25.15}$$

In the discharge mode, the generated voltage and frequency magnitudes are directly proportional to the flywheel speed. In the charge mode, the motor speed is spun up by increasing the frequency and voltage such that the V/f ratio is maintained constant. Maintaining the constant V/f ratio is necessary to avoid the saturation in the magnetic core of the machine. This is an open loop speed control.

Electrodynamically, the PM synchronous machine can become unstable if the load angle exceeds its maximum permissible value determined by the synchronous and transient reactances. This can happen under sudden load swings or rapid frequency changes during acceleration and deceleration. Two such machines working in parallel for redundancy or capacity boost can also become unstable with each other under load swings and/or uneven load sharing. With such complexities in operation, the synchronous machines could be unsuitable with widely swinging loads.

When two or more machines are working in parallel, as is always needed for reliability in the space application, merely maintaining a constant V/f ratio in an open loop system is not enough. The synchronous operation among all machines is necessary. Therefore, the speed of all machines must be maintained exactly constant, otherwise they would fall out of step and lose synchronism. The synchronism among all machines working in parallel is maintained by a closed loop system, in which the shaft position is sensed and fed back to control the frequency, and hence the speed. This is achieved by exciting the winding whose magnetic axis aligns with the PM pole axis. This, in essence, makes the machines work as synchronous machines.

The power electronic converters convert the variable voltage, variable frequency a.c. power to d.c. and vice versa. The L-C filter is used between the converter and the d.c. bus to minimize the harmonics generated by the PWM converters. The 3-phase machine windings are connected in ungrounded Y, which inherently eliminates all triple harmonics (3rd, 6th, 9th, ... etc.)

The power transfer between the d.c. bus and flywheel during the charge operation can be achieved in tapered power, constant power or constant current mode. The first two may require high current in the beginning, making it necessary to use a machine of high rating for a short time. The constant current mode at three current levels, i.e., high, moderate, and low, may be more mass efficient.

The synchronous machine can be electrically represented by a source emf and series resistance and reactance as shown in Figure 25.16(a). The relation between the internally generated emf and the terminal voltage is given by the phasor diagram shown in Figure 25.16(b) and (c) for the generating and the motoring operations, respectively. From the diagram, the power transfer is derived as[17]

$$P = P_{\max} \sin \delta \tag{25.16}$$

where $P_{\max} = E_f V_t / X$, E_f = internal field excitation voltage (emf), V_t = terminal voltage, X = synchronous reactance, and δ = power angel. Physically, δ is the angle between the magnetic axes of the stator phase coil and the rotor magnet. The power transfer also depends on the terminal voltage, the internal emf, and inversely with the synchronous reactance. It can be increased by choosing a machine with low synchronous reactance, which is made of the sum of the armature reaction and the leakage reactance. An ironless machine would result in a low reactance machine.

25.9.2 Induction Machine

This is inherently a brushless a.c. machine. On the ground, it is the most widely used machine as a motor because it is simple, of low cost, and has a rugged construction. In principle, it can work as a generator under a leading power factor, but its electrodynamic stability can be a difficult issue to resolve under varying operating conditions. Moreover, it gives lower specific power. Its speed can be changed by changing the frequency or by changing the applied voltage. The latter method of speed control is very inefficient as it produces losses in the rotor at high slip frequencies. Overall, this machine is unsuitable with the flywheel in space.

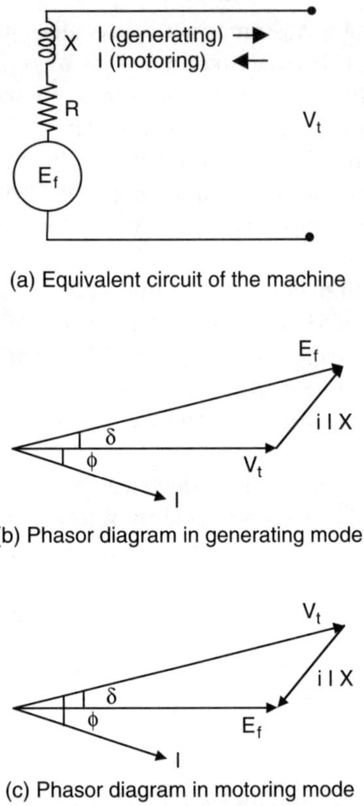

FIGURE 25.16 Synchronous machine equivalent circuit and phasor diagram.

25.9.3 Direct Current Machine

The d.c. machine has two varieties — excited and permanent magnet (with or without brushes). The speed is controlled by varying the applied voltage. The speed would change and settle where the resulting back emf plus the armature *IR* drop equals the applied terminal voltage, i.e.,

$$\text{speed} = \frac{V_t - I_a R_a}{K_e} \tag{25.17}$$

where V_t = applied voltage, I_a = armature current, R_a = armature resistance and K_e = electromagnetic constant of the machine.

The d.c. machine is inherently stable under load swings or voltage variations. With PM in the rotor, it is brushless, and has no fundamental I^2R loss in the rotor (only eddy and hysteresis losses due to parasitic flux pulsations).

Thus, the PM d.c. machine and the PM synchronous machine are the only two candidates in flywheel technology for space. These two machines look very much alike with almost the same specific power and no significant difference in efficiency or cooling requirements. Sometimes the two are not clearly distinguishable unless their control methods are examined. The synchronous machine speed can be controlled only by varying the applied frequency, whereas the d.c. machine speed is controlled by varying the applied voltage. Therefore, these two machines significantly differ in their speed control method and hence the power electronics design.

Being the heritage motor in the momentum and reaction wheels at present, the permanent magnet (PM) d.c. machine would be the lowest nonrecurring cost approach. In addition, the PM d.c. machine has certain desired performance characteristics suitable for space applications.

Since the d.c. machine speed is controlled by voltage and it has linear speed–torque–current relations, its controls and power electronics are simpler and lighter than with the synchronous machine. For this reason, the PM brushless d.c. machine is the most suitable candidate for electromechanical power conversion to and from the flywheel. The selection of the PM d.c. machine for flywheel technology in space is, therefore, a sound one for the reason given here. There are further options in the PM d.c. machine, namely the axial gap and the radial gap machines. Considering the flywheel geometry, the conventional radial gap machine is the most natural fit, which is also easily scaleable for different power levels.

25.10 Motor–Generator Design Issues

Some of the constructional features of the PM d.c. machine that may benefit the design are the ironless armature to eliminate iron losses, Litz wire to minimize the skin and proximity effects, and a composite carbon/graphite bobbin for improved heat conduction from the coil. Drawing from industry experience in designing high-frequency motors for aircraft and navy submarines and the superconducting generator design, the ironless armature of Litz wire in a graphite bobbin is a sound baseline approach. However, the technical challenge of fully developing a high speed, efficient, and entirely conduction cooled motor–generator design remains. Based on the ironless airgap armature and Litz wire construction, the estimated mass of a space worthy 1-kW, 20 kHz machine is 1.5 kg. A 5-kW, 2-pole PM machine is estimated to weigh 4 to 6 kg/kW at 3600 rpm, 2 to 3 kg/kW at 12,000 rpm and 0.4 to 0.6 kg/kW at 100,000 rpm. The scaling law for machine mass is as follows:

$$\text{mass} = \left(\frac{\text{kW}}{\text{speed}}\right)^{a} \tag{25.18}$$

where the exponent α equal to 0.5 applies to 1 to 5 kW machines, and 0.7 applies to 100 to 1000 kW machines.

25.10.1 Constant Torque Versus Constant Power Design

In the motor–generator (M–G) design, one approach is to deliver constant torque and the other is to deliver constant power. Table 25.6 shows that in the constant power design, the motor has to be rated 3.0 times that in the constant torque design. On the other hand, the charging power delivered to the flywheel at minimum speed under constant torque operation is only one third, the remaining two thirds must be diverted to the shunt dissipators. Although the constant torque approach optimizes the machine performance, it would lead to poor performance in the satellite power system as a whole. The constant power M–G design approach saves mass in the photovoltaic array several times the mass it adds in the motor-generator. Since proportionally similar savings would be seen in other missions as well, the constant power M–G design is the preferred design approach in the PV–flywheel power system for space applications.

One of the critical design issues in the motor–generators design for constant power would be the much higher (9 times) armature I^2R losses in the stator windings when operating at the minimum speed. Since the space motors must be cooled by conduction only, removing such a high level of heat may be a fundamental design issue not faced by their terrestrial counterparts. In space motors, the cooling would most probably require use of heat pipes. However, placing heat pipes in the extremely high frequency magnetic field (rotating at 100,000 rpm) would pose new design issues. For this reason, a motor–generator designed for terrestrial applications in general would not be applicable for space use. The underlying reason for this is that, in space, power not used in charging the flywheel is wasted, which is not the case in an automobile or in UPS applications on the ground. This system-level operating difference between the two applications introduces significant differences in motor–generator design.

Table 25.6 Motor–generator performance at one third speed relative to full speed

M–G performance parameter	Constant torque design	Constant power design
Voltage (%)	33	33
Current (%)	100	300
Power (%)	33	100
Armature loss	100	900
Eddy loss (%)	10	10
M–G rating (%)	100	300

25.11 Magnetic Bearings Design Issues

Conventional roller bearings in high-speed flywheels result in high transmitted vibrations and high drag torque resulting in high self-discharge rates. The magnetic bearings provide contactless suspension with far reduced drag torque and vibrations. They have flown on the Russian space station MIR for years. However, the design challenges with magnetic bearings are in the sensors and complex control system.

Actively controlled bearings require complex controls systems. The control current is proportional to the square of the gap. Passive magnetic bearings, on the other hand, give high system efficiency and reliability, but it typically gives lower stiffness and lower damping. A typical passive bearing design for 60-krpm operations may have radial stiffness of 1800 lb/inch and the modal damping of 5%. Earnshaw's theorem states that in any inverse square law system following the Laplace equation, such as the magnetic field, it is impossible to have a completely 3-dimensional stable passive magnetic bearing using only the forces of a static field of permanent magnets. The theorem, however, does not apply in active feedback systems, time-varying fields, ferrous liquids, and superconductors. The active feedback control system is the most commonly used method for stabilizing the magnetically suspended bearings. The operating speed, hence the specific energy of the flywheel, increases with the ratio of the polar to the transverse moment of inertia (I_p/I_t). The ratio also produces larger changes in the momentum for more effective attitude control. For a stable active bearing design, the I_p/I_t ratio should be preferably below 0.8 or above 1.2, although its value of about 1.0 has been used in a 60 krpm design.[7]

Several concepts for designing magnetic bearings are possible. The relative merit of any one configuration over another can be compared only after their designs are developed in sufficient details and traded for optimizing the performance, cost, and technical and schedule risks. However, any configuration that is selected must have the following five performance attributes:

- Low eddy current and hysteresis losses (magnetic drag)
- Sufficient restoring forces in all five degrees of freedom
- Spring constants, both in translation and torsion, that assure the mechanical stability and avoid resonance at the spacecraft excitation frequencies
- Electrical control loop stability
- Sufficient dynamic damping to attenuate disturbances

The following paragraphs compare a few design concepts offered on the market at present.

25.11.1 Low-Speed Bearings

Teldix, in Germany, and others have been offering magnetic bearings in their momentum wheels for many years. The speed range of these wheels is 3000 to 6000 rpm. In their design, the magnetic flux travels through the pole shoes on the stator and the magnetic feedback ring on the rotor. The reluctance lock between the pole shoes and the magnetic feedback ring provides the vertical restraint. The control coils are segmented, hence they would suffer excessive magnetic drag at high speed, making them unsuitable for the flywheel energy storage.

25.11.2 High-Speed Bearings

Among homopolar and other design concepts, one shown in Figure 25.4 uses a permanent magnet (PM) for passive restrain and electrical coils for active control. However, since the coils are segmented, it would create pulsating flux in the shaft as it rotates, thus generating magnetic drag. At speeds above 60 krpm, drag loss would limit the speed. One can reduce the drag by using thin laminations, but that would introduce noise and vibrations. This issue needs to be satisfactorily resolved for speeds approaching 100 krpm.

There are several other configurations of magnetic bearing commercially available which may be suitable for high-speed operation. Regardless of the configuration chosen, detailed electro-mechanical design for a magnetic bearing would need the following tasks to be performed:

1. The magnetic bearing must be restrained in five degrees of freedom (all except the one we want to rotate freely). The restoring forces and spring constants in the five degrees of freedom (three translational and two rotational) are computed from the finite element electromagnetic flux plots and using either the pressure approach and/or the energy approach. In free space, the magnetic pressure (force per unit surface) and the magnetic energy stored per unit volume are given by is $\frac{1}{2}B^2/\mu_o$. The change in energy per unit displacement is force, and the force per unit displacement is the spring constant. This would give a translational spring constant for a linear displacement, and a torsional spring constant for a rotational displacement. A positive spring constant would make the system naturally stable in that direction, while a negative spring constant would need a feedback control to stabilize the system.
2. The above spring constants, along with the dynamic damping and the mass and angular inertia constants, are used to develop a complete electro-mechanical model of the rotor-bearing system.

3. The resonance frequencies (square root of the spring constant over inertia) are computed and kept away from the system excitation frequencies.
4. The feedback control system model is then developed to study the dynamic electro-mechanical response and stability. The gains and damping are adjusted to fine-tune the design performance as required.
5. The finite element flux plots are used to calculate the magnetic drag at rated speed. There would still be significant stray (parasitic) loss at high speed that would defy all analyses. The magnetic loss components are therefore estimated using engineering judgment at the design stage, and then from carefully planned tests to separate eddy loss (varying with speed squared) and hysteresis loss (varying with speed) for use in subsequent designs.

25.12 Flywheel System Controller

The rotating flywheel has six degrees of freedom, the x and y positions at two axial locations which keep the shaft straight, one in the Z direction at either end, and one angular position. The angular position is unrestrained. That leaves five degrees of freedom that must be restrained by the bearings — mechanical or magnetic. The magnetic bearing works on magnetic forces — attraction and/or repulsion — created by passive permanent magnets or active electromagnets with feedback control systems. A typical clearance gap in the radial direction is 0.25 to 0.50 mm, and is maintained within 0.025 mm. The shaft position in the five degrees must be sensed and the errors amplified and fed back to the system for corrective response in terms of the coil currents in the electromagnets. Figure 25.6 earlier showed the location of the five restrains and the corresponding sensors.[8]

Eddy current sensors are generally used in magnetic bearings. The theoretical basis for these sensors is Faraday's law of electromagnetic induction. The induced voltage in any coil is proportional to the rate of change of the flux through the coil. A change in the shaft position in any direction causes a proportional change in the magnetic flux in the gap in that direction, which translates into the induced voltage in the eddy current sensor. The sensitivity of the sensor is therefore measured in terms of the induced voltage per micrometer of the position change. For greater sensitivity, the sensors are operated at high frequency. The eddy current sensor, therefore, has three major components: a radio frequency oscillator to generates a signal, the sensor that radiates the signal through the probe tip, and the demodulation circuit to convert the returned signal into usable form. A conducting material placed in the proximity of the sensor tip would generates eddy currents, which opposes the flux from the sensor tip in accordance with the Lenz law. The resulting loss of magnetic strength in the return signal is measured, demodulated, and converted into a voltage signal. This signal is proportional to the gap between the sensor tip and the conducting surface, which is fixed to the shaft. Such sensors are available commercially with a sensitivity of 10 to 15 mV/μm. The major design issues are the control system and the EMI cross-talk between the sensors, since some sensors are located close to each other.

A simplified schematic of the control systems is shown in Figure 25.17[18] The bearing control code generates command signal to allow shaft levitation. These command signals are converted to drive currents by the PWM amplifier, and passed on to the control system, which alters the shaft position. The eddy current sensor signal is processed by the signal conditioning system and fed back to the bearing controller. The mode controller controls the pulsing of the converter switches as required for operating the systems in the charge, discharge or idle mode. Truong, Wolff, and Dravid[18] have developed the modeling technique for a flywheel, a 3-phase PM synchronous motor-generator and a 3-phase bi-directional half-bridge converters (Figure 25.18). The operation is described below starting from the minimum DOD. The controllers set the system in the charge mode. Initially the motors take in the maximum charging current within the set limit. As the motor speed rises, the counter emf builds up and the current drops. When the speed reaches the maximum value, the system enters idle mode with zero power transfer. The wheel maintains its speed until commanded into the discharge mode. The synchronous machines as the generator feeds the bus with a constant current set by the maximum limit, or using a constant power criteria. The latter is perhaps a better approach.

Significant work has been conducted in the flywheel test facility at NASA Glenn. A successful operation with speed and position feedback control of the unit has been demonstrated.[19] A 4-pole permanent magnet synchronous machine suspended on magnetic bearings was controlled with a field orientation algorithm.

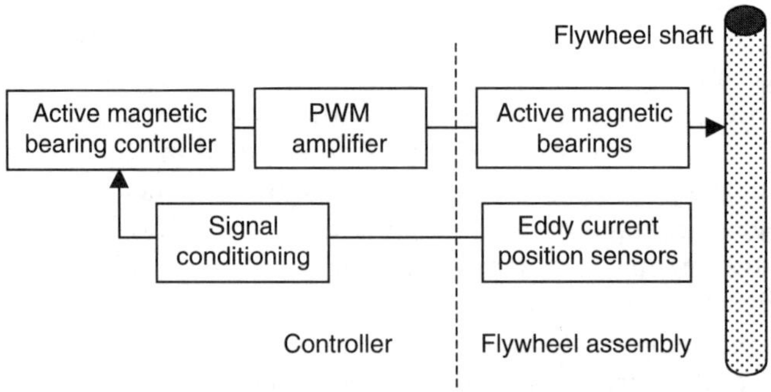

FIGURE 25.17 A simplified control system for active magnetic bearings.

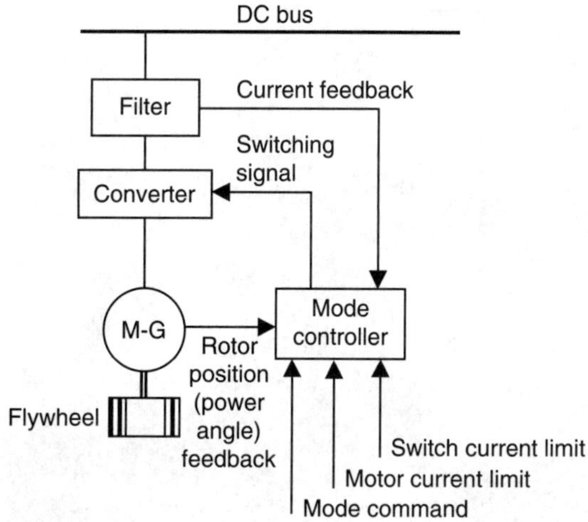

FIGURE 25.18 Flywheel control system configuration.

25.13 NASA Flywheel Program

Flywheel momentum storage has been in use for spacecraft navigation and attitude control system since the beginning of the space program. A prototype flywheel energy storage combined with attitude control system had been built and tested at Draper Laboratory at 30,000 rpm in the early 1990s (Figure 25.19). The concept is now under active developments at NASA Glenn for replacing the electrochemical battery in spacecraft electrical power systems. In the current NASA design, flywheel operating speeds vary from 60,000 rpm maximum to 20,000 rpm minimum, representing 90% DOD. Flywheel inertia is set by the energy storage requirement, and safety and reliability considerations set the maximum operating speed. The friction and windage losses are negligible. The voltage magnitude and frequency are linearly proportional to the flywheel speed. The phase current magnitude and the phase angle determine the magnitude and direction of the power transfer between the flywheel and the bus. The actual voltage and power controls are implemented by the PWM power converters. The switching action is affected by the power angle of the rotor. Capacitive filters are used on the bus side of the converters. A series inductive filter is also used to reduce the harmonic current in the phase coils.

The projected goals of the NASA and AFRL programs[13] at Honeywell and Lockheed Martin are:

- Specific energy >200 Wh/kg
- Cycle life >75,000

FIGURE 25.19 Energy storage and attitude control flywheel system for spacecraft under tests. (©The Charles Stark Draper Laboratory, Inc. All rights reserved. Reprinted with permission.)

- Round trip energy efficiency >90%
- Systems cost reduction >25%

NASA Glenn has also been developing flywheel energy storage systems (FESSs) for the International Space Station.[20] At present, the ISS power system consists of solar arrays and NiH_2 batteries. The estimated life of a battery is 5 years. However, the design life of the ISS is 15 years. NASA plans to replace one battery along with the battery charge/discharge converters with a flight-qualified FEES in 2006. The flywheel will replace the battery at the same time, thus requiring no modifications to the ISS. The projected life of the FESS is 15 years (92,000 C/D cycles and 1100 startup and shut-down cycles), thus requiring no replacements during the ISS mission life. NASA estimates the life cycle cost savings of about 150 million dollars.

The design features of the ISS flywheel unit (Figure 25.20) are:[11,20]

- Operating speed 53,000 rpm to 41,500 rpm, which results in a tip speed of 916 m/s and 39% DOD (versus 35% for the NiH_2 battery. The laminar stress at the rotor-hub interface is not an issue
- Contingency DOD 89% (53,000 rpm to 17,800 rpm) for 30 cycles over the life
- Specific energy of 27 Wh/kg versus 10 Wh/kg with the NiH_2 battery
- Radial and axial magnetic bearings in 10^{-6} torr vacuum and mechanical touchdown bearing
- Two counter-rotating rotors made of carbon fiber composite with titanium shaft
- Eclipse power 4.1 kW, peak power 5.5 kW (comparable with the NiH_2 battery)
- Energy storage 5.5 kWh, versus 4.6 kWh for the NiH_2 battery
- Round trip energy efficiency 83% versus 65% for the NiH_2 battery

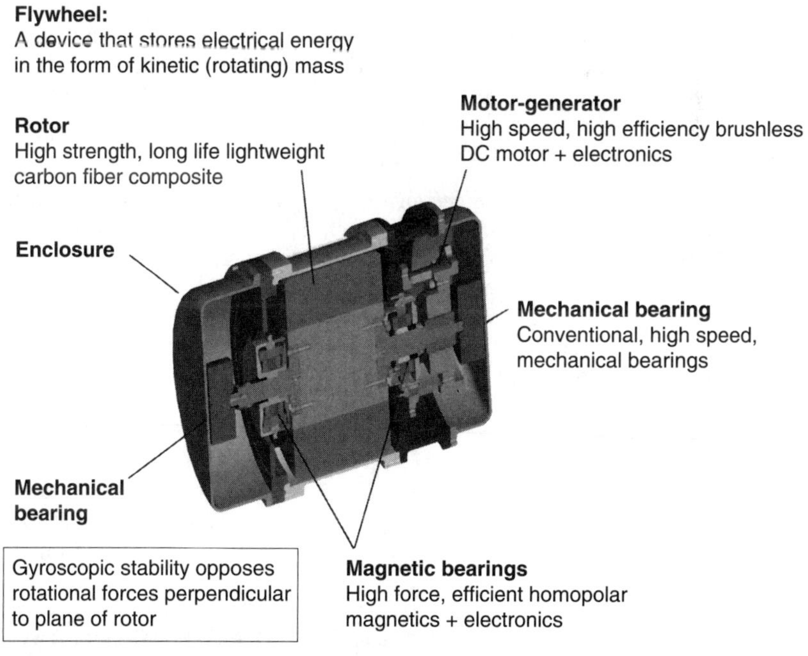

FIGURE 25.20 Cross section of NASA flywheel module.
(Source: E.B. Gietl et al. NASA Glenn TM 2000-210209.)

- Life expectancy 15 years versus 5 years for the battery
- Wheel size: 33 cm diameter × 48 cm long with 5 kWh capacity, initially in compression (near zero stress in operation)
- Electrical machine: 3-kW, 4-pole, 3-phase, PM synchronous machine
- Protection: RPC with less than 6 V dead-facing for crew safety
- Each FESS assembly: 0.9 m wide × 0.85 m long × 0.625 m high and 205 kg mass
- Each wheel assembly with signature for failure detection
- Life tested to 60 years (4 × operating life) as required by the manned space flight fracture design for critical components
- Ratio of polar to transverse moment of inertia I_p/I_t of 0.43
- Operating temperature 22 to 82°C, actively cooled
- Permanent magnet with homopolar magnetic bearings
- Three-phase a.c. permanent magnet motor–generator
- Housing to support the flywheel and protect from orbital debris (not hermetically sealed, the space naturally provides vacuum for the rotor)
- Wheel not spinning during launch
- Tested up to 60,000 rpm with safety enclosure containing water in the test bed.

In fully developed and space-qualified hardware, the specific energy of the flywheel system is estimated to be around 5 times that of the NiH_2 battery.

References

1. NASA/*Aerospace flywheel development program document*, Web site http://space-power.grc.nasa.gov/ppo/project/flywheel/May 15, 2001.
2. Harris, C., Flywheels for spacecraft energy storage and attitude control, in *Flywheel Energy Storage Workshop*, Oak Ridge, TN, National Technical Information Service, Washington, D.C., November 1995, Conference Records 9510242.
3. Wehmer, J., Spacecraft flywheel system requirements, in *Flywheel Energy Storage Workshop*, Oak Ridge, TN, National Technical Information Service, Washington, D.C., November 1995. Conference Records 9510242.
4. Uherka, K., Experimental test facility for flywheel energy storage utilizing superconducting magnetic bearings at the Argonne National laboratory, in *Flywheel Energy Storage Workshop*, Oak Ridge, TN, National Technical Information Service, Washington, D.C., November 1995, Conference Records 9510242.
5. Maass, D., Development in polar woven/RTM composite flywheel rotors, in *Flywheel Energy Storage Workshop*, Oak Ridge National laboratory, TN, National Technical Information Service, Washington, D.C., November 1995, Conference Records 9510242.

6. Siebert, M. et al. A passive magnetic bearing flywheel, in *Proceedings of the 36th Intersociety Energy Conversion Engineering Conference*, ASME, 2001, Paper No. 100-x.
7. Palazzolo, A. R., Beach, R., et al. Magnetic Suspension for flywheel batteries, in *Proceedings of the 36th Intersociety Energy Conversion Engineering Conference, ASME, 2001,* Vol. 1, pp. 97–103.
8. Dever, T.P., Palazzolo, A.B., Thomas, E.M., and Jansen, R.H., Evaluation and Implementation of Eddy Current Position Sensors in Magnetically Suspended Flywheel Systems, in *Proceedings of the 36th Intersociety Energy Conversion Engineering Conference*, ASME, 2001, Vol. I, pp. 689–694.
9. Harris, C., Flywheels for spacecraft energy storage and attitude control, in *Flywheel Energy Storage Workshop*, Oak Ridge, TN, National Technical Information Service, Washington, D.C., November 1999.
10. Aerospace Flywheel Workshop Records, U.S. Air Force Research Laboratory and NASA Glenn Research Center, Cleveland, OH, October 1999.
11. Thomson, R., Flywheel energy storage system in *Proceedings of the 35th Intersociety Energy Conversion Engineering Conference*, AIAA, 2002, Paper No. 100-x.
12. Patel, M.R., Flywheel energy storage for spacecraft power system, in *Proceedings of the 34th Intersociety Energy Conversion Engineering Conference*, SAE, 1999, Paper Number 183.
13. McLallin, K.L., Fausz, J., Jansoen, R.H., and Bauer, R.D., Aerospace flywheel development for IPACS applications, in *Proceedings of the 36th Intersociety Energy Conversion Engineering Conference*, ASME, 2001, vol. I, pp. 105–110.
14. Patel, M.R., Energy-momentum wheel for satellite power and attitude control systems, in *Proceedings of the 35th Intersociety Energy Conversion Engineering Conference*, AIAA, 2000, Paper No. 200-x.
15. Patel, M.R., Integrated energy-momentum wheels for spacecraft, in *Proceedings of the 35th Intersociety Energy Conversion Engineering Conference*, AIAA, 2002. Paper No. 200-x.
16. Chen, H., Walter, T., and Lee, N., Recent development of magnetic bearings for flywheel energy storage, in *Proceedings of the 1st International Energy Conversion Engineering Conference*, AIAA, 2003. Paper No. 6105.
17. Say, M.G., *Alternating Current Machines,* John Wiley & Sons, New York, 1983.
18. Truong, L.V., Wolff, F.J., and Dravid, N.V., Simulation of flywheel energy storage system controls, in *Proceedings of the 36th Intersociety Energy Conversion Engineering Conference*, ASME, 2001, Paper No. 33-38.
19. Kenny, B. H. et al. Advanced motor control test facility for NASA GRC flywheel energy storage systems technology developments unit, in *Proceedings of the 36th Intersociety Energy Conversion Engineering Conference*, ASME, 2001, vol. 1, pp. 91–96.
20. Grantler, J.H., Wiegmann, B.M., and Wagner, R., Flywheel energy storage system for the International Space Station, in *Proceedings of the 36th Intersociety Energy Conversion Engineering Conference,* ASME, 2001, Vol. I, pp. 263–267.

Chapter 26

Superconductors in Space

26.1 Introduction

This chapter deals with a technology that is fully developed on the ground, and naturally fits even better in the low temperature environment of space. Working at the liquid nitrogen temperature prevailing in space, the newly discovered high-temperature superconductors are expected to make rapid inroads into a variety of space applications. For example, it has a high potential for storing large amounts of energy for multi-megawatt burst power systems for space defense. While some barriers are yet to be overcome, these materials have enormous potential for new space and ground-based applications, or merely to expand proven technologies that currently use liquid helium superconductors. Cryogens carried on board for propulsion may be judiciously routed to augment the cooling of new superconductors at selected places on the spacecraft platform.

The energy density in the magnetic field is much higher than in an electrical field. The air can support a uniform E-field of 3.1 kV/mm in a standard atmosphere. Most liquids break down at 10 kV/mm. A liquid-filled capacitor operating at 3.5 kV/mm has a stored energy density of $U = \frac{1}{2}\varepsilon E^2 = \frac{1}{2}(2 \times 8.88 \times 10^{-12}) \times (3.5 \times 10^6) = 108\ \mathrm{J/m^3}$ between the electrode plates. The energy storage density in magnetic field is given by $U = \frac{1}{2}BH = \frac{1}{2}B^2/\mu o$. An air-cooled copper coil with a current density of $1.5\ \mathrm{A/mm^2}$ producing a 2 tesla B-field would store $U = \frac{1}{2} \times 2^2/(4\pi \times 10^{-7}) = 1.6\ \mathrm{MJ/m^3}$ in the hollow of the coil. This is four orders of magnitude higher than the energy density in the E-field. A superconducting coil carrying much higher current density would result in an even higher energy density in the square proportion of the magnetic field density. Commercial superconducting magnets are available at present to produce a 5 to 15-T field over 1/100 to 1 liter volume for a variety of applications. They yield an energy density of $40\ \mathrm{MJ/m^3}$ at 10 T and $90\ \mathrm{MJ/m^3}$ at 15 T.

26.2 Magnetic Energy Storage

The energy stored in the magnetic field of a coil in air or in free space is given by

$$U = \tfrac{1}{2}B^2/\mu$$

in joules/m^3, or

$$U = \tfrac{1}{2} I^2 L \tag{26.1}$$

in joules, where B = magnetic field density produced by the coil (T), μ = magnetic permeability of free space = $4\pi \times 10^{-7}$ H/m, I = current in the coil (A), and L = inductance of the coil (H).

The relation between the coil current I and the voltage V is

$$V = RI + L\frac{\mathrm{d}I}{\mathrm{d}t} \tag{26.2}$$

where R and L are the coil resistance and inductance, respectively. For steady-state energy storage at constant d.c. current, the second term in Equation 26.2 becomes zero. The voltage required to circulate a d.c. current is then $V = RI$. The resistance of the coil is temperature dependent. The electrical resistance of most conductors decreases with decreasing temperature as shown in Figure 26.1. In certain conductors, as the temperature is reduced to the cryogenic range, the resistance abruptly drops to precise zero at some critical temperature T_c. Below this temperature, no voltage is required to circulate current in the coil, and the coil terminals can be shorted to continue free circulation of the current. The circuit time constant L/R is now infinite, meaning that the current will continue to flow in the coil indefinitely. The coil is said to have attained the superconducting state with zero resistance. The energy in the coil then *freezes,* and remains stored indefinitely.

The superconductor requires a supercool operating temperature to attain the critical superconducting state. Early superconductors operating around 4 K were required to be immersed in liquid helium at almost absolute zero. Recently discovered materials that become superconducting at much higher

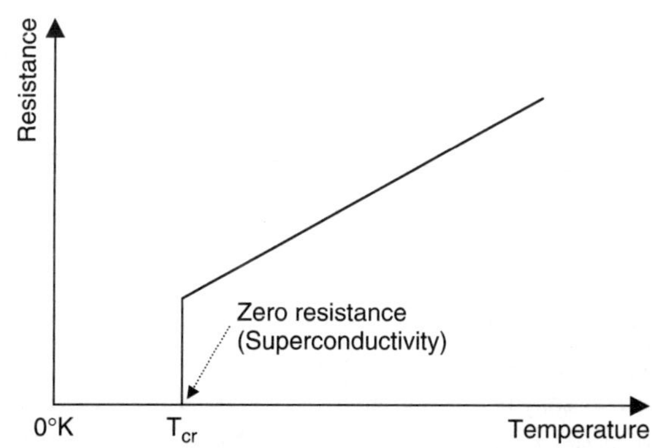

FIGURE 26.1 Conductor resistance versus temperature approaching absolute zero.

temperatures are leading to new technological breakthroughs. They can be operated at liquid nitrogen temperature (80 K), or even at temperatures exceeding 100 K. Raising the critical temperature of new materials to 100 K allows the use of liquid nitrogen as the coolant. That has a number of advantages, such as:

- Liquid nitrogen costs much less than liquid helium
- It has greater heat capacity, which means less is needed
- It can be transported in a portable container, and
- It has an ideal temperature range for optimum operation of semiconductor devices, making hybrids circuits of semiconductors and superconductors attractive

Although the superconducting phenomenon was discovered decades ago, industry interest in developing practical applications started in the early 1970s. In the USA, the pioneering work was done in this field by the General Electric Company, the Westinghouse Research Center and the University of Wisconsin. During the 1980s, a grid-connected 8 kWh superconducting energy storage system was built with funding from the Department of Energy, and was operated by the Bonneville Power Administration in Portland, Oregon. The system demonstrated over one million charge/discharge cycles, meeting its electrical, magnetic and structural performance goals. Conceptual designs of large superconducting energy storage systems up to 5000 MWh energy for utility applications have been developed in the past.

On the ground, the superconducting coil is suitable for long-term energy storage. It requires no electro-mechanical conversion, hence has high efficiency. Coils that can store 4 to 6 MJ are commercially available in low volume by American Superconductors. The Wisconsin Power Company has a large superconducting ring used for stabilizing the grid. High-temperature superconductors are expected to further accelerate the following ongoing research and development.

Utility load leveling by storing electrical energy at night for use at times of peak demand, eliminating brownouts and reducing the amount of generating capacity needed. Magnetic bearings and levitating trains suspended on magnetic fields could move from prototype to commercial use. In medicine, diagnostic machinery that uses superconductive equipment to probe the body's soft tissue could become more sensitive and less expensive.

Highly efficient generators and motors developed by companies like General Electric, Westinghouse, Siemens, and American Superconductors may become commercially viable products. The superconducting generator gets rid of all magnetic steel, giving much higher specific power by 4- to 5-fold (excluding the cryo overhead). Toshiba Corporation of Japan and

GEC-Alsthom along with Electricité de France are actively persuing developments in this field.[1,2]

Improved superconductors may be used in computers to replace the wires between transistors and chips. This could increase computing speed while reducing the amount of power that computers dissipate in heat. Superconducting interconnects between semiconducting devices could increase circuit speed as devices become smaller. Critical current densities around $1\,MA/cm^2$ are needed for interconnects between devices on chips, and $100{,}000\,A/cm^2$ densities are necessary on printed circuit boards between chips.

In space, superconductive switches — being very sensitive to certain kinds of radiation — could be used by the military to locate enemy satellites by detecting minute amounts of infrared radiation. The superconducting coil is also a potential candidate for energy storage for space-based missile defense weapons to produce brief pulses of extremely high power needed to drive antimissile lasers, particle beams, and electromagnetically propelled canons. The energy storage coil could be switched to a discharge mode in 30 ms, compared to 15 min to prepare a conventional rotating generator. The defense department has also funded programs at the utility companies to develop a large superconducting energy storage ring. In normal use, it would store energy generated at low cost during off-peak hours for their own use during peak hours. However, the Pentagon could turn to them in the event of a missile attack and instantly divert the stored energy to the defense needs of the moment.

Superconducting converter technology is also being developed for Air Force applications. It is for either the d.c.–d.c. converter or a.c.–d.c. converter depending on the power source. This technology results in very high conversion efficiency of about 99%, coupled with high specific power, around 1 kW/kg for d.c.–d.c. converters and 5 kW/kg for a.c.–d.c. converters in large power ratings typical of defense weapons. These power densities are about double the value that can be achieved with conventional designs.

And, of course, superconducting energy storage may one day be considered as a replacement for the battery, but that is some time in the future.

26.3 Critical *J*, *B*, and *T*

Three parameters, namely, the temperature, current density, and the magnetic field intensity jointly determine when a superconductor attains the precisely zero resistance. The critical value of each parameter — marking the transition to superconductivity — varies with changes in the other two. For example, as the temperature and magnetic field increase, the critical current density declines. Hence, superconductivity thrives below but vanishes above a three-dimensional critical surface, which is unique to each superconductor. Among the early superconductors, niobium titanium (NbTi) and niobium tin (NbSn) have been extensively used. Both have the critical temperature around 9 K, which requires liquid helium cooling to 4 K to allow some margin. The maximum flux density of around 9 T can be produced using NbTi at 4 K, and around 16 T using NbSn at 4 K. The critical current density is $1000\,A/mm^2$ at 10 T, both being orders of magnitude higher than those in the conventional copper coil magnets ($5\,A/mm^2$). NbSn is mechanically weak and brittle compared to NbTi, posing some difficulty in manufacturing.

The 1986 discovery of high-temperature superconductors has made three types of new superconducting materials available now, all made from bismuth or yttrium cuprate compounds. Current density and field requirements for various superconductors are shown in Figure 26.2. In the case of yttrium barium copper oxide, the critical surfaces for thin film and

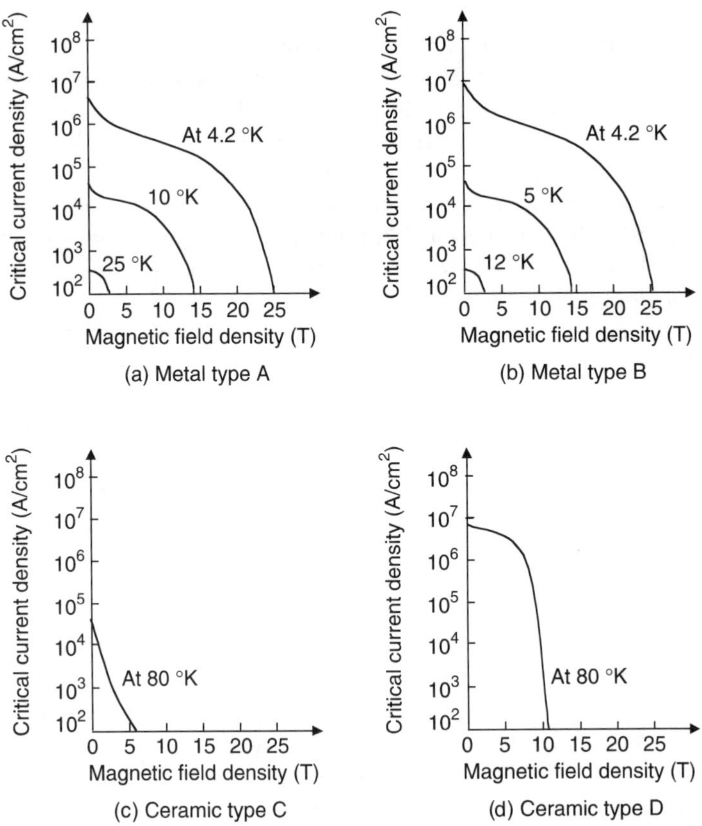

FIGURE 26.2 Critical current density versus magnetic field density and temperature for various superconductors.

bulk material made with AT&T's melt-textured growth process extend past 80 K, the liquifying temperature of nitrogen. The critical surfaces of the earlier superconductors, niobium titanium and niobium tin, are smaller, which force them to operate near 4 K, the liquifying temperature of helium. When metals (A and B), and ceramics (C and D) operating at 4 K and 80 K, respectively, are compared for current density and field relationships, the melt-textured bulk ceramic (C) is inferior, but if normally processed, would barely show up on the graph.

The superconducting coil is wound with wires made of numerous superconducting filaments in a copper or aluminum matrix. Copper and aluminum have lower resistance than the new materials in their non-superconducting state. Thus, if the critical current were accidentally exceeded in the superconductor, electrical current would switch from the ceramic to the surrounding metal matrix, reducing ohmic heating. The goal of the present development programs funded by the DoE and DoD are to produce such wires with engineering current carrying capacity of 1000 A/ mm^2 based on the entire cross sectional area of the wire, including the copper matrix and the insulation. This is about 70 to 100 times that of the traditional coil in standard atmosphere. The manufacturing process includes the coating of noble metals like silver to turn brittle high-temperature superconducting ceramics into flexible wires. It applies not only to the powder-based wires made of a bismuth strontium compound, but also to the next generation of wires using yttrium barium copper oxide, which many companies are developing in the USA, Europe, and Japan. The current density and magnetic field requirement for various superconducting applications are shown in Figure 26.3.

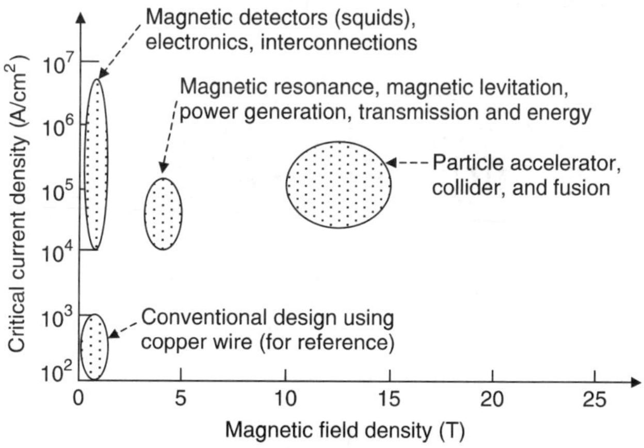

FIGURE 26.3 Current density and magnetic field requirement for various superconducting applications.

26.4 Magnet Coil Design

In spacecraft, a strong magnetic field emanating from a solenoid coil would interfere with the attitude control and other equipment sensitive to the magnetic field. The toroidal coil design eliminates this problem, as the flux is self-confined to a closed ring-shape volume as shown in Figure 26.4. The toroid with major radius R and the number of turns N carrying current I produce an internal magnetic field $B_\phi = \mu_o NI/(2\pi R)$, which varies inversely with the major radius. The circular current produces hoop stress in the conductor, and the circumferential current produces repulsive force between the two sides of the toroid. The resulting mechanical stresses vary around the circumference, thus causing strong hoop and bending stresses. The bending stress needs a center support structure. Once it is supported, the remaining force produces a pure hoop stress, which is internally supported by the tensile strength of the conductor. The superconducting coil is generally manufactured with high pre-stress, so that cooling to the operating cryogenic temperature does not produce damaging stress. However, mechanical creep during room temperature storage and high temperature excursions during forced warm up are of concern.

26.5 System Configuration

A typical superconducting energy storage system is shown in Figure 26.5. The coil is charged by an a.c.–d.c. converter in the power supply. Once fully charged with the required current and cooled below the critical temperature, a short is placed, thus making the solenoid coil a permanent magnet. The converter continues providing the low voltage needed to overcome losses in the room temperature parts of the circuit components. This keeps the constant d.c. current flowing (frozen) in the superconducting coil. In the storage mode, the current is circulated through the normally closed switch.

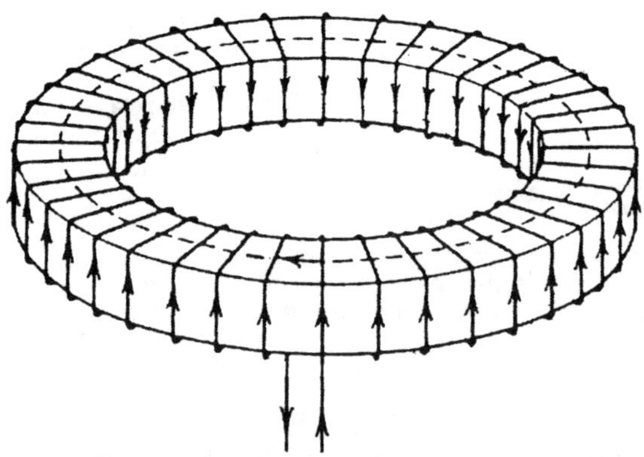

FIGURE 26.4 Self-contained magnetic field in toroidal coil.

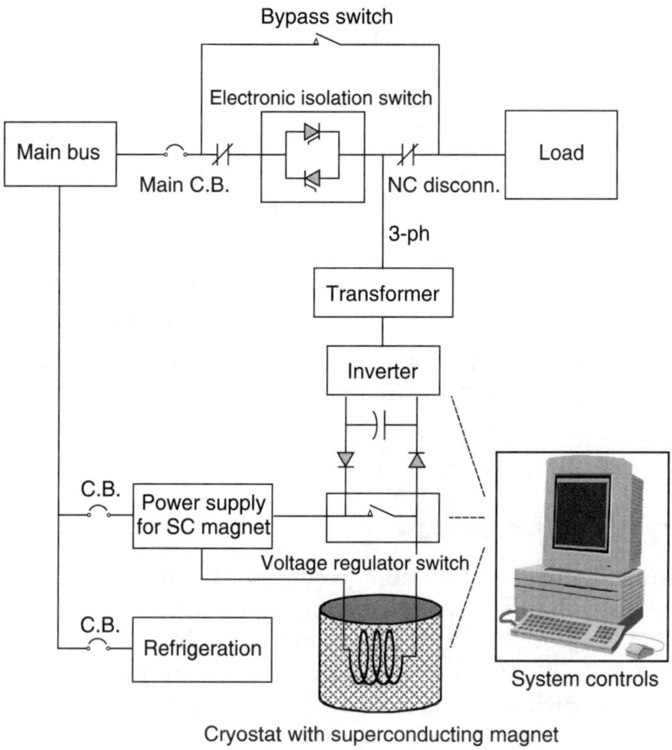

FIGURE 26.5 A superconducting energy storage system configuration.

The power supply itself for charging can be a superconducting d.c. generator, such as a Faraday disk.

The system controller shown in the figure has three main functions:

- Control the solid state isolation switch
- Monitor the load voltage and current
- Interface with the voltage regulator that controls the d.c. power flow to and from the coil

If the system controller senses the line voltage dropping, it interprets that the system is incapable of meeting the load demand. The switch in the voltage regulator opens in less than 1 ms. The current from the coil now flows into the capacitor bank until the system voltage recovers to the rated level. The capacitor power is fed directly to the load or after inverting into a.c. if necessary. The bus voltage drops as the capacitor energy is depleted. The switch opens again, and the process continues to supply energy to the load continually. The system is sized to store sufficient energy to power the load for specified duration.

Superconducting energy storage has several advantages over other technologies:

- The round trip energy efficiency of the charge/discharge cycle is high around 95%. This is higher than that attainable by any other technology
- Much greater charge/discharge cycle life
- Much longer life, up to about 30 years
- The charge and discharge times can be extremely short, making it attractive for supplying large power for short time if needed
- Have no moving parts in the main system, except in the refrigeration components, hence high reliability

26.6 Cryogenic Temperature

In a superconducting energy storage system, the main cost is to keep the coil below the critical superconducting temperature.

The NbTi and NbTi alloys are stable at high *B* and high *J* without quenching (i.e. losing superconductivity). It usually operates in a 4 K bath of liquid helium, which has a boiling point of 4 K at one atmospheric pressure. At 4 K, cooling becomes difficult. Recovering 1 W of heat loss from 4 K to room temperature cost about 500 W of refrigeration power, even at perfect Carnot cycle efficiency of $(T_h - T_c)/T_h$. Therefore, heat leakage from the leads connected to the external circuit must be controlled by proper thermal design, as they determine the capital and running cost of the cryogenic equipment.

On the other hand, room temperature magnets lose more power in the magnet itself, and little in the cooling power. The room temperature power in conventional magnet amounts to roughly 100 times that for the superconducting magnet's room temperature power. Thus, the trade is highly favorable even for the liquid helium superconducting magnet. The new superconductors having critical temperatures around 100 K are cooled by liquid nitrogen, which needs orders of magnitude less refrigeration power. As a result, numerous programs around the world have started to develop new commercial applications using liquid nitrogen cooled superconductors.

A copper matrix avoids local thermal transient heating and subsequent quenching. Also, when the coil is grossly quenched, it provides an additional path for the normal currents to flow without excessive heating. During quenching, the stored magnetic energy is converted into heat. Generally, quenching starts from a small local hot spot due to friction under vibration or any other reason. The spot may become over-heated more than the bulk surroundings. The hot spot temperature can be controlled by rushing coolant to avoid local burning. If the coil is quenched, a large potential of hundreds, or even thousand, of volts is induced in the coil, which may cause arcing between the turns. Switching an external dump resistor or a short-circuited coil across the magnet to inductively transfer part of the stored energy can provide the quenching protection.

26.7 Cryogenic Behavior of Components

Semiconducting devices, inductors, transformers, capacitors, resistors, and structural materials used in the power system components behaves differently at cryogenic temperatures around 100 K (near liquid nitrogen) than at the conventional temperature around 100°C.

26.7.1 Semiconductors

The electrical behavior of a semiconductor device is dominated by several key parameters, as follows.

26.7.1.1 Carrier mobility

The ease with which conduction electrons and holes move through the semiconductor crystal significantly varies at low temperatures. The carrier mobility increases, peaking between 100 K and 200 K depending on the doping concentration, and is typically a factor of 2 or 3 times that at the room temperature.

26.7.1.2 Number of dopant atoms ionized

The percentage of dopant atoms ionized and contributing to the conduction process begins to decrease at approximately 125 K. At 80 K the effective dopant concentration is lower and some conduction processes are affected.

26.7.1.3 Minority carrier lifetime

The probability of a trapping center to capture both a hole and a conduction electron increases, thus reducing the minority carrier lifetime. For all semiconductor devices, the leakage current, being thermally induced, is an exponential function of temperature, and therefore decreases rapidly as the temperature is lowered. Since the leakage current and the breakdown voltage vary in a similar fashion, the breakdown voltage also decreases as temperature is reduced, although not as rapidly. The reduction is relatively small, about 10% at 80 K.

26.7.1.4 PN junction diode

As the temperature is reduced, the bulk resistance due to the increase in mobility decreases, but it also increases the offset voltage. At 80 K, the offset voltage is approximately double the room temperature value (1.4 V versus 0.7 V). The reverse recovery time is reduced at lower temperature because of the lower minority carrier lifetime. The net result is a higher conduction loss and faster recovery. Even with these changes, the PN junction diodes functions satisfactorily at 80 K.

26.7.1.5 Schottky barrier diode

As in the PN junction diode, the offset voltage of the Schottky diode increases as the temperature is lowered. The ratio is about 2:1 from room temperature to 80 K. The bulk resistance has a greater importance in Schottky diode and it decreases with temperature. The net result is still an increase in conduction loss, but not as severe as in the PN junction device.

26.7.1.6 Bipolar transistor

The bipolar transistor operates through the injection of the minority carriers from the emitter region through the base to the collector region. The current gain depends on both the injection efficiency of the emitter–base junction and the lifetime of the injected carriers. Both these phenomena grow less efficient as the temperature is lowered, so the performance of the transistor decreases as it is cooled. The present generation of transistors is not designed to operate at low temperatures. As a result, their performance becomes relatively poor at 80 K. However, it is possible to design devices that will operate at such low temperatures. Therefore, such devices could become available when a significant market of low temperature electronics develops.

26.7.1.7 Power MOSFETs

The MOSFET operation does not depend on the injection of carriers across a junction, nor is the minority carrier lifetime a factor. As long as sufficient majority carriers are present in the source region, drain current will flow when a bias is present on the drain and a voltage is applied to the gate. Carrier freezeout in the source and drain regions does not impact device performance until the conduction electron (or hole) concentration is reduced to a small percentage of its original value. The main factor affecting MOSFET operation is the carrier mobility, which increases at low temperature. As a result, device on resistance is reduced by a factor of 2 to 3. With such a great improvement, the MOSFET actually works better at 80 K than at room temperature.

26.7.2 Magnetics

Ferromagnetic metals generally behave poorly at low temperatures, displaying reduced permeability and drastically increased losses. Nickel iron in tape, laminations or powder form is much better. The saturation flux density increases at low temperature in several magnetic materials. Increased flux density is particularly useful in filter inductors with large dc currents. For a given induced emf in the magnetic core, the e^2/R eddy loss increases at low temperature because the resistance decreases. The actual increase in core loss, however, is small. The ohmic loss in the winding conductor drastically decreases due to decreased resistance. Even if the windings are not superconducting, the low temperature is still advantageous. The resistivity of copper drops by a factor of 13 from +100°C to 80 K.

Some insulating materials used in magnetics show substantial improvements in dielectric strength. The dielectric strength of 1-mil thick Kapton film is about 6500 V/mil at room temperature and 11,000 V/mil at 80 K. The life expectancy also rises dramatically at lower temperatures. High-voltage power converters can benefit from both these trends. However, insulating films become brittle at low temperatures, so sharp radius bends must be avoided and adequate mechanical support provided. On balance, magnetics should demonstrate improved performance at 80 K, especially if the winding is immersed in liquid nitrogen.

26.7.3 Capacitors

Some capacitors are usable at cryogenic temperatures, and some are not. Film and mica capacitors maintain their capacitance values well at low temperature. Solid tantalum types also hold up very well down to 4 K, but wet tantalums are not usable at cryogenic temperatures due to large capacitance change and very high ESR. Inadequate data are available on the ESR of film capacitors at low temperatures, except that the dissipation factor of the dielectric increases and the resistivity of the foil decrease. The reduced ESR, coupled with the increased thermal conductivity of the dielectric, translates into a much higher current rating for low-temperature operation.

26.7.4 Resistors

Resistors are primarily used in the control circuits only, not in the power trains. Most metallic resistance decreases at low temperature. The performance of metal film resistors is more than adequate at 80 K. For precision dividers or similar uses, the bulk metal resistors are generally good and should be adequate even at 10 K (liquid helium temperature). Wire-wound resistors also would work well. However, the ubiquitous carbon composition resistor is not suitable for low-temperature operation. It has a large temperature coefficient, and the resistance value displays considerable hysteresis during temperature cycling.

26.7.5 Structural Materials

All materials become brittle at low temperature, including magnetic stainless steel. Nonmagnetic stainless steel remains relatively ductile at liquid helium temperature. Insulating materials, such as Kapton, KelF, and glass–epoxy break at lower elongation. Resins when cooled approach the glass transition temperature, below which they become brittle.

The thermal conductivity of most materials used in the power converter construction increases as the temperature is reduced. For example, the thermal conductivity of silicon increases by a factor of 10 from room temperature to 8 K. One of the greatest limitations to increasing power densities is the problem of getting heat out of the electronic parts. The heat transfer is very efficient if the device is operated in a boiling liquid nitrogen bath. Operating power converter circuits at 80 K reduces the total power dissipation and improves the heat transfer. This should allow achieving much higher power densities with no technical barriers.

To summarize this section, cryogenic power equipment can be designed to operate at liquid nitrogen temperatures. In many cases the performance can actually be improved over the room temperature operations. As the use of 80 K superconductors becomes practical, higher efficiency, and compact power converters could become possible with performance well beyond the present state of the art.

References

1. DeWinkel, C.C. and Lamopree, J.D., Storing power for critical loads, *IEEE Spectrum*, June 1993, pp. 38–42.
2. Balachandran, U., Super power, progress in developing the new superconductors, IEEE *Spectrum*, July 1997, pp. 18–25.

Chapter 27

Microwave Beam Power Satellite

27.1 Introduction

The traditional power system in space uses a solar array and battery for each satellite. The concept of using a centralized large power-satellite (powersat) has been receiving some seed funding at NASA for some time. In this concept, depicted in Figure 27.1, one high-power satellite would generate bulk power, which is then transmitted to multiple satellites by laser or microwave beams. The electric propulsion of spacecraft using beam power from a powersat was discussed in Section 23.8. The beam power can also be used for a rescue operation on a remote Earth surface or anywhere in the ocean. Advantages of the central power satellite transmitting beam power to multiple user spacecraft are:

- Replaces PV array by much smaller power receiving antenna in the user spacecraft
- Eliminates battery by proper orientation with the power satellite(s) that avoids satellite-to-satellite eclipse

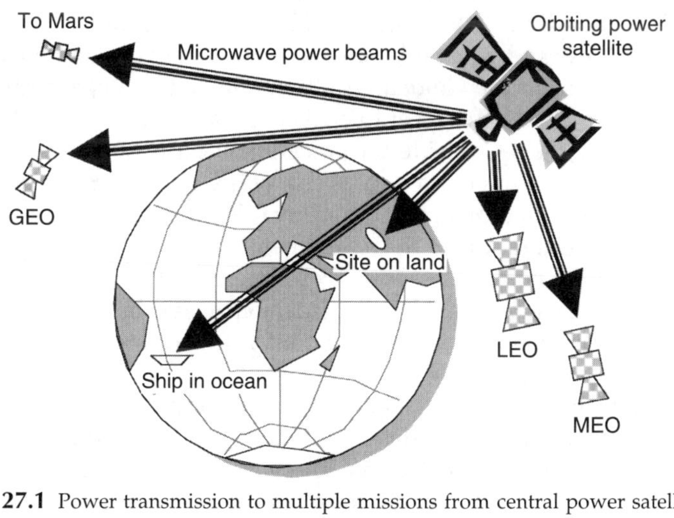

FIGURE 27.1 Power transmission to multiple missions from central power satellite.

- Extends the mission life where the spacecraft life is solar array or battery limited
- Provides large peak power if and when needed, such as for electric propulsion, thus significantly decreasing the on-board fuel mass
- Orienting the spacecraft to the powersat beam may be simpler than orienting to the sun

The power generated by the powersat can be transmitted to user satellites in two ways: by a microwave beam or by a laser beam.

27.2 Microwave Beam

The power transmission efficiency in space by microwave beam is given by the following expression,[1]

$$\eta = 1 - \exp(-A_t A_r / d^2 \lambda^2) \tag{27.1}$$

where

η = fraction of the transmitted power captured by the receiving antenna

A_t and A_r = transmitter and receiver antenna area, respectively

d = distance between the transmitter and receiver

λ = wavelength of the microwave power beam

The distance between the powersat and the user satellite may vary. Microwave transmission may be effectively used over short distances. The frequencies considered for such systems[2] are from 3 GHz to 300 GHz with the corresponding wavelengths from 100 mm to 1 mm.

In 2003, a NASA funded experiment was designed to transmit power by a microwave beam from the Goldstone 100-m antenna to a space sail at 800-km altitude. The Goldstone steerable dish radiates up to ½ MW power. However, the sail, being outside the range, was estimated to receive only 1700 W. The resulting microwave pressure was estimated to accelerate the sail by $10^{-7}g$ and demonstrate the principle of beaming power to a spacecraft. The acceleration of the sail depends only on power and not on the frequency of the beam. In another experiment at JPL and the University of California at Irvine, a 10-kW, 7-GHz microwave beam in a vacuum chamber produced a power density of 1 kW/cm^2 to heat a sail surface to 2000 K.

Microwave transmitters have been under development for much longer than lasers. They are far more efficient and cost much less to build. They do not damage the receiving surface as lasers can, and do not refract while passing through air. However, microwaves require much larger antennae for the same focusing ability.

27.3 Laser Beam

Laser beams may be more efficient over distances in the 10,000 to 50,000 km range.[3] The laser system consist of three parts:

- First the solar energy collected on the powersat is converted into a laser beam using a solid-state solar pumped laser. It consists of a crystal placed in the focus of a parabolic solar concentrator.[4]
- The monochromatic laser radiation is then focused into a beam using an optical mirror of appropriate size. The wavelength of such laser is equal to 1.06 μm, and the conversion efficiency 20 to 25%. A focused power laser beam then transmits the power to the user satellite or a rescue site on the Earth at 80 to 90% efficiency.
- Finally, the received energy is converted into electrical power using the conventional PV cell. The conversion efficiency of PV cell under laser illumination is around 50%. The maximum electrical power output of the silicon PV cell using such systems can be 3000 W/m^2 from laser radiation compared to 300 W/m^2 from natural solar radiation. The solar array requirement on the user satellite is thus greatly reduced, and the battery can be eliminated by avoiding powersat-to-user satellite eclipses.

The powersat, once developed and built, can make the user satellite much lighter and less expensive, so that many can be launched at low cost per launch. This could open up entirely new kinds of missions in the Earth orbit and in interplanetary space at much less incremental cost. The concept is like the 19th century railroad system: once the tracks are laid, the train itself is a small added expense.

27.4 Space-to-Ground Power

Driven by the environmental and depletion concerns of the fossil energy sources, both the DoE and NASA have funded space-based generation of power for ground use. The equivalent mass density of air on Earth with typical moisture and pollution is 1.5 times that of pure air. Solar radiation on the normal Earth surface with air mass 1.5 (AM 1.5) is about 1000 W/m^2. With 12-h nights, on average, seasonal sun variations, and overcasts, the annual average energy yield of a ground based PV system may be around 6 kWh/m^2 per day. For a 400 MW power station on the ground using 20% efficient PV cells would need 40 million square meters of PV array. Storing sufficient energy to last 5 days without sun would require 60,000 MWh of energy storage assuming 80% round trip energy efficiency. A space-based system can reduce the collector area to one quarter and eliminate the need of energy storage by making the beam power dispatchable on demand.

Performance of the solar array placed on the Earth's surface versus being in LEO and GEO is summarized in Table 27.1. It shows that the energy collection per square meter in space is several fold higher than that on Earth. Furthermore, it is 50% higher in GEO than in LEO. The conceptual study made in the past has considered the medium Earth orbit around 10,000-km altitude for such powersats for ground use.

NASA and DoE have explored ambitious proposals for harnessing solar energy for terrestrial use by deploying solar powersats in geosynchronous orbits that could generate power 24 hours a day, 365 days a year. One concept study has shown that one such satellite with 146-km^2 solar array could deliver power equivalent to 10 nuclear power plants on the Earth. Its 1 km^2 antenna would transmit power in a sharp 2.4 GHz microwave beam to an Earth-receiving station. Here, it would be converted into d.c. and inverted into 10,000-MW 60-Hz a.c. and delivered to the distribution system of the electrical power utility. The basic converter would use inductors — perhaps superconducting — to store energy and boost the voltage.

In the proposed concept, the beamed power is converted to 60 or 50 Hz utility power using high-voltage converters. Massive series–parallel connections of numerous converters would be needed to beam gigawatts of power to Earth from the outer space.[5] The space-to-space power transmission would be in watts/m^2, while the space–to-ground power transmission would be needed in hundreds of watts/m^2. However, U.S. government regulations limits microwave radiation to 110 W/m^2. Therefore, the hundreds of watts/m^2 beam intensity would require a change in the regulations or special permission in a remote location from where it can be transmitted to populated load centers. This high-beam intensity is still a few-fold lower than the natural sun intensity of 1000 W/m^2 on the ground. Therefore, it may not be a cancer risk to humans, as some may fear.

Several innovative concepts are being studied for collecting solar energy in space and transmitting microwave beams to another spacecraft that may be orbiting the Earth, or on a planetary surface, or on its way to an interplanetary mission. NASA's Solar Space Power Exploratory Research

Table 27.1 The performance of solar array placed on the ground and in LEO and GEO

Parameter	Earth	LEO	GEO
Air mass	1.5 with average moisture and pollution	0	0
Solar radiation W/m^2	1000 in full sun 500 in partial sun	1350	1350
Useful sunlight hours/day	6 on average	16	24
Incident energy Wh/year	1643	7884	11826
Launch and maintenance cost in orbit	0	Low	High

and Technology program is investigating systems at power levels ranging from 100 kW to 1200 MW. The building blocks of such a system are:

- A large sun-oriented solar array that tracks the sun and generates power at high voltage in the 400 to 1000 V range
- Rotating microwave transmitter in space that tracks a receiving antenna (rectenna)/rectifier station on Earth
- Microwave beam at several GHz frequency, using solid-state power converters, magnetrons or klystrons
- A rotary joint between the solar collector and the transmitter
- High-voltage cables
- d.c. to microwave power converters in space and microwave to d.c. power converters and d.c. to 50 or 60 Hz a.c. inverters on the ground

A concept study for delivering 1200 MW from GEO orbit to the ground grid has developed the following estimates:[6]

- Solar array with 4 to 5 concentration ratio with futuristic conversion efficiency of 39% generating at specific power of 1000 W/kg and 550 W/m^2 at 1000 V
- A 5.8 GHz microwave beam of Gaussian power density distribution with 10 dB taper from the transmitter center to the edge
- Solid-state transmitters operating at 80 V and the solar array at 1000 V
- The power distribution from solar array to the SSPC is at 100 kV, so converters are needed at both ends of the distribution lines. These converters substantially add into the system mass. Using 6000 V magnetrons and reducing the distribution voltage to 6000 V could eliminate the transmission voltage converters
- Rectenna size on Earth about 7450 m in diameter.
- Overall wireless power transmitting efficiency around 35%.

A system for 1200 MW power to the grid requires a solar array of 7,300,000 m^2 area, amounting to 2700 $\times$ 2700 m^2 or about 3000 m diameter array in space. The rectenna on the ground is estimated to be 44,000,000 m^2 area or 7500 m diameter. The ground receiver would be about six times the solar array area, because of: (a) power losses in various components, and (b) the beam power density limited under the federal regulation on microwave power.

The mass of the above concept satellite is estimated to be 22,500 to 30,000 metric tons at launch and 17,000 to 22,000 metric tons in orbit. A great many technological developments and demonstrations are needed to bring the cost per kWh energy delivered to the ground competitive with grid power.[7,8]

27.5 Solar Dynamic Power Generation

Solar to electrical power conversion in the above schemes can be done using two primary options:

- Thin-film PV cells on a flexible array, and
- Solar dynamic system using a concentrator and a Brayton, Stirling or Rankine engine driving an electrical generator

Between the two options, the solar dynamic using the Brayton cycle offers the following advantages:

- Moderate technological developments
- Higher conversion efficiency
- No batteries are needed, as the heat can be stored as the latent heat of working fluids
- Minimal performance degradation with age for longer life and low maintenance
- Higher specific power to minimize the launch cost
- Scalability to high power up to hundreds of megawatts.

The solar dynamic power system was discussed in Chapter 21. The solar dynamic concept described there was developed and demonstrated for space station design during the mid-1980s. The development work then was centered on a 300-kW, 3-phase, 460-V, 20-kHz power system. At the hundreds of megawatts power levels of interest here, the proposed dynamic scheme uses a refractive Fresnel concentrator to collect solar energy, a heat receiver, a Brayton cycle turbine, and a rotating generator to generate electrical power. The waste heat is dissipated back to space via a radiator.

The electrical generator is a 3-phase synchronous generator operating at high frequency in the range from 400 Hz to 20,000 Hz. The higher the frequency, the lower the mass of materials needed in the system magnetics, such as the generator and the transformer. This is because lower magnetic flux is required at higher frequency for a given voltage. The power can be generated at a low voltage (200 to 1000 V) and transformed into a high voltage (5 to 25 kV) for transmission and distribution to the transmitting antennas. Alternatively, the generation and transmission both can be at the same voltage around 15 kV.

Table 27.2 Performance projections for the 10-MW solar dynamic module

Parameter	State of the art	Near term	Far term
System specific power, W/kg	21	116	129
System specific area, W/m^2	283	389	450
Brayton cycle efficiency, %	39	42	47
PV efficiency, %	30	33	35
Concentrator area, m^2	24,800	22,300	20,800
Concentrator diameter, m	178	168	163
Radiator area, m^2	10,135	3200	1400

(Data source: Mason, L.S., *Proc. 34th IECEC*, p. 2601, SAE, 1999.)

Based on this earlier groundwork, Mason[4] of NASA Glenn Research Center has developed a conceptual design for a 10 MW module for a solar power satellite. The Mason study has taken into account a projected launch vehicle capability (< 20,000 kg) and solar concentrator sizes (< 200 m diameter). Table 27.2 shows the system level performance estimates that could possibly be achieved in a couple of decades. It assumes certain new enabling technologies identified in Table 27.3 that could be available between 2015 and 2020. A 1.6-GW power generation can perhaps be met using 160 modules in a 40 × 4 configuration. Mason estimates that such a station would cover 9.5 km × 0.65 km area with mass of 4.5 million kg.

Lynch[9] as proposed an alternate approach of using sun tower architecture for PV power generation as shown in Figure 27.2. It requires 100 kV transmission. The antenna in the figure is the microwave transmitting antenna.

Table 27.3 Solar dynamic component advances

Component	State of the art	Near term	Far term
Concentrator	Rigid panel, Al coating, 88% reflectance 10 kg/m^2	Rigid, Ag coating, 92% reflectance 2 kg/m^2	Inflatable, Ag coating, 92% reflectance 0.2 kg/m^2
Power system	500 V, 1 kHz, 3-ph generator, 60°C Si electronics, 10-kV distribution, 100-m cable, 93% efficiency, 2 kg/kW	5 kV, 1 kHz, 3-ph generator, 150°C Si electronics, 10-kV distribution, 100-m cable, 93% efficiency, 1 kg/kW	10 kV, 1 kHz, 3-ph generator, 300°C Si electronics, 10-kV distribution, 100-m cable, 95% efficiency, 0.2 kg/kW

(Data source: Mason, L.S., *Proc. 34th IECEC*, p. 2601, SAE, 1999.)

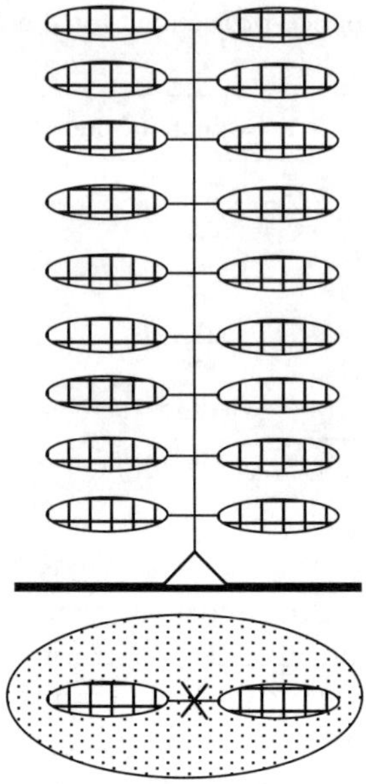

FIGURE 27.2 Solar PV power generation using the sun tower concept.

27.6 Powersat Developments

The proponents of space-based solar power seek a big science response to a problem that needs a more subtle solution. However, generating such power with gigantic panels in space, transmitting by microwave or laser beam, and distributing on Earth can be expensive and perhaps unsafe at present. The on-going research may find answers, but may take decades. However, the concept has captured the imagination of many engineers around the world. The studies include multi-megawatts solar powersats and microwave wireless power beaming to other spacecraft or to the Earth. On the other hand, technological implementation of the concept is exceptionally challenging. NASA estimates that tens of billions of dollars may be needed before transmitting the first watt of power from a GEO based powersat. The key technologies required before such concept can be materialized are:

- Low-cost thin film PV cells, concentrators, or high-temperature solar dynamic power generation
- Wireless power transmission by high-frequency power beams
- High-voltage, high-current, high-frequency power converters
- Structural, assembly, and thermal aspects of the massive station design
- Space transportation to the powersat and robotic assembly, maintenance, and servicing

- Resolution of the environmental, regulatory, and safety concerns
- Resolving the complex system integration issues

For the powersat beam power from space to ground to be economical, the cost at the 60 Hz side of the distribution station should be below 5 cents per kWh. For this to happen, the following performance would be required:

- PV efficiency of 45% with specific power of 1 kW/kg (about 40 times the presents state of the art)
- Thin film PV with vapor deposition on polymer substrate
- Potential PV candidates are $CuInGaSe_2$, $CuInS_2$, and CdTe, all of which work at voltages around 300 to 500 V
- High-temperature (400°C) polymer substrate or low-temperature deposition method.
- d.c.-to microwave beam-to a.c. conversion efficiency of 80%.
- Launch cost comparable with that of the airplane transportation at present.

All studies at present conclude that such powersats could not be cost effective in the near future, but may become economically viable in a few decades.

References

1. Lineberry, J.T. and Chapman, J.N., MHD Augmentation of rocket engine for space propulsion, in *Proceedings of the 35th Intersociety Energy Conversion Engineering Conference*, AIAA, 2000, Paper No. 3056.
2. Koert, P. and Cha, J.T., Millimeter wave technology for spec power beaming, *IEEE Transactions on Microwave Theory and Technology*, Vol. 40, No. 6, June 1992, pp. 1251–1258.
3. Grechnev, A.B. et al. Centralized power as basis of new philosophy of space power engineering, in *Proceedings of the 34th Intersociety Energy Conversion Engineering Conference*, SAE, 1999, Paper No. 2436.
4. Mason, L.S., A solar dynamic power option for space solar power, in *Proceedings of the 34th Intersociety Energy Conversion Engineering Conference*, SAE, 1999, Paper No. 2601.
5. Kusic, G., Conversion of beamed microwave power, in *Proceedings of the 35th Intersociety Energy Conversion Engineering Conference*, AIAA, 2000, Paper No. 3071.
6. SAIC and Futron Corporation, *Space Solar Power Concept Definition Study*, NASA Report No. SAIC-99/1016, February 1999.
7. Mankins, J.C. and Howell, J., Overview of the space solar power exploratory research and technology program, in *Proceedings of the 35th Intersociety Energy Conversion Engineering Conference*, AIAA, 2000, Paper No. 3060.

8. Carrington, C. et al. The abacus/reflector and integrated symmetrical concentrator concept for space power collection and transmission, in *Proceedings of the 35th Intersociety Energy Conversion Engineering Conference*, AIAA, 2000, Paper No. 3067.
9. Lynch, T.H., Sun tower PMAD architecture, in *Proceedings of the 34th Intersociety Energy Conversion Engineering Conference, SAE, 1999*, Paper No. 2441.

Index

A

D

E

G

H

I

J

K

L

M

N

O

Q

R

S

U

V